Neuropsychoanalysis in Practice

Neuropsychoanalysis in Practice

Brain, Self, and Objects

Georg Northoff
Canada Research Chair
Mind, Brain Imaging and Neuroethics;
Michael Smith Chair for
Neuroscience and Mental Health;
University of Ottawa Institute of Mental Health Research
Ottawa, Canada

OXFORD
UNIVERSITY PRESS

OXFORD
UNIVERSITY PRESS

Great Clarendon Street, Oxford, OX2 6DP,
United Kingdom

Oxford University Press is a department of the University of Oxford.
It furthers the University's objective of excellence in research, scholarship, and education by publishing worldwide. Oxford is a registered trade mark of Oxford University Press in the UK and in certain other countries

First published 2011

Published in the United States of America by Oxford University Press
198 Madison Avenue, New York, NY 10016, United States of America

British Library Cataloguing in Publication Data
Data available

Library of Congress Control Number: 2011925359

ISBN 978–0–19–959969–1

Foreword

More than 10 years have passed since we introduced to the world, at the Royal College of Surgeons of England, the "infant" discipline that Georg Northoff addresses in this remarkable new book. But the inaugural International Neuropsychoanalysis Congress in London was not so much a birthday as a "coming out" ceremony, of the kind traditionally held to introduce young ladies to the eligible bachelors of good society. Accordingly, among the invited guests at that congress in July 2000 were some leading scientists that we hoped might take an interest in Neuropsychoanalysis. Since then, she has had many suitors, not all of whom have pleased us. A regularly disappointing feature, even among those who displayed enthusiastic interest, has been a certain lack of substance. It goes without saying that what concerns us here is *intellectual* substance, an approach to science that reflects serious immersion in all relevant dimensions of intellectual "breeding."

You can imagine our delight, therefore, when Professor Dr med Dr phil Georg Northoff first declared the true extent of his interest in Neuropsychoanalysis, approximately five years ago. His pedigree was impeccable. A physician who had specialized first in psychiatry and then in philosophy and then in neuroscience, and who was, moreover, pursuing a highly productive research career, promised to meet our beautifully maturing adolescent's every need. We would have preferred him to have trained in psychoanalysis, but nobody is perfect. And few before Georg Northoff had shown anything like the depth of understanding of our daughter's background, her intellectual lineage if you like, both scientific and philosophical. Although such things still mean a great deal to us, let us not deny that times have changed. We are under no illusions as to what the future might hold. Even if this relationship does end in marriage, for example, still it might not last forever. There may yet be others. Notwithstanding these modern realities, we wish to declare, taking a deep breath, our enthusiastic support for Georg Northoff's proposal outlined in this book.

Before discretely withdrawing, however, and letting them get on with it, we would like to make one point clear to his readers. Neuropsychoanalysis takes a very particular approach to the mind, and thereby to the brain. It is remarkable how few scientists these days even bother to reflect critically on their approaches to the mind. Most disturbing of all, they do not always know what the thing actually *is* that they are approaching in this way. Yes, believe it or not, mental scientists today (cognitive neuroscientists, psychiatrists, psychologists, and psychoanalysts alike) for the most part, do not seem to know what a mind is! How can they ever hope to lift the top-down veils that still obscure this most beautiful, mysterious and profound part of nature if they do not know what it is? Few seem to realize that human mental processes, especially affective ones, are not exactly unique. Everything in nature evolves bottom-up. Accordingly, many features of the mind are almost impossible to understand by studying human cognition in isolation.

We would therefore like to give readers of Georg Northoff's proposals a clear, simple and succinct statement of the Neuropsychoanalytic approach to the mind. This provides a good basis for evaluating what he sets out in the pages that follow of this seminal, breathtakingly complex and wide-ranging book.

The mind, according to us, has three fundamental properties.

Foremost among them is *subjectivity*. Anyone who does not take the subjective nature of the mind seriously, as the starting point of their investigations, has lost the plot before they begin, and is therefore not suited to Neuropsychoanalysis. This is sadly true of many (if not most) mental scientists ... and here we are not only referring to behaviorists. The so-called cognitive revolution did not entail a crossing of this Rubicon. The "representations" of cognitivists are far from being similar in kind to the things we actually experience. They are usually conceptualized as just-so circuitry of some type, arising from complicated networks of information-processing machinery, not very different to the flow of information in silicone-based computers. Machinery, no matter how complex or abstract, is still machinery. Machinery is almost the opposite of subjectivity.

The reason for this neglect of subjectivity is simple: the subjective nature of the mind is an embarrassment to investigators who want to retain the mantle of objectivity. In order to retain their credentials as mental *scientists* therefore, they think they have to jettison the *mental* itself—the most essential property of the mind. But every part of nature has its own empirical challenges that may appear insurmountable to the custodians of pure reason. What is called for, as Northoff well realizes, is something more practical.

The mind does not consist in subjectivity alone. Although everything may be said to *be* something, not everything has a mind—a quality of being that *experiences* itself in the world. This touches on many philosophical problems, not least of them the problem of other minds. Since the mind is subjective, I can only ever know my own in the sense of directly experiencing it; that is what subjectivity means. It means "first-person perspective." How, then, can I know whether anything other than me has a mind? How can I ever know *what it is like* to be something other than me?

In our view, it works like this (and this, incidentally, is where the "Neuro-" enters Neuropsychoanalysis). I look inward, and experience myself. I look outward and I am attached to a body, a physical thing. In fact I am so deeply attached to this body that I call it 'me', using the same name that I use to refer to my subjective self (my mental being). I gain a better view of this thing by looking in a mirror, but when I close my eyes I notice that I am still my mind. I then realize that my body is *simply what I look like from the outside*. This body is 'me' from the objective, "third-person" point of view, even thought the first-person view reveals things that cannot be seen through the eyes of another.

We therefore may provisionally infer that all *things that look and behave like me from the outside feel like me from the inside*. Philosophers (like Descartes) might say that one can never really know such things—all the other things might really be zombies who just *look* like me but actually feel like nothing—but to think like that is mad. That is why we prefer to approach the problem scientifically. We do not seek to intuit ultimate truths; we are satisfied with testable hypotheses based on the best available evidence. We hypothesize that all things that look and behave like me from the outside feel like me from the inside, and we hold this to be true, unless someone adduces evidence to the contrary (evidence to the effect that all others are really zombies).

Approaching the problem this way encourages us to go further, and recognize that some aspects of the body are more centrally implicated than others in being a mind. When some parts of the body (like limbs) are removed, human beings typically tell us they still feel like themselves, and they behave accordingly. When other parts of the body (like the brain) are removed, the *being* of the human being seems to disappear. All the evidence points to that conclusion. If we continue in this vein we learn, including from research on other animals, that some parts of the brain are more centrally implicated than others in generating the being of a mind. So we proceed gradually, on this basis, to identify precisely *which parts of our anatomy are the objective manifestations of the state of being that we know subjectively as our minds*. This has led us to conclude that two parts are absolutely fundamental. These brain regions coincide with two further fundamental properties of what we call 'mind', and thereby characterize its unique subjectivity.

The first of these is the extended reticulothalamic activating system (ERTAS), a core emotional component of which is the periaqueductal gray (PAG). When these parts of the brain are removed, in the course of preclinical (animal) research or by experiments of nature, the mind disappears completely. To put it more precisely, *consciousness* disappears. *Consciousness, then, is a second fundamental property of the mind.*

Because only things that possess an ERTAS and PAG are capable of consciousness, and because rocks and plants (for example) do not possess such brain matter, we are unwilling to accept they have minds. The same applies to bacteria, and may well apply to some insects and molluscs, but evidence is harder to obtain. There is every reason to believe, on the other hand, that all birds and mammals are conscious like we are. On these grounds, in fact, all vertebrates are conscious. They are all possessed of the core brain tissues for experiencing oneself in the world. This conclusion is also consistent with the behavioral evidence. Stimulating and ablating the PAG, for instance, in other vertebrates causes exactly the sorts of behavioral changes one would predict on the basis of its role in human consciousness. To be sure, a *human* mind is not merely core consciousness; humans are also possessed of reflexive awareness of awareness. But that is another matter; it is not a necessary feature of mind.

Hopefully our approach is becoming clear. We base our conclusions about the mind on a triangulation of three types of evidence. From our starting point, which is (1) the *subjective* evidence, we make inferences about (2) the *anatomical* realization of subjective experience, manipulation of which then enables us to confirm via a third type of evidence—namely (3) *behavior*—whether the anatomical objects in question do indeed possess the subjective attributes we infer. Crucially, in humans, the latter type of evidence includes verbal reports of subjective experience, which brings us back to our starting point.

In other animals the quality of experience has to be inferred from behavioral indices alone (among which, incidentally, are emotional vocalizations that all mammals share). For example, since stimulation of dorsal PAG generates separation cries in both humans and other mammals (and birds), in the absence of any evidence to the contrary we may assume that all such animals experience mental pain in response to loss, even when they cannot tell us about it in words. Readers may be surprised to learn that very few neurological and mental scientists outside of Neuropsychoanalysis share this view. Some are unwilling to attribute consciousness to other animals on purely philosophical grounds; others on the basis of disciplinary intellectual traditions. We base our conclusions on the evidence. For example: brains with (1) intact ERTAS structures coupled with (2) behaviors indicative of consciousness, allow us to infer (3) the existence of consciousness itself. This conclusion is no more or less secure than the inference that the object reflected in the mirror is "me."

But the mind consists in more than subjectivity plus consciousness. Good evidence for this conclusion comes from sleep studies: the mind does not disappear during sleep. Consciousness on the other hand *does* disappear during (dreamless) sleep. So what is it that holds our selves together during sleep and other mental states devoid of consciousness? (The evidence for unconscious mental acts is now overwhelming.) To put it differently, what do we mean by unconscious mental activity; what is 'mental' about it?

Abundant evidence in this respect points to a constellation of nuclei in the upper brainstem, the core part of which appears to be the ventral tegmental area (VTA).[1] When this part of the brain is removed by experiments of nature, although the patient remains conscious, a mental attribute no less fundamental than consciousness is lost. This fundamental attribute is *volition*. (What we are

[1] Just as it was artificial to isolate the PAG from the rest of the ERTAS, so too it is artificial to isolate the VTA from the other nuclei with which it interacts (such as hypothalamic ones); we do not have space here to consider anything but the *most* elementary facts.

here calling 'volition' also goes by many other names: will, spontaneity, interest, expectancy, desire, agency, intentionality, wanting, and SEEKING.) The bottom line is that patients with damage in this part of the brain often *do* behave like zombies, and may legitimately be described as mindless. *Volition, then, is the mind's third fundamental property.* All animals that possess VTA-related networks, in other words all mammals (but not only mammals) may be assumed to possess at least the rudiments of a mind as we know it.

Since there is abundant evidence to the effect that volition is independent of consciousness (since unconscious intentionality is ubiquitous) we conclude that the mind consists in all three of these properties combined and inextricably intertwined: *subjectivity* and *consciousness* and *volition*. These properties unfold over many levels of the brain, including regions (like the neocortex) whose information-processing activities would not be recognizably mental if they were not infused from below with the primal qualities we have just described.

Having thus stated our point of view in the simplest possible terms, we leave it to Georg Northoff's readers to decide whether the marriage he proposes is *standesmäßig* —consistent with the Neuropsychoanalytic values we have outlined—and therefore likely to survive. Whatever readers conclude, however, we—the nervous parents—give Professor Northoff our unqualified support. We enthusiastically recommend his ideas as products of great intellectual substance, and assure you that any amount of effort expended in mastering them will be richly rewarded.

Mark Solms and Jaak Panksepp

Preface and Acknowledgments

This book is about neuropsychoanalysis and how we can practice and establish it as a distinct discipline that links psychoanalysis and neuroscience. By studying the work of Sigmund Freud and subsequent developments in psychoanalysis, I became increasingly aware of the scope of Freud's endeavor and his vision. The most revolutionary aspect of Freud's endeavor seems to me to be the fact that he aimed to bring together such diverse disciplines such as psychology, anthropology, philosophy, psychiatry, and even neuroscience. I greatly admire the scope and breadth of his vision. This book aims to pull together some of the different threads, and in doing so to link and integrate philosophical, anthropological, psychological, neuroscientific, and psychiatric issues.

The approach that is adopted in this book is a truly transdisciplinary one. I consider such an approach essential to allow neuropsychoanalysis to mature from the infant that it currently resembles into an adult discipline that can stand on its own two feet. In order to achieve this I suggest that neuropsychoanalysis must not only focus on specific contents, psychodynamic contents, and how these are linked to neuroscientific issues, but must also develop a specific neuropsychoanalytic methodology that sets it apart from both psychoanalysis and neuroscience. This book aims to outline such a truly transdisciplinary methodology, and at the same time to apply it. I am well aware that any such outline can only be sketchy and preliminary at this stage, and will raise many questions and critical comments from the disciplines concerned and from their respective intradisciplinary perspectives. However, even if many aspects are criticized or rejected, so long as this book offers the reader, whatever his or her background, some ideas for future conceptual or empirical investigation, it will have fulfilled its purpose.

I want to thank the editor at Oxford University Press, Martin Baum, who was very open to and supportive of the idea of such a book on neuropsychoanalysis. Further, I also want to thank his assistant editor, Charlotte Green, and especially my copy-editor Jo Hargreaves and production editor Abigail Stanley for excellent assistance in making the book considerably more readable. Thank you so much for an excellent job. A big thank you also to Laura Eckert and Victoria Howard for sorting out the references. In addition, I wish to thank the Neuropsychoanalytic Society, including Mark Solms, who strongly supported my work and ideas. For neuroscientific discussion and helpful advice I shall always be grateful to Jaak Panksepp, who probably knows more about the brain than anyone else, and for illuminating discussion and collaboration I thank Todd Feinberg. On the European side I want to thank Heinz Boeker in Zürich, who introduced me to the field and is now not only an excellent collaborator but a close friend. I also want to thank my neuropsychodynamic collaborators, Marina Farinelli, Oliver Lyttelton, and Jean Gagnon, for their patience with my often rather confused ideas, Moritz deGreck, who introduced me to Kohut and self psychology, and Stephan Doering, for stimulating discussions and for reading earlier drafts of some of the chapters. Finally, I want to thank my beloved John Sarkissian, who rather often has to endure the mental absence and musings of his partner before, during, and after the long hours of writing.

Contents

Part II **Neural Equipment**

Part III **Mental Equipment**

List of abbreviations

ACB	autoepistemic capacity of the body
ACC	anterior cingulate cortex
ADC	altered difference-based coding hypothesis in psychosis
AL	autoepistemic limitation
ALB	autoepistemic limitation of the brain
APD	antisocial personality disorder
AVH	auditory verbal hallucinations
BCI	brain–computer interfaces
BDI	Beck Depression Inventory
BHS	Beck Hopelessness Scale
BOLD	blood oxygen level dependent
BPQ	Body Perception Questionnaire
CMS	cortical midline structures
DMN	default-mode network
DMPFC	dorsomedial prefrontal cortex
DMT	dorsomedial thalamus
DSM	*Diagnostic and Statistical Manual*
EEG	electroencephalography
fMRI	functional magnetic resonance imaging
FPP	first-person perspective
GABA	gamma-aminobutyric acid
ICD	*International Classification of Diseases*
KAPP	Karolinska Psychodynamic Profile
LPFC	lateral prefrontal cortex
MDD	major depressive disorder
MEG	magnetoencephalography
MLR	mesencephalic locomotor region
MMN	mismatch negativity
MOFC	medial orbital prefrontal cortex
MPC	medial parietal cortex
MRS	magnetic resonance spectroscopy
MTG	middle temporal gyrus
NBR	negative BOLD response
NCC	neural correlates of consciousness
NCP	neural correlates of psychodynamics
NPD	narcissistic personality disorder
NPI	Narcissistic Personality Inventory
NPP	neural predispositions of psychodynamics

NPPC	neural predispositions of principal consciousness
NPS	neural predisposition of the self
NREM	non-rapid eye movement
OPD	Operationalized Psychodynamic Diagnostics
PACC	perigenual anterior cingulate cortex
PAG	periaqueductal gray
PBR	positive BOLD response
PCC	posterior cingulate cortex
PEP	pre-ejection period of the heart
PET	positron emission tomography
PGO	ponto-geniculo-occipital
REM	rapid eye movement
RSA	respiratory sinus arrhythmia
RSC	retrosplenial cortex
SACC	supragenual anterior cingulate cortex
SC	skin conductance
SCL-90	Symptom Checklist-90
SCMS	subcortical–cortical midline system
SMA	supplementary motor area
SN	substantia nigra
SRP	self-related processing
SSEP	steady-state evoked potential
STG	superior temporal gyrus
ssAEP	steady-state auditory-evoked potential
ssVEP	steady-state visual-evoked potential
TAS	Toronto Alexithymia Scale
TPJ	temporo-parietal junction
TPP	third-person perspective
VMPFC	ventromedial prefrontal cortex
VTA	ventral tegmental area

Introduction

> Psychoanalysis is not, like philosophies, a system starting out from a few sharply defined basic concepts, seeking to grasp the whole universe with the help of these and, once it is completed, having no room for fresh discoveries or better understanding. On the contrary, it keeps close to the facts in the field of study, seeks to solve the immediate problems of observation, gropes its way forward by the help of experience, is always incomplete and always ready to correct or modify its theories. There is no incongruity (any more than in the case of physics or chemistry) if its most general concepts lack clarity and if its postulates are provisional; it leaves their more precise definition to the result of future work. (Freud, 1923a)

Freud and the quest for neuropsychoanalysis

A combination of psychoanalysis and neuroscience? At first glance that looks like a rather strange marriage. How can psychoanalysis, which focuses so intensely on the meaning of psychological contents, be united with a discipline like neuroscience, which centers on the brain's neuronal states rather than meaning and contents? It is often forgotten that Sigmund Freud, the founder of psychoanalysis, trained initially as a neuroanatomist, and investigated nerve cells and the exact lesion localization of speech disorders such as aphasia. However, his early neuroscientific origins later became completely overshadowed by his focus on exclusively psychodynamic issues, such as the ego, our dreams, and the unconscious.

There is one specific point in his own writings when his search for psychodynamic mechanisms converges with the quest for corresponding neuronal mechanisms in the brain. This is his famous 1895 writing on a *Project for a Scientific Psychology*, in which he tries to link neuronal mechanisms to psychodynamic concepts (see Part II of this book for details). Although in that book he attempted to link specific neuronal mechanisms to his psychodynamic concepts, Freud later regarded his first neuropsychoanalytic endeavor as a failure, and so the manuscript was not published during his lifetime.

When it was eventually published in 1895, Freud's *Project for a Scientific Psychology* generated much discussion as to whether or not psychoanalysis can in principle be linked to neuroscience (Brook, 1998; Levin, 2003; Peled, 2008). Most recently, the *Project* writing has been regarded as evidence that psychoanalysis can indeed be linked to neuroscience. This has led to the birth of a new discipline, known as neuropsychoanalysis (Kandel, 1998; Solms and Solms-Kaplan, 2000; Solms and Turnbull, 2002; Fonagy, 2003; Shore, 2003; Mancia, 2004; Solms, 2004; Mancia et al., 2006;

Northoff et al., 2007a,c). Broadly defined, neuropsychoanalysis aims to link psychodynamic concepts and neuroscientific mechanisms, and thus to integrate the psyche and the brain. However, as with any peculiar birth, the rather long pregnancy and subsequent painful delivery of the discipline of neuropsychoanalysis, after its initial conception in 1895, have given rise to much controversy. In particular, this concerns what constitutes the most appropriate care for the new infant, with proponents and opponents being much divided about the kind of remedy that is required for the gestational complications.

The proponents have focused predominantly on linking psychodynamic concepts such as dreams, the unconscious, the ego and the self to specific psychological functions (e.g. cognitive and affective functions), which in turn may be localized in specific brain regions. One focus has been on the unconscious and its relationship to memories (Kandel, 1998; Mancia, 2004; Mancia et al., 2006), while others have searched for the neuronal mechanisms underlying drives (Solms, 1996; Panksepp, 1998; Fonagy, 2003), dreams (Solms, 1997, 2000; Solms and Turnbull, 2002; Mancia, 2004; Hobson, 2009), the ego (Northoff, 2007; Carhart-Harris and Friston, 2010), primary and secondary processes (Carhart-Harris and Friston, 2010), and defense mechanisms (Fonagy, 2003; Northoff and Boeker, 2006; Northoff et al., 2007a,c; Feinberg, 2010).[1] Since these neuropsychoanalytic investigations focus on searching for the brain regions whose neural activity correlates with the psychodynamic concept in question, it is possible to speak of the "neural correlates" of psychodynamic concepts.

In contrast, the opponents of such neuropsychoanalytic endeavors argue that any such linkage between neuronal mechanisms and psychodynamic concepts cannot account for their meaning and thus the hermeneutical dimension of the latter (Green, 2001; Schneider, 2006). More specifically, the quantification and objectification that are required in order to investigate neuronal mechanisms eliminate the qualitative and subjective features of the meaningful psychodynamic concepts. Moreover, those same qualitative and subjective features make it impossible to localize psychodynamic concepts in the quantitative and objective neuronal activities that are associated with certain regions in the brain (Green, 2001; Schneider, 2006). The opponents suggest that this makes any neuropsychoanalytic attempt futile, and they predict a rather difficult life if not premature death of the new infant neuropsychoanalysis, by comparison with its much older and more mature sister, namely psychoanalysis.

How would Freud have responded to such debate if he was alive today? Would he have embraced the newborn with the name "neuropsychoanalysis"? Or would he have left the new infant alone in the same way that he rejected and abandoned his 1895 writing, which was supposed to die a silent death and end up in the graveyard of unpublished writings? We do not know the answers to these questions, and we can only speculate. However, we do know (at least partially) why Freud abandoned his initial neuropsychoanalytic attempts. He considered that, during his lifetime, the level of knowledge of the brain and thus of neuroscience was insufficient to allow the linking of neuronal mechanisms to psychodynamic concepts.

What is the situation now? Is our current level of knowledge and insight into the brain's neuronal mechanisms sufficient to allow what Freud eluded, namely the link between neuronal mechanisms and psychodynamic concepts? Leading proponents such as Mark Solms (2004) and Jaak Panksepp (1998) do indeed consider neuropsychoanalysis to be the continuation and completion of Freud's project (Freud, 1895) (see also Pugh, 2006) and, more generally, of his endeavor to establish a scientifically based psychology of the human mind. Is neuropsychoanalysis thus nothing more than a continuation and completion of Freud's attempt to develop a scientific

[1] This list is by no means complete, as I wish to make clear to those authors who are not mentioned here.

psychology of the mind, which is now merely extended from the psyche to the brain? Let us now finally turn to the brain itself, and see whether neuropsychoanalysis can indeed overcome the deficits in knowledge of the brain that Freud identified during his lifetime.

Function- and localization-based approach to the brain

Current neuropsychoanalysis draws heavily on recent developments in cognitive, affective, and social neuroscience (Panksepp, 1998; Gazzaniga, 2008). For instance, different cognitive functions (e.g. attention, working memory, episodic memory, etc.) are taken as the initial starting point and are then related to supposedly corresponding concepts (e.g. memory, dreams) within the psychodynamic context (for examples of such an approach, see Solms and Turnbull, 2002; Mancia, 2006). Cognitive neuroscience has been most recently complemented by affective and social neuroscience (Panksepp, 1998) since psychodynamic concepts such as introjections, narcissism, self-objects, drives, etc. may be closely related to specific affective and social functions (Solms, 2004; Northoff et al., 2007a,c; Northoff and Panksepp, 2008). It is thus possible to speak of what I call a "function-based approach," which is the first characteristic of current neuropsychoanalytic approaches.

How can such a function-based approach be related to the brain? Broadly speaking, cognitive, affective, and social neuroscience aims to link cognitive, affective, and social psychological functions to the brain, and more specifically to the neural activity in particular regions of the brain. For instance, cognitive functions such as working memory, attention, etc. have been associated with the neural activity in specific cortical regions such as the lateral prefrontal cortex and the parietal cortex (Gazzaniga, 2008), whereas affective and basic social functions are more strongly associated with neural activity in subcortical regions such as the tectum, the periaqueductal gray (PAG), the dorsomedial thalamus, the colliculi, and so on (Panskepp, 1998; Panksepp and Northoff, 2009). Following the road maps of its guiding disciplines, neuropsychoanalysis aims to link specific psychodynamic mechanisms to the neuronal activity in particular regions of the brain.

One of the main frontrunners of such an approach was Mark Solms. He observed particular psychodynamic changes in his neurosurgical patients who had specific lesions in their brains. Extrapolating from the pathological to the healthy, this allowed him to link psychodynamic mechanisms to specific brain regions (Solms and Solms-Kaplan, 2000; Solms and Turnbull, 2002), and thus to establish the discipline of neuropsychoanalysis.[2] For example, relying on his observations in neurosurgical patients, he has argued that the ventromedial prefrontal cortex may be crucial in constituting the ego (or self) (for a detailed discussion see Chapter 9), whereas the parietal cortex may have a central role in constituting "one's body" as first self-object (Solms, 1999; for a detailed discussion, see Chapter 7). This approach has recently been extended to the healthy brain, so neuropsychoanalysis can be characterized by what I call a "localization-based approach," which is its second characteristic feature (see also Figure A).

Can neuropsychoanalysis in this sense, as characterized by a function-based approach and a localization-based approach, account for what Freud eluded in his early 1895 writing? Can the function- and localization-based approach to the brain in current neuropsychoanalysis make up for the deficits in our knowledge of the brain that Freud identified? Would such function- and localization-based approaches allow Freud to reverse his decision to abandon and then reject his

2 However, this was preceded by others, such as E. A. Weinstein (see Weinstein and Kahn, 1959) and H. Shevrin, who pursued even earlier neuropsychoanalytic investigations either in healthy subjects or in patients with lesions (for details about Weinstein in particular, see Feinberg, 2009, 2010).

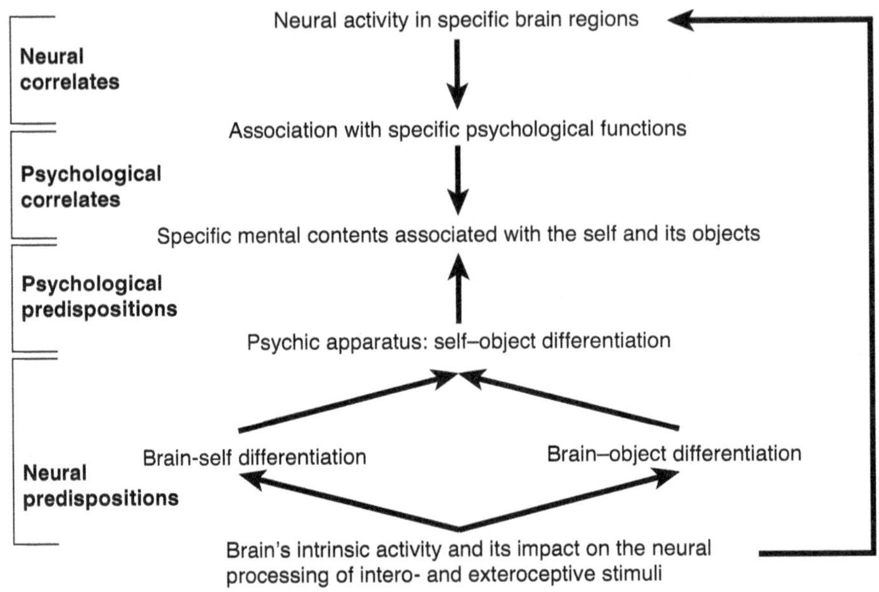

Fig. A. Neural correlates versus neural predispositions.

Project for a Scientific Psychology (Freud, 1895)? To answer these questions, we have to go back to Freud himself and see how, if he could have done so, he would have approached the brain.

Freud's search for psychological structure and organization

How did Freud approach psychological functions? And how did he relate them to what he called the "psychic apparatus"? Freud listened to his patients and investigated their mental contents, which he then associated with specific psychological functions. These psychological functions in turn were assumed to mirror a specific psychodynamic mechanism. For instance, the dreams showed his patient's mental contents to be predominantly sexual which, except in dreams, could not otherwise surface. From this he inferred a specific psychological mechanism, namely repression, which suppressed sexual contents during the day but allowed them to surface in dreams at night.

What does Freud presuppose here? He starts with observations of mental contents, such as sexual desire, and attributes a specific meaning to them in that they are supposed to express something specific about the person him- or herself. The next step is the crucial one, as Freud infers from the mental content and its personal relevance a specific psychological function, namely repression. Here he assumes that the mental content and its specific meaning are possible only on the basis of the specific psychological function of repression. One may therefore want to refer to a "function-based approach" in Freud.

Is there any difference between the "function-based approach" in the context of cognitive, affective, and social neuroscience and that assumed by Freud? In contrast to Freud, cognitive, affective, and social neuroscience does not start with specific individually meaningful contents that are subjective. Instead it starts with certain objective behaviors, such as social interaction, cognition, or emotions. Despite such a difference in the starting point, the extrapolation to psychological functions is the same. In the same way, cognitive, social, and affective neuroscience infers specific psychological functions to account for the observed behavior. Freud infers his psychological functions from the mental contents that he observed. One may therefore refer in both cases to a "function-based approach."

There is even further convergence. Many of the cognitive, affective, and social functions that are investigated in cognitive, affective, and social neuroscience are nowadays related to the psychological functions that Freud described. For instance, the cognitive function of memory, and especially autobiographical memory, may be closely related to Freud's psychological function of repression (Kendell, 1998; Manzia et al., 2006). Freud's "function-based approach" may consecutively converge with that presupposed in current cognitive, affective, and social neuroscience.

Does the convergence of both function-based approaches provide a common platform from which to make the second step, namely that to the brain? If the cognitive, affective, and social functions can be localized in specific regions of the brain, the respective associated psychological functions as described by Freud must be related to precisely these brain regions. The "localization-based approach" would then be seen as a natural extension of Freud's psychological functions to the brain. As previously indicated, this is the view of Mark Solms and Jaak Panksepp.

However, Freud did not extend and complement his function-based approach with a localization-based approach. Why was this? Let us listen to Freud himself: "Every attempt to discover a localization of mental processes . . . has miscarried completely. The same fate would await any theory that attempted to recognize the anatomical position of the system [consciousness]—as being in the cortex, and to localize the unconscious processes in the subcortical parts of the brain. There is a hiatus here which at present cannot be filled, nor is it one of the tasks of psychology to fill it. Our psychical topography has for the present nothing to do with anatomy" (Freud, 1915).[3]

Why was Freud so sceptical about the localization of his psychological functions? Was it only because he lacked the psychological inventory of cognitive, affective, and social neuroscience that we are so fortunate to have today? This is the position taken by Mark Solms, Jaak Panksepp, and many proponents of neuropsychoanalysis. However, let us return to Freud himself and see how his focus shifted, especially after his *Project* writing (Freud, 1895).

After his 1895 writing, Freud abandoned the reference to the brain altogether, and focused exclusively on psychological functions. However, he went one step further than mere psychological functions, as he aimed to put the assumed psychological functions into a larger psychological context. This context refers to psychological structure and organization[4] rather than to specific psychological contents. For instance, he introduces the tripartite structure of the psychic apparatus with its division between the id, the ego, and the superego. His focus thus moves beyond mere psychological functions to their underlying psychological structure and organization. Unlike cognitive, affective and social neuroscience, Freud does not complement his function-based approach with a localization-based approach. Instead, he takes a different direction, namely that of psychological structure and organization. I therefore refer to what I call Freud's "structure-based approach."

How are psychological structure and organization related to psychological functions? Psychological structure and organization enable and predispose to specific psychological functions as, for example, the ego makes possible the repression of sexual desire. Let us consider the difference between psychological functions and structure/organization in terms of necessary and sufficient conditions. Psychological functions are assumed to correlate with the mental contents

[3] Mark Solms (1998, 1999) may want to argue that this only concerns a specific form of localization in the brain, namely static localization, while it does not exclude what he calls "dynamic localization" as based on the work of the Russian neuropsychologist A. R. Lurija. However, even dynamic localization still presupposes localization, which I therefore consider to be associated with the neural correlates approaches (e.g. NCP), rather than mirroring neural predispositions (see Parts I and II of this book for a more detailed discussion of Solms' concept of dynamic localization).

[4] See in particular Part I of this book for a more detailed definition of "psychological structure and organization."

(or the behavior) in question. The former is thus a sufficient condition of the latter. This is different in the case of psychological structure and organization. The ego is not assumed to correlate with specific mental contents, and can therefore not be regarded a sufficient condition. However, it is assumed first and foremost to make possible all kinds of mental contents. Without the ego, there would be no mental contents at all. The ego is thus a necessary but non-sufficient condition of (possible) mental contents.

What implications does this have? Freud and current cognitive, affective, and social neuroscience appear to share the function-based approach. However, although they share the first step, they depart from each other in the subsequent step. Cognitive, affective, and social neuroscience go forward towards the brain when assuming a localization-based approach, whereas Freud decided to go into another direction, namely that of psychological structure and organization, thereby presupposing a structure-based approach. If we want to extend Freud to the brain, we thus have to extend his structure-based approach into a neural context. For this we must leave Freud and move forward to the neuroscience of the present time.

"Neural correlates" and "neural predispositions"

One of the main research fields in current neuroscience is the search for the neuronal mechanisms underlying consciousness. This is described and subsumed under the heading "neural correlates of consciousness" (NCC) (Koch, 2004). The NCC aim to identify the neuronal mechanisms that are sufficient to induce consciousness. Thus they aim to describe the relationship between consciousness and specific psychological functions such as attention, working memory, and so on, as well as how consciousness is related to the neural activity in specific brain regions (Koch, 2004). The function- and localization-based approach is thus prevalent in the NCC (albeit in a somewhat modified form).

Analogous to the case of consciousness and the NCC, current neuropsychoanalysis also aims to search for the sufficient neural conditions of specific psychodynamic mechanisms. For instance, this is the case when Solms locates early defense mechanisms such as introjections and other forms of internalization in the region of the ventromedial prefrontal cortex. Therefore, analogous to the NCC, one may refer to what I call "neural correlates of psychodynamics" (NCP). The NCP share with the NCC the focus on sufficient neural conditions as well as a (rather implicitly presupposed) function- and localization-based approach. However, there is one crucial difference. Instead of being restricted to consciousness as the NCC, neuropsychoanalysis targets both conscious and unconscious states. The NCP do consecutively cover a more extensive territory, including consciousness and unconsciousness, by comparison with the NCC and their exclusive focus on consciousness.

The NCP focus on the sufficient neural conditions of the psychological functions that Freud described. With regard to their necessary neural conditions, Freud extended and complemented psychological functions by his search for psychological structure and organization, and thus by a structure-based approach. Instead of sufficient conditions, psychological structure and organization concern the necessary but non-sufficient conditions of mental contents. How can we extend the psychological structure and organization to the brain? The mere neural correlates, the NCP, are not appropriate, as they refer only to the sufficient conditions, not the necessary ones. Instead we need to target those neural conditions that are necessary but non-sufficient by themselves. Since those neural conditions may enable and predispose to the psychological functions and mental contents, I here refer to them as "neural predispositions."

Let us examine the concept of neural predispositions in more detail. The concept of neural predispositions describes the necessary but non-sufficient conditions provided by the brain itself.

These conditions are supposed to enable and predispose to psychological functions and their respective mental contents. One may therefore refer to what I call "neural predispositions of psychodynamics" (NPP), which refer to the necessary but non-sufficient (i.e. enabling and predisposing) neural conditions of mental contents as described in psychodynamic concepts. As such the NPP must be distinguished from the NCP, which refer to the sufficient but non-necessary (i.e. operating and executing) conditions of the mental contents as mentioned in psychodynamic concepts.

Having charted out the conceptual territory, we can now finally turn to the empirical grounds of the brain. What do neural predispositions and thus NPP refer to in terms of empirical brain function? Whereas the neural correlates and the NCP, being function and localization based, can be associated with the neural activity in specific brain regions, the situation remains unclear in the case of the neural predispositions. In order to gain a glimpse of the empirical functions associated with neural predispositions, we may need to turn to the brain itself independent of the psychological functions and their particular localization. In his metapsychology (see Part I of this book for details), Freud aimed to target the psychic apparatus and its structure and organization independent of specific mental contents. Similarly, we need to see what happens in the brain prior to the constitution of the various psychological functions and their localization targeted in the NCP and current cognitive, affective, and social neuroscience.

The brain's intrinsic activity as neural predisposition

The brain is not only active when we present it with specific stimuli and observe the associated activity changes, so-called "stimulus-induced activity." Even when a person is resting without any specific external stimulation, their brain is still active and shows high resting-state activity. This is, for instance, reflected in the fact that the brain consumes energy even when in the resting state. Although it comprises only 2% of the body's mass, the resting awake brain consumes 20% of the body's total oxygen. Oxygen is essential for any kind of neuronal activity, and the abundant use of the body's oxygen during rest indicates that something must be happening in the brain during rest (Raichle, 2009). Although the concept of an active brain showing high resting-state activity is not a new one, and can be traced back to historical predecessors (Raichle, 2009), the purpose of such activity remains unclear. In short, we need to understand why the brain uses such a large amount of oxygen in the resting state, and what purpose this serves.

Let us first briefly consider the resting-state activity itself. Single cells in the cortex of the brain use tonic activity for constant firing, with baseline firing rates ranging from a few spikes per second to tens of spikes per second. The resting-state activities of different cells may be coordinated and harmonized by synchronization and oscillation. At a more macroscopic level, specific regions show high resting-state activity, so that when confronted with a stimulus they can only lower (rather than elevate) their activity level. This seems to occur in particular in the regions of the so-called default-mode network (DMN). Brain imaging has shown that the DMN displays high resting-state activity and predominant so-called deactivation, and a corresponding negative BOLD response (NBR), during stimulus-induced activity (Raichle et al., 2001; Buckner et al., 2008; Raichle, 2009). The DMN includes the anterior and posterior medial cortical regions, such as the ventro- and dorsomedial prefrontal cortex, the perigenual, supragenual anterior and posterior cingulate cortex, the lateral parietal cortex, and the medial temporal lobe, including the hippocampus. However, even sensory regions such as the auditory and visual cortex, as well as other regions such as the thalamus, show resting-state activity that is too high (Llinás, 2001; Hunter et al., 2006; Wang et al., 2008).

What does the brain do with this high level of resting-state activity? One may regard this activity as mere noise in the background of stimulus-induced activity, the latter being "the real thing."

But why then does the brain expend so much energy and effort on mere noise? There must be more to the brain's resting-state activity than this, so it must somehow affect stimulus-induced activity, with the latter being dependent upon the former. Therefore there must be what I shall hereafter refer to as "rest–stimulus interaction" (for an overview of this subject, see Northoff et al., 2010).

Recent studies do indeed lend empirical support to such rest–stimulus interaction. For example, Mandaag et al. (2007) used anesthetic drugs to manipulate the resting-state activity level in rats, and investigated their neural activity changes during subsequent execution of similar movements. The different resting-state levels led to different neuronal activity patterns in the cortex during the subsequent movements, with some regions being active only at one specific resting-state activity level, and others being recruited at another one. Human studies have shown that the degree of stimulus-induced activity in the perigenual anterior cingulate and the visual cortex is dependent upon the resting-state concentration of gamma-aminobutyric acid (GABA) in precisely the same regions (Boly et al., 2007; Northoff et al., 2007a,c; Muthukumaraswamy et al., 2009; for a review, see Northoff et al., 2010).

In addition to such rest–stimulus interaction, there is also reverse traffic, with stimulus-induced activity affecting and modulating subsequent resting-state activity, amounting to what I call "stimulus–rest interaction." For example, Lewis et al. (2009) demonstrated that visuospatial learning led not only to activation changes in the visual cortex but also to connectivity changes of that region with DMN regions during the subsequent resting state (for a review, see Northoff et al., 2010). Similar observations were made in studies on working memory and self-relatedness. Higher degrees of working memory and self-relatedness were found to lead to stronger connectivity and activity changes within the DMN in subsequent resting-state periods when compared with lower degrees of working memory and self-relatedness (Schneider et al., 2008; Pyka et al., 2009).

What do these examples tell us about the brain and its involvement in specific cognitive, affective, or social functions as described by particular psychodynamic concepts and localized in specific brain regions? When investigating the brain's resting-state activity and how it affects stimulus-induced activity (and consequently psychological functions) (Northoff et al., 2010), we focus on the necessary but non-sufficient conditions and thus on those neural conditions that enable and predispose to psychological functions and their respective associated mental contents. One may then describe the brain's resting-state activity and its impact on subsequent stimulus-induced activity as necessary but non-sufficient conditions, and thus as enabling and predisposing rather than executing and operating conditions. In other words, investigation of the brain's resting-state activity may give us a glimpse into what I described as "neural predisposition," as distinct from mere neural correlates.

Brain–self and brain–object differentiation

Earlier I characterized current neuropsychoanalysis by neural correlates amounting to what I called NCP. Although the NCP could be considered to be analogous within the brain's neural context to what Freud was aiming for when targeting psychological correlates, the NCP fall short when it comes to psychological structure and organization. I therefore introduced the concept of NPP. Furthermore, I postulated that one such neural predisposition consisted of the brain's intrinsic activity and its impact on the stimulus-induced activity as associated with the psychological functions and their respective mental contents. However, I left open how such neural predisposition, the brain's intrinsic activity, may be related to the kind of psychological structure and organization that Freud envisaged.

Freud considered the distinction between the id, the ego, and the superego to be a characteristic feature of the structure and organization of the psychic apparatus. Thereby the ego or self (here I use the terms "self" and "ego" synonymously for the sake of operational simplicity; for a detailed discussion of their conceptual and empirical distinction, see Chapter 9) describing the person's inner world must be distinguished and differentiated from objects in the outer world. I refer to this as "self–object differentiation" (for further details, see Chapter 6). The concept of self–object differentiation refers to the processes that enable and predispose to the constitution of self and object as distinct from each other. Such constitution of self and objects is possible only on the basis of energy. The id is considered to provide and invest the energy (i.e. cathexis as described by Freud; for further details, see Chapters 4 and 5) that is necessary to enable and predispose to the constitution of self and objects as distinct from each other (i.e. self–object differentiation).

What does such self–object differentiation entail for the neuropsychodynamic context? Prior to their constitution in the gestalt of self–object differentiation, both self and objects must first and foremost be distinguished (i.e. differentiated) from the brain. How are we able to experience and perceive a self that is different from our brain (and our body)? How can objects be distinguished from our brain and its neuronal processes that enable and predispose to the constitution of objects in our mental states?

So long as self and object are not distinguished from the brain in our perception and experience, and thus in our mental states, we could not even speak of self and object. This means that they must be distinguished from the brain. This amounts to what I call "brain–self differentiation" and "brain–object differentiation." The concepts of brain–self differentiation and brain–object differentiation refer to and describe those processes that enable and predispose to the constitution of both self and objects as distinct from the brain. As such, brain–self differentiation and brain–object differentiation may also be implicated in allowing for the subsequent distinction between self and object, namely self–object differentiation (see also Figure A).[5]

Metaphorical excursion: brain, self, and objects

What are the implications of the concepts of brain–self and brain–object differentiation for the empirical (i.e. neuronal) context and thus for the NPP? The NPP may need to target the brain's neuronal mechanisms that enable and predispose to the differentiation of self and object from the brain (i.e. brain–self and brain–object differentiation). We therefore need to investigate how the neural predispositions that are provided by the brain itself are related to the constitution of self and objects as distinct from the brain. Based on the preceding section, I consider the brain's intrinsic activity and its specific impact on stimulus-induced activity, i.e. rest-stimulus interaction, as necessary (but non-sufficient) condition and thus as neural predisposition to brain-object and brain-self differentiation.

---

[5] What I here call brain–self and brain–object differentiation, within the neuropsychodynamic context, may resurface in different gestalts in different contexts (see Chapter 3 for details of the concept of context). For instance, within the philosophical context of mind and brain, brain–self and brain–object differentiation may resurface as "brain–mind differentiation," thereby referring to the necessary conditions that enable and predispose to the possible differentiation of what we call (and associate with) the mind, including its mental states from the brain and its mere neuronal states. Within the current neuroscientific context, one may speak of "brain–consciousness differentiation." Brain–consciousness differentiation may be regarded as a specific subset of the concept of brain–mind differentiation with regard to the specific instance of consciousness. As such it targets the processes and mechanisms that enable and predispose to the differentiation of consciousness and thus phenomenal or mental states from the brain's mere neuronal states.

Let me draw an analogy with a Scottish castle, in order to illustrate exactly what I mean by featuring the brain's intrinsic activity as a neural predisposition in the sense of NPP. The Scottish Highlands are very hilly, and so that it is located in a prominent position that is clearly visible from all around, the castle stands on a high hill, where a special platform had to be constructed on which to build the walls of the castle. People coming from different directions (e.g. neuroscience, psychoanalysis, and philosophy) will therefore perceive the same castle in different ways (e.g. as consciousness, self/objects, or mind). Most importantly, due to the hilly landscape, none of them see the platform underlying the castle, so they all think that they are talking about different things when they are communicating with each other. However, they all ask the same question: how could such a beautiful castle be built on such hilly terrain?

They therefore raise the issue of the platform underlying the castle. Some, like the neuroscientist, may deny any principal difference between the platform and the castle, thus only taking into account what can be observed, namely the castle itself, including its wall and its roof, which the neuroscientist consider to be its constructional correlates. The philosopher, in contrast, may reason that there must be more to the castle than its walls and its roof. He therefore assumes that there is something hidden inside, outside, or beneath the castle itself, which he calls "mind," that allows for the transformation of the mere walls and roof into a castle. Finally, the psychoanalyst may also assume that there is something more than the castle's walls and roof, but unlike the philosopher he "locates" this additional factor within the actual walls and roof of the castle by attributing a specific structure and organization to them. The structure and organization that distinguish the mere walls as the castle's walls then correspond to what Freud assumed to be the structure and the organization of the psychic apparatus.

From where do the organization and structure of the castle's walls and roof originate? They originate from the platform on which the castle is standing, and the platform's specific shape and structure are determined by its immediate environment, namely the hilly landscape. Although the platform should not be confused with the castle itself, the construction of the latter may have been impossible without the former. Therefore the platform is not a sufficient (i.e. executing and operational) condition, but rather an enabling and predisposing (and thus necessary but non-sufficient) condition of the possible construction of the castle and its walls. More specifically, it is the design, shape, and configuration of the platform that enable and predispose the castle's walls to be organized, shaped, and structured in a certain way, while at the same time excluding other ways. For instance, if the platform is rectangular, this may make the construction of a round castle impossible. Similarly, if the structure of the psychic apparatus was not tripartite, but was instead uni- or bipartite, our mental states and their respective contents would be different.

The platform that I have described corresponds well to the brain's resting-state activity, which provides the platform for subsequent stimulus-induced activity in the same way that the platform provides the foundation for the walls of the castle. Stimulus-induced activity may correspond to the walls of the castle and the various rooms, rather than to the platform. Although the self (ego) may be analogous to the castle as a whole (as it can be seen even from a distance in the Highlands), what can be seen from the castle itself may correspond to objects in the psychodynamic context. Therefore, when opting for neural predispositions rather than neural correlates, I shift my focus from the castle itself (including its various rooms) to its underlying platform, without which the castle's construction and its specific form and shape could never have been achieved.

In order to fully understand the brain's differentiation from self and objects, we need to go back to the "neural platform" (the brain's resting-state activity). We need to investigate how this neural platform affects the construction of self and objects, thus targeting the brain's own neural processing of intero- and exteroceptive stimuli from the body and the environment. Therefore, to understand the NPP as distinct from mere NCP, we need to consider the brain itself and more specifically its resting-state activity and how it enables and predisposes to the differentiation of self and objects (i.e. brain–self and brain–object differentiation).

Focus of the book: general aims and hypotheses

The general aim of this book is to link neuroscience and psychoanalysis and to demonstrate how the psychodynamic concepts of self and objects can be linked and related to specific neuronal mechanisms in the brain as their neural predisposition. The aim is thus to reveal which neuronal mechanisms our brain employs, and how it does so, to enable and predispose to the differentiation of both self and objects from the brain itself (i.e. what I call brain–self and brain–object differentiation). This leads me to four specific hypotheses.

First, I postulate that we need to refine our conceptual and methodological approaches to the brain in order to be able to properly investigate brain–object and brain–self differentiation. This will be the focus of Part I of the book (entitled Conceptual Equipment). Secondly, I postulate that a specific neural code (i.e. difference-based coding; see Chapter 4) is employed by the brain, which may be closely related to the psychodynamic concepts of cathexis (i.e. the investment of energy) and defense mechanisms (i.e. mechanisms that originally constitute and later defend self and objects). I then postulate that the brain's specific neural code (i.e. difference-based coding) enables and predisposes to brain–self and brain–object differentiation. This will be the focus of Part II of the book (entitled Neural Equipment).

Thirdly, I postulate that neuronal mechanisms which are closely related to the brain's neural code (i.e. difference-based coding) enable and predispose to the constitution of originally psychodynamic concepts, such as self, narcissism, and unconscious. This will be the focus of Part III of the book (entitled Mental Equipment). Finally, I postulate that the neuronal mechanisms are redirected and thus altered in neuropsychiatric disorders such as psychosis and depression, and can therefore lead to abnormal changes in brain–self and brain–object differentiation as psychodynamic key features in patients with these disorders. This will be discussed in Part IV of the book (entitled Disordered Equipment).

Plan of the book: overview of contents

Part I of the book (see Figure B), on conceptual equipment, discusses the theoretical issues relating to the concept of the brain. In the first chapter I take up some of Freud's metapsychological ideas, place them within the philosophical context of Kant, and discuss how the latter's transcendental approach can be fruitfully extended with regard to the concept of the brain. As has been observed, for example, by Brooks (1998), Freud's concept of the structure and organization of the psyche is related to Kant's concept of mind. The similarities and differences between Freud's approach to the psyche and Kant's approach to the mind are discussed and extended to the concept of the brain. The conceptual implications and empirical relevance of such a transcendental approach to the brain, as first suggested by the philosopher Schopenhauer, are discussed and placed in the context of neuropsychoanalysis.

Chapter 2 follows up on the transcendental approach to the brain and discusses the relevant epistemic implications (e.g. what we can and cannot know about our own brain). This can be regarded as an extension of Freud's concept of the "psychic apparatus" and his claim that we may remain unable to fully cognize and know it independent of ourselves and our knowledge of it. Here I focus on possible similarities and analogies between Freud's claims about what we can (and cannot) know about the psychic apparatus and our possible (and impossible) knowledge about our brain. I then distinguish between different concepts of the brain, namely the brain as

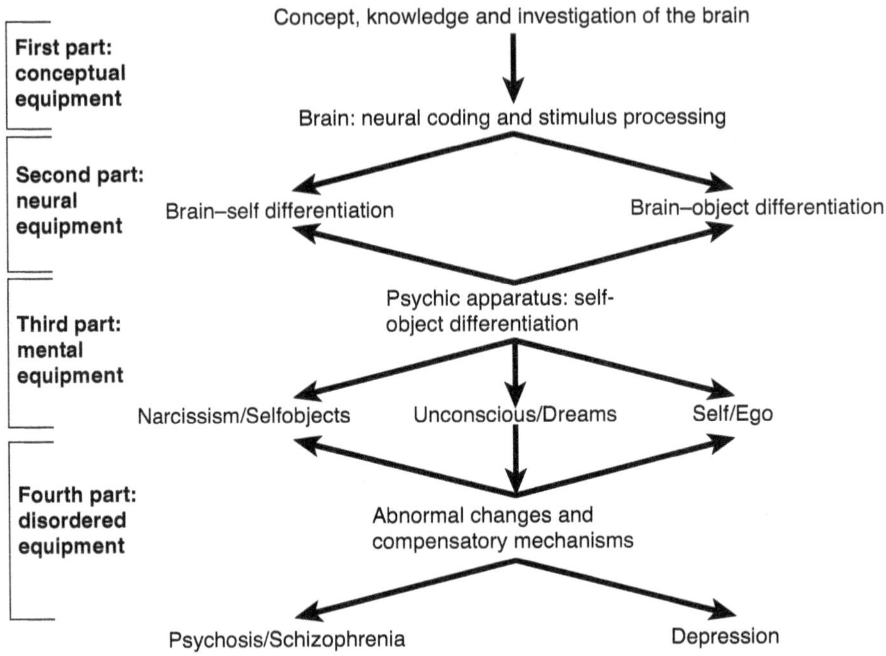

Fig. B. Plan of the book.

observed, the brain as functioning, and the brain as experienced. After a discussion of the empirical, epistemic, and conceptual relevance of these different concepts of the brain, it will become clear that we need to target the brain as functioning rather than the brain as observed in order to better understand the neuronal predispositions of brain–self and brain–object differentiation.

Chapter 3 raises the methodological question of how we can link and connect psychodynamic concepts and neuroscientific facts in a valid and reliable way. Investigation of brain–self and brain–object differentiation requires a methodology that goes beyond mere correlation between and mutual enrichment of psychodynamic concepts and neuroscientific facts. Rather than aiming to reveal mere neural correlates (e.g. consistency between or correspondence of specific psychodynamic contents) (see, for instance, Cathart-Harris and Friston, 2010, pp. 1–2, 13 and Carhart-Harris et al., 2008, p. 15), I go further by searching for those neuronal mechanisms and the kind of neural code that enable and predispose the brain to differentiate self and object from itself (i.e. brain–object and brain–self differentiation). For this we need to develop a specific methodological strategy that allows us not only to put psychodynamic concepts within the brain's neural context but also, conversely, to consider the brain's neuronal mechanisms within the psychodynamic context of self and objects. This makes it necessary to mutually compare and match psychodynamic concepts and neuroscientific facts with each other, resulting in what I call "neuropsychodynamic iterativity." I develop an outline of neuropsychodynamic iterativity as transdisciplinary strategy in Chapter 3, and this will be the methodological approach that guides the empirical Parts II, III, and IV of the book.

Part II of the book, on neural equipment, investigates the brain's neuronal predisposition and determines a specific form of neural coding (i.e. difference-based coding) for this. Difference-based coding describes how all incoming stimuli (i.e. intero- and exteroceptive) are coded and processed in relation to the brain's resting-state activity. Chapter 4 focuses on the psychodynamic concept of cathexis and its relationship to the brain's energy, including how it invests its own resting-state activity in stimulus-induced activity, thereby enabling and predisposing to difference-based coding. This leads on to a consideration of recent findings on rest–stimulus and stimulus–rest interaction mirroring the brain's employment of difference-based coding of stimulus-induced activity.

Chapter 5 moves on from the purely neuronal ground and describes in detail how the brain's difference-based coding enables and predisposes to the transformation of neuronal states into mental states, which I call "neuronal–mental transformation." The concept of mental states is here understood in a phenomenological sense as referring to the subjective experience of phenomenal or mental states from the first-person perspective, as distinct from the observation of neuronal states from the third-person perspective. Here I shall specifically focus on two key features of mental states, namely their "going beyond" the information that is provided by the stimulus itself, and their intrinsic or intentional relationship to the world. Finally, difference-based coding as supposedly enabling and predisposing the constitution of mental states is discussed within the context of Freud's own account of the neural genesis of mental states in his *Project for a Scientific Psychology* (Freud, 1895).

Chapter 6 goes on to discuss how difference-based coding and the associated neuronal mechanisms of rest–stimulus and stimulus–rest interaction enable and predispose to the constitution of self and objects as distinct from the brain. I specifically focus on early defense mechanisms, such as internalization (e.g. introjection) and externalization (e.g. projection), as they enable and predispose the brain to first constitute and later defend self and objects. Since both internalization and externalization are crucial when constituting self and objects, they are postulated to enable and predispose to brain–self and brain–object differentiation. This chapter thus has an essential role in that it bridges the gap between the neuroscience of the brain and the psychodynamic concept of the psychic apparatus as characterized by objects and a self.

Part III of the book focuses on the mental equipment of the psychic apparatus, self-objects, and narcissism, the unconscious as it surfaces in dreams, and finally the self and its ego. Chapter 7 deals with self-objects in the context of the psychodynamic concept of narcissism as introduced by Kohut and self psychology. Thus narcissism is to be understood not merely as a pathological condition or as a personality trait in healthy individuals, but rather as a basic and universal feature of the human psyche. Neuronal mechanisms that focus on affective assignment and intero-exteroceptive linkage are discussed, and are related to the psychodynamic concept of self-objects that is regarded as central to narcissism.

Chapter 8 focuses on yet another piece of mental equipment, namely the unconscious and its specific manifestation in dreams. I demonstrate that the same neuronal mechanisms that enable and predispose to brain-object and brain-self differentiation in the awake state are also at work in the dreaming state, although in a slightly different gestalt. This accounts not only for the various mental features that are often observed in dreams, but also for the transition between unconsciousness and consciousness that was so well highlighted by Freud himself.

Chapter 9 discusses the psychodynamic concept of the self and its relationship to that of the ego within the neuropsychodynamic context. Different concepts of the self in neuroscience and psychoanalysis are directly compared with each other with regard to their plausibility and compatibility with the current empirical (i.e. neuronal) findings with regard to the self. I then develop and discuss what I call the "relational concept" of the self, as distinct from the concepts of the ego and the id.

Finally, Part IV of the book concerns the application of the conceptual, neural, and mental equipment as raised in the first three parts of the book to neuropsychiatric disorders such as depression and psychosis. Chapter 10 is about depression, where the aforementioned psychodynamic key features of internalization and introjection resurface in a distorted gestalt. Rather than focusing on abnormalities in specific brain regions, I here highlight the abnormal redirection of otherwise normally functioning neuronal mechanisms and their consecutive induction of the symptoms and psychodynamic key features that characterize depression.

Chapters 11 and 12 focus on psychosis. Due to the complexity of psychosis, I have devoted the whole of Chapter 11 to a discussion of its psychodynamic key features, including object loss, abnormal projection, and the splitting and fragmentation of the self. The corresponding neuropsychodynamic hypotheses are then developed in Chapter 12. As in the case of depression, rather than focusing on single neuronal deficits, I highlight the employment of normally functioning neural coding (i.e. difference-based coding) within an abnormal neuronal context. And it is the employment of the otherwise normally functioning difference-based coding that is supposed to enable and predispose to the complex symptoms and psychodynamic features which are so typical of psychosis.

This is followed by an Appendix that discusses what we can learn from these neuropsychiatric disorders about the healthy brain. Freud regarded the various neuropsychiatric disorders as the starting point for learning something about the healthy psyche (i.e. how the healthy psychic apparatus must function if it can produce the kinds of symptoms that are observed in such pathological conditions). The Appendix adopts this approach and claims that depression can be regarded as a paradigmatic example of altered brain–self differentiation, whereas in psychosis brain–object differentiation may be altered in a paradigmatic way. These alterations are only possible if specific neuronal mechanisms (e.g. a specific kind of neural coding, such as difference-based coding) are at work. Thus depression and psychosis indirectly tell us something about the neural predispositions of brain–object and brain–self differentiation in the healthy brain which we apparently remain unable to tackle directly.

I conclude the book with an imaginary and slightly humorous dialog between a psychoanalyst, a neuroscientist, and a philosopher. By debating how neuroscience and psychoanalysis can and cannot be linked together, they examine many of the issues raised in this book from their respective intradisciplinary perspectives. In so doing they highlight some of the possible difficulties, obstacles, and counterarguments that such a transdisciplinary approach to neuropsychoanalysis will almost certainly encounter.

Guidance for the reader

I am aware that my approach to neuropsychoanalysis in this book is truly transdisciplinary in that it invokes such diverse disciplines as philosophy, neuroscience, psychoanalysis, and psychiatry (see Figure C). In addition to the potential strengths of such an approach, it is of course prone to criticisms and counterarguments, which often stem from intradisciplinary perspectives and requirements (see the Epilogue for further details). This may also make it difficult for the reader to follow the book in all its detail. I therefore wrote each part of the book (and where possible each chapter as well) in such way that it could stand on its own more or less independent of the other parts (or chapters).

How should one go about reading such a complex book, touching as it does upon topics and issues that are dealt with in different disciplines? Although, as the author of the book, I of course recommend that the whole book should be read from beginning to end, readers with different disciplinary backgrounds may wish to focus only on particular aspects. The philosopher and neurophilosopher may be most interested in Chapters 1 and 2, where I discuss the notion of the brain and its conceptual and epistemic ramifications, thereby drawing heavily on Kant's transcendental framework. However, the philosopher of science may want to focus on Chapter 3, where I develop a novel methodological strategy, namely neuropsychodynamic iterativity,

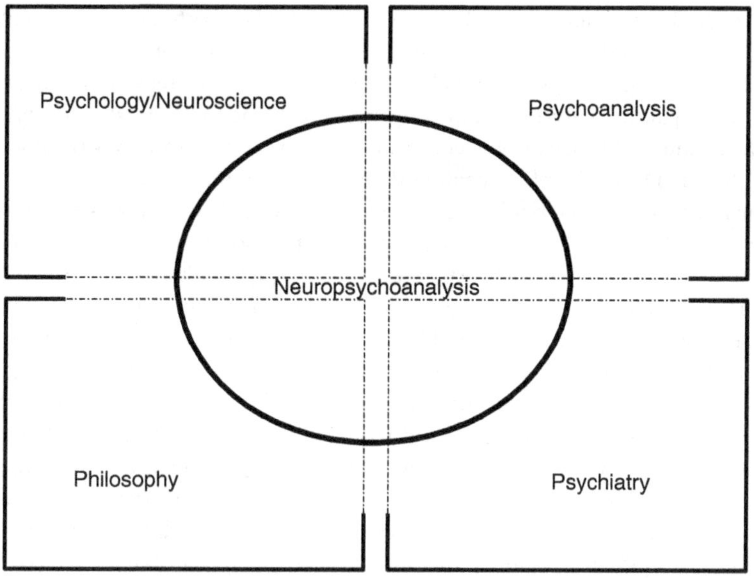

Fig. C. Transdisciplinary approach to neuropsychoanalysis.

for transdisicplinary approaches that may also be applied to the links between disciplines other than neuroscience and psychoanalysis (e.g. between neuroscience and anthropology) (Northoff, 2010).

The neuroscientist may be most interested in Chapter 4, in which I discuss rest–stimulus interaction and postulate that a specific form of neural coding (i.e. difference-based coding) is at work in the brain's neural processing of intero- and exteroceptive stimuli. In Chapter 5, the application of difference-based coding to the question of how neuronal states are transformed into mental states may be especially relevant for those neuroscientists with an interest in consciousness. Finally, Chapter 6, which focuses on defense mechanisms, may be of particular interest to neuropsychoanalysts and psychoanalysts, as it is here that I make the transition from the neuronal context of the brain to the psychological (e.g. the psychodynamic context of the psychic apparatus).

Part III of the book, on self, narcissism, and unconsciousness, may be the domain of the neuropsychoanalysts and interested affective and social neuroscientists. Part IV, on depression and psychosis, may be particularly relevant for neuropsychoanalysts focusing on disorders, and for psychiatrists and clinical neuroscientists (and psychotherapists). Finally, the Epilogue is written for readers from all disciplines as well as for those who are not specialists in one of the aforementioned fields. The format, consisting of a lively and humorous dialog, should give an impression of the struggles and difficulties that are involved in transdisciplinary work in general and neuropsychoanalysis in particular. The reader may therefore, after having read this Introduction, wish to read the Epilogue next, and then turn to the chapters of interest.

Finally, it should be noted that the transdisciplinary approach that has been adopted here makes it impossible to spell out all of the philosophical, neuroscientific, and psychodynamic details with which the respective specialists will be familiar. Instead I focus on the links between neuroscientific mechanisms and psychodynamic concepts, with these links being the main topic. Hereafter I discuss the different meanings and definitions of specific psychodynamic concepts only where it is necessary to do so in order to link them to specific neuroscientific mechanisms.

Otherwise I disregard psychodynamic details. The same applies to philosophical details, especially in Part I of the book, and to the neuroscientific findings and mechanisms that are discussed in Parts II, III, and IV. Although I am well aware that such a strategy may not be altogether welcomed by the psychoanalyst, the philosopher, and the neuroscientist, and may give rise to criticism on grounds of being overly selective and exclusive, one has to draw the line somewhere, and the line here is a transdisciplinary rather than an intradisciplinary one.

Particular limitations of the book concern psychodynamic key features such as memory and psychotherapy. I allude to memory at various points while not discussing it in full detail. This is not due to a lack of interest in memory, but rather to the fact that the neuropsychodynamic processes targeted here are more basic and are supposed to occur prior to the instantiation of memory, although at the same time enabling and predisposing to it. The situation is slightly different for psychotherapy, which is completely neglected in this book. Such neglect does not mirror lack of interest and ignorance. In fact, being a practising psychiatrist and psychotherapist myself, I am extremely interested in psychotherapy in general, and in psychodynamic psychotherapy in particular. I envisage in the future a form of psychodynamic psychotherapy that is based on specific neuronal mechanisms and their relationship to particular psychodynamic mechanisms. One could refer to this as "neuropsychodynamic psychotherapy" or, as I prefer to call it, "brain-based psychodynamic psychotherapy." However, if brain-based psychodynamic psychotherapy is to develop, we need to be clear about the relationship between neuronal and psychodynamic mechanisms, and the present book aims to contribute to this. Once we are clear about the details of the neuronal–psychodynamic relationship, we shall be able to develop an empirically and conceptually plausible brain-based psychodynamic psychotherapy, about which I and others will be able to write another book.

Acknowledgments

I am grateful to Stephan Doering and Oliver Lyttelton for reading and providing very helpful comments on previous drafts of the introduction.

Part I

Conceptual Equipment

This part focuses on theoretical issues relating to Freud's metapsychology and his approach to what he calls the "psychic apparatus." Analogous to Kant's approach to the mind, Freud approached the psychic apparatus in a transcendental way (see Chapter 1). In addition, Freud assumed that the psychic apparatus itself, and its real nature, remain hidden from us, thus indicating that our knowledge (i.e. our epistemic ability or lack of it) has major limitations (see Chapter 2). My first aim here is to put these features of the psychic apparatus into a philosophical framework, specifically a transcendental one. This paves the way for my second aim, which is to extend Freud's transcendental approach from the psychic apparatus to the brain.

Based on Kant and Schopenhauer, I shall suggest a transcendental approach to the brain in Chapter 1, and diagnose an epistemic limitation in our possible access (e.g. perception and experience, our own brain as brain) in Chapter 2. These more theoretical considerations are accompanied by a discussion of the implications of the brain's transcendental and epistemic features for empirical and experimental investigation of the brain and its relationship to psychodynamic mechanisms. I shall argue that consideration of the brain's transcendental and epistemic features is necessary to properly account for brain–self and brain–object differentiation on conceptual grounds. And most importantly, I shall argue in Chapter 2 that this will also allow access to those empirical features in the brain that may enable and predispose to brain–self and brain–object differentiation.

Chapters 1 and 2 on Freud's metapsychology of the psychic apparatus and its extension to the brain are followed by a chapter on the appropriate methodological strategy in neuropsychoanalysis. Following Freud's own positioning of psychoanalysis in an intermediate position between philosophy and the humanities on the one hand and science on the other, I here propose a method that allows psychodynamic concepts and neuroscientific observations to be systematically linked and mutually validated. Since such truly transdisciplinary methodology requires continuous and iterative movement between psychodynamic concepts and neuroscientific facts, I here speak of "neuropsychodynamic concept–fact iterativity."

Chapter 1

Transcendental Approach to the Brain

Philosophical concepts

The psyche's input to cognition: Kant and Freud

Freud was interested in the psyche or the "psychic apparatus" as he called it. More specifically, he focused on how the human psyche works, how it is organized and structured, and how in turn its structure and organization determine the contents that it can process and represent. For instance, consider Freud's text about the ego and the id (Freud, 1923a). The first two chapters are about the structure and organization of the ego, and its functions and characterization. The ego and its tripartite structure describes the organization of the psychic apparatus which it applies to the processing of stimuli. However, at the end of the second chapter there is a shift, as Freud starts to describe the ego in terms of specific contents such as bodily contents as distinct from its functions and structure. This moves the focus from the structure and organization of the ego to its contents, as has been pointed out by Brook (2003).

This double focus on both structure/organization and content led Freud to suggest a twofold distinction with regard to the psychic apparatus. The human psyche can be characterized first by a certain structure and organization, and secondly by the stimuli and their associated contents that it processes. The structure and organization are, for instance, well manifested by the tripartite structure (ego, superego, and id) and the various defense mechanisms (e.g. projection, introjection, reaction formation, etc.) (for further details, see Parts II and III of this book). These features of the structure of the psychic apparatus are applied to any kind of incoming stimulus, and this interaction results in what I shall refer to as "contents." By exerting the effect of its structure and organization upon the stimuli and their processing, the psychic apparatus can modulate and influence the latter according to its own specific needs and purposes. In other words, the psychic apparatus imprints itself on the processing of incoming stimuli and the resulting contents.

One may further refine the two features of the psychic apparatus. The structure and organization characterize the psychic apparatus itself, which is therefore not an "empty box," but rather a box containing various tools that are designed to be used for the kinds of stimuli and contents that are to be processed. One may thus liken the psychic apparatus to a toolbox, and the tools within it can be regarded as its input to its processing of stimuli from the outside world and our cognition and perception of them. In contrast, the stimuli and contents originate not from the psychic apparatus itself (the toolbox), but from the outside world, so must be regarded as the world's input. There is thus a double input to be considered. The input from the psychic apparatus imprints itself on the processing of stimuli as the input from the world. The two inputs interact with each other, resulting in the output, which is our perception and cognition of contents in the world. Thus the output of the psychic apparatus may be regarded as a mixture or hybrid of the input from the psychic apparatus and the input from the world.

How can we further illustrate the distinction between the input and the output of the psychic apparatus? Consider a specific defense mechanism, such as projection. Projection describes the externalization of personal contents when perceiving the other person in terms of one's self (for details, see Part II). This occurs, for instance, when one attributes certain features, such as being malicious, to the other person, although these features may in fact be an aspect of oneself. What is the input and what is the output in this instance? The output is clearly the attribution of malice to the other person. This output (i.e. the content), may clearly vary from case to case and thus from one person to another (e.g. being malicious, or being virtuous). The psyche's output thus describes the content of projection.

How can we describe the input from the world and the input from the psychic apparatus in more detail? This is easily done in the case of the world's input, as this refers to the stimuli coming from and associated with another person. However, in contrast to the world's input, and the output of the psychic apparatus, we do not have direct access to the input of the psychic apparatus. Thus we can only access it indirectly. We can infer from the defense mechanisms such as projection the kind of structure and organization that the psychic apparatus must employ in order to enable and predispose to projection.

How did Freud come to presuppose such methodology when inferring the structure and organization of the psychic apparatus in this way? Although he was not particularly fond of philosophy, due to its speculative nature, its strong focus on consciousness and its corresponding neglect of the unconscious (see Chapter 3 and also Brook, 2003; Tauber, 2009), Freud clearly presupposes the philosophy of Kant. Brook (2003) and Tauber (2009) have recently highlighted the importance of Kant for Freud in relation to, among other aspects, the structure–content distinction, the tripartite structure of the mind/psychic apparatus, the relationship between percepts and concepts, and the concept of unknowability as the noumenon.

In this chapter I shall focus on the structure–content distinction that mirrors the distinction between the input and the output of the psychic apparatus. Analogous to Freud's concept, Kant distinguished between input and output of cognition. However, due to Kant's philosophical and thus logical and epistemic context, he attributed the input to the mind rather than to the psychic apparatus, whereas Freud presupposed a rather psychological and thus empirical context. This difference in context means that Kant focused more on knowledge whereas Freud referred to empirical functions such as cognition, perception, desires, and so on. The contextual difference between Kant and Freud is important, as it makes direct one-to-one translation of Kant's concept of the mind into Freud's concept of the psychic apparatus impossible (more detailed information about such one-to-one translation can be found in Chapter 3). However, it does not exclude the possibility of indirect translation, which is the path I shall pursue here.

How did Kant specify and characterize the mind's input to cognition? The most important issue here is what we can learn from this with regard to the psychic apparatus and, most importantly, the brain. My aim is to extend Kant's characterization of the mind and Freud's characterization of the psychic apparatus to the brain, by revealing the brain's input to our cognition and perception. In my view we need to determine the brain's input to our perception and cognition, as distinct from the brain's output, in very much the same sense that Kant did with regard to the mind and Freud postulated for the psychic apparatus. I therefore intend to extend Kant's philosophical context of the mind and Freud's psychological context of the psychic apparatus to the brain and its neuronal context, in order to establish and develop a neuropsychoanalysis in the spirit of Freud.

The mind's input to cognition: transcendental approaches in Kant and Freud

Let us begin with Kant and the way in which he conceived the mind's input to cognition.

Kant was primarily interested in the way that we as humans perceive the world, which he traced back to reason and understanding. In order to reveal the specific human mode of cognition, he investigated the contribution of reason and understanding by posing the following question: "What must reason and understanding be like in order for us to be able to cognize the world in our specific human mode?" Kant suggested that reason puts certain constraints on our cognition, and Kitcher (1990, pp. 221–227) speaks of the "demand of reason." Our cognition and especially our mode of cognition must show a specific structure that can only be traced back to the contribution of reason, and reason puts certain "demands" on our cognition of the world by predisposing it to a specific organization and structure.[1] In other words, we perceive the world according to the design of reason, and thus according to the contribution that reason has made to our cognition of it.[2]

Reason and understanding are thus the faculties or functions of the mind that provide the mind's input to our cognition of the world. More specifically, reason and understanding are responsible for what Kant called the "categories," namely certain principles according to which the mind structures and organizes our cognition. By applying the categories that originate in the faculties or functions of reason and understanding to the stimuli that we receive from the outside world, the latter are organized and structured in such way that they become available as contents in our cognition, thus denoting the output of cognition.

How can we investigate the mind itself and its input to our cognition of the world?

If the mind's input to cognition must be distinguished from its output, we may also need different methods to account for input and output. Kant distinguished between an empirical and a transcendental view of cognition. The empirical view focuses on the contents of cognition and hence what we can perceive, such as certain objects or events in the world (Kant (1998) (as translated by Guyer and Wood), A56–7/B81), which is conceptualized here as the mind's output. The transcendental view of cognition focuses on the mind's input to cognition, namely the form or mode in which we perceive the objects or events in the world: "I call all cognition transcendental that is occupied not so much with objects but rather with our mode of cognition of objects insofar as this is to be possible a priori." (Kant (1998) (as translated by Guyer and Wood), A11–12).

The transcendental view focuses on those features that the mind itself, independent of the world and its input, puts into our cognition of the world. Kant characterizes such cognition of the mind's input as "a priori," and distinguishes it from the cognition of the mind's output, the events and objects of the world, which he designates as "a posteriori": "namely that we cognize of things a priori only what we ourselves have put into them" (Kant (1998) (as translated by Guyer and Wood), Bxviii). The question then arises of what is specific for the cognition of the mind's input as opposed to the mind's output. We describe the objects and events that we perceive in our cognition of the world by using terms or, better still, concepts. The fact that we use concepts in general and specific concepts in particular may be traced back to the mind itself and its specific input. If this is so, the cognition of the mind's input, which Kant characterized as transcendental and a priori, must concern the concepts that we use to describe our cognition, the "a priori" concepts of objects (understood in a Kantian rather than Freudian sense).

---

1 It should be noted that reason prestructures only cognition of the world, not the world itself. The contribution of reason is thus determined epistemically (i.e. cognition of the world) rather than ontologically (i.e. the world itself). This distinction is implicitly presupposed by Allison (1983, pp. 14–35) when he argues that we have to expect nature (i.e. the world as perceived by us) to show a certain structure since, given the way that reason constitutes our cognition of the world, that is the only way in which we could acquire knowledge.

2 This contribution of reason to our cognition may in turn be perceived by itself, and it is this cognition that Kant characterized as a priori: "namely that we cognize of things a priori only what we ourselves have put into them" (Kant, 1998, Bxviii).

Although I do not wish to pursue the exact role of concepts here (further details can be found in Chapter 3), I want to rule out some possible misunderstanding of what Kant meant by the transcendental view of the mind's input to our cognition. The transcendental view of knowledge does not aim to target any objects (in a Kantian sense) that lie beyond our possible cognition and thus also beyond the a priori concepts of such objects. The concept of the "transcendental" must thus be distinguished from the concept of "transcendent," which goes beyond or transcends any possible knowledge of humans into a world that lies beyond that which we inhabit.

Let me rephrase this important distinction. Whereas the concept of "transcendental" concerns possible knowledge of the objects within the world which we can possibly perceive,[3] the notion of "transcendent" goes beyond the objects that we can possibly perceive by postulating some objects in a world that lies beyond our possible cognition and knowledge (i.e. a transcendent world). The term "transcendental' thus refers to the mode in which we perceive objects in the world, whereas the term "transcendent" refers to objects in a non-natural world that we cannot perceive at all (see also Kant (1998) (as translated by Guyer and Wood), p. 717, footnote 6).[4] Thus there is nothing mysterious about the transcendental view of the mind's input to our cognition of objects and events in the world, whereas the search for transcendent objects beyond and thus outside our (cognition of the) world is mysterious.

What can we learn about Freud from Kant? Can we draw an analogy between what Kant characterized as the mind's input and Freud characterized as the psyche's input? In a similar way to that in which Kant presupposed a specific kind of cognition to account for the mind's input, Freud must (although rather implicitly) have assumed a specific view of the psyche (i.e. a transcendental view), in order to account for the psyche's input (as distinct from the psyche's output). Like Kant, Freud was not interested in postulating some transcendent feature or properties that lie beyond our (possible cognition of the) world, and thus adopting a transcendent view. Instead his focus was, in a similar way to that of Kant, on the structure and organization of the psychic apparatus. Although Freud does not appear to explicitly mention such a transcendental view of the human psyche, he must have presupposed it, otherwise he would not have been able to draw the distinction between the psyche's input (i.e. its structure and organization) and its output (i.e. its contents). However, before going on to discuss Freud's concept of the psychic apparatus and its possible extension to the brain, we need to learn more about the distinction between structure/organization and content. For this we again return to Kant, where this distinction appears in the gestalt of that between mode/form and content of cognition.

Form and content: Kant's concept of cognition as "composite" and Freud's concept of the psychic apparatus

What is meant by form or mode of cognition? As a preliminary definition I want to describe form or mode as organization and structure according to which it is possible for us to perceive and know objects or events in the world. By applying a specific organization and structure to all

3 The transcendental view thus concerns a reflection upon the objects and events that we can perceive which takes place within the same world and concerns the same objects as the empirical view that focuses on the cognition of the objects and events themselves (see also Svare, 2006, p. 150, who interprets Kant in this way).

4 In other words, the concept of "transcendent" has ontological–metaphysical implications, and thus belongs to the ontological–metaphysical domain, whereas the concept of the transcendental remains purely epistemic. This also makes it clear that the presupposed concept of the "transcendental" does not refer to the possible cognition of objects outside the limits of our cognition and a priori cognition (for a more refined post-Kantian concept of the transcendental, see McDowell, 2002).

incoming stimuli from the world, the mind provides an input that in turn makes it possible for us to perceive these as objects or events in the world. This organization and structure as the mind's input and how it influenced our cognition and knowledge of the world (including the mind as part of this world) was the target of Kant.

The form or mode of cognition describes the rules and principles according to which the cognition of objects is organized. The rules and principles determine how different objects are ordered and organized in our cognition and, most importantly, how they are related to each other when we perceive them. The order and organization in cognition are reflected in the relationship between the cognized objects, and must therefore be distinguished from cognition content that describes the objects of cognition. The relationship originates in the mind's input (i.e. the structure and organization that it employs when processing the incoming stimuli). The objects of cognition can be traced back to the stimuli from the world, and hence ultimately to what Kant in the *Prolegomena* refers to as "matter" (Kant (1998) (as translated by Guyer and Wood), pp. 282–283, p. 375).

In contrast, Kant locates the origin of order and organization (i.e. cognition form) in the logical domain as provided by the mind's faculties or function of reason and understanding. This is well reflected in the following quote: "I call that in the appearance which corresponds to sensation its matter, but that which allows the manifold of appearance to be ordered in certain relations I call the form of appearance. Since that within which the sensations can alone be ordered and placed in a certain form cannot itself be in turn sensation, the matter of all appearance is only given to us a posteriori, but its form must all lie ready for it in the mind a priori, and can therefore be considered separately from all sensations." (Kant (1998) (as translated by Guyer and Wood), A20/B34).

According to Kant, form and content (or better still, matter to which content can be traced back) are reciprocally related to each other. Form requires matter to employ its principles and rules of organization, and thus to constitute the objects and their contents in cognition and to relate them to each other. Without matter, form remains empty and thus without any impact on cognition. Conversely, matter requires form in order to become accessible to cognition by being transformed into objects and content. Without form, matter remains blind and cannot enter cognition. Form and matter are thus both not only necessary conditions of cognition,[5] but must also be reciprocally linked to each other to make cognition possible at all.[6]

The reciprocal linkage between form and matter implies that cognition is, in the terms used by Kant, a "composite" of form (what he here calls "cognitive faculty") and matter (what he here also calls "sensible impressions" or "fundamental material"): "For it could well be that even our experiential cognition is a composite of that which we receive through impression and that which our own cognitive faculty (merely prompted by sensible impressions) provides out of itself, which addition we cannot distinguish from that fundamental material until long practice has made us attentive to it and skilled in separating it out" (Kant (1998) (as translated by Guyer and Wood), A, A2/B5, B1–2).

The determination of cognition as composite of form and content shifts the focus in the investigation of cognition. Rather than focusing exclusively on content (as philosophical empiricists such as Locke and Hume did) or form (as philosophical rationalists such as Leibniz or Berkeley did), the investigation should concentrate on the ways and gestalts in which form and content are

---

[5] Collins (1999, p. 9) also refers to "subjective and objective conditions." Subjective conditions are the constitutional or internal features in our mind, which therefore reflect what Kant calls reason/understanding or form, whereas objective conditions are the external conditions, and thus correspond to what Kant calls sensibility or matter.

[6] For example, "The former (i.e. matter) signifies the determinable in general, the latter (i.e. form) its determination" (Kant, 1998, A266/B322).

linked to each other in cognition. The form should not be detached from the cognition content, otherwise one may easily slide into the realm of unknowable or transcendent objects that lie beyond our empirical world in a transcendent and rather mysterious world. One may then be at risk of confusing the transcendental view with a transcendent account. Conversely, if cognition content is detached from or confused with its form (as in empirical approaches), one may be tempted to neglect the mind's input itself, and the form and mode that it applies to our cognition of the world. One may then be at risk of confusing the transcendental with the empirical view.

Kant's assumption of a reciprocal linkage between form and content thus teaches us an important lesson. In contrast to empirical approaches to the mind, we should not focus exclusively on the content and the mind's output and thereby neglect the mind's input, including the form or mode of our cognition. Nor should we go to the opposite extreme, as in rationalist approaches that focus exclusively on the mind's input, including the form or mode of cognition. while neglecting the contents in our cognition of the empirical world (i.e. its objects and events that we can perceive).

Freud appeared to have taken this lesson to heart when he described the psychic apparatus in terms of both a specific form and mode on the one hand, and specific contents on the other. For instance, he characterized the psychic apparatus as having a threefold structure, consisting of the id, the ego, and the superego, which has some astonishing parallels with Kant and his description of the structure of the mind (for an excellent account, see Brook, 2003). Rather than discussing the details of the analogy between Freud and Kant with regard to the tripartite structure of the mind and the psychic apparatus,[7] I here want to point out that Freud, in the same way as Kant, saw the crucial importance of linking form and content.

Like Kant, Freud (implicitly) presupposed that our cognition of the world (including ourselves and our psychic apparatus) is a composite of the psyche's input (i.e. the form and mode that it employs) and the world's input (i.e. the stimuli that we receive from the outside world). Consider the example of the threefold structure of the psychic apparatus, consisting of the id, the ego, and the superego. Freud assumed that this threefold structure has a major role in organizing and structuring the psychic processing of the stimuli from the outside world, which in turn enables and predisposes us to access and thus perceive and cognize these stimuli as psychological contents (i.e. as corresponding to what Kant called objects and events). Due to the psyche's input to the subsequent psychic processing of stimuli from the outside world, we are able, for instance, to react to these stimuli with defense mechanisms as we are enabled and predisposed to do so by the psyche's input.

Interestingly, Freud appears to have gone even further than Kant with regard to the interaction between the psyche's input and the world's input. Kant assumed that the mind's input (i.e. reason and understanding) was innate, and that as such it must be fully developed and ready for use prior to the first encounter with and experience of the world. The form and thus the structure and organization that the mind applies to our cognition must exist prior to any possible experience of and encounter with the world. In other words, the psyche's input must be innate (or intrinsic).[8] Although Kant assumed that there was some interaction between the mind and the world, he does not appear to have regarded it as relevant in constituting the mind's input itself.

7 Freud's concept of the superego corresponds to the moral function of what Kant called reason when he characterized the mind by the categorical imperative. The ego in Freud as synthesizing the id and superego may be more or less paralleled by Kant's concept of understanding and its synthetic powers. Only Freud's concept of the id may have no clear analog in Kant, and at best may be related to what he called sensibility. Instead, Freud's id may closely correspond to what Schopenhauer called the will (Young and Brook, 1994).

8 However, there is some debate about what Kant meant by "innate," which may not necessarily exclude some kind of "original acquisition" (Kitcher, 1992).

This is obviously different from Freud's approach, as has also been pointed out by Brook (2003). Freud argued that even the tripartite structure of the psychic apparatus, and in particular the ego and the superego, are developed and constituted on the basis of the interaction with the world. Thus the ego and the superego do not exist prior to experience of the world, as in the case of Kant's concepts of reason and understanding, but rather they are acquired by being constituted and developed within the interaction between the psyche and the world. This allowed Freud to emphasize the importance of child development in constituting the ego and the superego.

The only innate element that Freud assumed in his concept of the psychic apparatus appeared to be the drives, the libido, and thus the id (which was later complemented by Eros as another innate element).[9] Another innate element, which Freud did not explicitly mention but implicitly presupposed, may be the predisposition to continuously interact with the world. Such psyche–world interaction must be necessarily presupposed to enable and predispose to the constitution of the ego and superego on the basis of the drives, the libido and the id (see Chapter 5 for further elaboration of this point in relation to the concept of embeddedness in the context of Freud). The psyche–world interaction as another apparently innate element will become crucially relevant later, when Freud's concept of the psychic apparatus is extended to the brain, as the brain's specific way of processing stimuli from the world may predispose and enable us to continuously interact with the world, and thus to develop an ego and a superego.

What does this mean for the relationship between Kant and Freud? Freud not only shifted Kant's higher cognitive faculties of reason and understanding to the lower level of drives and libido, but also placed even more emphasis on the crucial role of the interaction between the psyche and the world. Kant restricted the interaction between the mind and the world to our cognition of the world, while apparently sparing the mind's input from that interaction. Freud went one step further by not only exposing the cognition of ourselves and the world to the interaction between the psyche and the world, but also most parts of the psyche itself, including the psyche's input. To use Kantian terminology, one may characterize as "composite" not only our cognition of the world, but also the psychic apparatus itself (i.e. its structure and organization), resulting from the psyche–world interaction.[10]

Transcendental approach

Transcendental psychology versus cognitive psychology: identification or distinction in Kant and Freud?

How does Kant's concept of mind and its characteristic features, the mind's input and the transcendental approach, translate into the current discipline of psychology in general and cognitive psychology in particular? This question is of interest not only for relating Kant's approach to the current

---

[9] Subsequent psychoanalysts shifted this to other elements. For instance, Kohut in his later work assumed the narcissistic drive rather than the libidinal drive to be innate.

[10] I am well aware that, from the philosopher's point of view, this conclusion is logically circular. How can the psyche be a composite that results from the interaction between the psyche and the world? The psyche cannot be presupposed as one element of an interaction from which it is supposed to result. Although the philosopher is clearly correct in logical terms, he may be wrong in empirical terms, as Freud and many of his successors demonstrated. This may stipulate that the philosopher searches for concepts that may be able to describe such an interaction, where there is at least conceptually a strong overlap between output and input (i.e. the concept of psyche in our case, in a non-circular and thus logically compatible argument (see, for example, my concept of "co-occurrence and co-constitution," which represents a first attempt in this regard; Northoff, 2004a,b).

disciplinary landscape, but also in the current context of Freud. I characterized Freud's approach to the psychic apparatus as transcendental. If we are able to link Kant's transcendental approach to the mind to current psychology (e.g. cognitive psychology), we may also be able to extend Freud's approach from the psychic apparatus to cognitive psychology. Since cognitive psychology is nowadays closely associated with neuroscience (i.e. cognitive neuroscience), linkage of Freud and his concept of the psychic apparatus to cognitive psychology may open the door for psychoanalysis to be linked to the brain and thus to neuroscience. However, before discussing this in detail we need to return again to Kant and consider how his approach stands in relation to the cognitive psychology of our time, as has been discussed extensively by Patricia Kitcher (1990) in her book about Kant.

As has already been mentioned, content can be traced back to matter and therefore to the empirical domain and its sensible functions. In Kant's time, content was determined clearly and exclusively as sensible. However, in our time the recent discovery of cognitive functions such as working memory, attention, etc. raises the question of whether these should be characterized by either form/mode or cognitive content as opposed to mere sensory content. This problem has been well delineated by Kitcher (1990), who has characterized transcendental psychology in relation to both cognitive and empirical psychology. Kitcher (1990) considers cognitive tasks or functions such as attention, working memory, consciousness, etc. to be the subject of transcendental psychology: "Later I argue that transcendental psychology analyzes cognitive tasks to determine the general specifications for a mind capable of performing those tasks. That is how Kant is going to show that certain aspects of our knowledge are grounded in our faculties" (Kitcher, 1990, p. 13).

Although this quote does not explicitly distinguish between cognitive and non-cognitive (i.e. perceptual or sensory) tasks, it does seem to imply such a distinction. If this is true, transcendental psychology can be characterized by the investigation of cognitive tasks as higher faculties, as distinct from sensory or perceptual tasks as lower faculties, which are dealt with in empirical psychology. Kitcher supports her interpretation of the transcendental–empirical distinction in terms of higher–lower faculties by citing several examples in which she associates the transcendental view with cognitive tasks or capacities (see Kitcher, 1990, pp. 17–18). What Kitcher calls transcendental psychology may then correspond to what is nowadays called cognitive psychology, whereas what she calls empirical psychology appears to correspond to what may now be described as perceptual psychology. In other words, she seems to identify the notion of "transcendental" with the concept of "cognitive."

Such identification of transcendental psychology with cognitive psychology is further supported by Kitcher's definition of transcendental psychology as the investigation of cognitive faculties: "Transcendental psychology investigates the faculties required for the performance of basic cognitive tasks. . . . The only goal is to explore the requirements of various cognitive tasks. In this respect his work is centrally in epistemology and very different from empirical psychology" (Kitcher, 1990, p. 25). Taken together these quotes seem to suggest that Kitcher parallels her distinction between transcendental and empirical psychology with that between higher and lower faculties, which implies that transcendental psychology may be (more or less) identified with cognitive psychology.

However, the impression that Kitcher distinguishes between transcendental and empirical psychology along the lines of higher (i.e. cognitive) and lower (i.e. sensory or perceptual) faculties becomes questionable when she characterizes them as"different modes of addressing a common subject matter" (Kitcher, 1990, p. 25). She goes on to assume that transcendental psychology "guides and constrains empirical psychology in its attempts to determine the subjective conditions under which the mind actually performs those tasks" (Kitcher, 1990, p. 26; see also Kant (1998) (as translated by Guyer and Wood), A53–55, B77–79), although she apparently identifies "empirical psychology" with what Kant called "applied logic" (Kitcher 1990, p. 26). This is also suggested when she determines transcendental psychology by "disclosing universal and necessary features of human cognition" (Kitcher, 1990, p. 19), which may be considered as "abstract or general specifications" or "abstract descriptions" of a mind that is capable of performing cognitive tasks (Kitcher, 1990, pp. 13–14, p. 26).

What are the implications of this for the relationship between transcendental and cognitive psychology? Since the "abstract or general specifications" cannot be identified with the cognitive functions as their "concrete or particular specifications" (as one may be inclined to say in an analogous way), they must be regarded as transcendental rather than as empirical. To regard cognitive functions as transcendental would be to confuse "abstract or general specifications" with "concrete or particular specifications." Moreover, it would be to confuse what Kitcher calls "subjective conditions" with what they are supposed to condition, namely the cognitive and sensory functions. This means that cognitive psychology cannot be identified with transcendental psychology, but may rather be subsumed as one subset of empirical psychology besides perceptual psychology. Empirical psychology is then understood in a broader sense, including both cognitive and perceptual psychology, and thus higher-order cognitive and lower-order sensory functions.

One may now be inclined to argue that this discussion is of mere conceptual relevance, by shifting and playing with the concepts that we use to characterize transcendental psychology and empirical psychology. If transcendental psychology is characterized by cognitive functions, the latter are regarded as the form and mode, and thus the structure and organization, of our cognition. In this case the transcendental–empirical distinction corresponds to that between cognitive and sensory functions.

If, in contrast, transcendental psychology is not identified with cognitive psychology, the form and mode and thus the structure and organization of our cognition can no longer be determined by cognitive functions. This raises the question of how to characterize the form and mode and the structure and organization of our cognition, if not by cognitive function.

This is the point at which Freud enters the picture. By assuming, for instance, the tripartite structure and defense mechanisms as structural and organizational characteristics of the psychic apparatus, he considers them to be the form or mode that our psychic apparatus applies to stimuli from the world around us. More specifically, stimuli from the world around us are processed by the sensory and cognitive functions of the psychic apparatus, which in turn is structured and organized by its tripartite structure and defense mechanisms. And, most importantly, it is this structure and organization of the psychic apparatus that enables and predisposes to what stimuli it can process (and how), and the kind of sensory and cognitive functions. This in turn also determines what we remain largely unable to perceive (and how).

On the basis of his psychodynamic considerations, Freud gained some insight into the structure and organization of the psychic apparatus and how it influences our perception and cognition of the world, including ourselves and our psychic apparatus. He thus provided an outline of what I would call transcendental psychology, as distinct from both cognitive and perceptual psychology, which may rather be subsumed under empirical psychology. One might again claim that this is of mere conceptual relevance. However, this is not the case, as identifying the relationship between transcendental psychology and cognitive psychology will prove crucial in determining how we can bridge the gap between Freud's psychic apparatus and the brain, and thus between psychoanalysis and neuropsychoanalysis.

Transcendental psychology and cognitive neuroscience: Freud and the concept of neuropsychoanalysis

One might now want to argue that the structure and organization of the psychic apparatus involve only cognitive function. For instance, the superego requires moral reasoning as a specific cognitive function in the same way that defense mechanisms may correspond to different configurations between cognitive and affective functions. If so, the distinction between structure and organization on the one hand and cognitive functions on the other may turn out to be somewhat artificial and ultimately futile. This in turn would render meaningless the distinction between transcendental psychology on the one hand, and cognitive psychology as a branch of empirical psychology on the other. Must Freud's psychoanalysis therefore be extended to and complemented by the discipline of cognitive psychology which, as it is now closely associated with cognitive neuroscience, also opens the door to the brain and thus to neuropsychoanalysis?

Let us briefly return to the "composite" that characterizes our cognition. The concept of the "composite" describes how our cognition of the world and the psyche's input itself results from the interaction between the psyche and the world. This indicates that the structure and organization of the psychic apparatus can be traced back to the interaction between the psyche and the world. This interaction is in turn only possible if based on sensory and cognitive functions as dealt with in perceptual and cognitive psychology. However, the interaction between the psyche and the world itself presupposes the psyche, and specifically the psyche's input, otherwise any such interaction and consequently the recruitment of cognitive and sensory functions would be impossible. The psyche's input must thus consist of a specific psychological predisposition that makes the psyche–world interaction first and foremost possible. This interaction may then correspond to what Kitcher called "abstract or general specifications" or "subjective conditions" while the cognitive and sensory functions being conditioned mirror Kitcher's "concrete, or particular specifications."

Freud argued that the psychological predisposition lies in (among other things) the tripartite structure of the psychic apparatus and the defense mechanisms. By being structured and organized along these lines, sensory and cognitive functions allow the psyche to interact with the world. If this is the case, transcendental psychology targets those psychological features that enable and predispose our psyche to interact with the world. More specifically, the psychological predispositions in question must be able to structure and organize our cognitive and sensory functions in such a way that they allow the psyche to interact with the world. In fact this is exactly what Freud seemed to have in mind when he determined the psyche's input by the tripartite structure and the defense mechanisms, as both of these features enable and predispose us to continuously interact with the world (i.e. to engage in psyche–world interaction).

What are the implications of this for our concept of neuropsychoanalysis? Let us elaborate briefly. Transcendental psychology cannot be identified with cognitive psychology because that would mean confusing the "subjective conditions" of the psyche–world interaction with what they condition, namely the interaction itself as manifested in cognitive and sensory/perceptual functions. The distinction between transcendental psychology and cognitive psychology means that we cannot simply delegate the psyche–world interaction itself to cognitive neuroscience in order to link Freud's concept of the psychic apparatus to the brain, and thus to develop neuropsychoanalysis. This would mean to confuse the "subjective condition" of possible psyche-world interaction, e.g. Freud's transcendental characterization of the psychic apparatus, and the cognitive-perceptual function including their neuronal realization, e.g. the objective conditions of the empirical level.

If neuropsychoanalysis cannot be based upon cognitive neuroscience, how can we develop a road map from the psychic apparatus to the brain? One possible option is to revert to some mysterious and hidden properties in the brain that are not related to cognitive and perceptual functions. However, to suggest the existence of such non-neuronal properties is to introduce strong elements of mysteriousness, as for instance Descartes did with regard to the epiphysis, which he considered to be at the cross-road between mind and brain. More recent examples are the so-called mysterians, such as C. McGinn (1991), who assume the existence of additional non-neuronal properties in the brain (e.g. what McGinn calls "property P," that mediates between neuronal and mental states).

However, following Kant, this is to confuse the concept of the transcendental with that of the transcendent, which can only lead to unjustified ontological–metaphysical assumptions (see footnote 4). Therefore, assuming that the existence of additional properties or contents in the brain is not a solution to bridging the gap from the psychic apparatus to the brain, and to developing neuropsychoanalysis, we may need to reconsider our approach to the brain, which leads us to Schopenhauer.

Transcendental approach to the brain: Schopenhauer's interpretation of Kant

Let us now consider one of Kant's philosophical successors, Arthur Schopenhauer (1966),[11] who shifted the emphasis from the mind, as conceived by Kant, to the brain. Instead of taking a transcendental view of the mind, he took a transcendental view of the brain and described the brain as both the "subject and object of our knowledge."[12] He realized that our cognition can be traced back to our head and thus to our brain, rather than to reason and the mind as postulated by Kant: "With Kant, the critical philosophy appeared as the opponent of this entire method. It makes its problems just those veritates aeternae that serve as the foundation of every such dogmatic structure, investigates their origin, and then finds this to be in man's head. Here they spring from the forms properly belonging to it, which it carries in itself for the purpose of perceiving and apprehending an objective world. This here in the brain is the quarry furnishing the material for that proud dogmatic structure" (Schopenhauer, 1966, p. 421).

According to Schopenhauer, Kant investigated the "share of the brain functions (although not under this name)" (Schopenhauer, 1966, p. 418) in our cognitions: "Locke's philosophy was the criticism of the functions of sense; but Kant has furnished the criticism of the functions of the brain" (Schopenhauer, 1966, p. 11). Schopenhauer consecutively associates the a priori characteristics of the mind, the categories such as causality, that Kant attributed to the mind as the "faculty of knowledge," with the brain: "This proves that the whole of the material world with its bodies in space, extended and, by means of time, having causal relations with another, and everything attached to this—all this is not something existing independently of our mind, but something that has its fundamental presuppositions in our brain functions, by means of which and in which alone is such an objective order of things possible. For time, space, and causality, on which all those real and objective events rest, are themselves nothing more than functions of the brain; so that therefore this unchangeable order of things, affording the criterion and the clue to their empirical reality, itself comes first from the brain, and has its credentials from that alone. Kant has discussed this thoroughly and in detail; though he does not mention the brain, but says 'faculty of knowledge'" (Schopenhauer, 1996, p. 8).[13]

---

[11] Schopenhauer is close to Freud with regard to the characterization of the unconscious, which both authors described in terms of drives (and libido) that are beyond our control (Young and Brook, 1994). Therefore Freud's conception of the id may be close to Schopenhauer's concept of the will (Brook, 2003). However, I do not want to emphasize this aspect, but rather Schopenhauer's focus on the brain and his conceptualization of it which, as will be described below, may provide a bridge from Freud's concepts of the psychic apparatus and psychoanalysis to the brain and thus to neuropsychoanalysis.

[12] However, this amounted to what has been called "brain-paradox" (Northoff, 2001, 2004), since one and the same concept, namely the brain, cannot be designated and characterized by opposite and mutually exclusive features (i.e. subject and object). This contradictory and thus paradoxical determination of the brain has allowed philosophers to abandon Schopenhauer's position, and ultimately to neglect the brain and replace it with the concept of mind.

[13] One may consequently regard Schopenhauer as the first neurophilosopher to raise the possible epistemic and thus philosophical relevance of the brain. Although the term "neurophilosophy" was coined much

Following Kant and Schopenhauer, one may distinguish between an empirical and transcendental view of the brain. The empirical approach to the brain focuses on how the brain processes the stimuli that it receives from the world (the world's input). Thus it searches for the neural contents that correspond to the perceptual and cognitive contents by means of which we perceive and cognize the world's input. The empirical approach thus focuses on revealing what are called the "neural correlates," as the sufficient neural conditions of perceptual and cognitive contents (Koch, 2004). Since the neural correlates focus on neural contents, one may consider this approach, following Kant, to presuppose an empirical view of the brain.

In contrast, the transcendental view of the brain aims to reveal the brain's input to cognition and perception, rather than the neural contents that correspond to and correlate with perceptual and cognitive contents. The transcendental view focuses more on what the brain itself contributes to the neural processing of the cognitive and perceptual contents. Therefore the focus here is on those forms and modes, or structures and organizations, that the brain itself utilizes in its own neural (i.e. cognitive and sensory) processing of the stimuli from the world (the world's input). The transcendental view is thus not so much concerned with neural contents as neural correlates of perceptual and cognitive contents, as with the neural organization and structure that the brain itself utilizes in its own neural processing of the former. I shall refer to the brain's neural organization and structure as "neural predisposition" (see also Introduction), as it enables and predisposes to the neural processing of neural contents as their necessary but non-sufficient neural condition. This mean that my concept of "neural predisposition" must be distinguished both conceptually and empirically from that of "neural correlate," which targets sufficient but non-necessary neural conditions (see Introduction).

How is such a transcendental approach to the brain related to the concept of neuropsychoanalysis? I demonstrated that cognitive neuroscience cannot provide the road map to neuropsychoanalysis, as that would mean to confuse neural correlates with neural predispositions. However, this raises the question of how we can determine neural predispositions if not by cognitive functions. There is much at stake here. If we do not get a grasp of the specific characterization of the neural predisposition, we may remain unable to bridge the gap from Freud's transcendental concept of the psychic apparatus and its "psychological predisposition" to the brain and its "neural predispositions." In other words, the very possibility of neuropsychoanalysis is at stake here.

Where does this leave us? We now know that we need to adopt a transcendental approach to the brain rather than an empirical one. This is the lesson that we learned from Schopenhauer. And we further extended Schopenhauer's approach when we characterized his transcendental view of the brain by a focus on neural predispositions, as distinct from mere neural correlates, which tend to be associated with an empirical view of the brain. If the newborn discipline of neuropsychoanalysis is to live, grow, and flourish, we now need to nourish it with some input from the concept of the brain's neural predisposition. To further illustrate what we mean by the concept of "neural predisposition" within the conceptual context, we return again to both Kant and Freud in order to extend the features that they attributed to the mind and the psychic apparatus to the brain (see Figure 1.1). The precise empirical characterization of the brain's neural predisposition will be discussed in Chapter 2 and most importantly will explained in details in Parts II, III, and IV of this book.

later, by Churchland in 1986 in her book of the same name, Schopenhauer may be credited with first introducing the brain into the context of philosophy (see also Northoff, 2001, 2004; Birnbacher, 2005).

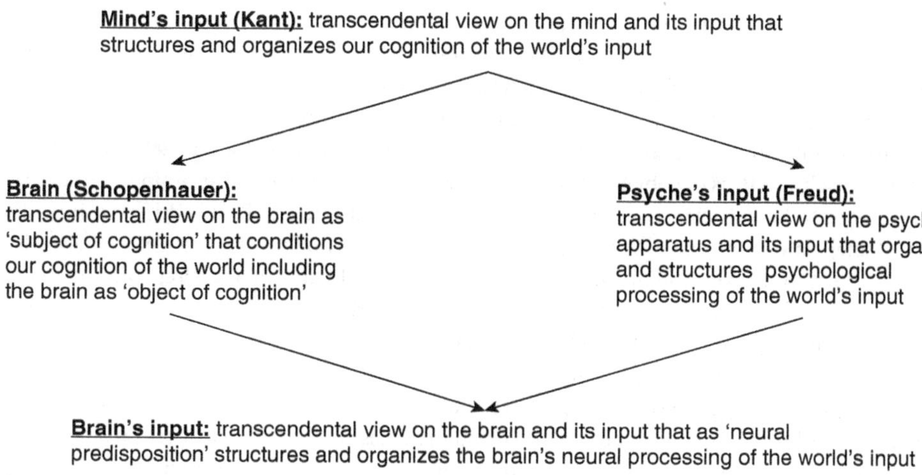

Fig. 1.1 The mind's input to the brain's input.

Approach to the brain

From the psychic apparatus to the brain: from Kant to Schopenhauer to Freud

Relying on Kant, I characterized the mind as having a specific input to our cognition of the world. Schopenhauer shifted Kant's concept of mind to the brain, and presupposed a transcendental view of the brain. He consequently assumed that the brain provides a specific input to our cognition of the world (the brain's input). However, he left open how we can specifically characterize the brain's input to our cognition.

The situation is almost reversed with regard to Freud and his concept of the psychic apparatus. Freud characterized the structure and organization of the psychic apparatus in detail, and was able to reveal the psyche's input to our cognition. In contrast, as documented in his 1895 paper, *Project for a Scientific Psychology* (Freud, 1895), he felt unable to extend his concept of the psychic apparatus, including the psyche's input, to the brain and to determine the brain's input to our cognition. This is the point at which, at least conceptually, Freud might benefit from looking to Schopenhauer and his transcendental view of the brain (whereas Schopenhauer might learn from Freud how to empirically specify and detail the brain's input).

What is the brain's input to cognition? Although Schopenhauer advocated a transcendental view of the brain, and distinguished it from an empirical view of the brain, he remained unclear about the differential features that both views target. Answering our question is easy if one adopts the empirical view of the brain. The empirical view focuses on those neuronal contents that underlie the constitution or representation, as current neuroscience would have it, of sensory and cognitive contents. This allows one to investigate the neural correlates of, for example, cognitive functions such as attention, working memory, and executive functions. It also leads to a search for the neural contents of consciousness, resulting in what I call "neural correlates of consciousness," which involves the search for the sufficient (but non-necessary) neural conditions of consciousness (Koch, 2004). By analogy, one may also search for the neural correlates of other mental states, such as free will (Walter, 2002) and the self (Northoff et al., 2006).

Although I do not wish to describe in detail the exact neural correlates of these mental functions at this stage (see Parts II and III for a detailed discussion), the important point here is that they all presuppose an empirical view of the brain. This means that they presuppose some input from the world which is supposed to induce (i.e. cause) neuronal activity in the brain. The subsequent neuronal activity changes as induced by the respective stimuli (i.e. stimulus-induced activity) are then considered to be the sufficient neural conditions and thus the neural correlates of the object or event in the world as associated with the respective stimuli (i.e. the world's input).

Most importantly, the observed neuronal activity in the brain is to be traced back completely to the stimulus itself and thus the object or event in the world. In other words, the brain's neural activity is supposed to be sufficiently accounted for by the stimulus itself. This is different from the transcendental view of the brain. In the latter, the target focus is no longer so much on the world's input (i.e. the stimuli from the objects and events in the world and their corresponding neural activity changes, namely stimulus-induced activity). Instead, the focus shifts to the brain itself and the kind of input that it contributes to the processing of the stimuli from the world (the world's input). In the same way that Kant was interested in the mind's input to our cognition, and Freud targeted the psyche's input to our psychological functions, the transcendental view of the brain aims to reveal the brain's input to the brain's own neural processing of stimuli from the world (i.e. the world's input). Thus the focus here shifts from the neural correlates of the world's input to the neural predisposition of the brain's input as provided by the brain itself and its specific input that enables and predisposes to the former. Using Kantian terms, one could say that the focus here is on what lies in the brain itself prior to its own neural processing of the world's input.

How can we determine the brain's input in more detail? Kant argued that the mind's input consisted of the application of a specific form or mode (a specific structure, order and organization) to the processing of the world's input. He considered the mind's input to consist of higher-order cognitive functions as reflected in reason and understanding. Schopenhauer shifted these higher-order cognitive functions from the mind to the brain when he introduced his transcendental view of the brain, as distinct from a merely empirical one. However, this left the nature of the brain's input open. Schopenhauer also introduced the concept of the will which, closely resembling Freud's unconscious, he considered to be even more basic than the brain itself (for further details, see Young and Brook, 1994).

Freud assimilated aspects of both Kant and Schopenhauer. He followed Kant with regard to the transcendental view of cognition but, due to his empirical (i.e. psychological) context as opposed to Kant's logical and epistemic context, he replaced Kant's concept of mind with his concept of the psychic apparatus. And he followed Schopenhauer by replacing Kant's emphasis on higher-order cognitive functions of consciousness such as reason and understanding with the lower-order functions of the unconscious and hence the drives and the libido. Freud did not follow Schopenhauer in tracing back Kant's concept of the mind to the brain itself, but only traced it back to the psychic apparatus.

Why did Freud not make the leap from his concept of the psychic apparatus to the brain itself? This is even more surprising given his own background as a neuroanatomist and his own idea of a scientific and thus brain-based psychology as outlined in his *Project for a Scientific Psychology* (Freud, 1895). He himself argued that the neurosciences of his time had nothing to offer that would have made it feasible to replace his concept of the psychic apparatus, including its psychological structure and organization, with the concept of the brain. We are now able to make this argument more specific. The neurosciences of Freud's time focused only on the localization of specific cognitive contents, and were thus exclusively interested in what one might nowadays call the neural correlates of the world's input. The input of the brain itself (the brain's input) was not yet the focus. The view of the brain was thus almost exclusively empirical, while a transcendental approach to the brain and its respective empirical features was not yet on the neuroscientific horizon.

It is not surprising that Freud did not consider the neurosciences of his time to be ready for his detailed account of the structure and organization of the psychic apparatus. He would have needed an account of the brain that targeted its empirical features from a transcendental rather than an empirical perspective. This would have allowed him to reveal the brain's input as the specific neural organization and structure that the brain itself applies to its own neural processing of the stimuli as the world's input. In other words, he would have been less interested in what we now call the neural correlates of sensory and cognitive functions, and consequently in searching for neural correlates (see, for example, Freud, 1900, p. 611), but rather he would have focused on the brain's input and its neural predispositions which make the former possible. However, to obtain further support for this argument, we need to specify the brain's input in an empirical context and thus describe in detail to what kind of neuronal processes the brain's neural predispositions refer.

From psychological to neural predispositions: Freud and current neuroscience

How can we specify the brain's input beyond the rather general concepts of form, mode, structure, order, and organization that Kant has already raised with regard to the mind's input? Let us return to Freud himself and his concept of the psychic apparatus. Freud argued that the drives, especially the libidinal drive, are an essential input of the psychic apparatus to our cognition. The libidinal drive (and later the ego) may thus be regarded as the psyche's input. Once it has been exposed to the world and its input, the libidinal drive generates conflict with opposing wishes and demands. Broadly speaking, this leads to the subsequent development and constitution of the ego, and ultimately the tripartite structure of the psychic apparatus, including its various cognitive and sensory functions. Accordingly, the libidinal drive has the effect of structuring and organizing the subsequent development of the ego and its functions.

Rather than describing some psychological correlate as a necessary and sufficient condition of a specific content, the libidinal drive can be regarded as a necessary (but not sufficient) psychological condition that predisposes the psychic apparatus to be structured and organized in a certain way (i.e. in a tripartite manner). Thus Freud was less interested in what are now called "psychological correlates" than in what I conceptualize as "psychological predispositions." What do I mean by psychological predisposition? I use the term to describe the psychological structures and organization that enable and predispose to the psychological processing of stimuli from the world, and thus what I earlier referred to as the psyche–world interaction.[14] I believe that Freud's primary focus was on psychological predispositions, in order to go beyond mere psychological correlates, for which he focused on the psychic apparatus itself. This view is supported by the following quote from his later work, *An Outline of Psychoanalysis*: "Every science is based on observations and experiences arrived through the medium of our psychical apparatus. But . . . our science has as its subject the apparatus itself" (Freud, 1940, p. 159).

What are the implications of this focus on psychological predisposition for Freud's approach to the brain and neuroscience? Freud would have been less interested in what we now call the "neural correlates" than he would have been in the neural structure and organization of the brain

[14] The rejection of possible neural correlates and the consecutive shift to neural predispositions is also well documented in the following quote: "We can avoid any possible abuse of this method of representation by recollecting that ideas, thoughts and psychical structures in general must never be regarded as localized in organic elements of the nervous system, but rather, as one might say, between them where resistances and facilitations provide the corresponding correlates. Everything that can be an object of our internal perception is virtual, like the image produced in a telescope by the passage of light-rays. But we are justified in assuming the existence of the systems (which are not . . . accessible to our psychical perceptions) like the lenses of a telescope, which cast the image" (Freud, 1900, p. 611). See also Talvitie and Ihanus (2006, p. 94), who argue that what I here call "neural predisposition," and what they call "function," "are not 'things' to be found in genes." What Freud here calls "correlates" corresponds clearly to what I describe as "psychological predispositions," which may be reflected in what he refers to as "resistance and facilitation."

that it applies to any kind of neural processing of stimuli to enable and predispose to their perception and cognition. In other words, his focus would have been not so much on the neural correlates of perceptual and cognitive contents as on the neural predispositions that enable and predispose to our specific modes of perception and cognition.

As he was more interested in neural predispositions than in neural correlates, Freud could not detect any hint of the latter in the neuroscience of his time, which focused on localizing specific sensory contents in specific regions of the brain, or in the interaction between different regions. Thus an empirical view of the brain, with the subsequent search for neural correlates of specifically perceptual contents, dominated neuroscience at that time. There were virtually no hints of a transcendental view of the brain, with a focus on neural predispositions rather than neural correlates, although in his writing on aphasia Freud (1891) referred to H. Jackson and his structural–organizational concept of positive and negative symptoms in the case of brain lesions (see Chapter 5 for a more detailed discussion of neural localization). It is not surprising that Freud, although he considered the idea of mere neural correlates (Freud, 1891, 1895), subsequently rejected it and limited himself to the psychic apparatus, rather than searching for the brain itself and its neural predisposition.

Are we ready at the present time for Freud's search for the brain's neural predisposition? Does current neuroscience go beyond the mere neural processing of the world's input and lay bare the contribution of the brain itself (the brain's input) to the processing of so-called neural correlates? We may not be ready for the extension of Freud to the brain as long as we insist on mere neural correlates rather than neural predispositions. This means that we shall not be prepared for neuropsychoanalysis if we continue to use cognitive neuroscience as our road map. This is because cognitive neuroscience itself is still unable to reveal the brain's input (independent of the stimuli and thus the world's input) and thus the neural predispositions that Freud would have searched for if he intended to bridge the gap between his concept of the psychic apparatus and the brain.

However, we may be able to give some encouraging signals to Freud, though first we need to briefly describe the different ways in which the brain's input and its neural predispositions could be determined. First, we may determine the brain's input in empirical terms. This would be analogous to the empirical characterization of the psychological predisposition by Freud as, for example, by the tripartite structure and defense mechanisms. Secondly, we may determine the brain's input in conceptual terms by associating different concepts of the brain with neural correlates and neural predispositions. This would be more or less analogous to Freud's distinction between metapsychology, which focuses on the psychic apparatus itself, and psychology, which deals with the psychological processing of stimuli from the world. Thirdly, we may determine the brain's input and thus its neural predispositions in an epistemic way by associating a specific type of knowledge or lack of knowledge with it. This is analogous to Freud's characterization of the unknowability of the psychic apparatus, as it is independent of us and our cognition of it, thereby presupposing Kant's transcendental framework.

All three determinations (epistemic, conceptual, and empirical) will be discussed much more fully in the rest of this book. In the next chapter I shall describe conceptual and epistemic characterization in more detail, while Parts II, III, and IV of the book focus on the empirical determination of the brain's input in much the same way that Freud did in the case of the psychic apparatus.

Let me just point out that, as I have already hinted in the Introduction, the brain's resting-state activity and its impact on subsequent stimulus-induced activity may correspond to the empirical determination of the brain's input. The brain's resting-state activity and the way in which it can (and cannot) affect stimulus-induced activity, and vice versa, the so-called rest–stimulus and stimulus–rest interaction (Northoff et al., 2010), may thus be considered a neural predisposition of the neural processing of stimuli (i.e. the world's input) in the gestalt of sensory

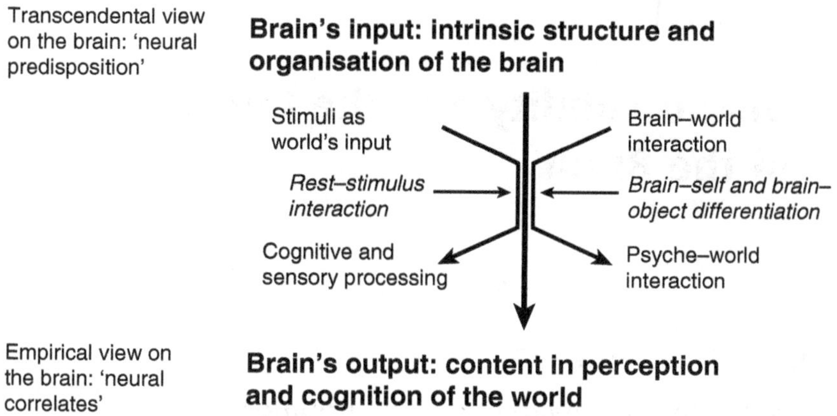

Fig. 1.2 Interaction between the brain's input and the world's input.

and cognitive processing. Once we understand rest–stimulus and stimulus–rest interaction, we may also get a grasp of the brain's neural coding. The neural code refers to the kind of algorithms that the brain uses to code and thus to transform the incoming stimuli into neural activity (see also Talvitie and Ihanus, 2006, p. 88), which remains unclear so far (see Chapter 5 for a detailed discussion of the brain's neural code).Thus, although it appears abstract at first glance, the transcendental view of the brain and the associated empirical determination of the brain's input can make a significant contribution to neuroscience by filling some of its current gaps. This will become even clearer in Parts II and III of the book, in which I discuss these empirical issues in much more detail.

My brief indication of the possible empirical, conceptual, and epistemic determination of the brain's input paves the way to extending Freud's concept of the psychic apparatus to the brain, and thus to developing the concept of neuropsychoanalysis. While extending Freud's concept to the brain, we may at the same time learn from him. His detailed description of the various psychological predispositions that characterize the input of the psychic apparatus may provide us with a road map which shows what the brain's input must look like. More specifically, the brain's input and its neural predispositions must be such that they enable and predispose to psychological characteristics such as the tripartite structure and defense mechanisms that Freud so cleverly described. Thus we may take Freud's psychological predispositions as a starting point and put them in the context of current neuroscience in order to learn more about the brain's input and its neural predispositions. This I shall do in Parts II and II of this book.

Once we have revealed the brain's input and the neural predisposition, we may better understand how the brain can (and cannot) interact with the world. This so-called "brain–world interaction" may allow for the constitution of objects (in the Freudian sense) and the self, and thus be crucial in enabling and predisposing to what I call "brain–object differentiation" and "brain–self differentiation" (see Introduction). This will also enable us to understand how the constitution of the psychic apparatus that Freud characterized by objects and the self as resulting from psyche–world interaction is enabled and predisposed to by the brain's input, which structures and organizes the brain–world interaction (see Figure 1.2). Thus, to extend Freud's concept of the psychic apparatus to the brain, we need to go back to the brain's input and show how it enables and predisposes to the brain–world interaction with the consecutive constitution of objects and self (i.e. brain–self and brain–object interaction). This will be the focus of Parts II and III of this book.

Chapter 2

Unknowability and the Concept of the Brain

Philosophical concepts

"Doctrine of the unknowability of the noumenal mind": Kant and the mind–brain problem

In the previous chapter I demonstrated that Kant and Freud shared a focus on structure and organization as distinct from mere contents, be they cognitive or sensory. The input to our mind, as highlighted by Kant, and the input to the psychic apparatus, as emphasized by Freud, contribute to our cognition and perception of the world's input (the stimuli we receive from the objects and events in the world, as we can cognize it). I then utilized Schopenhauer's interpretation of Kant and Freud to extend their concepts to the brain, thereby distinguishing the brain's input from the contents in our perception and cognition of the world's input in the form of contents.

The question that now arises is whether and how we can access and thus know the brain's input and distinguish it from the actual contents. This concerns what I called in the previous chapter the "epistemic determination of the brain's input," which will be the main focus of this chapter. How can we access the brain's input and thus the brain itself in our perception and cognition, independent of our perception and cognition of the contents as related to the world's input? The answer to this question reveals another analogy between Kant and Freud. Both argue that we cannot directly access and know the input of the mind or the psyche themselves, and remain unable to reveal what the input of the mind or the psychic apparatus looks like by itself independent of our cognition of the contents that it generates in interaction with the world's input. The input of the mind or the psychic apparatus and thus the mind and psychic apparatus themselves are thus supposed to remain a "blind spot" in our knowledge. This is because our knowledge of them is always already confounded by the contents of our cognition and perception, and thus by the world's input, which prevents us from directly accessing the mind's input or the psyche's input, as these are independent of the world's input.

Let me start with Kant. There has recently been a resurgence of interest in Kant's philosophy in the context of neurophilosophy and cognitive science. Brook (1994, 1998), one of the main proponents, has introduced many elements of the Kantian philosophy, including synthesis, mental unity, and self-consciousness, to current discussions in cognitive science and neurophilosophy. Moreover, Brook (1994, 1998) described one of Kant's central insights, namely that we cannot know the mind and the world in general as they are independent of our cognition of them. Brook calls this the "doctrine of the unknowability of the noumenal mind." All that we can know is how the mind appears to us in our cognition (i.e. in phenomenal terms), while we remain largely unable to know what the mind really is by itself independent of our cognition of it (i.e. in noumenal terms). This implies that we cannot know and determine the ontological reality and existence of the mind. Consequently, any ontological account of the mind, be it mental, physicalistic, or anything else, remains largely impossible for us and our limited cognition (see, for example, Schlicht, 2007).[1]

Why do we remain largely unable to cognize the mind itself as it is independent of our cognition of it? This question leads to the central insight of Kant. Following Kant, our mind itself contributes something to our cognition and knowledge (i.e. the mind's input), which imposes boundaries and limits upon our possible cognition and thus knowledge of the world (i.e. the world's input). However, the epistemic boundaries and limits due to the mind's input apply not only to our cognition of the world, but also to the cognition of the mind itself, including the mind's input. This means that our cognition and knowledge of our own mind and its specific input are constrained by the mind itself, and more specifically by the input that the mind itself contributes to our cognition and knowledge of both mind and world.

However, the epistemic boundaries and limits in our cognition and knowledge of the world need to be specified. The mind receives a double input, namely the mind's input and the world's input. The content in our cognition may thus be traced back not only to the world's input, but also to the mind's input and its interaction with the world's input. In order to cognize and know the mind's input itself, independent of the world's input, we would therefore need to disentangle the former from the latter when cognizing the mind. However, such disentangling of the mind's input from the world's input is impossible for us, because our cognition is always a "composite" or "hybrid" where it remains apparently impossible for us to tease the two inputs apart, as I shall discuss later in more detail. This makes it largely impossible for us to know the mind and its specific input, as they themselves are independent of our cognition of the world. Following on from Kant, the philosopher Brook therefore referred to what he called the "doctrine of the unknowability of the noumenal mind" (Brook, 1994).

If we cannot know the mind, including the mind's input to our cognition, as these are independent of the contents of our cognition, we may also be unable to determine and thus to know whether the mind really exists as such, distinct from and independent of the body. This raises the question of the relationship between mind and body in Kant, the so-called mind–body problem, which nowadays has resurfaced as the mind–brain problem. Independent of our cognition of the contents, they generate an interaction with the world's input. Different authors have suggested different mind–body conceptions in Kant, which will be discussed briefly here.

Karl Ameriks (2000, pp. 303–319) claims that immaterialism of the mind is Kant's solution to the mind–body problem. He argues that Kant's position is highly compatible with the assumption of immaterialism. According to Ameriks, immaterialism "means more than the restricted doctrine that not every concrete thing is material; I mean the unrestricted doctrine that every concrete thing is basically non-material." Immaterialism must be distinguished from spiritualism that implies immortality. Beings that are material and mortal might well be compatible with an immaterial mind that disappears with the death of the body. The mind may thus be immaterial, but it is not spiritual (Kant (1998) (as translated by Guyer and Wood), B420; Ameriks, 2000, p. 305).

The claim of possible immaterialism presupposes that we indeed have some knowledge of the noumena (the thing in itself), since otherwise we could not claim that it is non-material. However, this violates what Brook (1994, 263, note 23) calls the "doctrine of the unknowability of the mind as it is." Ameriks refutes this doctrine by claiming that some aspects of the mind, its appearance and its empirical features, can nevertheless be known. For instance, we can cognize the distinction

[1] Kant's position on the mind–body problem has been interpreted in different ways that do not always refrain from ontological assumptions, which range from functionalism (Kitcher, 1990; Brook, 1994,) through epistemic dualism (Schlicht, 2007) to immaterialism (Ameriks, 2000).

between appearance and the "thing in itself" with regard to the mind. This presupposes at least some indeterminate knowledge about the mind as it is, namely the noumenal mind. This provides us with the knowledge about the noumenal mind that is necessary to cognize the distinction between the noumenal and the phenomenal mind.[2]

Rather than assuming immaterialism as an ontological position, as Ameriks does, Schlicht (2007) interprets Kant' position on the mind–body problem as epistemic dualism. We have limited cognitive capacities, with two perspectives that mutually exclude each other. Relying on Colin McGinn, Schlicht (2007, pp. 241–242) argues that inner sense (i.e. introspection) provides access to consciousness but not to the brain, whereas outer sense (i.e. perception) captures the brain but not consciousness. The epistemic ability that is necessary to decide upon the ontological substrate of mind–brain would be a third capacity, which allows accessing of both brain and consciousness simultaneously. However, as we do not have such a third epistemic ability in addition to perception and introspection, we remain largely unable to make any ontological inferences from our limited epistemic abilities. Schlicht (2007, pp. 285, 287) emphasizes that we should also refrain from ontologizing both epistemic abilities, as neither inner sense as introspection nor outer sense as perception can be associated with specific ontological assumptions (e.g. existence and reality). Instead, due to our epistemic boundaries and the limits of our cognition and knowledge of the mind, we have to remain "ontologically abstinent" with regard to the mind, as we are restricted by our specific modes of perception, cognition, and introspection. Schlicht characterizes this position as "epistemic dualism."

In addition to immaterialism and epistemic dualism, functionalism has also been assumed by many authors to characterize Kant's mind–body position. Due to the fact that Kant describes cognitive processes while remaining indifferent to their underlying ontological substrate (according to Brook this is because of the "unknowability of the noumenal mind"), Brook (1994) considers Kant to be a functionalist "avant le mot." Kitcher (1990, pp. 111–112) also points out that Kant's relational or synthetic view of the contents of cognitive states resembles the functionalistic (i.e. extrinsic) account. Others who offer functionalistic accounts of Kant include Sellars (2007), Dennett (1978), Meerbote 1989), and Powell (1990). Functionalists claim that the mind is what it does and how it operates and functions, whereas its underlying substrate (e.g. a brain or a machine) does not matter. The way that the mind represents the world does not depend on its underlying ontological substrate. Kitcher emphasizes that Kant, in a similar way to functionalism, was not at all interested in the psychological and physiological embodiments of cognitive states (Kitcher, 1990, pp. 25, 28). Similarly, Brook (1994, p. 13) assumes that Kant was more interested in how the mind works, functions, and operates than in how it is built and structured and its underlying ontological substrate. This comes close to a general version of functionalism.

---

2 Ameriks claims that certain phenomena, such as apperception, are not expected to be explained materialistically because they defy in principle any such explanation (see Ameriks, 2000, p. 306). Schlicht (2007, pp. 280, 282) makes a similar argument with regard to the unity and spontaneity of self-consciousness. The unity and spontaneity of self-consciousness are not compatible with any empirical and thus materialistic account because they require a purely logical explanation. Unity and spontaneity must thus be attributed to what Schlicht calls logical self-consciousness as distinct from mere empirical self-consciousness (Schlicht, 2007, pp. 242–243, 253–276, 280–282). If however unity and spontaneity cannot be accounted for in empirical terms, a completely materialistic basis of logical self-consciousness remains impossible. Apperception and the unity and spontaneity of self-consciousness do thus indicate that some immaterial basis must underlie our mind.[2]

Functionalism implies that the mind's function is not dictated by its ontological substrate, but can stand independently by itself. Due to the unknowability of the noumenal mind, access to any possible ontological substrate remains blocked. This implies that we are left with the appearances of the inner sense. Brook (1994, p. 224) claims that the relationship between the mind as it appears to us and its noumenal substrate, as claimed by Kant, must be considered analogous to the relationship between mental functions and physical structures, where the latter realize and implement the former. In the same way, the noumenal substrate of the appearing mind remains undetermined, as the physical substrate of the brain does not determine its mental functions. This is because mental functions can be realized and implemented in different ontological substrates, including both physical and non-physical ones. In a similar way to functionalism, Kant's approach can therefore be characterized by what Brook (1994, pp. 23, 259) calls "ontological neutrality."

"Unknowability of the psychic apparatus": Freud and inner sense

Kant's characterization of the mind as noumenal and unknowable raises the question of whether Freud considered the psychic apparatus in the same way. Did he also consider the psychic apparatus, including the psyche's input, to be unknowable and inaccessible independent of the contents in our perception and cognition? Can we thus extend Kant's "doctrine of the unknowability of the noumenal mind" to the psychic apparatus and speak of a "doctrine of the unknowability of the noumenal psyche"?

Freud himself saw a clear analogy between his concept of the psychic apparatus and Kant's concept of the unknowability of the mind. In the same way, we cannot know the mind itself and its input, and we remain unable to know and directly access the psychic apparatus itself. All that perception and cognition can provide is access to contents and our consciousness of them which, following Kant, must be considered merely phenomenal rather than noumenal. What Kant calls the noumenal mind may then correspond to the (implicitly presupposed) concept of the "noumenal psyche" in Freud. The question now is why Freud considered the psychic apparatus to be inaccessible and unknown, and thus noumenal.

Based on Kant, Freud distinguished between inner and outer senses. The outer senses, or the external sensory organs as one would call them today, provide access to the outer world (e.g. during perception of objects and events). In contrast, the inner senses, which today would most probably be referred to as introspection, provide access to ourselves and our own inner state. If we want to access and know the psychic apparatus itself, we must recruit our inner senses. However, Freud argued that the inner sense does not provide direct access to the psychic apparatus itself, including its input, because it is clouded or confounded by the contents, the inner contents or, as one would say today, the mental contents. All that we can perceive and thus know in inner sense are mental contents, while the psychic apparatus itself, its structure and organization, including its specific input that allows the mental contents to be constituted, remains hidden behind them.

Freud specified this scenario with regard to consciousness and unconsciousness. The mental contents that we can perceive and cognize in inner sense (i.e. introspection) are tied to consciousness, while the processes themselves, which are thought to underlie the constitution of those mental contents, remain inaccessible to us and thus unconscious. This opens the door for Freud to parallel Kant's distinction between phenomenal and noumenal mind with that between conscious psychological contents and the unconscious psychic apparatus. What Kant described as the phenomenal mind within the philosophical (i.e. epistemic–ontological context) context may then resurface as conscious psychological contents within Freud's more psychological context, while

Kant's philosophical concept of the noumenal mind may correspond to the unconscious psychic apparatus within Freud's psychological context.[3]

How does the unconscious psychic apparatus relate to the inner and outer senses? Freud claimed that he extended Kant's concept of outer sense to one of inner sense, and applied it to the unconscious psychic apparatus (i.e. the noumenal psyche), just as Kant regarded the outer sense as crucial in his concept of the noumenal mind: "In psycho-analysis there is no choice but for us to assert that mental processes themselves are unconscious, and to liken the perception of them by means of consciousness to the perception of the external world by means of the sense organs. We can even hope to gain fresh knowledge from the comparison. The psycho-analytic assumption of unconscious mental activity appears to us as an extension of the corrections undertaken by Kant of our view on external perception. Just as Kant warned us not to overlook the fact that our perceptions are subjectively conditioned and must not be regarded as identical with what is perceived though unknowable, so psycho-analysis warns us not to equate perceptions by means of consciousness with the unconscious mental processes which are their object" (Freud, 1915, p. 171).

However, there is an error in Freud's reasoning here. Although he is certainly right in pointing out the analogy of his concept of the psychic apparatus as hidden and noumenal with Kant's concept of the noumenal mind, he is wrong when he attributes to Kant the outer sense but not the inner sense. As Brook (2003) has pointed out, Kant was fully aware that the inner sense, analogous to the outer sense, only provides us with consciousness and thus the appearance of the mind (i.e. the phenomenal mind), rather than the mind in itself (i.e. the noumenal mind). Freud seems to overlook this when he claims to extend Kant's assumption of outer sense and its access to the mind to the inner sense and how it can (and cannot) access the psychic apparatus. However, this should not distract us from the main commonality between Kant and Freud across their different contexts (i.e. philosophical and psychological). Both concur that the mind/psychic apparatus, including their respective inputs to our cognition, can neither be accessed directly nor known as such (i.e. independent of any contents, be they mental or non-mental) in either inner or outer sense (i.e. introspection or perception).

However, there are some discrepancies between Freud and Kant with regard to the possibility of gaining indirect knowledge of the mind/psychic apparatus. Whereas Kant restricted our knowledge to the phenomenal mind, and denied any kind of possible knowledge of the noumenal mind, Freud did indeed assume some kind of possible knowledge of the psychic apparatus (Brook, 2003). Following Freud, the impossibility of directly accessing the noumenal psychic apparatus does not prevent us from accessing it indirectly through those phenomena in which the unconscious becomes manifested, such as dreams, slips of the tongue, jokes, neurotic behaviors, etc. (see also Tauber, 2009, p. 5). Thus we are able to gain some knowledge of the unconscious and therefore the psychic apparatus itself (i.e. the noumenal psyche), although only indirectly through the contents in our cognition and our consciousness of them, rather than by directly accessing the unconscious and thus the noumenal psyche and its input.

Parallelism between inner and outer sense: Freud, Solms, and the "brain–mental apparatus dilemma"

As I have pointed out, Kant's concept of the noumenal mind as unknowable has led to some controversy about his exact stance on the relationship between mind and body/brain, which

[3] There is indeed some evidence from private conversations between Freud and Ludwig Binswanger (cited by Taube, 2009, p. 5) that Freud compared his concept of the unconscious directly with Kant's concept of the noumenal mind.

oscillated between immaterialism, epistemic dualism, and functionalism. Although one may regard the question of the mind–body/brain relationship as purely academic, it is of crucial importance for the development of neuropsychoanalysis. If the latter discipline is to be established, it will be necessary to clarify the exact relationship between mind and body, and more specifically that between mind and brain. Only if there is at least some kind of relationship between mind and brain, be it reductive or non-reductive, can neuropsychoanalysis get off the ground. If, in contrast, one assumes dualism or strict parallelism between mind and brain, the development of neuropsychoanalysis as a bridge between mind and brain may be doomed to failure.

Based on his Kantian framework, Freud was well aware that we cannot conceive and know the noumenal mind and its ontological reality and existence because we are always already tied to consciousness: "In our science, as in the others, the problem is the same: behind the attributes (qualities) of the object under examination (the unconscious) which are presented directly to our perception, we have to discover something else which is more independent of the particular receptive capability of our sense organs and which approximates more closely to what may be supposed to be the real state of affairs. We have no hope of being able to reach the latter itself, since it is evident that everything new that we have inferred must nevertheless be translated back into the language of our perceptions, from which it is simply impossible for us to free ourselves. But herein lies the very nature and limitation of our science. . . . Reality will always remain unknowable" (Freud, 1940, p. 196). Meissner (2003a,b, 2007) has provided an excellent account of Freud's somewhat ambivalent position on the relationship between mind and brain.

What are the implications of this for the noumenal psyche and our possible knowledge of the mind–brain problem? Freud was well aware that all we can do is to investigate the contents that inner and outer sense present to us in a phenomenal way, while the noumenal psyche itself and its relationship to the brain and body must remain elusive and thus unknown to us: "Like the physical, the psychical is not necessarily in reality what it appears to us to be. We shall be glad to learn, however, that the correction of internal perception will turn out not to offer such great difficulty as the correction of external perception—that internal objects are less unknowable than the external world" (Freud, 1915, p. 171).

Mark Solms picks up Freud's dualism between inner and outer sense and puts it into the present-day empirical context of neuroscience (Solms and Kaplan-Solms, 2000; Solms and Turnbull, 2002; Solms, 2006). He argues that, analogous to Freud and Kant, there are inner and outer phenomena that can be accessed and cognized in inner and outer sense. Outer phenomena concern our perception of the external world, and are considered to be objective, while inner phenomena target the inner world and thus our mental states that are subjective. Like Freud, Solms focuses on the inner world and its subjective mental states. For instance, he regards emotions and emotional feelings as subjective phenomena that reflect the appearances or phenomena in inner sense in the same way that our perceptions of the objects and events in the external world are appearances and phenomena in outer sense (Solms, 1996, 2006). Thus there is parallelism between inner and outer sense as mirrored by the parallelism between mental and physical states.

What are the implications of the parallelism between inner and outer sense for the concept of neuropsychoanalysis? Neuropsychoanalysis aims to link inner and outer sense. While the outer sense focuses on what we observe as neuronal states of the brain in perception, the inner sense targets the mental states of what appears as mind rather than as brain in introspection. Since they target different phenomena (i.e. neuronal and mental states, respectively), inner and outer sense cannot be reduced (or even eliminated) to each other. Instead, both must be considered in a parallel and complementary way. Thus, in order to link and connect neuronal and mental states, neuropsychoanalysis must devise a method of systematically linking inner and outer sense rather than reducing the former to the latter. In summary, Solms extends and goes beyond Freud's parallelism between the physical and the psychological as related to inner and outer sense by reframing it as (epistemic) parallelism between neuronal and mental states.

Solms further claims that both neuronal and mental states can be traced back to a common origin, which he calls the "mental apparatus" (Solms, 2006), that can be considered more or les analogous to Freud's concept of the psychic apparatus (and Kant's concept of the "noumenal mind"). For instance, the subjective experience of emotional feelings or affects in inner sense refers to what in outer sense is observed objectively as the brain's neural activity changes when experiencing affects (Solms, 1996, pp. 489–493). Although they target apparently different states (i.e. neuronal and mental), outer and inner sense refer to one and the same apparatus, namely the mental apparatus, from different perspectives (Solms, 1996, p. 500). Thus, relying on inner sense, psychoanalysis investigates the inside of something (what Solms calls the "mental apparatus"), whose outside is explored by neuroscience using outer sense.

How can we further characterize what Solms calls the "mental apparatus"? Applying Dennett's concept of the design level, Solms characterizes the "mental apparatus" by the design that makes the constitution of inner and outer sense as phenomenal possible. What Solms calls the "design level" is thus more or less similar to what Kant described as "noumenal" with regard to the mind (Solms, 2006). He therefore considers the design level of the mental apparatus to be the unifying link between inner and outer sense, and thus between neuronal and mental states. According to Solms, both "empirical realms of physicality and intentionality" originate and can be traced back to one and the same feature in the mental apparatus (Solms, 2006, p. 99). Interestingly, what Solms calls the design level seems to be more or less analogous to what Kant originally described as structure, organization, order, form, and mode that the noumenal mind applies to our cognition.

However, Solms appears to go beyond Kant, who rejected any ontological characterization of the mind–body/brain relationship, when he claims dual aspect monism as the mind–body theory (Solms and Turnbull, 2002, pp. 56–57). Inner and outer sense can be regarded as dual aspects of one more basic noumenal level, namely the design level of the mental apparatus. What we perceive in the outer sense is the brain, including its neuronal states, while in the inner sense we perceive the mind and its mental states. Body and mind are considered to be aspects of a more basic and unifying entity which Solms calls the mental apparatus or "human mental apparatus" (see also Solms and Turnbull, 2002, p. 57). Ontologically, the mental apparatus is thought to provide the monistic entity underlying both mind and body/brain as distinct aspects.

This account by Solms goes beyond Kant in that Solms must presuppose some degree of knowledge about the noumenal mental apparatus, and it is this degree of knowledge that Kant deemed to be impossible. Thus Solms seems to side here more with Freud who, as I pointed out did not claim complete lack of knowledge of the psychic apparatus, as otherwise he would have been completely unable to conduct his research program on the unconscious psychic apparatus. However, it remains unclear what is the exact relationship between the brain and what Solms calls the mental apparatus.

This discussion may sound academic, but it has serious implications for the way in which we conceive of the brain in neuropsychoanalysis. Central to neuropsychoanalysis is the linkage between neuronal and mental states. If both of these can be traced back to a common origin in the mental apparatus, the question of the relationship between the brain's neuronal states and the mental apparatus arises. If we identify the brain (i.e. our concept of the brain) with neuronal states as we can observe and perceive them in outer sense, the concept of the mental apparatus must be distinguished from the concept of the brain. Otherwise the mental apparatus would be merely neuronal (i.e. physical), and thus phenomenal rather than noumenal as assumed by Solms.

If, in contrast, the concept of the brain extends the characterization by neuronal states, the brain may indeed be identified with the mental apparatus. However, the brain in this sense does not refer to the brain as we can observe and perceive it in outer sense, which means that we may then need to distinguish between the brain as noumenal and the brain as merely phenomenal.

Thus we are confronted with a dilemma. Either we completely and exclusively determine the brain by neuronal states (which, however, means that the characterization of the mental apparatus remains open, because it cannot be defined by purely phenomenal features such as the brain's neuronal states), or we determine the brain by some states or features other than neuronal states that lie beyond our possible perception and observation in outer sense. Although the latter approach makes the characterization of the mental apparatus by the brain possible, at the same time it introduces some mysteriousness to our determination of the brain. This is because it remains unclear and rather mysterious what these non-neuronal states or features are and how we can access and determine them if they remain beyond our observation in outer sense. Therefore the price that is paid for determining the mental apparatus in terms of the brain may be the introduction of mysteriousness about the brain itself.

Why does this raise a dilemma? The dilemma is that in either case (identity or non-identity between the mental apparatus and the brain) we end up with another problem. If we assume non-identity between the brain and the mental apparatus, the determination of the mental apparatus remains open. In contrast, if we assume identity between the brain and the mental apparatus, the brain and its determination become rather mysterious. Since the dilemma concerns the determination of the brain and its relationship to the mental apparatus, I shall call it the "brain–mental apparatus dilemma." How can we escape the "brain–mental apparatus dilemma"? To answer this question we need to discuss and problematize something whose determination and definition are usually taken for granted in neuroscience, philosophy, and psychoanalysis, namely the brain and our concept of it.

Neurophilosophical concepts

Concept of the brain: application of Kant, Freud, and Solms to the brain

Following the distinctions made by Kant, Freud and Solms between inner and outer sense, I here want to distinguish between the concept of the "brain as observed" and that of the "brain as experienced." While the "brain as observed" is the brain that we can access in outer sense, the "brain as experienced" is the brain that we can access in inner sense. Both the concept of the "brain as observed" and that of the "brain as experienced" presuppose an empirical view of the brain, as they are based on our senses (i.e. inner and outer sense). However, if inner and outer sense only provide access to an appearance of our brain, or the phenomenal brain as Kant would probably have called it, we may also need to assume a concept of the brain that refers to the brain itself, as it is by itself independent of our perception and cognition of it in inner and outer sense. As discussed in Chapter 1, this requires a transcendental view of the brain that opens the door to the development of a concept of the brain as noumenal, which I call the "brain as functioning."

I shall now discuss the concepts of the "brain as observed" and the "brain as experienced" in detail, before providing further information about the "brain as functioning" in the next section.

The concept of the "brain as observed" refers to the brain that we can observe in outer sense and more specifically from the third-person perspective (TPP) as presupposed in observation. What do we observe from TPP when we observe the brain? We observe what we call neurons and neuronal states which we trace back to physical properties (or, to put it ontologically, in terms of existence and reality). Thus the "brain as observed" refers to the brain that we target in neuroscience which, relying on observation from TPP, aims to reveal the brain's neuronal states and their specific mechanisms.

In contrast to the "brain as observed," the concept of the "brain as experienced" presupposes inner rather than outer sense. More specifically, the concept of the "brain as experienced" refers to the brain that should be accessible in inner sense, and thus from what is today referred to as first-person perspective (FPP).[4] One may now be inclined to argue that we do not experience any brain from FPP, so the assumption of a "brain as experienced" does not refer to the brain. We introspect in inner sense neither neuronal states nor any kind of brain (as we can observe it from TPP), but rather what are called mental states, from which we infer some kind of mind as distinct from the brain. However, the absence of the brain in inner sense should not prevent us from assuming the "brain as experienced" as mere concept.

Unlike the "brain as observed," the concept of the "brain as experienced" seems to be plausible only within the conceptual context, whereas there is no such plausibility for it within the empirical context. Yet even if there is no evidence for any "brain as experienced" within the empirical context of inner sense and subjective experience from FPP, this should not prevent us from assuming such a concept within a purely conceptual (-logical) context. However, we should be well aware of the discrepancy between conceptual and empirical contexts in the case of the "brain as experienced," as otherwise we may be prone to make inferences from the "brain as experienced" to its empirical and ontological features. This seems to distinguish the concept of the "brain as experienced" from the one of "brain as observed," where such discrepancy between conceptual and empirical/ontological contexts does not seem to hold. To better understand both concepts (i.e. "brain as observed" and "brain as experienced"), we may compare them in more detail with regard to their similarities and differences.

Let me first discuss their similarities. Both the "brain as observed" and the "brain as experienced" are analogous in an epistemic sense, as both refer to the senses (i.e. inner and outer sense). Thus there is what one may describe as "epistemological similarity." How can we describe such epistemological similarity in more detail? Kant would probably have said that both concepts refer to phenomenal rather than noumenal features, and must therefore be regarded as appearances rather than things-in-themselves. Schopenhauer, who distinguished the brain as object of cognition from the brain as subject of cognition (see Chapter 1), would probably have subsumed both "brain as observed" and "brain as experienced" under his concept of the brain as object of cognition as distinct from the brain as subject of cognition. Finally, Freud and also Solms would probably have argued that the "brain as experienced" mirrors the inner sense in the same way as the "brain as observed" reflects the outer sense which distinguishes both concepts of the brain from that of the mental apparatus. In conclusion, this clearly indicates that the concepts of the "brain as observed" and the "brain as experienced" must be put on an equal footing in epistemological terms. They are epistemologically similar or analogous because both of them refer to phenomena or appearances rather than noumena or the thing-in-itself (Kant's concept),

[4] I am aware that the present-day distinction between FPP and TPP may not be completely analogous to the earlier distinction between inner and outer sense. This may be an issue for philosophers to discuss, but is not of crucial importance here.

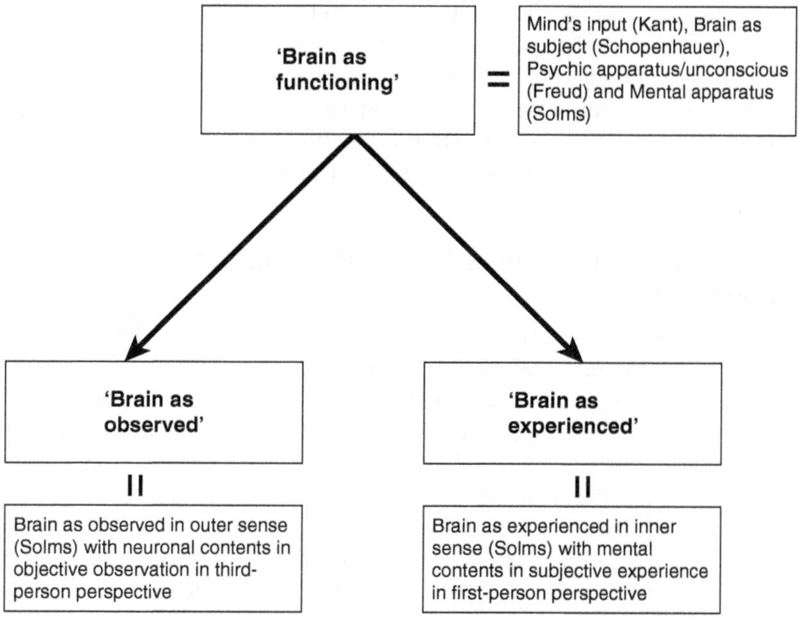

Fig. 2.1 Different concepts of the brain and corresponding concepts used by Kant, Freud, and Solms.

to objects rather than subjects of cognition (Schopenhauer's concept), and to the senses (i.e. inner and outer sense) rather than the mental apparatus itself (the concept of Freud and Solms) (see Figure 2.1).

However, as well as such epistemological similarity, there is also, as already indicated, some empirical discrepancy with regard to their respective contents (see Chapter 1 for the determination of the concept of "empirical" as presupposed here). Although the concept of the "brain as observed" refers to neural contents (i.e. the brain's neural states), the concept of the "brain as experienced" does not refer to such neural contents, but rather to some non-neural contents which we designate as mental contents. Due to the reference to different contents (neural and mental), the concepts of the "brain as experienced" and the "brain as observed" must subsequently be assumed to differ within the empirical context. There is empirical discrepancy between the concept of the "brain as observed" and that of the "brain as experienced."

So far I have considered the concepts of the "brain as observed" and the "brain as experienced" in conceptual, epistemological, and empirical contexts. I diagnosed conceptual difference (as otherwise I would not need to distinguish both concepts), epistemological similarity with regard to their characterization as phenomenal rather than noumenal, and empirical discrepancy with regard to their reference to different contents. However, the philosopher may be more interested in the ontological context that concerns the existence and reality associated with both concepts of the brain. Can the "brain as experienced" and its non-neuronal contents be traced back to physical existence and reality, and thus to neuronal states as associated with the "brain as observed"?

This question obviously touches upon what is referred to as the mind–brain problem in current philosophy of mind. There are two possibilities. It may be that the mental states as the contents of the "brain as experienced" are supposed to be traced back to the physical properties of the "brain as observed," as is assumed to be the case in materialistic–monistic mind–brain theories. Alternatively, the mental states as the contents of the "brain as experienced" may be supposed to be traced back to ontological properties that are different from physical properties, such as mental or other properties, with the subsequent assumption that the mind is ontologically different from the brain, as in dualistic mind–brain theories. However, this is a purely philosophical discussion which may not be relevant within the neuropsychodynamic context that is the focus here. I shall therefore refrain from entering deeper into this debate, and leave it for the philosophers to make their points. Instead, I shall now characterize the third concept of the brain, namely the "brain as functioning," in more detail.

"Brain as functioning": conceptual and empirical extension of Kant and Freud

What exactly do I mean by the concept of the "brain as functioning"? This cannot refer to the brain that we can access in our senses, be it the inner or the outer sense, as otherwise it could not be distinguished from both the "brain as observed" and the "brain as experienced." However, if the "brain as functioning" does not refer to the senses and thus to access from either FPP or TPP, it remains very unclear what it refers to and whether we can possibly distinguish it from the other two concepts of the brain. This is an especially important consideration given that we have no direct access to the brain other than through our senses. However, as we have already seen in the case of Freud, the impossibility of gaining direct access to the psychic apparatus because it is independent of our senses does not preclude the possibility of gaining indirect access. I shall now apply the same strategy to the concept of the brain, and argue that the concept of the "brain as functioning" may target the brain, as it is independent of our access to the brain in terms of our senses (i.e. inner and outer sense). I shall characterize the concept of the "brain as functioning" in further detail in this section, and will then go on to describe our access to it in terms of indirect access in the next section.

Does the "brain as functioning" refer to something that lies beyond our senses? Without plunging into the ontological mind–brain debate, this "something" is not supposed to refer to something ontological, as in the case of the so-called "mysterious philosophers" such as McGinn (1991) and Nagel (1998), who assume that a special non-physical and non-mental ontological property characterizes the brain. Such ontological characterization of the "brain as functioning" is rejected here because I do not intend to introduce any kind of mysteriousness in my concept of the brain by, for example, postulating non-neuronal brain properties. Although I do not presuppose any such mysterious ontological property here, I shall refrain from any further ontological characterization of the "brain as functioning," as that may be relevant only within the philosophical context, but not within the neuropsychodynamic context that is targeted here. Instead, I shall focus on the determination of the "brain as functioning" within conceptual, empirical, and epistemological contexts.

How can we characterize in more detail the "brain as functioning" within the conceptual context as distinct from both the "brain as observed" and the "brain as experienced"? Let us return briefly to Kant and Schopenhauer. Kant introduced the concept of the "transcendent object," which I suggest shows some similarity with what I here describe as the "brain as functioning." He assumes that both inner and outer sense can be traced back to the transcendent object as their common underlying ground: "But if one wants to broaden the concept of dualism as it is usually applied and take it in a transcendental sense, then neither it, nor the pneumatism that is opposed to it on the other side, nor the materialism on the other side, have the least ground, since then one's concepts would lack determination, and one would take the difference in the mode of representing objects which are unknown to us as to what they are in themselves, for a difference in these things themselves. I, represented through inner sense in time, and objects in space outside me, are indeed specifically wholly distinct appearances, but they are not thereby thought of as different things. The transcendental object that grounds both outer appearances and inner intuition is neither matter nor thinking being in itself, but rather an unknown ground of those appearances that supply us with our empirical concepts of the former as well as the latter" (Kant (1998) (as translated by Guyer and Wood), A379–380).

How is this relevant to determining the concept of the "brain as functioning" within the conceptual context? According to Jaspers (1962a, p. 37), Kant even said at some point that the brain itself may be what Kant describes as "transcendental object": "All sorts of statements can be found in Kant; he even said that the world is conditioned 'by our brain'" (Jaspers, 1962a, p. 37). Unfortunately, Jaspers does not provide any reference by Kant himself. If the world is indeed "conditioned by our brain," Kant cannot but presuppose the "brain as functioning," which may then assume the kind of role that Kant attributed to his concept of the transcendent object. Unlike both the "brain as experienced" and the "brain as observed," the "brain as functioning" is assumed to provide the common underlying ground to which both inner and outer sense ("outer appearance" and "inner intuition" as Kant himself states) can be traced back.

Although in his major writings, such as the *Critique of Pure Reason*, Kant did not refer explicitly to the brain, it was Schopenhauer (1966), a later philosopher, who traced back Kant's concept of the mind and its input to our cognition to the brain. He distinguished between the brain as "object of cognition" and the brain as "subject of cognition." The brain as "object of cognition" refers to the brain as we can perceive and cognize it in both inner and outer sense, thus involving the "brain as observed" and the "brain as experienced." The brain as "subject of cognition" corresponds to what Schopenhauer called the brain as "subject of cognition." Unfortunately, Schopenhauer left open the exact epistemological and empirical characterization of the brain as "subject of cognition." However, the brief excursion into Kant and Schopenhauer showed that conceptually the "brain as functioning" may be close to concepts such as the "transcendent object" and the brain as "subject of cognition," which provides some insight into its conceptual context.

How can we characterize the concept of the "brain as functioning" within the empirical context in more detail and distinguish it from the concepts of the "brain as observed" and the "brain as experienced"? In order to characterize the "brain as functioning" empirically, we may need to look to Freud and his opening quote in *An Outline of Psychoanalysis*: "We know two kinds of things about what we call our psyche (or mental life) (Psyche, Seelenleben): firstly, its bodily organ and scene of action, the brain (or the nervous system) and, on the other hand, our acts of consciousness. . . . Everything that lies between them is unknown to us, and the data do not include any direct relation between these two terminal points of our knowledge. If it existed, it would at the most afford an exact localization of the processes of consciousness and would give us no help towards understanding them." (Freud, 1940, pp. 145–146).

What are the implications of this for the empirical characterization of the "brain as functioning"? The empirical features that specifically characterize the "brain as functioning" must "lie between" what we can observe (i.e. neuronal states) and what we can experience (i.e. mental states or consciousness) in the brain (i.e. the "brain as observed" and the "brain as experienced"). This means that neither neuronal states as observed in outer sense nor mental states as experienced in inner sense can be considered candidates for empirical characterization of the "brain as functioning." This is because mental and neural states do not "lie between" brain/body and consciousness but rather on either side of them, belonging either to the brain/body or to consciousness.

The exclusion of neuronal states means that even the neuronal states underlying consciousness, including mental contents, the so-called neural correlates of consciousness, cannot be considered as empirical features of the "brain as functioning." That would simply mean recreating the dualism between brain/body and consciousness and projecting it onto and ingraining it into the brain itself (i.e. the "brain as functioning"), with the latter being false-positively confused and equated with the "brain as observed." Instead, one would do better to single out an empirical feature that truly "lies between" the brain and consciousness, as pointed out by Freud. Although we are still being abstract and vague at this point, we now at least know the criterion that the postulated empirical characterization of the "brain as functioning" needs to fulfil. The empirical feature in question should be neither neural nor mental by itself, and it should account for the alleged empirical difference between the "brain as functioning" and the "brain as observed."

How can we escape the dichotomy between inner and outer sense, including the associated dichotomies between mental and neuronal states and between the "brain as experienced" and the "brain as observed" in our empirical characterization of the "brain as functioning"? Rather than adopting an empirical view of the brain that remains stuck in inner and outer sense and hence in the dichotomy between mental and neural states, we might do better to adopt a transcendental view of the brain. Such a transcendental view may allow shifting empirical features of the brain into our view that are neither neural nor mental, but which "lie between" the two. As we learned in Chapter 1 of this book, such a transcendental view targets the organization and structure of the brain rather than its neural and mental contents and their respective neural and mental states. In that chapter I also briefly characterized the neural organization and structure of the brain by the brain's resting-state activity and the impact on the neural processing of stimuli and subsequent stimulus-induced activity (Northoff et al., 2010). However, it is not only the brain that imposes itself on stimulus-induced activity. The traffic also moves in the reverse direction, with the stimulus and stimulus-induced activity imprinting themselves on the brain and its resting-state activity. This amounts to what we characterized as stimulus–rest interaction (Northoff et al., 2010).

This double modulation, consisting of rest–stimulus and stimulus–rest interaction, may be considered an organizational and structural feature of the brain that can be revealed by presupposing a transcendental view of the brain rather than an empirical one. Moreover, as I shall suggest and then demonstrate in Part II of this book, rest–stimulus and stimulus–rest interaction are supposed to enable and predispose to the transformation of neuronal states into mental states. If they indeed enable and predispose to what I shall later call "neuronal–mental transformation" (see Part II for further details), rest–stimulus and stimulus–rest interaction must be assumed to truly "lie between" neuronal and mental states, and thus between brain/body and consciousness in exactly the same way as Freud described in the quote (see p. 47). Both features, namely the presupposition of a transcendental view of the brain and their intermediate position between brain and consciousness, thus make rest–stimulus and stimulus–rest interaction suitable and likely candidates for the empirical features that characterize the "brain as functioning" as distinct from both the "brain as observed" and the "brain as experienced."

What are the implications of this for neuropsychoanalysis? If rest–stimulus interaction and stimulus–rest interaction are indeed the empirical characterization of the "brain as functioning," then due to their intermediate "location" they should enable and predispose to those mental features such as the self and objects (and others, such as the tripartite structure of the ego and defense mechanisms) that are so well documented in psychoanalysis. Therefore, if rest–stimulus interaction and stimulus–rest interaction are relevant to the psychic apparatus as conceptualized by Freud, we need to investigate their relationship to some of the psychoanalytic key concepts and how they enable and predispose to these. I shall focus on the psychodynamic concepts of self and objects in particular, and how they are enabled and predisposed by the brain and its rest–stimulus and stimulus–rest interaction, which I describe in terms of the concepts of brain–self and brain–object differentiation. This will be the aim of Part II and especially Part III of this book.

First, however, we need to characterize the concept of the "brain as functioning" within the epistemic context, which may also help us to further refine its empirical characterization.

"Unknowability of the 'brain as functioning'": epistemic extension of Kant and Freud

So far I have characterized the "brain as functioning" in empirical terms by rest–stimulus and stimulus–rest interaction. However, it is also necessary to characterize the "brain as functioning" in more detail in epistemic terms, thus raising the question of what we can and cannot know and how we can and cannot access the "brain as functioning." As already described, both Kant and Freud considered the mind or the psychic apparatus to be unknowable. Extending the views of both authors to the brain, one may consider the "brain as functioning" to be unknowable, too, so that one may characterize it epistemically by what I call "unknowability of the 'brain as functioning.'" There is much at stake here. If the "brain as functioning" is indeed unknowable, its empirical characterization by, for example, rest–stimulus and stimulus–rest interaction is doomed to failure. The aforementioned organizational and structural features of rest–stimulus and stimulus–rest interaction may then be found to characterize the "brain as observed" rather than the "brain as functioning." Are we therefore confusing the "brain as observed" with the "brain as functioning," and thereby violating the "unknowability of the 'brain as functioning'"? Before answering this question, we may want to consider in more detail what exactly the "unknowability of the 'brain as functioning'" is describing. One limitation of our possible knowledge of our own brain has already been noted by previous philosophers, and will be briefly recounted below.

Schopenhauer (1966) was probably the first to point this out when he assumed that we cannot cognize the brain as a necessary condition of our cognition (the brain as "subject of cognition"), whereas we are able to cognize the brain as cognition content (the brain as "object of cognition," as he described it). Wittgenstein made the same observation, that we have no direct access to our brain as brain, which is implied in the following statement: "Strange coincidence, that all persons, whose skull was opened, had a brain" (Wittgenstein, 1994, p. 254; also see p. 249, pp. 199–200, 217–219). The phenomenological philosopher Husserl also explicitly mentioned that, in contrast to the case of our body as "lived body," we remain unable to perceive and cognize the brain as brain, as for instance some of its parts, like the frontal lobe: "But my frontal lobe is not the bearer of a field of touch and is for me not at all something appearing" (Husserl, 1989, p. 45).

Even some of the current neurophilosophical approaches have acknowledged a knowledge deficit with regard to our brain. For instance, Churchland describes how something appears as non-inferential and spontaneous when we cannot perceive the underlying processing: "Second, when one becomes aware of the result of nonconscious processing one has no introspective (conscious) access to the processing steps that went into producing the results. It is therefore entirely inevitable that there will be some discrimination—the results of nonconscious processing—that are spontaneously, noninferentially, and, one might say, directly experienced. That one cannot articulate how the discrimination was made is simply explained by the fact that there is a vast amount of nonconscious neural activity to which one does not—and perhaps cannot—have conscious access. For example, even feedback techniques will probably not allow me to be aware of what the amacrine cells of my retinae are now doing, any more than I can become aware of when hormones are released form the pituitary or what my blood pressure is. I do not have introspective access to the processing that yields a stereoptic, three-dimensional representation of the visual scene. I just simply see things in stereoptic depth. I cannot tell you how I identify a melody as 'Three Blind Mice'; I just do" (Churchland, 2002, pp. 118–119).

Thomas Metzinger, another current neurophilosopher, is also aware of this limitation: "In this context, it is also interesting to note that there is one decisive part of our bodies that is self-representationally blind. This part is the brain itself. It possesses no self-directed sensory mechanisms at all. For instance, we know from neurosurgery done on conscious patients that it is insensitive to pain. The body can feel itself with the help of the brain, but the brain itself is unable to directly feel itself. It is the blind spot of self-representation as it were. As a medium it is functionally transparent to the organism it supports, for instance, in functionally appropriating and owning itself with the help of a conscious self-representation" (Metzinger, 2003, p. 294; see also p. 177).

Even staunch neuroscientists who presuppose reduction and elimination of the mind in favor of the brain, such as Francis Crick, are aware of this limitation: "There is general agreement that people are not conscious of all the processes going on in their heads. . . . While you may be aware of many of the results of perceptual and memory processes, you have only limited access to the processes that produce this awareness" (Crick, 1994, p. 19). Finally, traditional philosophers such as Maxwell Bennett and Peter Hacker, who oppose conceptual interference between philosophy and neuroscience, are also aware that we remain unable to perceive our own brain: "And since normally we cannot perceive the activities of our own brain, we cannot become and then be conscious of them; but this does not mean that they are unconscious activities, either of ourselves or of our brains. The fact that we are not conscious of the activities of our brains does not imply that we are engaged in them unconsciously; but nor does it imply that the brain is engaged in these activities unconsciously. For only a creature that can do something consciously can also be said to do something unconsciously" (Bennett and Hacker, 2003, p. 270).

These examples make it clear that a limitation in our cognition and knowledge of the brain has been noted previously. However, we need to specify this limitation in both epistemic and empirical terms. I shall address the epistemic specification in this section, which will then be followed by the empirical specification of the "brain as functioning" in the next section.

How can we epistemically specify the limitation in our cognition and knowledge of the brain? In addition to the TPP, which may be considered to be more or less analogous to what in earlier times was described as outer sense, our (human) epistemic design can also be characterized by the FPP, which may more or less correspond to inner sense. For example, we can access our mental states from FPP in that we experience ourselves as such within the environment. In contrast, no one has ever subjectively experienced their brain as their brain and its neuronal states as such in FPP. Subjective experience remains "blind" to one's own brain, which never appears as such (i.e. as one's brain and neuronal states) from FPP. Although we can subjectively experience ourselves, including our own body, from FPP, this does not include our brain and its neuronal states as such. I call this inability to directly access the brain in one's skull as one's brain from FPP the "autoepistemic limitation of the brain" (ALB).

I define ALB as the epistemic inability to subjectively experience and thus to directly access one's brain as brain from FPP.[5] For example, we have no subjective experience of brain pain along the lines "My brain hurts" or "My brain aches." Even in the case of migraine we speak of headache, rather than of "brain ache." This inability to subjectively experience one's brain as such has been nicely described by Todd Feinberg: "Now here comes the tricky part. What is it about the

[5] At this point I presuppose a broad definition of FPP that includes both subjective experience and introspection. Some authors (see Northoff 2004a,b for details) distinguish the second-person perspective from FPP by relating the latter to subjective experience and the former to introspection. Since subjective experience in phenomenal terms presupposes consciousness and thus introspection (Metzinger, 2003), I subsume both under the FPP.

brain, what neurological factor or property allows the brain to create subjective neural activities? There is a simple fact about the brain that is often neglected that provides the answer to this question. The conscious brain has no sensation of itself. It has been known since the time of Aristotle that the brain is insensate. For instance, sticking a pin in the cortex itself evokes no pain that is referable to the brain itself. The brain has no sensory apparatus directed towards itself. As Globus outs it, the brain does not 'represent in any way its own structure to the subject.' There is no way that the subject can become aware of his own neurons 'from the inside.' They can be known objectively 'from the outside.' We have already seen that there is no 'inner eye,' no inner homunculus watching the brain itself, perceiving its own neurons, no 'brain-skin' which feels the neurosurgeon's knife. When I test a patient's pinprick sensitivity by applying a pin to the hand, and I ask them to localize where on the body the sensation is, no one has ever pointed to their head. Conscious neural activity refers to things, not to the brain itself. Conscious neural states are about things, not about the neurons themselves" (Feinberg, 2001, p. 147).[6]

I shall now compare our epistemic inability with regard to the brain (e.g. ALB) with our epistemic abilities with regard to the rest of the body, such as our muscles, heart, liver, etc.[7] In contrast to the case of the brain, we are well able to subjectively experience our own body and its distinct parts. For example, we actually say "My stomach hurts," "My muscles are sore," "My liver aches," "My heart is jumping," and so on.[8] What exactly, then, do we subjectively experience? We can, for example, feel and experience the process of digestion and our heartbeat. Due to the fact that our body has sensory receptors for the stomach and the other internal organs, we can subjectively experience our body as such from FPP.

I call this epistemic ability to subjectively experience one's body as such from FPP the "autoepistemic capacity of the body" (ACB). Philosophically, terms such as "phenomenal body," "lived body," or "Leib" have been used to describe our subjective experience of our own body from FPP (Merleau-Ponty, 1962). The "lived body" and its appearance in subjective experience from FPP must be distinguished from the "objective body" that we observe from TPP. The main difference between the "lived body" and the "objective body" is that the former is specifically related to the person showing what is called "mineness" (Metzinger, 2003), whereas the latter remains independent of the specific person. Such "mineness" is, for instance, not given in the case of the brain and its different regions, such as the frontal lobe which, to draw the analogy to the body, we can access (i.e. observe) as "objective brain" but not as "lived brain." In summary, one's body is accessible by both subjective experience from FPP and objective observation from TPP, which I here describe as ACB. This distinguishes the body from the brain, which can be accessed from TPP but not from FPP, for which I use the term ALB.[9]

---

6 Metzinger (2003, pp. 177, 294) also recognizes our inability to subjectively experience our own brain as brain.

7 In the following account I distinguish between brain and body, although I am of course well aware that the brain is part of the body. Putting such empirical concerns aside, I here distinguish between brain and body, and this is intended in a purely epistemic way, without making any empirical implications.

8 As such, there are other organs in addition to the brain which cannot subjectively experience. For example, we cannot subjectively experience our liver or our spleen. These organs are thus in the same situation as the brain, as neither have any interoceptive sensory system on their own.

9 One may be inclined to argue that autoepistemic limitation (AL) of the brain can be resolved by either technical or social means. Technically, the recent introduction of brain–computer interfaces (BCI) (Birbaumer et al., 2003) has made it possible to visualize and thus to access one's own brain states and to consecutively modulate them. Would that not overcome AL and make the "brain as functioning" directly accessible to us? Even if one acknowledges that BCI provides access to one's brain and its neural states,

What are the implications of ALB for the epistemic characterization of the "brain as functioning"? Although we can certainly observe the brain's neuronal states and thus the "brain as observed,"

---

the access would still be indirect via the perception of the visualization of one's brain states. The only novelty about BCI is that it provides online access to one's own brain as distinct from access to other brains (as in neuroscientific observation). However, in both cases it is only the "brain as observed" that comes into focus since, even in BCI, I access my own brain only as I can observe and perceive it in my observation of its visualization. Thus BCI still concerns the "brain as observed" rather than the "brain as functioning."

Another way to circumvent AL of the brain may be to revert to social means, and specifically to recruit a community of researchers rather than individual ones, who would then correct each other and would thus overcome AL. Such a scenario is, for instance, outlined by Metzinger in his example of neglected patients: "It is, of course, an intriguing question whether there are, even for 'normal' persons simple facts—for instance, about the relationship between mind and body or about the nature of the phenomenal self—which are in principle cognitively unavailable, because our brains are not able to supply the frame of reference that would be necessary to carry out the cognitive simulations leading to convincing theoretical solution. Are we, like Ellen, suffering from a theoretical version of the looking-glass syndrome? Some philosophers have argued that the puzzle of conscious experience itself belongs to this class of problems (McGinn). Are we cognitively closed, because we are autoepistemically closed? As I pointed out, "autoepistemic closure" does not necessarily refer to cognitive closure in terms of the unavailability of theoretical, propositionally structured self-knowledge. And we have in the past solved many theoretical puzzles (e.g. in theoretical physics) in ways that were unexpected and probably unexpectable by anyone—simply by employing mathematical and other intersubjective, nonphenomenal media of representation. A community of neglect patients could certainly find out about the nature of mirrors and the representational roots of the looking-glass syndrome" (Metzinger, 2003, p. 227; see also p. 389, where he speaks of "violating the principle of autoepistemic closure").

Metzinger's analogy to the community of neglected patients fails because it aims only to demonstrate that these patients find out from the mirror that they are neglected, whereas it does not discuss whether this community of neglected patients would also be able to resolve the nature of neglect as a neurological disease. I believe that the community of neglected patients would remain unable to do so if one extends the example to what I call the "neglect world." Imagine a world in which all individual subjects suffer from neglect from their birth onward. They never get to know any subject who is not suffering from neglect. This community of neglected patients living in their neglect world would remain unable to resolve neglect as a neurological disease. For that to be possible, they would have to be able to critically distance themselves from their own neglect. However, that is only possible if somebody knows or at least has some positive (i.e. not only negative) ideas about the opposite condition (i.e. subjects without neglect). Then their own neglect could be contrasted with the non-neglect and the former could be resolved. I regard the community of neglected patients living in a neglect world as analogous to our situation with regard to AL. Due to the latter, we live in a world where our own brain is neglected, and this cannot be resolved because we remain unable to contrast it with a positive idea of what it would be like to live in a world without AL. Since every single one of us suffers from AL, even the social community of all individual subjects will remain largely unable to resolve AL. This is due not only to epistemic self-circularity, as discussed below, but also to the lack of positive ideas about the opposite or contrasting case.

However, the reference to intersubjectivity and thus social means of resolving AL (see also Churchland, 2002 for a similar argument) confuses the general nature of AL with an individual and contingent limitation in accessing the brain. If the whole community and not only individuals suffer from AL, we may be largely unable to overcome AL even by intersubjective and social means. We may now want to argue that in such a setting we may remain largely unable to know about AL as such, because of the lack of a contrasting case (i.e. one without AL). However, this is not an argument against AL, but against our possible knowledge of AL as AL. Thus the only way in which we can make a statement about AL is in the form of a hypothesis, so that AL may be considered a neuro-epistemological hypothesis rather than knowledge (in the philosophical and thus epistemological sense) (for a detailed discussion, see Northoff, 2004a,b; Northoff and Musholt, 2006).

ALB prevents us from directly accessing the "brain as functioning." This means that epistemically the "brain as functioning" may be characterized by a limitation in our cognition and knowledge of our own brain as brain (e.g. ALB). Does this limitation and epistemic inability mean that we cannot know anything about the "brain as functioning" via the brain's empirical features, including its empirical distinction from the "brain as observed"? Although this is certainly true for the epistemic case of direct access, it may not necessarily hold when accessing the "brain as functioning" in an indirect way, as for instance when inferring from mental and neural states back to their commonly underlying organization and structure. This leads to another question. How must the "brain as functioning" be structured and organized in order to enable and predispose to the kind of neuronal states that we observe from TPP (i.e. the "brain as observed") and the mental states that we experience from FPP (i.e. the "brain as experienced")? We may need to indirectly infer from ALB back to the kind of empirical processes that must occur in order to enable and predispose to the kind of neuronal and mental states that we can observe and access from FPP and TPP and thus in inner and outer sense.

"Earlier processing stages" and the "unknowability of the 'brain as functioning'"

So far I have described ALB as the principal limitation to our possible knowledge of the "brain as functioning." The question now is why there is such limitation to our knowledge of the brain. We remain unable to directly access the brain as brain from FPP, whereas we can access the brain as "brain as observed" from TPP. Why, though, do we remain unable to directly access the brain as brain from FPP? One may postulate that ALB can be traced back to the brain itself and the way that it organizes and structures our perception and cognition (and thus our knowledge) of both the world and the brain from FPP and TPP. If so, the question about the necessary conditions for and thus the origin of ALB and its underlying empirical features leads us back to the brain itself, and more specifically to the way that our brain's input structures and organizes our perception and cognition of the world, including our perception and cognition of the brain itself as part of that world. Thus the mechanisms underlying ALB may turn out to be precisely those that may characterize the "brain as functioning" in the empirical context, which may also distinguish it empirically from both the "brain as observed" and the "brain as experienced."

How can we further specify the brain's empirical or neural features that enable and predispose to ALB? Let me compare what I call ALB with what the philosopher Metzinger calls "autoepistemic closure." Metzinger (2003) also speaks of an epistemic limitation with regard to one's self, which he describes as autoepistemic closure: "I have called this structural characteristic of the neurophenomenological caveman's conscious mind 'autoepistemic closure,' referring to it as an 'inbuilt blind spot,' a structurally anchored deficit in the capacity to gain knowledge about oneself. It is important to understand that autoepistemic closure as used in this book does not refer to cognitive closure or epistemic 'boundedness' in terms of the perhaps principled unavailability of theoretical, propositionally structured self-knowledge. Rather, it refers to closure or boundedness of attentional processing with regard to one's own representational dynamics. It is a limitation in mental resource allocation expressed on the level of nonconceptual content. Autoepistemic closure, in the current context, consists in humans beings in ordinary waking states, using their internal representational resources—that is, by introspectively guiding attention—not being able to attentionally penetrate into *earlier processing stages* [my italics] in the ongoing construction of their conscious self-model" (Metzinger, 2003, p. 564; see also p. 57).

What is the consequence of our apparent lack of insight into what Metzinger calls the "earlier processing stages"? Attentional unavailability of these earlier information-processing stages leads to what Metzinger calls "transparency." Transparency means that we "look through something," such as the content, without seeing its underlying properties, such as the vehicle that is carrying the content (see the original definition by G. E. Moore, 1903, quoted in Metzinger, 2003, p. 163). Transparency can thus be regarded as "synonymous to a missing of information" (Metzinger, 2003, p. 175). Metzinger speaks of "phenomenal transparency,"[10] when we miss information in the context of subjective experience. He cites the following example, where we miss experiencing the window: "We don't see the window, but only the bird flying by" (Metzinger, 2003, p. 169).[11]

How can we characterize the "earlier processing stages" in more detail? According to Metzinger, phenomenal transparency is a matter of attentional availability in introspection of earlier processing stages, with each being inversely proportional to the other. The lower the attentional availability of earlier information-processing stages, the more subjective experience can be characterized by phenomenal transparency. Conversely, the higher the attentional availability of earlier information-processing stages, the less subjective experience can be characterized by phenomenal transparency. Subjective experience can then be characterized by the opposite of phenomenal transparency, namely phenomenal opacity: "Phenomenal opacity is simply the degree of attentional availability of earlier processing stages, and the degree depends on how adaptive it was to make these earlier processing stages globally available" (Metzinger, 2003, p. 175). Once our attention focuses on the window itself, it becomes clear that the bird is flying outside the room: "If the window is dirty or has a crack, we realize that we view the bird flying by through a window" (Metzinger, 2003, p. 170).

Although Metzinger puts the concept of "earlier processing states" into the conceptual context of phenomenal transparency, attentional availability, and phenomenal opacity, he leaves open their exact empirical characterization.[12] It seems as if we have no direct access to them as such, and their existence can be only inferred and hypothetically assumed on an indirect basis. However, this is all that he tells us about them. He neither specifies any empirical processes nor makes any hypothetical assumptions by at least indirectly inferring their possible empirical features and characteristics.

What exactly are these "earlier processing stages" and more specifically what kind of empirical processes are going on in these "earlier processing stages"? I postulate that these "earlier processing stages" are characterized by rest–stimulus and stimulus–rest interaction where the brain and its intrinsic activity (i.e. the brain's input) directly interface with the stimuli that the brain receives from the world. This encounter between the brain and the world (which I called the "brain–world interaction" in Chapter 1), seems to correspond within the empirical context to the linkage of the brain's resting-state activity to stimulus-induced activity (i.e. rest–stimulus and stimulus–rest

[10] Metzinger distinguishes phenomenal transparency from epistemic transparency (which concerns missed conceptual and propositional information), semantic transparency (which he describes as a property of extensional contexts), and referential transparency (which he considers to be a property of media as used in the theory of telecommunication) (Metzinger, 2003, pp. 170, 339–340, 436).

[11] This understanding of transparency, in which the representational character of something that is represented remains hidden, contrasts with that suggested by Tye, who argues that transparency is revealing that phenomenal experience is nothing but introspection and representation, because it reveals only its representational content: "Generalizing, introspection of your perceptual experiences seems to reveal only aspects of what you experience, further aspects of the scenes, as represented. Why? The answer, I suggest, is that your perceptual experiences have no introspectable features over and above those implicated in their intentional contents. So the phenomenal character of those experience is identical with, or contained within, their intentional contents" (Tye, 1995, p. 136; see also Legrand, 2005, p. 8).

[12] I claim that two central characteristics of early information-processing stages are sensory coding and further processing of sensory information in self-related processing.

interaction. However, such rest–stimulus and stimulus–rest interaction means that we remain unable to access the brain's resting-state activity directly by itself independent of its interaction with stimulus-induced activity. More generally this means that epistemically we remain unable to tease the brain and the world apart from each other in our perception and cognition of them, and it thus involves what one might call "hybrid knowledge" that describes such a mixture between the brain and the world.

It should also be clear from this that neither ALB nor the "earlier processing stages" refer to any kind of transcendent or mysterious properties. Such properties would refer to features in the brain whose ontological existence and reality lie beyond the domain of physical properties. This is, for example, assumed by the Mysterian philosophers such as Colin McGinn (1991) and Thomas Nagel (1998), who postulate special properties in the brain (e.g. a property P that can account for the linking of neuronal and mental states). This must be distinguished from my account here. Both ALB and "earlier processing stages" and thus the "brain as functioning" refer to an epistemic inability, namely our inability to directly access the "brain as functioning," including its earlier processing stages, rather than to some specific ontological properties and features that are distinct from physical (and mental) ones. To put it in philosophical terms, I determine the concept of the "brain as functioning" and thus ALB in a purely epistemic sense and not in an ontological sense, thereby, in Kant's terms, presupposing a transcendental view of the brain rather than postulating transcendent properties within the brain itself.

What are the epistemological implications of this empirical mixture of resting-state activity and stimulus-induced activity, and thus of brain and world? First, it means that the perception and cognition of ourselves and the world is a composite or hybrid of both brain and world, in very much the same sense that Kant characterized our cognition as being a composite of form originating in ourselves (i.e. our mind) and content derived from the world. Secondly, it means that we may remain unable to directly perceive and cognize our brain and its input to our cognition independent of the stimuli from the world. The hybrid and composite nature of our brain's neural activity (as we can observe it), being a mixture of both the world's input and the brain's input and their interaction, makes direct observation and experience of the brain's input itself, independent of the world's input, impossible. Thus, as ironic as it sounds, the "unknowability of the 'brain as functioning'" and thus ALB may be traced back to the brain itself and the interaction of its own input with its own neural processing of the world's input[13] (see Figure 2.2).

[13] Does this mean that we cannot know anything about the "earlier processing stages"? Metzinger denies this when he argues that "autoepistemic closure" does not mean "cognitive closure": "Are we cognitively closed, because we are autoepistemically closed? Autoepistemic closure does not necessarily refer to cognitive closure in terms of an unavailability of theoretical, propositionally structured self-knowledge" (Metzinger, 2003, p. 227). Metzinger argues that on higher levels, relying for example on what he calls the attentional and cognitive selves, we can escape our epistemic limitation and go beyond it by "violating the principle of autoepistemic closure" (Metzinger, 2003, p. 389). Due to the fact that it can be resolved by higher-order cognitive and social functions, Metzinger's "autoepistemic closure" is not epistemically but only empirically relevant. As such it must be distinguished from ALB, as set out here, which describes a principal epistemic limitation that cannot be resolved by either higher-order cognitive or social means, making ALB not only empirically but also epistemically relevant.

This is in contrast to ALB, which (unlike Metzinger's autoepistemic closure) is supposed to be impossible to resolve by technical, social, or cognitive means. Even if our higher-order cognitive functions may be able to obtain some knowledge about the "earlier processing stages," this knowledge will always (or, as the philosopher might say, necessarily and principally) remain inferential and indirect. ALB

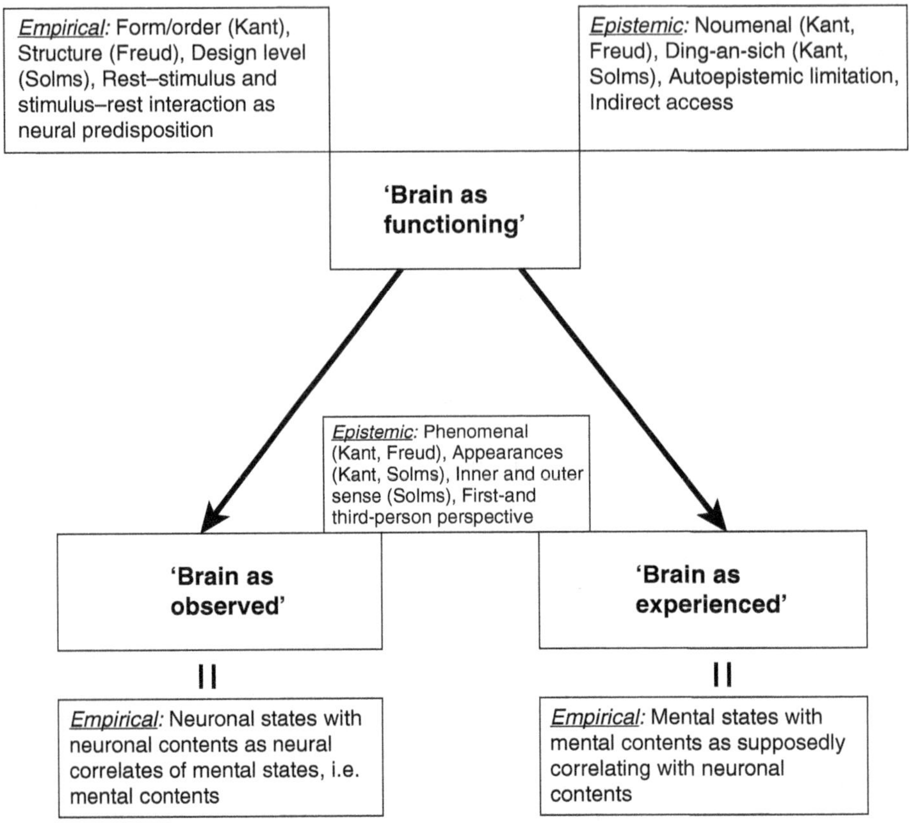

Fig. 2.2 Empirical and epistemic characterization of the different concepts of the brain.

What are the implications of this for neuropsychoanalysis? Neuropsychoanalysis must target the "brain as functioning" if it really wants to make the leap from Freud's concept of the psychic apparatus to the brain. The fact that we cannot directly access and know the "brain as functioning" as such means that we may need to devise indirect methodological strategies that allow us to indirectly infer from what we know about the "brain as observed" and the "brain as experienced" back to the "brain as functioning." We therefore need to develop a methodological strategy that links mental and neural states and thus the "brain as observed" and the "brain as experienced" in order to make some inferences about the kind of empirical features that commonly underlie them (e.g. the earlier processing stages as empirical characteristics of the "brain as functioning"). The development of such an indirect methodological strategy will be the focus of the next chapter, while the investigation of the empirical details of the earlier processing stages, including rest–stimulus and stimulus–rest interaction, will be addressed in Parts II and III of this book.

describes exactly the inability to directly access these "earlier processing stages" in our brain and thus to experience or observe them as such. Interestingly, these "earlier processing stages" were apparently also referred to and alluded to in the aforementioned quotes by Churchland and Crick when describing our inability to perceive and cognize the brain (i.e. the "brain as functioning").

Chapter 3

Transdisciplinary Methodology and Neuropsychodynamic Concept–Fact Iterativity

Freud's duality between science and hermeneutics of mind: "concept–fact linkage"

Freud was originally a neuroanatomist and intended to develop a scientific account of the mind in empirical terms. This is well reflected, for instance, in his mechanistic view of the mind (i.e. the psychic apparatus), which he compared with a hydraulic machine by using terms such as energy, suppression, etc. He thereby relied on observation from the third-person perspective (TPP), which allowed him to approach the mind in an empirical way and thus to develop what can be called "science of mind" (i.e. "science of the psychic apparatus").

However, Freud aimed to achieve more than just a science of mind. Going beyond mere observation from TPP, he strived to understand his patients' subjective experience from the first-person perspective (FPP). More specifically, he raised the question of how the mind can generate and constitute the kind of meaning, personal values, and norms that he observed to be inherent in his patients' subjective experiences. For instance, he raised the question of the sense and meaning of dreams and how the mind must be functioning in order to enable and predispose to the constitution of the kind of meaningful contents that are so prevalent in dreams. This allowed him, as Tauber (2009) has pointed out, to go beyond pure TPP-based observation and objectivity by complementing it with FPP-based experience and subjectivity.

The consideration of his patients' meanings, personal values, and norms involves a shift in focus from the mind's pure mechanics to its constitution of meaningful contents and behavior. The "science of mind," which targets mere mechanisms, is thus complemented by what one may call "hermeneutics of mind," which focuses more on the meaningful contents associated with the psychic apparatus. However, the explicit inclusion of, reliance on, and reference to meaning and personal values as elucidated from his patients' subjective experience stands in the way of the ideal of a science of mind that is solely based on TPP-based observation and objectivity. A science of mind that is based on TPP-based observation and objectivity must exclude any FPP-based subjective experience and the strong traces of subjectivity that are associated with personal values and meaning. However, Freud was apparently not willing to succumb to either scientific TPP-based observation and objectivity or FPP-based experience and subjectivity.

One may therefore identify a methodological duality in Freud, with the consideration of both FPP-based subjective experience and TPP-based observation with objective facts and subjective values standing side by side. This methodological duality is nicely described by Tauber (2009): "Freud's positivist project, built on the ideals of the physical sciences, breaks apart on those normative shoals, for the neat division of objective facts and subjective values only admits the separation of certain kinds of facts from certain kinds of value. The interpretative nature of psychoanalysis disallowed such a simple prescription, although formally, Freud held to his earlier scientific ideal to separate scientific objectivity from personal value. A positivist philosophy that had served a newly emerged scientific medicine so well was not only inappropriate for psychoanalysis; its aspirations were unattainable as well" (Tauber, 2009, p. 34).

Brook (1998) also identifies an almost analogous duality in Freud when considering his psychology and its relationship to metapsychology. Psychology describes concrete psychological contents and uses intentional concepts that describe goal-oriented and meaningful behavior. Such intentional concepts presuppose what is generally described as *Verstehen*, which is the main methodological tool in any kind of hermeneutics, including the hermeneutics of mind. Such *Verstehen* is, for instance, presupposed when Freud aimed to understand the sense and meaning of his patients' dreams, the personal relevance and values of their contents for the respective person itself and biography. In contrast, metapsychology concerns psychological structures and organizations which can be described in purely mechanistic terms by using concepts (e.g. energy, etc.) that describe causal processes. This presupposes *Erklären* rather than *Verstehen*, and thus a science of mind. Hence, following Brook, Freud's metapsychology corresponds to what I here described as TPP-based objective science of the mind, whereas his psychology must rather be considered to target an FPP-based subjective "hermeneutics of the mind."

However, to merely identify duality between science and hermeneutics of mind (e.g. between *Verstehen* and *Erklären*) is not sufficient. Freud aimed for something more radical, and was not satisfied with merely considering both subjective (i.e. hermeneutical) and objective (i.e. scientific) aspects of the mind. Rather than allowing both aspects to stand side by side in parallel, he aimed to integrate them and show how the mind's mechanical structure and organization enable and predispose to the constitution of its meaningful contents, including personal values and norms. In other words, he was interested in the transition from the mind's objective mechanical structure to its subjective meaningful contents. By focusing on this transition, he moved way beyond mere parallelism between science and hermeneutics of mind.

For instance, he described the mind and what he, in his metapsychology, called the "psychic apparatus" in terms of the mechanistic concepts of the id, the ego, and the superego. However, at the same time he associated meaningful contents with them (e.g. drives/wishes, personal values, and norms). Most importantly, he aimed to understand how the concepts of the id, ego, and superego can enable and predispose to the constitution of the drives/wishes, personal values, and norms. For that purpose he introduced more dynamic terms, such as energy (i.e. cathexis), in his description of the mechanistic concepts (for further details, see Part II of this book). This allowed him to better understand the transition from the objective structures of the id, ego, and superego to subjective meaningful contents (e.g. drives/wishes, personal values, and norms).[1]

How can we account methodologically for the transition from science to hermeneutics of mind? Reduction or even elimination of the hermeneutics of mind in favor of a science of mind is not an option. Such a reductive–eliminative strategy, as philosophers call it, is often presupposed in current neurophilosophy (see, for example, Churchland, 1986, 2002; Bickle and Mandik, 2006). The subjective meaning of philosophical concepts is reduced to and sometimes even replaced by the objective facts of neuroscience, with the latter being regarded as the neural correlates of the former (the concept of "neural correlates" is discussed in more detail below and in the Introduction). However, this is not an option here as, metaphorically speaking, that would mean

[1] This presupposes a wider meaning of science, as is nicely expressed by Alfred Tauber: "Freud struggled with fitting the scientific mode of knowing within the broader therapeutic agenda he set himself. For him, science did not function solely as some kind of separate intellectual or technical activity to study the natural world, but rather became an instrument to help define human realities in the human quest of knowing the world in order to place ourselves within it" (Tauber, 2009, p. 41).

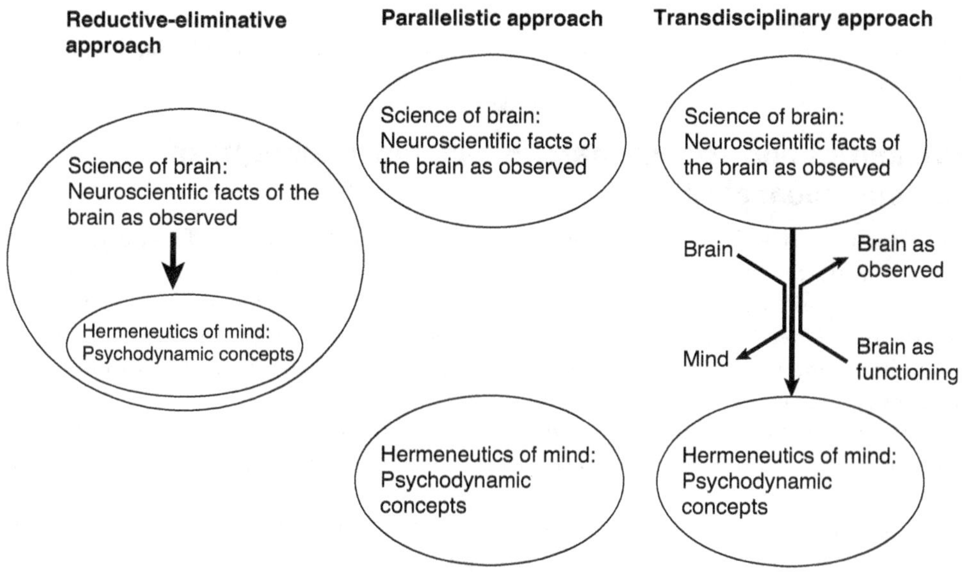

Fig. 3.1 Different approaches to science of the brain and hermeneutics of the mind.

to cut off one side of the bridge (the subjective meaning in the mind's contents) from the other side across the river (the objective structure of the mind). The mind itself may thus be likened to a bridge between objective structures and subjective meaning. To reduce the latter in favor of the former would mean locating the bridge on only one side. However, this would strip it of its very nature as a bridge. Thus a reductive–eliminative strategy cannot be considered an appropriate methodological approach for focusing on the mind, as that requires bridging of the gap between science and hermeneutics of mind.

Do we have to succumb to mere parallelism? A merely dualistic and parallelistic approach to science and hermeneutics of mind is not sufficient either, and so is not an option. Presupposing mere duality or parallelism between science and hermeneutics of mind would mean, speaking metaphorically again, considering both sides of the river while neglecting the bridge that crosses the river. Thus both approaches, reductive–eliminative and parallelistic, must be deemed inappropriate for Freud's aim of understanding the transition from science to hermeneutics of mind. Thus, despite their differences, both approaches (i.e. reductionism and parallelism) have in common the fact that neither provides us with a bridge which crosses the river, since either the bridge is shifted to just one side of the river, or it is not built at all (see Figure 3.1).

What does the inadequacy of both reductive–eliminative and parallelistic approaches mean for our methodological strategy in positive terms? TPP-based observation of objective facts, as presupposed in science of mind, needs to be not only complemented but also linked and integrated with FPP-based concepts that describe subjective meaning. This means that we need to devise a methodological strategy that allows us to directly link and connect facts and concepts without either reducing the one to the other or allowing them to stand side by side in a parallel manner. In other words, we need a transdisciplinary methodological strategy (see Figure 3.1) that allows for what I call "concept–fact linkage."[2] The main aim of this chapter is to describe the development

---

[2] I am of course aware that the description of facts in neuroscience also presupposes concepts so that, literally understood, the term concept–fact linkage is misleading. I therefore understand the concept of concept–fact linkage and the subsequent one of concept–fact iterativity in a rather figurative sense, as distinct from

and theoretical outline of such a non-reductive/non-eliminative and non-parallel methodological strategy. Its application and concrete realization will then be demonstrated in Parts II and III of this book.

Humanities and science: narrow and wider concepts of neuropsychoanalysis

There is much at stake here. The required non-reductive/non-eliminative and non-parallel methodological strategy links not only science and hermeneutics of mind but also more generally the rather opposite camps of science and philosophy. Therefore in order to understand the full implications of Freud's enterprise, we may need to briefly shed some light on the conceptual presuppositions and more specifically on the role of philosophy and its possible (or impossible) relationship to science.

What was Freud's stance with regard to philosophy? Although he had received formal training in philosophy, Freud almost completely rejected the method of philosophy, the purely and exclusively logical investigation of concepts: "Philosophy is not opposed to science, it behaves like a science and works in part by the same methods; it departs from it however by clinging to the illusion of being able to present a picture of the universe which is without gaps and is coherent, though one which is bound to collapse with every fresh advance of our knowledge. It goes astray in its methods of over-estimating the epistemological value of our logical operations and by accepting other sources of knowledge such as intuition" (Freud, 1933, pp. 160–161).

Tauber (2009, p. 36) singles out three reasons why Freud rejected philosophy. First, philosophy cannot respond to new empirical discoveries, and therefore remains unable to integrate empirical findings in the concepts that it uses to describe the world. Thus there is what I would call a "blindness of philosophical concepts to empirical facts." Secondly, beyond the blindness of its concepts to empirical facts, philosophy is closed in the purely logical approach and operation of its concepts, thereby presupposing the domain of logic to be distinct from the empirical domain. Therefore there is what may be described as a "closed logical domain of philosophical concepts." Thirdly, philosophical investigation assumes that the merely logically investigated concepts mirror and reflect human reality, and that they thus provide us with knowledge about ourselves. In other words, non-scientific forms of knowledge are considered epistemologically relevant in the same way as scientific forms. However, this view may be considered flawed, and could be described as "false-positive epistemological relevance attribution to non-scientifically based concepts."

However, Tauber convincingly demonstrates that Freud's rejection of philosophy concerned only a particular kind of philosophy, namely that which adheres to the aforementioned claims. In contrast, a kind of philosophy that focuses on the human itself and considers scientific knowledge was of great interest to him, and may even have been his goal from the very beginning (Tauber, 2009, p. 37). Tauber (2009, p. 42) describes Freud as exemplifying "the 'natural philosopher' in the broadest and most noble tradition: to know nature; to define ourselves within nature; to control nature in order to direct human destiny" (Tauber, 2009, p. 42).

What kind of method is required to pursue Freud's ideal of philosophy? Let us return to the three reasons why Freud rejected philosophy. First, the "blindness of philosophical concepts to empirical facts" would need to be replaced by a method that allows for "openness of philosophical

a literal one. I am also aware that the concept of fact is itself a problematic one, discussion of which I leave to the philosophers. I use it here in an operational way to describe those processes, mechanisms, substances, etc. that were observed on the basis of TPP when conducting experimental investigations of the brain.

concepts to empirical facts," without eliminating the former in favor of the latter. Secondly, the "closed logical domain of philosophical concepts" would need to be replaced by a method that allows not only the opening of philosophical concepts to the empirical domain, but also the direct interfacing and linking of empirical and logical domains. This may amount to what I call the "systematic empirical–logical interface in philosophical concepts." Thirdly, the "false-positive epistemological relevance attribution to non-scientifically based concepts" would need to be replaced by a method that allows one to base epistemological characterization on the empirical discoveries of science. However, scientific facts provide only the basis, thus leaving some room for concepts and their meaning as associated with these scientific facts. This amounts to the development of a method for an "empirically plausible epistemological attribution to scientifically based concepts."

Summarizing all of this, one may want to argue that Freud opted for a broader sense of philosophy, one that does open its doors to empirical facts, the empirical domain, and empirical plausibility. More specifically, such a broader sense of philosophy considers facts when determining its concepts. Most importantly, it does not see the empirical domain in opposition to the logical–conceptual domain, and it compares and matches its conceptual definitions with empirical results to make the former not only conceptually but also empirically plausible. Thus the focus here is on the transitions between facts and concepts, between empirical and logical–conceptual domains, and between conceptual and empirical plausibility. Rather than excluding philosophy from the sciences, as in a narrow sense of philosophy, the broader sense of philosophy focuses on the transitions between (narrow) philosophy and science.

How can the transitions between philosophy and science be investigated? Current proponents of neurophilosophy, such as Patricia Churchland (Churchland, 2002) and John Bickle and Pete Mandik (Bickle and Mandik, 2006), may want to argue that philosophical concepts can be reduced to and eliminated by the scientific facts, thus presupposing a reductive–eliminative approach. Meanwhile the opponents of such a reductive–eliminative approach insist on the principal difference between empirical and logical–conceptual domains, thus presupposing parallelism between philosophy and science (see, for example, Bennett and Hacker, 2003).

However, neither approach can account for Freud's presupposition of a broader sense of philosophy that focuses on the transitions between facts and concepts, between empirical and logical–conceptual domains, and between conceptual and empirical plausibility.[3] As in the case of science and hermeneutics of mind, both reductive–eliminative and parallelistic approaches to philosophy and science fail to take into consideration the transitions themselves. We therefore need a non-reductive/non-eliminative and non-parallelistic approach to philosophy and science in order to account for the transitions between them and to make possible a broader concept of philosophy. More specifically, such a non-reductive/non-eliminative and non-parallelistic approach must operate across or in between facts and concepts, empirical and logical–conceptual domains, and conceptual and empirical plausibility to focus on their transitions.

[3] Freud himself was well aware of such an intermediate position when, for example, he aimed to avoid the localization of the mind, and specifically the hermeneutics of mind, in the brain (i.e. the science of the brain): "Freud's vision was to search for the truth about mental functioning, but he wanted to position his insights into the right category of realization. At all costs he was trying to avoid the category mistake of trying to locate within the brain events and processes that belonged to the mind. By 1924, disillusioned by the lack of interest in psychoanalysis by philosophy and medicine, and by the "cultural hypocrisy" of society, Freud had lowered his sights somewhat: "So it comes about that psychoanalysis derives nothing but disadvantages from its middle position between medicine and philosophy" (Mancia, 2006a, p. 39).

Let me be more specific about this with regard to the case of neuroscience and philosophy. Neuroscience focuses on the brain, whereas philosophy presupposes the concept of mind. If we now want to focus on the transitions between neuroscience and philosophy, we need to understand the transitions from brain to mind, which here in the methodological context I call "brain–mind transition" (and in the Introduction I called "brain–mind differentiation"). Since such "brain–mind transition" requires direct linkage and connections between facts and concepts, between empirical and logical–conceptual domains, and between empirical and conceptual plausibility, one needs to devise a methodological approach that allows this. However, such an approach would need not only to consider the methodological features of both neuroscience and philosophy, but also to devise a specific methodology to link them in a non-reductive and non-parallelistic way. This amounts to nothing less than a truly transdisciplinary approach, that operates across the boundaries between philosophy and science, which will then enable us to understand brain–mind transition.

What are the implications of this for the concept of neuropsychoanalysis as one of the most recently delivered babies that fall in between psychoanalysis and neuroscience? Neuropsychoanalysis aims to link the characterizations that Freud attributed to the mind (i.e. the psychic apparatus) to the brain and its neural apparatus. Does this mean that everything which comes with the mind now needs to be transferred to the brain? One item of baggage that comes with Freud's approach to the mind (i.e. the psychic apparatus) is his focus on the transitions from the mind's objective mechanic structure to the constitution of its subjective meaningful contents, including personal values and norms. The implications of this for neuropsychoanalysis are that it needs to focus on the transitions between brain and mind and thus between the neural and psychic apparatus, rather than simply transferring the latter to the former. This book targets one such neuropsychodynamic transition between the brain's neural apparatus and the mind's psychic apparatus in the gestalt of brain–self and brain–object differentiation (as I referred to this in the Introduction). And here, in this chapter, I suggest that in order to do this we need to devise a specific methodology which allows us to consider such a transition from the science of the brain to the hermeneutics of the mind (i.e. the psychic apparatus).

Building on and extending the concepts of Freud, neuropsychoanalysis must aim to understand the transition from brain to mind (i.e. "brain–mind transition"). Thus the concept of mind that is presupposed in "brain–mind transition" refers to subjective meaning, personal values and norms, and thus to the hermeneutics of mind. Such a hermeneutical meaning of the concept of mind in "brain–mind transition" must be distinguished from the scientific meaning of the concept of mind, where it is characterized by mere objective facts as they can be observed from TPP. If one presupposes a merely scientific meaning of the concept of mind, neuropsychoanalysis degenerates into a search for the sufficient neural conditions of psychic features (i.e. neural correlates of psychodynamic features). However, to adopt such a neural correlate approach is to neglect the mind's hermeneutical features, namely its ability to constitute meaningful contents, including personal values, meaning, and norms.

Based on the different concepts of mind in "brain–mind transition," one may want to distinguish between narrow and wider concepts of neuropsychoanalysis (see Figure 3.2). The narrow concept of neuropsychoanalysis seeks neural correlates of the mind's objective features and thus complements Freud's science of mind with a science of the brain. In contrast, the wider concept of neuropsychoanalysis aims for something more radical, namely to understand how the brain must be neurally structured and organized in order to enable and predispose to the constitution of the mind's subjective features, its ability to constitute meaning, personal values, and norms. Rather than seeking mere "neural correlates" of the mind's structure and organization, the wider concept thus searches for the necessary (but non-sufficient) conditions and thus for what I call

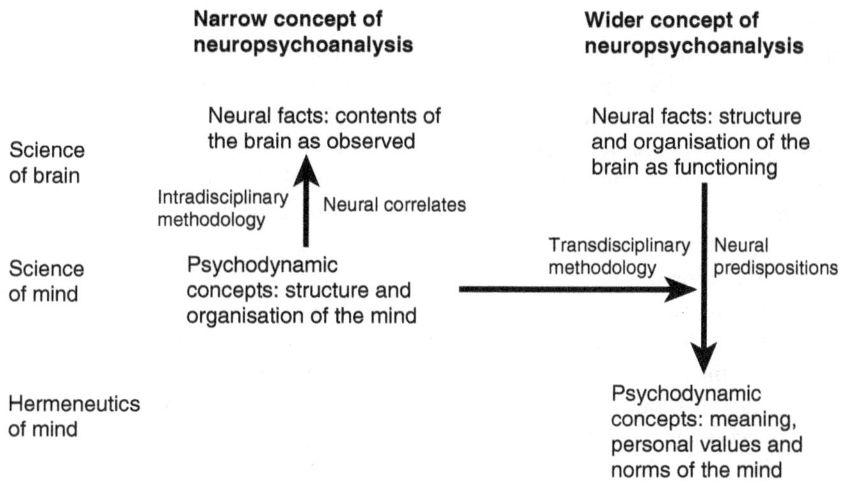

Fig. 3.2 Narrow and wider concepts of neuropsychoanalysis and their methodological approaches.

"neural predisposition" (see Figure 3.2; further details can be found below and in the Introduction). The focus on "neural predispositions" rather than "neural correlates" also involves conceptual and empirical differentiation between the "brain as observed" and the "brain as functioning," something which is not considered in the narrow concept of neuropsychoanalysis (see Figure 3.2).

What are the implications of the narrow and wider concepts of neuropsychoanalysis for the relationship between the science of brain and mind and hermeneutics of mind? The narrow concept of neuropsychoanalysis must be located between a science of the brain (i.e. neuroscience) and a science of the mind (i.e. psychology, and especially cognitive psychology). In contrast, the wider concept of neuropsychoanalysis must be situated in between a science of the brain (i.e. neuroscience) and hermeneutics of mind (i.e. philosophy), and thus more generally between science and the humanities.

What are the implications of this for the methodological strategy? The narrow concept can readily rely on a purely scientific and therefore intradisciplinary strategy that is focused on TPP-based observation and facts. In contrast, the wider concept of neuropsychoanalysis must presuppose a non-reductive/non-eliminative and non-parallelistic approach, which is able to focus on the transitions from the brain's neural structure and organization to the mind's hermeneutical features (i.e. meaning, personal values, and norms). This, as demonstrated previously, is only possible when presupposing a truly transdisciplinary approach. This will be specified and outlined in the remaining sections of this chapter. I thus presuppose here and throughout the rest of this book the wider concept of neuropsychoanalysis, rather than the narrow one.

Solms' quest for method: neuropsychodynamic concept–fact iterativity

In a similar way to Freud and Kant, Solms regards both inner and outer sense and thus the "outer world of objective observation" and the "inner world of subjective experience" as merely phenomenal, and thus as appearance (for details, see Chapters 1 and 2). Both may be traced back to a common ground, a common explanatory framework, which Solms associates with what he calls the "mental apparatus" as being analogous to Freud's concept of the "psychic apparatus." The mental apparatus cannot by itself be directly accessed by inner and outer sense, for which an additional perspective, besides inner and outer sense and thus beyond FPP and TPP, would be necessary. However, this additional perspective (if it would be a perspective at all, or rather what Kant and others have called "God's point of view") is (due to what I called "autoepistemic limitation" in Chapter 2) not available to us (Kaplan-Solms and Solms, 2000, p. 235). Since this additional perspective and thus direct access to the mental apparatus itself are not available to us, we have no choice but to stick to inner and outer sense and combine and link them, or correlate them, to use Solms's words (Solms and Turnbull, 2002, pp. 56–57; Solms, 2006), in order to make some indirect inferences about the mental apparatus itself (and thus ultimately what I called the "brain as functioning" in Chapter 2).

This raises the question of how we can link and combine inner and outer sense in a way that best allows some inferences to be drawn about the mental apparatus itself and thus the "brain as functioningt." Solms himself expressed this as follows: "So now we are faced with two descriptions of the metapsychology of consciousness (and emotion, memory, dreams, and so on)—two descriptions of everything that makes up the chapter headings of the book. This is neither a happy nor a healthy situation. Clearly, there is only one thing to be done about it: the two perspectives have to be integrated and combined, and thereby reconciled with one another" (Solms and Turnbull, 2002, p. 294).

Following Solms, we cannot do other than link and combine the data from inner sense with those from outer sense. Solms assumes that both data sets can be traced back to the mental apparatus and its specific design, and thus to what he calls the "design level of explanation" (Solms, 2006, p. 99). He also assumes some "law-like connections" between neuronal data and psychodynamic data (as he would probably call them, due to their association with the observation of inner sense): "The design level—the level upon which the 'mental apparatus' exists—is therefore the link between the empirical data of psychoanalysis and those of neuroscience. Hence its profound importance for our field: the design level must provide a unifying explanatory account of the data of both sciences. This is the scientific task of neuropsychoanalysis" (Solms, 2006, p. 99) (see also Kaplan-Solms and Solms, 2000, p. 244).

We have to link and integrate the psychodynamic and neuroscientific data in an ingenious way to provide some indirect access to the mental apparatus and ultimately to the "brain as functioning." However, such integration faces a problem in that different concepts are used to describe psychodynamic and neuroscientific data. As Solms and Turnbull (2002, p. 306) themselves remark, "when analysts talk about 'drives' or 'arousal' or 'inhibition' they seldom mean the same thing as neurobiologists using the same words." We therefore need to translate the different meanings of psychodynamic and neuroscientific concepts and to seek correspondences, overlaps, and ultimately novel and integrated concepts: "Scientific models cannot be integrated in theory—one needs to determine (by laborious scientific observation and experimentation) where and whether they refer to the same things and then codify the dual referents into a new, integrated language" (Solms and Turnbull, 2002, pp. 306–307).

According to Solms and Turnbull, this "new, integrated language" will no longer be characterized by either psychodynamic concepts or neuroscientific concepts, but specifically by neuropsychodynamic concepts. However, we need to develop a specific method that allows us to do this: "What is needed is a method by means of which one and the same thing can be studied simultaneously from both the psychoanalysis and the neuroscientific perspectives, so that one can be sure that the two sets of observations (and the resultant theoretical accounts) refer to the same piece of reality. Only this enables us to link the two theories in reality rather than in words" (Solms and Turnbull, 2002, p. 307).

Solms and Turnbull go on to suggest specific empirical–experimental methods, such as the clinico-anatomical method and the psychopharmacological challenge (Solms and Turnbull, 2002, pp. 307–312), but the exact procedure and methodology for linking neuroscientific and psychodynamic concepts and transforming them into a new, integrated language remain unclear. Therefore a precise methodological approach needs to be found for developing such truly neuropsychodynamic concepts (in the wider sense of the term) as opposed to psychodynamic and neuroscientific ones.

It is true that we cannot derive concepts other than from inner and outer sense. This is what the epistemological reality tells us. However, the failure to directly approach the mental apparatus independent of inner and outer sense should not prevent us from developing a third rather indirect approach, which consists of combining and linking the concepts as yielded on the basis of inner and outer sense in a specific way. We have two options. First, we can link neuroscientific and psychodynamic concepts in an arbitrary and intuitive way according to how we think they could fit together in the case under consideration. Alternatively, we could develop special rules and regulations with regard to how psychodynamic and neuroscientific concepts can and cannot be linked and connected to each other. This would mean developing a systematic rather than intuitive methodology for the links that are and are not possible between neuroscientific and psychodynamic concepts.

Since neuroscientific concepts are predominantly based on facts, whereas psychodynamic features rely on the concepts that we use to describe our subjective experience, I call such systematic methodology "neuropsychodynamic concept–fact iterativity." The development of such a truly neuropsychodynamic method may, for intradisciplinary orientation, be subsumed under what may be described as "philosophy of neuropsychoanalysis."[4] Here it is my intention to sketch a preliminary outline of such "neuropsychodynamic concept–fact iterativity" as can tentatively be determined as the continuous movement and systematic interaction between neuroscientific facts and psychodynamic concepts. This continuous movement and systematic interaction between neuroscientific facts and psychodynamic concepts is supposed to take place in a series of steps, which I shall describe in detail in the following sections.

Neuropsychodynamic concept–fact iterativity

"Feedforward translation": direct versus indirect translation

We are simultaneously confronted with concepts from psychoanalysis and with neuroscientific concepts that refer to specific facts.[5] Psychodynamic concepts (e.g. the concept of the ego) describe psychodynamic mechanisms that are associated with subjective experience as accessed in inner sense (e.g. FPP). In contrast, neuroscientific concepts (e.g. the concept of cortical midline structures, or CMS, which are a specific set or network of regions in the medial part of the brain's cortex)

[4] The concept of "philosophy of neuropsychoanalysis" is also used by Talvitie and Ihanus (2006, p. 96), although they appear to use the term in a much broader sense that includes not only methodological questions as originally discussed in philosophy of science, but also ontological issues with regard to the psychic apparatus and metapsychology as corresponding (more or less) to the discussion of the mind, as in philosophy of mind. I here use the concept of the "philosophy of neuropsychoanalysis" in the narrow sense that restricts it to methodological questions and procedures while neglecting ontological issues.

[5] As already mentioned in footnote 2, the concept of concept–fact iterativity is not to be understood in a literal sense, since that would not take into account that the facts yielded in neuroscience are also described in terms of concepts. Thus, in literal terms, one would need to speak of concept–concept iterativity rather than concept–fact iterativity. However, to indicate the principal differences between empirical and conceptual domains, I here speak of concept–fact iterativity, thereby presupposing this concept in a figurative rather than literal way.

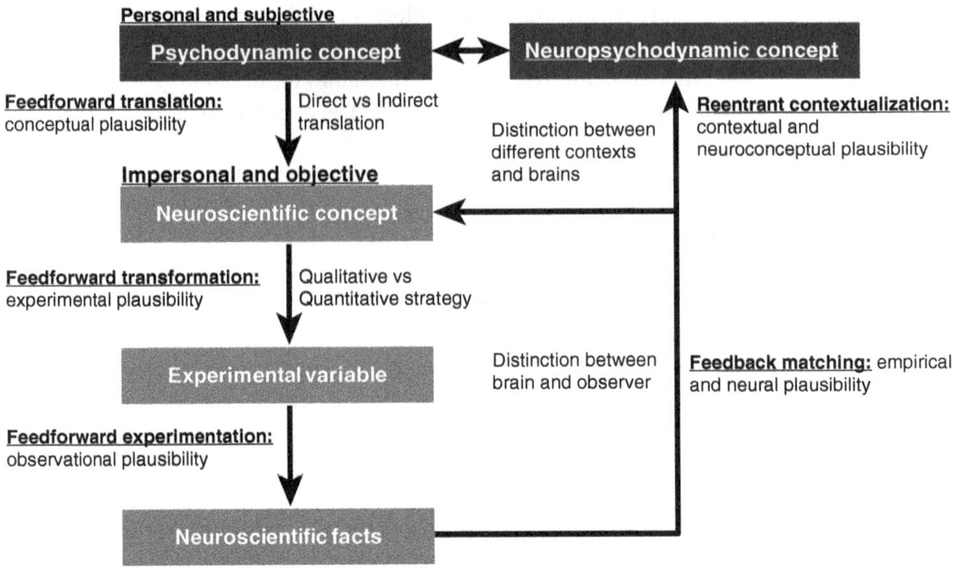

Fig. 3.3 Neuropsychodynamic concept–fact iterativity.

describe a specific neuroscientific fact (e.g. region, process, mechanism) that is assumed to occur in the brain as accessed by TPP-based observation and experiments.[6] Although both psychodynamic and neuroscientific concepts refer to contents, they differ from each other in the kind of contents that they describe (i.e. psychodynamic and neuroscientific contents), and in the way that they access those contents (i.e. from FPP and TPP). Therefore, if we want to link psychodynamic and neuroscientific concepts, we need to connect the different contents and their different perspectives. The first step in my neuropsychodynamic concept–fact iterativity, namely "feedforward translation," focuses on the linkage between the different contents, while the second step, namely "feedforward transformation," targets the linkage between FPP and TPP (see Figure 3.3).

How can we link psychodynamic and neuroscientific contents as described by their respective concepts? In order to link psychodynamic concepts to neuroscientific concepts and their respective contents, we need to translate the former into the latter, thereby generating a feedforward system. How can we translate psychodynamic concepts into neuroscientific ones? Such translation can be either direct or indirect. In direct translation the psychodynamic concept and its contents are directly translated into the neuroscientific concept and its contents, whereas in indirect translation some kind of mediating concept is required for the psychodynamic concept to be translated into the neuroscientific concept.

Let me start with direct translation. Direct translation presupposes that the psychodynamic concept can be directly linked to and translated into the neuroscientific concept. Ideally such direct translation is supposed to be possible in a one-to-one manner, so that the psychodynamic content can be "mapped onto" the neuroscientific one. For instance, one could assume that the concept of the ego can be directly translated in a one-to-one manner into the neuroscientific concept of CMS. All of the contents described by the concept of the ego are then supposed to be

[6] I am well aware that the concept of fact is by itself a problematic one. However, I leave discussion of this to the philosophers. I here use the concept of fact to describe those processes and states that can be traced back to observation from TPP (i.e. outer sense).

directly related to or correlated with the contents associated with the CMS. Thus there is what the neurophilosopher Churchland (1986, 2002) called "intertheoretic reduction," whereas neuroscientists such as Koch (2004) would probably speak of "neural correlates," with the CMS being the sufficient neural condition of the psychodynamic concept of the ego. Therefore direct translation involves the assumption of what I call "neural correlates of psychodynamic contents" or the "neural correlates approach."

Philosophically, one may now want to debate whether or not such direct one-to-one translation in the gestalt of correlates is possible. This has indeed been done and, importantly, doubts have been expressed.[7] I leave discussion of this to the philosophers. However, even if such direct translation is in principle possible, we may not be able to fully access and thus know (i.e. perceive and cognize) its outcome. In order to evaluate the outcome of translation, one would need to have a perspective that was independent of those by which both concepts were generated (i.e. FPP and TPP). Such an additional, independent perspective is not available to us, so we can never be sure whether or not the outcome of the translation between psychodynamic and neuroscientific concepts corresponds to a one-to-one translation,[8] and thus whether or not the translation was successful. In other words, we seem to remain largely unable to establish any kind of outcome measure of even a successful one-to-one translation and also of the neural correlate approach. Our possible knowledge (i.e. perception and cognition) may be too limited to account for such an outcome measure. The mere possibility of such a limitation is often neglected, and I describe this as "limitation neglect."

Even if, for example, ego and CMS can be directly translated to each other on a one-to-one basis, we may remain unable to access and know this. Thus there may always remain some gap because we can never be sure whether the psychodynamic content really fully corresponds to the neuroscientific content (and vice versa), as in the case of a successful one-to-one translation. However, this raises the question of whether, for example, the CMS can be considered the neural correlate of the psychodynamic concept of the ego or not (see Chapter 9 for a detailed discussion of this issue).

Another feature that makes direct translation problematic is the almost total neglect of the context in the neural correlates approach. Psychodynamic and neuroscientific concepts differ in their respective contexts. Psychodynamic concepts presuppose a personal and subjective context as characterized by meaning, personal values, and norms, whereas neuroscientific concepts remain rather impersonal and objective. For instance, the concept of the ego is ultimately a personal and subjective context (although that view may be debated; see Part III of this book), whereas the concept of CMS is clearly impersonal and objective. Since these contextual differences are almost completely neglected in the neural correlates approach, which focuses exclusively on contents independent of their respective contexts, I here speak of "context neglect."

In conclusion, the consideration of both knowledge limitation and the context sheds some doubt on the feasibility of direct translation between psychodynamic and neuroscientific concepts in a one-to-one manner. This has serious implications for the methodological strategy of neuropsychoanalysis. If direct and one-to-one translation remains problematic, the neural correlates approach may turn out to be problematic as well. In that case, neuroscientific contents cannot be

[7] For example, Quine (1976) does indeed argue (from the philosophical side) that any translation between different concepts remains impossible in principle.

[8] I am aware that this argument would require much more philosophical and specifically epistemological elaboration. However, this cannot be the focus here, and I have discussed it elsewhere (Northoff, 2011).

regarded as neural correlates of psychodynamic contents, because that would mean overlooking possible limitations in our knowledge and possible contextual differences.

Do we therefore have cause to doubt whether neuropsychoanalysis is possible at all? Opponents of neuropsychoanalysis dwell on the contextual differences when arguing that neuropsychoanalysis remains impossible in principle because it cannot account for the meaning and the respective personal values that are inherent in psychodynamic concepts but absent from neuroscientific concepts (Hennigsen, 1998, 2000; Blass and Carmeli, 2007). However, this targets only a specific form of neuropsychoanalysis, one that presupposes direct translation and the neural correlates approach, which corresponds to what I earlier described as the narrow concept of neuropsychoanalysis. Meanwhile their argument does not touch upon those forms of neuropsychoanalysis that do not presuppose direct translation and the neural correlates approach, as is the case in the wider concept of neuropsychoanalysis. Thus, by rejecting neuropsychoanalysis altogether on the basis of the argument of context neglect, the opponents themselves suffer from the neglect of different forms and concepts of neuropsychoanalysis.

However, this puts the responsibility back with the proponents of neuropsychoanalysis, who need to devise other forms that do not rely on direct translation and the neural correlates approach. They may suggest what I call "indirect translation," in which there may be a mediating concept between neuroscientific and psychodynamic concepts that can serve as the bridge between their contents and their respective contexts. For example, the concept of self-related processing could be such a mediating concept in the case of the ego and the CMS. Self-related processing (see Part III of this book for further details) describes a specific process by which stimuli can be related to the self of a person. At the same time, self-related processing may correspond on the neural side to the activity observed in CMS (see Northoff et al., 2006; Qin and Northoff, 2011). As such the concept of self-related processing appears to link the personal and subjective context of the ego with the impersonal and objective context of the CMS. This means that the neuroscientific and psychodynamic concepts of the ego and the CMS can be linked and translated into each other via the mediating concept of self-related processing, which then serves as what one might call a "bridge concept."

In contrast to the views of the opponents of neuropsychoanalysis, such indirect translation does not involve any neglect of contextual differences. Due to the mediating concept (i.e. bridge concept), the transition from the impersonal and objective context of the CMS to the personal and subjective context of the ego can be accounted for. Therefore any arguments that meaning, personal values, and norms are lost in neuropsychoanalysis can be refuted. However, on the neuroscientific side it is the neural correlates approach that needs to be refuted. Rather than describing neural correlates of psychodynamic contents, neuroscientific concepts can then only be considered to refer to what I call "neural predispositions" which, via the mediating concept, enable and predispose to the constitution of the features referred to in the psychodynamic concepts. Thus the preference for indirect rather than direct translation is associated with a shift from neural correlates to neural predispositions.

How can we account for and measure the outcome of such indirect translation? I have already demonstrated that any such outcome measurement remains impossible in the case of direct translation, and thus in the neural correlates approach. How can we measure the success or failure of the assumed indirect translation and thus whether the mediating concept in question really underlies the psychodynamic concept, and is at the same time associated with the facts described by the neuroscientific concept? The association of the mediating concept with the facts described in the neuroscientific concept must ultimately be tested experimentally, thereby yielding what I shall later refer to as "empirical plausibility," while the relationship of the mediating concept with the psychodynamic concept must be elucidated by conceptual comparison. More specifically, we need to compare the meaning of the psychodynamic concept with the meaning of the supposedly mediating concept. This will tell us whether there is an overlap or even similarity in the meaning of the two concepts as dependent on both their contents and contexts. The test for conceptual overlap or similarity will reveal whether or not the mediating concept is related to the psychodynamic concept in question in a plausible way. If it is, one may speak of "conceptual plausibility," and if it is not, one may describe this as "conceptual implausibility."

For example, one may compare the meaning of the concept of the ego with the meaning of the concept of self-related processing. To what extent does the concept of the ego in the psychodynamic concept correspond to the concept of self-related processing? Is there an overlap in the meaning or the conceptual extension between both concepts? If self-related processing can, for example, be determined by personal and subjective features, it may indeed be conceptually related to the concept of the ego, which makes their relationship conceptually plausible. A high level of conceptual plausibility of the relationship between the psychodynamic concept (i.e. ego) and the mediating concept (i.e. self-related processing) is one necessary precondition for successful indirect translation from the psychodynamic concept (i.e. the ego) to the neuroscientific concept (i.e. CMS).

Conceptual plausibility may be considered the outcome measure of indirect translation, but it is only half the story, as although it accounts for the measurement of the outcome of the linkage between psychodynamic and mediating concepts, it does not account for the measurement of how the mediating concept is linked to the neuroscientific concept. For this we need to turn to the second step of neuropsychodynamic concept–fact iterativity, namely feedforward transformation.

"Feedforward transformation": quantification and experimental variables

The neuroscientific concept is essentially based on observation in outer sense (i.e. TPP), as presupposed in scientific experiments that manipulate experimental variables. This means that if we want to associate the supposedly mediating concept in question with the neuroscientific concept, we need to transform the former concept into an experimentally accessible variable that can be observed from TPP. Thus we are confronted with two main tasks, namely to transform the concept into an experimentally accessible variable that can be observed from TPP (step IIa), and to link FPP and TPP (step IIb) (see also Figure 3.3).

Let me start with the first task, which involves the transformation of the mediating concept into an experimentally accessible variable. One central difference that confronts us here is the difference in quantification. An experimental variable is supposed to be quantifiable. This makes it possible to perform experimental manipulation, the effects of which can then be measured and thus observed. How can we quantify the mediating concept?

This is a crucial step, because the opponents of neuropsychoanalysis argue that this is the stage at which the meaning, personal values, and norms inherent in psychodynamic concepts become lost. However, we need to be careful that we do not transform the psychodynamic concept itself into a quantifiable experimental variable, but only the mediating concept, which is related to the psychodynamic concept in a conceptually plausible way. To argue that the meaning of the psychodynamic concept has been lost is to confuse the mediating concept with the psychodynamic concept itself. It would mean identifying the concept of self-related processing with the concept of the ego, which implies that major differences in the context between the two concepts have been ignored. Therefore the opponents' argument, if it is to be successful, must rely on exactly the same mistake (i.e. neglect of contextual differences with subsequent context confusion, which is manifest as context neglect by the proponents in the neural correlates approach and of neuropsychoanalysis in general).

How can the mediating concept be transformed into an experimental variable? One could select some experimental variables from those available in current neuroscience and link them to the mediating concept. For example, one could take a cognitive variable such as attention, with a subsequent behavioral measurement such as reaction time to denote self-related processing. If so, one presupposes self-related processing to be dominated by a specific cognitive process rather than some basic process such as affective or sensory processing. Now the ball is back in the conceptual park. One may discuss whether one really wants the concept of self-related processing to be characterized in predominantly cognitive terms, or in affective and sensory terms (for further details, see Part III of this book). If one chooses cognitive terms, the choice of attention as the experimental variable seems to be appropriate, so that there is a good match between concept and experimental variable. Alternatively, if one decides against the cognitive characterization, the choice of attention must be considered inappropriate, with consecutive mismatch between concept and experimental variable.

One may then look for other possible experimental variables for the mediating concept (e.g. self-related processing). If it is assumed to be characterized by strong affective processing,[9] one may for example choose an experimental variable for arousal, such as skin conductance (since arousal is supposed to be an indicator of emotional processing; Panksepp, 1998). However, arousal indicates only affective processing, not its specific linkage with self-related processing, as is required. One may then look for an experimental variable that captures the interaction between emotions and self which, if presupposed in this way, may characterize the concept of self-related processing. For example, one may develop some visual analog scales for the quantified and subjective evaluation of self-relatedness (see, for example, Northoff et al., 2009) and combine this with experimental variables for affect (e.g. skin conductance).

The example of self-related processing makes it clear that the linkage between the mediating concept and its possible associated experimental variables is itself an iterative procedure. More specifically, one has to move to and fro between the conceptual assumptions implicitly or explicitly made in the mediating concept and the conceptual presuppositions associated with the possible (and impossible) experimental variables. Only when there is a certain degree of conceptual overlap between the conceptual assumptions of the former and the conceptual presupposition of the latter does the selected experimental variable have an opportunity to match and thus be appropriate for the mediating concept in question. This also makes it clear that the success or failure of feedforward transformation can be measured by the degree of conceptual overlap between the mediating concept and the conceptual presuppositions associated with the experimental variable.

"Feedforward transformation": quantitative versus qualitative strategies

So far I have described the first aspect of the transformation, namely the linkage of the mediating concept with an experimental variable, while I left open how the FPP, as inherent in the mediating concept, can be linked to the TPP in which we observe the respectively associated

[9] One can already see that even on the side of the neuroscientific concepts we run into deep conceptual problems. Concepts such as attention and emotion/affect are highly ambiguous and have been defined in different ways by different authors. Consequently they are also associated with different experimental variables. Although this subject is more the domain of the critics of neuroscience than of a book about the practice of neuropsychoanalysis, it tells us that the apparently meaning-free neuroscience is not in fact as devoid of meaning (and thus conceptual issues) as it often claims to be.

experimental variable. The problem here is how to link the different perspectives (i.e. FPP and TPP) and thus inner and outer sense, including their opposite features. TPP-based observations can be quantified on an intersubjective or objective basis, and are thus accessible to all other individuals, whereas this is impossible in the case of FPP-based subjective experience.

At this point the opponents of neuropsychoanalysis might want to raise another objection. Due to the fact that subjective experience from FPP cannot be quantified as objective observation from TPP, the linkage of the former to the latter remains theoretically impossible. However, if FPP cannot be linked to TPP, any attempt to link neuroscientific and psychodynamic concepts must necessarily fail. The failure to link FPP and TPP then in turn leads to the failure of the whole project of neuropsychoanalysis.

The answer to this argument may lie in developing different approaches to linking FPP and TPP. One may argue against the opponents and claim that quantification of FPP-based subjective experience is possible. This is the strategy that many proponents of neuropsychoanalysis, such as Peter Fonagy, have pursued, which I call the "quantitative strategy" (Fonagy, 2003; Northoff et al., 2007a,c). Alternatively, one may concur with the opponents' claim and nevertheless link TPP-based observation of neural states to the meaning of even non-quantifiable and thus qualitative features of FPP-based subjective experience. This is the strategy that is advocated by Mark Solms, which I call the "qualitative strategy." Both strategies will now be described in more conceptual detail.

Let me start with the quantitative strategy. The systematic examination and evaluation of subjective experience must preserve its richness and complexity on the one hand, and objectively quantify its main characteristics on the other. The objectification and quantification of subjective first-person data allow scientific investigation and consequently the establishment of what can be called a "science of experience" (Varela and Shear, 1999).

Based on a science of experience, a "science of psychodynamic processes" needs to be developed, which should place great emphasis on patients' mental life or inner experience in order to preserve the richness and complexity of subjective experience and clinical description (Gabbard, 1992). At the same time, these subjective features must be objectified in order to provide reliable and quantifiable data.

This can be achieved by asking the subjects to complete rating scales. For example, visual analog scales (Northoff et al., 2009) with regard to personal identity or idiographic instruments such as the repertory grid test might be applied in order to allow the subjects themselves to evaluate their experiences. One might also apply structured interviews with valid and reliable instruments to allow evaluation of the subjects' relevant psychodynamic features by an experienced investigator. General instruments include, for example, the Karolinska Psychodynamic Profile (KAPP), which assesses different psychodynamically relevant dimensions of a person's structure (Weinryb and Rossel, 1991). Another instrument is the Operationalized Psychodynamic Diagnostic (OPD) system (Arbeitskreis, OPD 2004), which, based on the *International Classification of Diseases (ICD)* and the *Diagnostic and Statistical Manual (DSM)*, develops five psychodynamically relevant axes (illness experience and expectations, relationship, conflict, structure, and psychic and psychosomatic disturbances). More specific instruments are scales for quantifying defense mechanisms, such as the Defense Mechanism Rating Scale (Perry et al., 1989), the Defense Style Questionnaire (Kipper et al., 2004), and the Shedler–Westen Assessment Procedure (Shedler and Westen, 2004) for assessing psychodynamic psychopathology in personality disorders.

One potential criticism of such a "science of psychodynamic processes," relying on ideographic approaches, is that the individual content from FPP cannot be considered "data" in the way that this term is understood in science. For example, Metzinger (2003) argues that the concept of data implies their generalizability to groups, their possible validation in an intersubjective process, their extraction by technical measures, and their linkage to public procedure. He subsequently describes first-person data (i.e. the individual contents as subjectively experienced from FPP) as confusion of "semantic elements with a concept borrowed from the theory of science" (Metzinger, 2003, p. 591). However, I do not consider this argument to be applicable to the "science of psychodynamic processes," for the following reason. Individual contents that are central to psychodynamic processes can indeed be generalized to groups by comparing the similarities and differences of individual contents between different individual subjects.

One of the main features of psychodynamic theory is that it was able to reveal the same kind of individual contents and structures of psychological processes across different individuals. As the examples of operationalization demonstrate, it is quite possible to validate and objectify individual contents in psychological processes. Although this does not require technical measures, it can be regarded as a public procedure. Thus, in view of these arguments, individual contents of psychological processes can be understood as data in the scientific sense. This enables us to speak of first-person data that describe the subjective experience of individual contents in a general way (i.e. across different individuals).

Solms and Turnbull (2002, p. 307ff)) make a somewhat similar point to Metzinger when they argue that FPP-based subjective experience and thus psychodynamic processes cannot be operationalized and standardized, thereby refuting the idea that their quantification is theoretically possible. However, in contrast to the opponents of neuropsychoanalysis, Solms argues that this does not mean that neuropsychoanalysis is an impossibility. One can still alter the neuronal processes in a quantitative way by, for example, neurological or technically induced lesions in certain brain regions (what Solms calls the "clinico-anatomical method") (Solms and Turnbull, 2002, pp. 307–308) or psychopharmacological challenge (Solms and Turnbull, 2002, p. 309), and observe the effects of this on the qualitative features of FPP-based subjective experience. The outcome measure is then not some quantified and standardized variables, but qualitative reports from different individual subjects who are specially instructed to target a specific process or content when introspecting their subjective experience. As mentioned, I call this the "qualitative strategy."

Are the quantitative and qualitative strategies mutually exclusive? Although they are certainly incompatible with regard to the question of whether or not FPP-based subjective experience can be quantified, they may not be mutually exclusive when it comes to the issue of linking FPP and TPP. Rather than being mutually exclusive, quantitative and qualitative strategies may be complementary and compatible. Consider our example of self-related processing. One may, for example, develop quantified instruments such as subject- or observer-based scales for measuring self and self-relatedness, including its potentially affective, sensory, and also cognitive aspects (see Chapter 9 for further details). These scales may be related to the neuronal states in, for example, the CMS as observed from TPP (Northoff et al., 2009). Based on these results, the observed midline regions may be manipulated by, for instance, psychopharmacological means or neurological lesions (Solms and Turnbull, 2002, p. 307ff). The respective effects on the FPP-based subjective experience of one's self (and the ego) can then be investigated in a purely qualitative way by, for example, instructing the subjects to focus on specific psychodynamic features. One may then compare the qualitative information from the subjective reports with the information obtained from the quantitative scales. Thus both quantitative and qualitative strategies to account for FPP-based subjective experience may well be compatible and complementary when it comes to linking FPP and TPP.

Independent of whether one chooses the quantitative or qualitative strategy to account for FPP-based subjective experience, the success of the linkage between FPP and TPP depends on their degree of conceptual overlap. More specifically, the meaning that is targeted in the concepts that are used to describe the FPP-based subjective experience should overlap with the meaning that is presupposed in the concepts to describe the experimental variable as observed from TPP. Thus the degree of conceptual overlap between FPP-based subjective experience and the experimental variables, as accessible to objective observation from TPP, can be taken as the outcome measure for the success or failure of linking FPP and TPP. If one fails to link FPP and TPP and claims, for instance, that FPP can be replaced by TPP, one may confuse both perspectives. This amounts to what I call "FPP–TPP confusion" (as often seems to occur in the neural correlates approach).

The consideration of the experimental variable is a crucial step. If the experimental variable is well linked to the mediating concept (and thus to the neuroscientific and psychodynamic concepts), one may speak of high "experimental plausibility." High experimental plausibility indicates that the experimental variables mirror and reflect the mediating concept in a plausible way that is highly compatible with both general experimental constraints and the specific meaning of the mediating concept. If experimental plausibility is low, the experimental variable is not really linked to the mediating concept, making the former's association with the psychodynamic concept in question rather weak, and thus intuitive.

The feedforward transformation of the mediating concept is probably one of the most difficult steps in neuropsychodynamic concept–fact iterativity. Now one must design an experiment that allows the presupposed experimental variable to be experimentally tested. Although I admit that this in itself is difficult, I shall not discuss this step in my methodological account of neuropsychodynamic concept–fact iterativity, because it pertains to standard methodology and is therefore not a challenge from a purely methodological point of view. Therefore I shall not discuss the next steps in concept–fact iterativity in detail. These include the step from the experimental variable to the observation of data, and the step from the neuroscientific data to neuroscientific facts (for further details of these two steps, see Northoff, 2011 and Figure 3.3). The outcome measures of conceptual and experimental plausibility are complemented in these steps by technical plausibility and observational plausibility, which describe how valid and reliable the experimental variable is, and thus how plausibly it can be accounted for, given the limitations of technology and the observer. Rather than focusing on these steps in detail, which is provided elsewhere (see Northoff, 2011), I shall now focus on what happens to the data and facts that are obtained.

"Feedback conceptualization": matching of data/facts[10] and concepts

We have now acquired the empirical data (e.g. the observed changes in the neuronal states during experimental manipulation). The question now is what to do with the experimental data. We can stop here and argue that we now have all that we need. The experimental data as observed from TPP can be considered to be facts, and that is the end of the story. The concept–fact iterativity comes to an end here before any iterative movement from concepts to facts and back again from facts to concepts gets off the ground. However, this fails to take into account the need to interpret the experimental data, which means that they need to be put back in the context of the initial concept (i.e. the neuroscientific concept) that was the starting point for designing the experiment and obtaining experimental data (i.e. the second step).

The third step of the concept–fact iterativity describes how the obtained experimental data/facts can be put back in the context of the original neuroscientific concept, and thus how they can

[10] I am aware that the concepts of "facts" and "experimental data" cannot be regarded as identical from a purely conceptual point of view. However, this is a conceptual and thus philosophical issue, rather than being crucially relevant in the present context. For practical purposes, I here use the terms "data" and "facts" synonymously.

be linked to the latter. The direction is thus the opposite of that in the second step. The movement is now no longer a feedforward movement from a neuroscientific concept to experimental variables and then to facts, but rather the reverse, from the experimental data, which are considered to be facts, back to the original neuroscientific concept. However, the original neuroscientific concept, as presupposed by and transformed into the experimental variables, may need to have its meaning changed and modified according to the experimental data/facts that have been obtained. Thus the meaning and consequently the conceptual extension of the neuroscientific concept in the third step may no longer be exactly the same as in the second step. In other words, there may be an iterative change in the neuroscientific concept's meaning in relation to the acquired experimental data (i.e. the facts).

The question now is how that change in meaning can come about, and what are the criteria that determine whether or not the meaning of the neuroscientific concept should be modified. Any form of one-to-one mapping of the experimental data onto the neuroscientific concept with the former being the neural correlate of the latter remains impossible because of the fundamental difference between facts and concepts. Therefore we need to match and compare rather than correlate the experimental data with the original neuroscientific concept. The third step of neuropsychodynamic iterativity describes this matching process between the experimental data (i.e. the facts) and the original neuroscientific concept, with possible modification of the meaning of the latter.

Let me start by describing the process of matching between experimental data/facts and the original neuroscientific concept. Suppose, for example, that we did indeed observe some changes in the CMS (our experimental variable) when comparing self-stimuli (e.g. one's own face) with non-self-stimuli (e.g. the faces of other people) in functional brain imaging. This will lead us to assume that our concept of self-related processing, the mediating concept, may indeed be related to the CMS. This means that our hypothesis is confirmed by the experimental data/facts, and that we do not need to change and modify (one aspect of) the meaning of the neuroscientific concept of CMS as being associated with and functionally determined by self-related processing. Thus it could be argued that the (functional and processual) determination of the neuroscientific concept of CMS by self-related processing is empirically plausible (see Chapter 9 for details).

Empirical plausibility must be regarded as the outcome measure of the third step, namely the matching of the experimental data/facts with the original neuroscientific concept. If the experimental data are consistent with the original hypothesis and thus with the neuroscientific concept, the supposed meaning of the latter can be considered to be empirically plausible. However, if the experimental data are not consistent with the original hypothesis and thus with the neuroscientific concept, the originally assumed meaning of the latter may need to be modified and adapted to the experimental data/facts.

The latter may be the case if, for example, the results do not show any change in the CMS during self-specific stimuli when compared with non-self-specific stimuli. Such a finding puts the functional characterization of the CMS by self-related processing in doubt, thus influencing the meaning of the neuroscientific concept of CMS. We may therefore need to modify the meaning and conceptual extension of the neuroscientific concept of CMS. This allows us to match the meaning of the concept with the novel facts in order to generate an empirically plausible neuroscientific concept. Thus empirical plausibility must here be considered to be the outcome measure that describes the degree of matching between the experimental data/facts and the neuroscientific concept.

However, a different situation may occur. It may turn out that the changes observed in the CMS during the comparison of self-specific with non-self-specific stimuli are more related to the experimental need to include an evaluation task, as is often the case in functional magnetic resonance imaging (fMRI), rather than to the self-specific stimuli themselves. When one compares two conditions one may generate some kind of result by default. As it may be generated by default, the result that is obtained (e.g. the changes in the midline structures) can then not be interpreted as indicating self-related processing. Instead, processes other than self-related processing, such as familiarity or the task itself (e.g. evaluation), may also yield activation in the same regions, as has indeed been suggested by several authors (Gillihan and Farah, 2005; Legrand and Ruby, 2009). However, this casts doubt on the empirical plausibility of the midline structures being characterized by self-related processing.

One might want to argue that the concept of self-related processing needs to be modified and redefined, and thus to have its meaning changed. The concept of self-related processing may, for instance, be detached from the processing of self-specific stimuli as distinct from non-self-specific stimuli. Rather than presupposing an "all-or-nothing processing" of self-specific stimuli and associating this with self-related processing, one may instead assume a more continuous form of processing, with multiple and continuous transitions in the degree of self-relatedness of the stimulus. Self-specific and non-self-specific stimuli would then describe only the opposite and extreme ends of the process (i.e. the self-related processing that is supposed to underlie such a continuum) (see Chapter 9 for details). Such conceptual modification can then be taken as a starting point for the design of novel experiments to test for its empirical plausibility (see, for example, Northoff et al., 2009). The changed meaning of the mediating concept may then be fed back with the original psychodynamic concept so that one may need to take an iterative sidestep back to the original psychodynamic concept to achieve higher empirical plausibility of the mediating concept. This will be described in the fourth step of our neuropsychodynamic concept–fact iterativity. First, however, it is necessary to make an important methodological comment that concerns the distinction between brain and observer in our experimental data/facts.

"Feedback conceptualization": distinction between brain and observer

One may also raise another methodological question, namely whether the previous results showing self-specific activity changes in the CMS may be related to the experimental method (i.e. the need to compare and contrast two conditions, such as self- and non-self-specific stimuli), rather than to the midline regions and thus the brain itself. We therefore need to determine which parts of the experimental data may be more related to the experimental design (e.g. the comparison), and which are truly related to the brain itself. This means that conceptually we need to distinguish the observer and his experimental input (e.g. the experimental design and variables) from the neural activity and thus the brain itself. Certain changes in the neural activity may be related to the input applied by the observer rather than reflecting the neural processing of the brain itself, as it is independent of us as observers, including our respective experimental input. I use the term "observed-induced artifacts" to refer to these observed changes in the brain's neural activity that are more related to the observer's experimental input than to the brain itself (as it is independent of our observation and experimental input). The detailed investigation of such observer-induced artifacts is beyond the scope of this book, and has been discussed elsewhere (Northoff, 2011), but here I give some empirical examples and discuss them in a footnote.[11]

[11] Let me offer an example of an observer-induced artifact, namely perception. We might make conceptual distinctions on the basis of our perception. For instance, I can perceive objects with specific contents such as the table in front of me, which is made of wood and has a brown color, these characteristics distinguishing it from the floor, the window, the bed, etc. This ability to perceive objects and to differentiate them from other objects led to the distinction between content and context. So far so good—no

Why do we need to distinguish between the observer and his experimental input on the one hand and the brain and its neural activity as it is independent of us on the other? One could, for instance, assume that the way we investigate and observe the brain exactly matches the way that the brain functions, and thus it is independent of ourselves and our observation. This argument may be further strengthened by the fact that our observation and experimental input themselves rely on the brain and its method of neural processing. One would therefore expect that the brain-based observation and the observer's experimental input would match the brain itself, as it is independent of us and ourselves, including our observation.

However, such a match between the observer (e.g. his experimental input) and the brain itself as it is independent of our observation does not appear to be available to us. This is because we cannot exclude observer-induced artifacts signaling a possible discrepancy between observer and brain. Such a discrepancy can only be related to the observer himself and how he can (and cannot) perceive the brain, which in turn determines how he can (and cannot) set up the experimental design and variables and what we can and cannot observe.

The possibility of observer-induced artifacts may indicate that there is indeed a discrepancy between the concept of the brain as observed, as related to the observer and his experimental and epistemic input, and the concept of the brain as functioning, remaining independent of the observer and his experimental input. Thus the difference between the brain as observed and the brain as functioning may indicate not only a pure conceptual difference but also a possible epistemic and (even more important) empirical difference. The epistemic difference consists, as described previously, of the difference between inner and outer sense (i.e. FPP and TPP) as perspectival cognition on the one hand, and a perspectival cognition that would be necessary to perceive the brain as functioning on the other. The empirical difference may consist of specific neural processes that are assumed to be predominantly stimulus based (i.e. based on the stimuli that are applied as experimental variables by the observer) in the brain as observed. This may contrast with the neural processes in the brain as functioning, where the brain's own activity (its resting-state activity) may be regarded as dominating the observer's input (i.e. the stimuli that he applies in order to test the brain's neural processes) (see Part II of this book for empirical details of resting-state activity and stimulus-induced activity).

What does this mean for the outcome measure? I assumed that empirical plausibility was the main outcome measure in the process of matching experimental data and neuroscientific concept. However, if the empirical plausibility is potentially confounded by the observer's experimental input and thus by possible observer-induced artifacts, we may need to introduce another conceptual distinction. Empirical plausibility must then be distinguished from what I call "neural plausibility." Empirical plausibility describes the degree of accordance between experimental data

observer-induced artifact is looming yet. However, this situation changes once the perceptual content–context distinction is projected upon the brain and is supposed to be neurally realized, with separate brain systems underlying content and context, respectively. I do not wish to discuss in detail whether this is true or false. What is important here is that the observer's conceptual distinction between content and context is directly mapped onto the brain, and although this may be true, Baars (2005) cannot exclude at least the possibility that the neural distinction may be more likely to be due to his conceptual distinction than to the brain's own processes. Others, such as Freeman (1997), deny that there is such a neural distinction between content and context in our brain's neural organization. An organizational feature like re-entrant or cyclic processing also argues against a strict distinction between content and context in the brain's neural processing. If this is the case, we cannot exclude the possibility that the neural distinction between content and context may turn out to be an artifact that can be traced back to the observer and his concepts, rather than to the brain itself.

(the facts) and the original neuroscientific concept. However, as the experimental data (the facts) may be confounded by possible observer-induced artifacts, the concept of empirical plausibility must be distinguished from the concept of neural plausibility. The latter thus describes the degree of accordance of the original neuroscientific concept with the neural processes of the brain itself as they are independent of our observation, including our experimental data.

Although one may assume that in most cases there is considerable if not complete overlap (and in the best possible scenario even exact correspondence) between empirical and neural plausibility, we may always have to consider at least the possibility of divergence between the two. This means that we remain unable to completely exclude the possibility of observer-induced artifacts in our experimental data. This should therefore be considered as a major confounding factor when matching the experimental data with the original neuroscientific concept. Otherwise we run the risk of confusing observer and brain and thus, to use psychodynamic terminology, of projecting ourselves and our ideas and wishful fantasies as observers onto the brain itself. By ignoring the possibility of what I call "brain–observer confusion," we may ultimately end up with our own projections (i.e. ideas and fantasies) of the brain, rather than with the brain itself as it is independent of us. In other words, by ignoring possible brain–observer confusion, we may run the risk of confusing the brain as observed with the brain as functioning.

"Re-entrant contextualization": consideration of different contexts and concepts of the brain

We are now ready to put the experimental data (the facts) and the possibly modified and thus empirically and neurally plausible neuroscientific concept back into the context of the original psychodynamic concept. More specifically, the neuroscientific concept re-enters the psychodynamic context which I describe as "re-entrant contextualization" (see Figure 3.3).

Let me briefly revisit what I mean by psychodynamic context. As described in the initial sections of this chapter, the psychodynamic context can be characterized by subjective experience of meaning, personal values, and norms. In order to investigate meaning, personal values, and norms, we need a "hermeneutics of mind" (see this chapter for details) (Tauber, 2009). This contrasts with the neuroscientific context, which can be characterized by objective observation, impersonal neural states, and facts, and thus remains devoid of meaning as crystallized in what Tauber (2009) called "science of mind." Although in the first step of the neuropsychodynamic iterativity we translated the psychodynamic concepts into neuoscientific concepts, searching for a mediating concept, we are not confronted with the need to put the latter back into the context of the former. More specifically, we need to find out how to link objective observation of impersonal neural states with subjective experience of meaning, including personal values and norms.

Let us return to our example of the ego and the CMS. I assumed self-related processing to be the mediating concept that is associated with the ego and the CMS. The concept of the CMS is based on objective observation, impersonal neural states, and mere facts, whereas the concept of the ego is supposed to account for subjective experience, personal values, and norms. If we now want to put the concept of the CMS back into the context of the ego, the mediating concept of self-related processing needs to provide the glue between the neuroscientific context and the psychodynamic context.

How can this be achieved? The answer is very simple. The concept of self-related processing must be characterized in such a way that it allows and accounts for the transition from the neuroscientific context to the psychodynamic context (i.e. from impersonal neural states to personal values, from objective observation to subjective experience, and from facts to norms). Does the concept of self-related processing provide such a bridge between the neuroscientific and psychodynamic contexts?

I leave this question open here, and refer the reader to Chapter 9, in which I shall discuss both conceptual and empirical details.

Methodologically, we are here also confronted with the problem of the rather general nature of psychodynamic concepts. The concept of the ego is a rather broad and general one, which is so basic that it is supposed to underlie almost any empirical observation. The rather general and broad meaning of psychodynamic concepts contrasts with the specific and restricted meaning of neuroscientific concepts, especially if they are evidenced by specific neuronal mechanisms, while in feedforward translation the broad and general meaning of psychodynamic concepts is pinned down to the specific and restricted one of neuroscientific concepts (via the mediating concept). We now need to reconsider the meaning of the latter if we want to put them back into the psychodynamic context. Bilateral interaction between the neuroscientific concept and the psychodynamic context may be needed. In the same way that we narrowed down (i.e. translated) the meaning of the psychodynamic context to the neuroscientific concept, we now need to broaden the latter when putting it back into the psychodynamic context, if we really want to account for the transition between both via the mediating concept (see also Fonagy, 2003 who similarly opts for a more specific and less metaphorical determination of psychodynamic concepts).

Let us return to my example of the ego. The concept of the ego is a rather broad and general one, and we may need to single out one particular feature that restricts, specifies, and narrows down the concept of the ego. One may, for instance, take the concept of the self in psychoanalysis, which has been described by Freud (who, however, remained ambivalent about it; see Chapter 9 for details) and others (e.g. Kohut and Milrod) as the subjective–experiential side of the psychologically objective structure and organization of the ego. One may even more specifically assume a specific feature of the self, namely Kohut's concept of self-objects, which roughly describes the subjective experience of objects in close relation to the self (see Chapters 4 and 9 for details). What the concept of self-related processing must then achieve is to allow for the transition from CMS to self-objects, rather than from the CMS to the ego. Although this may account for self-objects and consecutively for the self as the subjective–experiential side of the ego, it may fail to consider other features of the ego that may need to be transferred to other mediating concepts (see Chapter 9 for details).

This makes it clear that the re-entrant contextualization of the neuroscientific concept into the psychodynamic concepts may involve modifications of both the neuroscientific and the psychodynamic concept. The neuroscientific concept of CMS may need to be modified in functional terms in orientation on the experimental data on the relationship between self-related processing and CMS. At the same time, the concept of the ego may also need to be modified and adapted to the mediating concept of self-related processing and its specific features. This psychodynamic slant on the concept of self-related processing may in turn involve different experimental variables and thus novel experimental designs and testing. The results obtained may then trigger subsequent modifications of the neuroscientific concept (e.g. the concept of the CMS). In short, there may be iterative dynamics at work that lead to continuously ongoing feedforward and feedback movements between the psychodynamic and neuroscientific concepts, with possible modifications of both via the mediating concept.

What is the output measure of the fourth step, namely the re-entrant contextualization? So far we have investigated whether the experimental data and the subsequently matched neuroscientific concept are plausible with regard to the psychodynamic context of the original psychodynamic concept. One may consequently speak of neuropsychodynamic plausibility that describes the degree of accordance of the neuroscientific concept with the psychodynamic context or, conversely, of the psychodynamic concept within the neuroscientific context. Neuropsychodynamic plausibility may thus be characterized by what may be called "contextual plausibility" (as distinct from mere conceptual plausibility) as the outcome measure.

For instance, the concept of CMS may not show a high degree of neuropsychodynamic plausibility in the context of the psychodynamic concept of the ego, due to their discrepancy in conceptual extension. This is because the two concepts may not match well with each other in the mediating concept of self-related processing. In contrast, the concept of CMS may show a high degree of neuropsychodynamic plausibility with regard to the psychodynamic concept of the self, both converging and matching well in the concept of self-related processing (see Chapter 9 for details).

The outcome measure of neuropsychodynamic plausibility as contextual plausibility will guard against any mismatch between neuroscientific and psychodynamic concepts as well as against the possible neglect of the differences in their respective contexts. Neuroscientifically oriented approaches (e.g. the neural correlates approach) may want to stop after the third step of our neuropsychodynamic iterativity, omitting the fourth step on the grounds that it is unnecessary, superfluous, and redundant. However, this means that the meaning and the different contexts that characterize psychodynamic concepts are neglected and ultimately reduced to and eliminated in favor of the context associated with the neuroscientific concepts (i.e. the neuroscientific context). One may consequently diagnose "context confusion" in such cases, because the psychodynamic context is here false-positively taken to be identical and reducible to the neuroscientific context. The "hermeneutics of mind" are then indeed reduced to a "science of mind" as is claimed by the opponents of neuropsychoanaylsis. Thus the fourth step of the neuropsychodynamic iterativity, namely the re-entrant contextualization, will guard against such context confusion while at the same time allowing the linking of psychodynamic and neuroscientific contexts.

However, not only do we need to be careful that we do not confuse psychodynamic and neuroscientific contexts, thus avoiding "context confusion," but we must also consider the "correct" concept of the brain (i.e. the brain as functioning rather than the brain as observed). We shall only be able to achieve high neuropsychodynamic plausibility if we presuppose the "correct" concept of the brain (i.e. the brain as functioning rather than the brain as observed). Possible mismatch between the neuroscientific and psychodynamic concepts and thus low neuropsychodynamic plausibility may not only be due to a mismatch between the psychodynamic and neuroscientific contexts, but may also stem from confusion with regard to the different concepts of the brain. If the neuroscientific concept is more related to the brain as observed and does not fully account for the brain as functioning (which of course we cannot always know), it may be mismatched with the psychodynamic concept, because that presupposes the brain as functioning rather than the brain as observed. There may thus be confusion between the different concepts of the brain (i.e. "brain confusion"), with possible mismatch in the presupposed concept of the brain within the different contexts, amounting to what I call low "neuroconceptual plausibility." Thus, in addition to context confusion, low neuropsychodynamic plausibility may not only stem from low contextual plausibility but also from low "neuroconceptual plausibility" when presupposing different or "wrong" concepts of the brain.

"Re-entrant contextualization": neuropsychodynamic concepts as "hybrid concepts"

Finally, we need to briefly characterize neuropsychodynamic concepts themselves in more detail. As outlined here, neuropsychodynamic concepts must be considered "hybrid concepts." They are hybrid in that they include both neuroscientific and psychodynamic features, and thereby allow for the transition from neuroscientific to psychodynamic contexts. However, it is important to be aware that neuropsychodynamic concepts are not simply a mixed bag of both psychodynamic and neuroscientific features that stand side by side (and are at best intuitively linked). As well as containing features of both psychodynamic and neuroscientific contexts, the neuropsychodynamic concepts themselves should refer to and thus describe the transition between the two contexts. More specifically, the neuroscientific concept should be linked to another concept, namely the mediating concept, which describes how the neuronal mechanisms in the former can enable and predispose to the transition from the neuroscientific to the psychodynamic context.

Let me give an example. The mediating concept of self-related processing may be considered to describe those features that allow for the transition from the neuroscientific context of the neuroscientific concept of CMS to the psychodynamic context of the concept of self and self-objects. The concept of self-related processing may then be characterized as a truly neuropsychodynamic and transdisciplinary concept that, as such, must be distinguished from the intradisciplinary concepts of CMS (neuroscience) and self/self-objects (psychoanalysis). Another example of such a hybrid concept may be brain–self and brain–object differentiation, which contain not only neuroscientific concepts (i.e. brain) and psychodynamic concepts (i.e. self and object), but also quite directly a concept that describes their transition (i.e. differentiation). I therefore consider brain–object and brain–self differentiation to be truly neuropsychodynamic concepts that as such must be considered hybrid in an intrinsic sense (i.e. as a feature of the concept itself) rather than an extrinsic one (i.e. as imposed by our conceptual deficits).

This also makes it clear that neuropsychodynamic concepts cannot be considered mere conceptual correlates of neuronal concepts. The concept of self-related processing is not the conceptual correlate and thus the neural correlate (in conjunction with the CMS) of self-objects. This would amount to at best an interdisciplinary concept where features of both psychodynamic and neuroscientific contexts are simply put together in a "mixed bag." Instead, neuropsychodynamic concepts describe predispositions rather than correlates. For instance, self-related processing describes the processes that enable and predispose the CMS to contribute to constituting the subjective experience of self-objects. Neuropsychodynamic concepts thus describe those processes that enable and predispose to the transition from the neuroscientific to the psychodynamic context. As such, neuropsychodynamic concepts must be considered to reflect neural predispositions rather than neural correlates, and thus to be truly transdisciplinary rather than remaining intra- or interdisciplinary.

The hybrid nature of neuropsychodynamic concepts has also been alluded to by Brook (1998). He speaks of two different kinds of concepts, namely intentional concepts that describe some goal-oriented and meaningful behavior or functions, and mechanistic concepts that refer to causal processes. He associates Freud's psychological concepts with intentional concepts, which he considers to be extensions of common-sense psychology, while he associates the mechanistic concepts (e.g. energy, forces, flow, etc.) with metapsychology and Freud's description of the psychic apparatus. However, the mechanistic concepts of metapsychology are penetrated by intentional concepts of psychology when, for example, Freud attributes objects and aims to the blind flows of energy in his writing on sexuality (Freud, 1905).[12] And it is this penetration, as Brook describes it,

[12] Brook (1998) considers this to be a violation of the intentional–mechanistic dichotomy. However, it may have been regarded as appropriate by Freud himself in order to account for the hybrid nature of the metapsychological processes in question, which must in turn account for and thus predispose to the link between seemingly purely mechanistic processes and rather intentional and meaningful psychological contents. Interestingly, Brook (1998) puts Freud's dichotomy between mechanistic and intentional concepts and its violation into a wider framework of the duality between *Verstehen* and *Erklären*, as mirroring the duality advanced here between a hermeneutics and a science of mind. He also associates this duality with current thinking in philosophy of mind and neurophilosophy with, for example, Dennett resisting the reduction of intentional concepts to purely mechanistic ones, as is for instance advocated

that may well correspond to what I here describe as the "hybrid" nature of neuropsychodynamic concepts.

The opponent of neuropsychoanalysis may nevertheless try to refute neuropsychodynamic concepts, arguing that the hybrid nature is nothing but a confusion between different contexts and thus between different concepts, which makes it impossible to pin them down and determine them. Therefore neuropsychodynamic concepts may be regarded as mere confusion and category errors, rather than as valid concepts. The opponent is right if he employs his purely intradisciplinary perspective and its respective context (i.e. the psychodynamic context), which he obviously still presupposes in his sceptical attack. Taking such a purely intradisciplinary perspective and thereby presupposing a purely psychodynamic context, the neuropsychodynamic concepts must indeed be diagnosed as mere confusion and category errors.[13]

However, once one switches to a truly transdisciplinary perspective, as Freud himself obviously had in mind, things look different. Then it all depends on the method and the systematic nature of the linkage between the distinct aspects within the neuropsychodynamic concept whether it can be regarded as a valid concept or as mere confusion and thus invalid. Rather than being dependent on the content of the concept and thus the result obtained, the validity of the neuropsychodynamic concept may then depend on the systematic (i.e. valid and reliable) nature of the method that served to generate the respective concept. So long as the hybrid nature of the neuropsychodynamic concept is not extrinsic, stemming from us as investigators and our possible ambiguities and deficiencies, but rather is related to the linkage of the contexts themselves, thus being intrinsic to the concept itself, refutation of the hybrid nature of neuropsychodynamic concepts is possible only when presupposing a purely psychodynamic context. In order to exclude any traces of extrinsically imposed ambiguities and deficiencies arising from us as investigators, we need to develop and employ a systematic methodology to yield neuropsychodynamic concepts.

---

by Churchland. This suggests that the description of neuropsychodynamic concepts as "hybrid," as suggested here, taps into a duality that until now has remained unresolved, as Brook himself acknowledges (see Brook, 1998, pp. 75–77).

[13] Opponents of neuropsychoanalysis may also want to argue that this characterization of neuropsychodynamic concepts as hybrid and neural predispositions may be an ideal, but is impossible to achieve. They may thus say that we are here confronted with another gap between the methodological ideal and epistemological reality. They are right in that we may never ultimately be able to know whether the neuropsychodynamic concept in question can really provide the link from the neuroscientific to the psychodynamic concept. This is because we cannot go beyond inner and outer sense in our possible perception and cognition, which would be necessary in order to directly perceive and cognize those neural predisposition that are described in neuropsychodynamic concepts. Due to the impossibility of gaining direct epistemic access to the neural predispositions themselves, our neuropsychodynamic concept must (necessarily, as the philosopher would say) remain hypothetical. Here the opponents of neuropsychoanalysis are right.

However, they are wrong in that the (seemingly necessary) hypothetical nature of the neuropsychodynamic concept prevents their investigation on the basis of a systematic method that takes into account our epistemic limitations. This is exactly what the neuropsychodynamic iterativity that is suggested here aims to achieve (although, as I am well aware, only in a preliminary way), namely the suggestion of a systematic method that, by considering our epistemic boundedness to inner and outer sense, allows us to minimize the hypothetical nature of neuropsychodynamic concepts in an iterative procedure. Yes, the opponents of neuropsychoanalysis are right in that we shall never be able to eliminate the hypothetical character of neuropsychodynamic concepts and turn them into knowledge (as the philosopher might say). However, they are wrong in that we do not have the methodological means to at least minimize and narrow down the hypothetical and thus the ultimately speculative nature of neuropsychodynamic concepts.

This is precisely the purpose of neuropsychodynamic concept–fact iterativity. I am well aware that what I have outlined here needs to be elaborated in more theoretical and empirical detail. The empirical application of such neuropsychodynamic concept–fact iterativity and the fruitfulness of this are demonstrated in Parts II, III, and IV of this book. First, however, I want to raise one last point about the concept of neuropsychoanalysis.

"Method-based neuropsychoanalysis" versus "result-based neuropsychoanalysis"

This last point concerns the need to develop what I call "method-based neuropsychoanalysis." Only if we have a clearly and systematically defined method can we counter the sceptic's objection about the speculative and erroneous nature of neurosychodynamic concepts. This means that neuropsychoanalysis must develop its own specifically transdisciplinary methodology that must be distinguished from merely intrainterdisciplinary methodologies in both neuroscience and psychoanalysis. Part of any methodology must be the indication of potential pitfalls and, most importantly, outcome measures that enable us to test the validity and reliability of the results. I have already indicated the possible pitfalls and outcome measures in each step of my proposed methodology of neuropsychodynamic concept–fact iterativity. The proposed methodology that I outline below should be considered a first step in the direction of developing a specifically transdisciplinary and thus neuropsychodynamic methodology as the basis of what I call a "method-based neuropsychoanalysis."

Such method-based neuropsychoanalysis must be distinguished from what may be called "result-based neuropsychoanalysis." From the very beginning, result-based neuropsychoanalysis pre-empts its preferred output in two ways. First, it assumes that there is a relationship between the psychodynamic and neuroscientific concepts, with the latter being assumed to be the neural correlates of the former. Their possible correlation is not in question. The only question remaining is which neuroscientific concept provides the neural correlates of the psychodynamic concept. However, this is to know the child before it is born. How can the neuropsychoanalyst know from the very beginning that there is such relationship between the neuroscientific and psychodynamic concepts?

Pre-empting the outcome (i.e. the existence of a relationship) allows the concept–fact iterativity to degenerate into mere concept–fact circularity, in which neuroscientific facts and psychodynamic concepts no longer show any real (i.e. iterative) interaction with each other with possible bilateral modification. This is because concept–fact iterativity also involves possible failure (i.e. mismatching) when relating psychoanalytic and neuroscientific concepts. This, however, is excluded from the very beginning in such result-based neuropsychoanalysis.

Most importantly, result-based neuropsychoanalysis with neuropsychodynamic concept–fact circularity poses the risk of what I call a "locked-in syndrome." This refers to a neurological condition in which patients suffer from a brain lesion that prevents them from communicating with the outside world, while their consciousness of themselves and their inner world is fully preserved. In the following account I shall use this concept in a metaphorical way to describe the relationship between neuroscience and psychoanalysis. One may, for example, determine the initial psychodynamic concept in such way that it remains impossible to be linked to any kind of neuroscientific concept. The psychoanalyst purist as the sceptic of neuropsychoanalysis may feel validated, but this in fact merely indicates that he is locked into the world of his psychodynamic concepts and can no longer see any doors to the world outside. I shall refer to this as "psychoanalytic locked-in syndrome."

Alternatively, one may define the initial neuroscientific concept in such way that any linkage to psychodynamic concepts remains impossible, while simultaneously claiming that neuropsychoanalysis (and also psychoanalysis) remains at best irrelevant and at worst pure speculation. However, this merely indicates that the neuroscientist is, in the same way as his psychoanalytic counterpart, locked into his world of neuroscience and thus the brain as observed, which he false-positively identifies with the brain as functioning, as it is independent of him and his own observation of the brain. I shall refer to this as "neuroscientific locked-in syndrome."

The diagnosis of neuroscientific locked-in syndrome may be extended to the result-based neuropsychoanalyst. Unlike his psychoanalytic and neuroscientific counterparts, he is at least aware that there is a door which leads outside the room in which he is located. However, because he is paralyzed, he remains unable to walk through the door, and therefore closes it and tries to relate the paintings on that door (i.e. the psychodynamic concepts) to the rest of the room (i.e. the neuroscientific context). Since he does not step through the door and see the other side of it, he cannot of course do other than pre-empt the relationship between what he sees on this side of the door and his room. He therefore searches for neural correlates while ignoring the neural predisposition when employing neuropsychodynamic concept–fact circularity rather than neuropsychodynamic concept–fact iterativity. This amounts to what I call "neuropsychodynamic locked-in syndrome."

Part II

Neural Equipment

I now abandon the theoretical and conceptual issues that were addressed in Part I and plunge straight into the empirical issues of the brain itself. Rather than searching for the relationship between specific psychodynamic concepts, specific psychological functions, and particular brain regions (e.g. neural correlates), I here focus on basic neuronal mechanisms and neural coding as neural predisposition. I assume that the brain itself employs specific mechanisms and a particular way of coding the stimuli that it processes, which in turn enable and predispose to the constitution of the psychodynamic processes that Freud and others described. Most importantly, I postulate that the brain's neuronal mechanisms and the neural coding it applies to its own neural processing of stimuli are crucial in enabling and predisposing to the differentiation between the brain's neural apparatus and the psyche's psychic apparatus, and thus to what I called in the Introduction "brain–object and brain–self differentiation."

I shall start with the psychodynamic concept of cathexis, which can be located at the very interface between the psyche's psychic apparatus and the brain's neural apparatus (see Chapters 4 and 5 for details). The concept of cathexis describes the investment of energy. I here focus on the brain's own energy, namely its intrinsic or resting-state activity, and what it does and even more important does not do with it. I shall demonstrate that the brain's energy, or its intrinsic or resting-state activity, is invested in stimulus-induced activity and thus the neural activity changes that we observe during experimental stimulation of the brain as investigated in neuroscience (see Chapter 4). I assume that this is only possible due to the existence of a specific kind of neural coding, namely difference-based coding. This difference-based coding is assumed to have a crucial role in transforming the brain's neuronal states into mental states as characteristic of the psyche's psychic apparatus, thereby accounting for cathexis as a "neuro-mental bridge concept" (see Chapter 5).

This has major implications for our understanding of the transition from neural to mental states and thus from brain to mind, which is in turn essential for understanding how our brain enables and predisposes to brain–self and brain–object differentiation. These implications and the neuropsychodynamic hypotheses of brain–self and brain–object differentiation are discussed within the context of defense mechanisms in Chapter 6. I here consider two aspects of defense mechanisms. In early life they are essential for constituting self and object as distinct from brain and body, whereas in later life they are defensive when they serve to protect the self and object from foreign intrusions (e.g. trauma) (see Chapter 6). By constituting self and object as distinct from brain and body, early and more immature defense mechanisms such as internalization and externalization operate directly at the interface between the brain's neural apparatus and the psyche's psychic apparatus. My focus is therefore on these early and immature defense mechanisms, and I postulate that they enable and predispose to brain–self and brain–object differentiation.

A brief comment about my methodological strategy is relevant here. The reader may be expecting to be shown the neural correlates of certain psychodynamic processes. As discussed in the Introduction and Part I of this book, this is not the kind of strategy that I pursue here. Instead of searching for certain brain regions and specific cognitive function as neural correlates of psychodynamic concepts, I shift my focus to neural predisposition, by which I mean the kind of neural mechanisms and neural coding that the brain must employ to enable and predispose to the features described in psychodynamic concepts. Rather than taking both neural and psychodynamic mechanisms as a fait accompli and correlating them, I here start earlier and thus prior to their completion by searching for the kind of mechanisms that allow for their differentiation from the brain's neuronal mechanisms, which more generally amounts to brain–object and brain–self differentiation.

In order to do this, I shall have to delve deep into the brain itself and its neuronal mechanisms and neural coding. Only if we reveal and understand these will we be able to understand how the brain enables and predisposes to the more complex psychodynamic concepts (e.g. self, narcissism, self-objects, unconscious, etc.). These psychodynamic concepts will be subsumed under the heading of "mental equipment" and discussed in Part III of this book, while here in Part II the main focus will be on the brain's neuronal equipment that enables and predisposes to the differentiation of the brain from both self and object (i.e. brain–self and brain–object differentiation). Chapters 4 and 5, which discuss cathexis, focus more on the neuronal mechanisms that are assumed to occur prior to brain–self and brain–object differentiation. Chapter 6, which discusses defense mechanisms, directly addresses both kinds of differentiation. That chapter is therefore crucial not only in that it addresses the transition from the brain's neural apparatus to the psyche's psychic apparatus, including self and object, but also in allowing us to make the step from the neural equipment discussed in Part I of this book to the mental equipment discussed in Part III.

One final comment is needed here. As already indicated at the end of the Introduction, I shall not be describing all of the empirical details of the proclaimed neuronal data and mechanisms, or giving a complete account of all the conceptual subtleties and intricacies associated with the psychodynamic concepts discussed here. In both cases I can only recount those empirical and conceptual features that are relevant to the current neuropsychodynamic contexts, while disregarding those that only appear to be relevant within the respective intradisciplinary contexts (i.e. neuroscientific and psychoanalytic contexts) (see also the Epilogue, which provides an account of the tension between a transdisciplinary approach and intradisciplinary expectations).

Chapter 4

Cathexis and the Energy of the Brain

Determination of cathexis

Strachey (1967, p. 62) used the Greek term "cathexis" to translate what Freud described as *Besetzung* in German. Although Strachey pointed out that the English terms "occupation" and "filling" capture some of the meaning of what Freud referred to as *Besetzung*, the term "cathexis" has suffered from ambiguities subsequently (Holt, 1962; McIntosh, 1993; Hoffer, 2005). However, these ambiguities have been caused not only by the translation of *Besetzung* as "cathexis," but also by Freud's multiple usage of the former term. Before considering these ambiguities in detail, I shall briefly introduce the term "cathexis" and discuss some of its key features.

What does *Besetzung* mean? This term is used in German in different contexts for which the English translation uses different terms, as Hoffer has pointed out: "'Besetzen' means 'taking over something and using it in a certain way' which may happen in different contexts. For instance, a restroom is occupied, a washing machine is in use, a lake is stocked with trout, the military is capturing territory, a telephone line is busy, and a role in a play is cast. While the English uses different terms, occupied, in use, stocked, capturing, busy, and cast to describe these different contexts, the German language expresses this by one and the same term, besetzen" (Hoffer, 2005, p. 1127).

To overcome this difference between the English and German languages, various authors have suggested that the term "investment" might be more useful to describe cathexis, as opposed to "occupation" or "filling" (Macdiarmid, 1989; McIntosh, 1993; Milrod, 2002a,b; Hoffer, 2005, p. 1127). So how can we define cathexis as investment? Investment means that "something is or has been invested in" and, most importantly, the commodity that is invested is energy. Therefore the term "cathexis" may preliminarily be defined as the investment of energy which consecutively leads to occupation or *Besetzung*.

The definition of cathexis as investment of energy mean that the energy must be invested in something. That "something" consists of objects in the most general sense, be they people, ideas, words, objects, the self and the ego, the other, the drive and libido, events, or mental states (see also Ornston, 1985, pp. 395–397; Macdiarmid, 1989; MacIntosh, 1993). The energy is thus invested in or directed towards objects in this very general sense, so that those objects then bind the energy. In the absence of such objects the energy would remain free and unbound. Thus there is a distinction to be drawn between "bound and free cathectic energy" (Holt, 1962, pp. 519–520). This means that cathexis is determined not only by energy but also by its relationship to objects, as either free or bound energy.

Thus the energy that is a characteristic feature of cathexis is invested in objects. This sounds straightforward enough, but it becomes more complex when one considers the implications for the object (which is conceptualized here in a wide and non-specific sense rather than in terms of one particular school of psychoanalysis). The object itself is changed and transformed by the energy that is invested in it. The cathected object, as one may call it, can thus be distinguished from objects in which no energy is invested (i.e. non-cathected objects) (see, example, Hoffer, 2005, pp. 1028–1029). More specifically, when the cathected object is occupied by energy, it is being held or reserved for a certain purpose. Let us return to the aforementioned examples. The restroom is occupied for a certain purpose by the person using it, the washing machine is in use for a certain purpose, and the territory is captured by the military for a certain purpose. Similarly, the object that is bound to and thus cathected by energy is used for a certain purpose (as will be determined later). Thus the investment of energy in a specific object not only distinguishes that object from other non-cathected objects, but also is consistent with the existence of a specific purpose (i.e. the use of that object). While the specific purpose itself remains unclear at this point, it is clear that the attribution of purpose to objects by the investment of energy may be considered a third characteristic feature of cathexis, alongside energy and objects.

Following Hoffer (2005, pp. 1028–1029), the introduction of purpose (or usage) may hold for a certain time span, so the purpose can change over time. Most importantly, the constitution and attribution of purpose are associated with exclusivity, which means that a specific purpose and no other is exclusively assigned to this particular object and no other. Cathexis, as Hoffer says, "entails a certain element of exclusivity, which implies that it (the object) cannot be used for anything other than its designated purpose, by any other than the designated agency. When the line is busy, no third party can get in on the conversation; when the bathroom is occupied, no one but the current occupant can use it" (Hoffer, 2005, p. 1028).

The exclusivity that accompanies the binding of energy to specific objects leads to a certain organization of objects, as the energy is bound in certain ways to specific objects and serves particular purposes which do not apply to other objects. Cathexis thus allows different objects to be distinguished from each other, thereby involving the constitution of organization and structure. More specifically, the binding of energy to particular objects that serve a specific purpose involves a constant shift and movement in the streams of energy that are shifted backward and forward between different objects, with the balance between bound and free energy continuously changing. Thus various relationships between different objects can be built and constituted (Holt, 1962, pp. 519–523), which ultimately is considered to account for the structure and organization of the psyche (see the next section). These different movements and shifts of energy have been described by the terms "anticathexis" (the countering of energy investment), "decathexis" (the withdrawal of energy), and "hypercathexis" (the reinforcement of energy) (Holt, 1962).

In summary, the concept of cathexis can be preliminarily defined in terms of three characteristics, namely energy, investment of that energy in non-specified objects (thereby endowing the latter with purpose and exclusivity), and constitution of structure and organization in the psychic apparatus.

Ambiguities in the meaning of cathexis

The first ambiguity in the meaning of cathexis concerns not so much the concept of cathexis itself as an associated concept, namely the concept of the object. I have defined the objects of cathexis in the broadest possible way, including people, words, erotic and non-erotic objects, one's own ego and self, libido, the other, etc. (see also Ornston, 1985, pp. 395–397). For instance, recounting the history of psychoanalysis, Macdiarmid (1989) distinguishes between two basic objects of cathexis, namely one's self, amounting to what he calls "self-cathexis," and the other person, amounting to what he describes as "other-cathexis." In the case of self-cathexis, energy is invested predominantly in the self (ego) (the two terms are used synonymously here for the sake of operational simplicity; for their exact determination and differentiation, see Chapter 9) which in extreme cases can lead to narcissistic, schizoid, and introverted personalities. In the case of "other-cathexis," investment of energy in the other person predominates over investment in the self, which ultimately results in rather extrovert, anaclictic, and hysterical personalities (see Macdiarmid,

1989, p. 850). Usually, there is a balance between self-cathexis and other-cathexis, which if lost can lead to extreme cases and personality types.

Must the objects of cathexis be characterized by the self–other polarity and thus by either self- or other-objects? This would fail to take into account the fact that not every object can be classified as either self- or other-object (e.g. objects of love). Instead of self-object–other-object polarity, one may rather assume the existence of a multitude of objects in which energy can be invested, these objects being found along a continuum between self- and other-objects. The concepts of self- and other-objects must thus be regarded as extreme ends of a continuum of different objects with different degrees of relationship and balance to both self and other, mirroring what I shall later call "relatedness" or "self-relatedness" (see Chapter 9). Most importantly, this relatedness or self-relatedness applies to all kinds of different objects, resulting in, for example, libidinal or more non-libidinal objects, and narcissistic or less narcissistic objects, etc. This clearly shows that the concept of the object in the context of cathexis is not determined by its specific contents, but rather by its relatedness to the self and the other, while in contrast the object's contents remain non-specific. I argue that the concept of cathexis as the investment of energy in objects allows the establishment of such relatedness independent of the specific contents of the objects. Thus cathexis remains independent of or precedes the determination of the objects by their contents.

The same applies to mental and non-mental objects. McIntosh (1993, pp. 9–10) argues that the object of cathexis is not a mere mental representation of an object, but rather the real (and thus non-mental) object itself. Even fantasies as apparently entirely mental states in which energy can be invested are, according to McIntosh, not purely internal but concern objects in the outside world and thus real rather than purely mental objects. This means that even objects that seem to be entirely mental can be characterized by some degree of relatedness to both self and other, which implies that they may not be as purely mental as is often presupposed. However, it also means that the investment of energy (i.e. cathexis) may itself be independent of and precede the characterization of objects as either mental or non-mental (with the mental–non-mental distinction eventually being more related to us as observers than the process of cathexis itself). In other words, the concept of cathexis remains independent and thus non-specific with regard to the distinction between mental and non-mental objects (see also Holt, 1962, pp. 519–520).

Is the non-specificity and thus ambiguity of the concept of the object in the context of cathexis an intrinsic or extrinsic feature? Is the concept of cathexis by itself non-specific and independent with regard to its objects? The concept of cathexis describes the investment of energy in objects, and the focus here is on the investment of energy and thus on the concept of energy, rather than on the objects themselves. The concept of energy itself remains non-specific with regard to the objects to which it is directed and in which it is invested. To specify the energy with regard to a specific object would mean ignoring the fact that the concept of energy itself would be overspecified and therefore unnecessarily limited if tied to a specific object.

Let us use oil as an analogy. Oil describes a form of energy that can be used for (i.e. invested in) a wide range of purposes (e.g. as a fuel for cars, heating, cooking, etc.). To limit the concept of oil to one of these particular uses (e.g. as a fuel for cars) would be to confuse the general purpose of energy itself (the provision of power) with the more much specific purpose of providing fuel to power a car. Similarly, to limit the psychodynamic concept of energy to one particular object (e.g. the libidinal drive or narcissistic drive) would be to confuse the purpose of the former (the provision of power for any object) with the much more specific purpose of the libidinal drive (e.g. the sexual drive).

What are the implications of this for the concept of the object within the context of cathexis? The fact that the concept of the object remains non-specified, broad, and general must be considered essential to the concept of energy and thus to the concept of cathexis. To specify the concept of the object further would mean to disregard the purpose of energy, and thus cathexis, in providing mere power or energy independent of the specific object (i.e. its content). Even worse, it would mean confusing the purpose of cathexis with the purpose of the object in which it is invested. To avoid such confusion, we must keep the concepts of energy and cathexis "object-open," as I call it, rather than "object-closed." This means that the ambiguity in the concept of the object within the context of cathexis must be considered intrinsic (i.e. related to the concepts of energy and cathexis themselves) rather than being extrinsic, thus mirroring our inability to define objects properly and unambiguously.

The second ambiguity concerns the concept of cathexis itself that is both static and dynamic at the same time. The investment of energy in the object means that the object is reserved or being held by that energy and a specific purpose, while at the same time making other forms of investment in the same object, including other possible purposes of the latter, impossible. This corresponds to a static state in which the object is held and the energy is bound, thus attributing a static meaning to the concept of cathexis. At the same time, however, the investment of energy in an object and its specific purpose is not only static but also a highly dynamic process that accompanies the redirection of energy streams from one object to another, changes in purpose, and shifts in the organizational structure of objects.

As it is a concept that describes both a static state and dynamic processes, cathexis can be characterized by a dual meaning. Hoffer describes cathexis as "an action that is simultaneously static and dynamic. When something is besetzt, it is being held or reserved for some action or process that takes place within its parameters" (Hoffer, 2005, p. 1029). Hoffer (2005, pp. 1029–1030) assumes that Strachey, who translated Freud's works into English, was well aware of such a dual meaning when he characterized cathexis in terms of both process and state, which mirror its dual characterization as dynamic and static.

However, one may make a further distinction. The static state may be considered the result or outcome of a preceding dynamic process (see also Ornston, 1985, p. 392). Both the static state and dynamic processes are intrinsically related to each other while at the same time being distinct aspects or manifestations of one and the same process (i.e. the investment of energy in an object). Although conceptually (i.e. logically) they appear to be incompatible because they describe opposite features (i.e. static and dynamic), the static state and dynamic processes are considered within the empirical domain to be two distinct aspects of one and the same feature, namely the investment of energy in an object. This means that what Hoffer (2005, pp. 1029–1030) describes as the "duality of meaning" and thus ambiguity is intrinsic to the concept of cathexis itself, rather than being related to our difficulties or inability with regard to providing a clear-cut definition of the concept of cathexis that is non-ambiguous.

In summary, I here identify two ambiguities in the concept of cathexis. First, there is the ambiguity with regard to the specification of objects in which energy is invested, with the objects remaining non-specified, including self- and other-objects as well as mental and non-mental objects. Secondly, there is ambiguity with regard to the kind of processes in which cathexis is involved, including both dynamic and static processes. Both ambiguities must be considered as inherent or intrinsic rather than as extrinsic to the concept of cathexis, as without them cathexis would lose its meaning and definition as the investment of energy.

Neuropsychodynamic hypotheses of cathexis

So far I have considered the nature of the concept of cathexis. As discussed, three characteristic features of the concept of cathexis can be identified. First, it involves the investment of energy. Secondly, all possible and thus non-specified objects, including mental and non-mental as well as self- and other-objects, are the targets of energy investment that assigns purpose and exclusivity to those objects. Thirdly, it involves the constitution of organization and structure by energy investment, including both static and dynamic processes.

How can we now put the concept of cathexis and its three characteristic features into the neural context (i.e. the context of the brain)? In his 1895 writing, *Project for a Scientific Psychology*, Freud considered the concept of cathexis within the neural context of the brain, and regarded it as the discharge of electrochemical energy. However, he then abandoned the neural context and replaced with a purely psychological one, thus associating cathexis with motivation, drive, libido, etc. If one now wants to develop a truly neuropsychodynamic hypothesis, one must shift the concept of cathexis back from the psychological context of the psyche's psychic apparatus to the neural context of the brain's neural apparatus, while at the same time maintaining its characteristic features as previously outlined. This approach may be described as "shifting cathexis into the neural context of the brain."

Before outlining any neuropsychodynamic hypotheses, a brief methodological comment should be made here. Shifting the concept of cathexis back into the neural context of the brain should not be confused with what may be described as "neuralizing cathexis." How does "shifting cathexis into the neural context of the brain" differ from "neuralizing cathexis"? The concept of cathexis would be neuralized if one equated, identified, and ultimately replaced it with a specific neuronal process (e.g. the influx of specific molecules or the activation of specific regions in the brain). In contrast, "shifting cathexis into the neural context of the brain" means that neuronal mechanisms are targeted that enable and predispose to cathexis, although the latter is not identified with or reduced to them. One would consequently look for neuronal mechanisms that may enable and predispose to the kind of features that are addressed by the concept of cathexis as previously outlined, rather than searching for identical neural processes that replace the concept of cathexis. In short, my aim here is to target the neural predispositions rather than the neural correlates of cathexis (for an explanation of this distinction, see the Introduction). This means that I am more interested in the enabling and predisposing condition of the transition from the neural context of the brain to the psychodynamic context of the psychic apparatus, rather than in neuronalizing and thus reducing and ultimately eliminating the latter in favor of the former.[1]

In order to develop a neuropsychodynamic hypothesis, I therefore aim to shift the concept of energy back into the context of the brain and its specific methods of neuronal processing and coding stimuli from the outside world. The concept of cathexis as investment of energy is now set in the neuronal context of the brain, where it resurfaces as investment of the brain's energy. And in the same way that the psychodynamic concept of cathexis is supposed to describe the investment of energy in all kinds of possible objects, the concept of object resurfaces in the neuronal context of the brain in the concept of stimuli from outside the brain, be they from the body or the environment, that are processed by the brain. Taking all of the above into account, one can now formulate the

[1] Why is it important to consider the difference between "shifting cathexis into the neural context of the brain" and "neuralizing cathexis"? This difference is important because, as outlined, the concept of cathexis defies any characterization as being either physical/neural or psychological/mental by itself, which means that it remains independent of any such distinction. However, such neural–psychological independence would be violated in the case of neuralization, as then the concept of cathexis would indeed be characterized as being physical/neural, and thus no longer independent. The concept of cathexis would therefore be unable to demonstrate one of its central features, namely its independence of the nature of the objects that remain non-specific. Therefore, if we want to maintain the conceptual features of the concept of cathexis, we have no choice methodologically but to pursue the strategy of "shifting cathexis into the neural context of the brain" rather than "neuralizing cathexis."

general neuropsychodynamic hypothesis of cathexis. The psychodynamic concept of cathexis may correspond neuropsychodynamically to the investment of the brain's energy in the brain's own neural processing of stimuli from the body and the environment.

However, this general neuropsychodynamic hypothesis leaves both the brain's energy and the processed stimuli undetermined, so that it remains unclear how the conceptual features of cathexis already outlined can be maintained within the neural context of the brain. We therefore need to specify our general neuropsychodynamic hypothesis, and in order to do this I shall now state three specific neuropsychodynamic hypotheses.

What do we have to search for in the brain if we want to find a neural process that corresponds to the concept of energy? The concept of energy itself remains independent of any objects. Transferred to the context of the brain, this means that the brain must show some neural activity that remains independent of any kind of stimulus. In other words, such neural activity must be intrinsic to the brain itself which, defined in an experimental (i.e. operational) context, describes the absence of any kind of stimulus-induced activity and thus what may be called "resting-state activity." Such resting-state activity must be somehow invested in the brain's own neural processing of bodily and environmental stimuli. This leads me to the first specific neuropsychodynamic hypothesis. The brain must show some intrinsic neural activity (i.e. resting-state activity) that it invests in its own neural processing of bodily and environmental stimuli, with such investment in the brain's neural context supposedly corresponding to cathexis as the investment of energy in objects as described within the psychodynamic context.

As discussed previously, the second characteristic feature of the psychodynamic concept of cathexis is that energy can be invested in any kind of object, be it internal or external, self- or other-related, mental or physical. The brain may thus invest its intrinsic activity (the resting-state activity) in the neural processing of various stimuli regardless of their origin in either the body or the environment. This leads me to the second specific neuropsychodynamic hypothesis. There must be some interaction between the brain's intrinsic resting-state activity and the brain's own stimulus-induced activity as related to extrinsic stimuli, which empirically and operationally may be called "rest–stimulus interaction." In the same way that the objects were non-specified in the psychological context of cathexis, the stimuli are non-specified in the neural context of rest–stimulus interaction.

Another feature of the psychodynamic concept of cathexis is that the investment of energy in objects has an impact not only on the objects themselves but also on the general structure and organization of the psyche as assumed by Freud in his later work. Transferred to the brain's neural context, this means that the investment of the brain's intrinsic activity in the neural processing of bodily and environmental stimuli may structure and organize not only the neural processing of the stimuli themselves, but also the brain itself, its neural organization and the structure of its own intrinsic activity (i.e. its resting-state activity). This leads me to the third specific neuropsychodynamic hypothesis. The brain's neural structure and organization and thus its own intrinsic activity (i.e. its resting-state activity) may be modulated and shaped by its own neural processing of stimuli (i.e. stimulus-induced activity), involving what I call "stimulus–rest interaction." Stimulus–rest interaction is believed to enable and predispose to the constitution of the brain's neural structure and organization, which in turn are supposed to correspond to what Freud described as the "structure and organization" of the psychic apparatus within the psychodynamic context (see Figure 4.1).

Energy, investment, and the brain's intrinsic activity

The brain is a strange organ. It consumes a large amount of energy in the form of oxygen and glucose. In fact it accounts for about 20% of the total oxygen consumption by the body, although

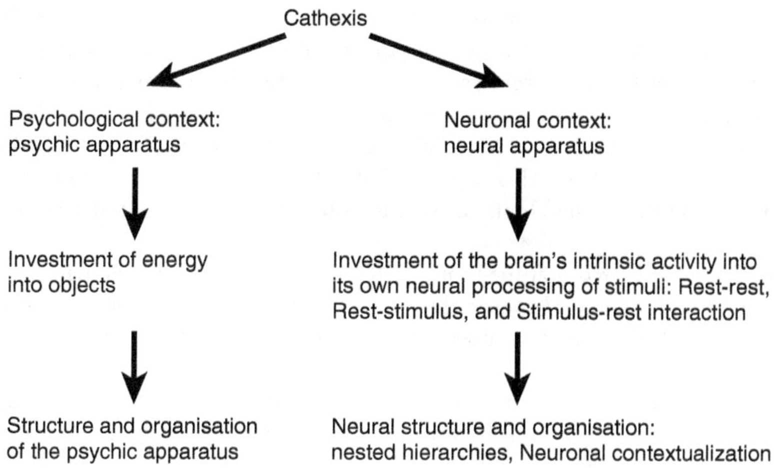

Fig. 4.1 Concept of cathexis in psychological and neural contexts.

it represents only 2% of the body's total mass. Therefore the brain can be described as extremely "energy-hungry."

What does the brain do with all of this energy? It contains nothing more than a bundle of neurons, but the neurons themselves appear to use a lot of energy. They show high activity, with baseline firing rates ranging from a few spikes per second to tens of spikes per second. This suggests that our neurons are always active and never at rest. Our brain is like a small child who cannot sit still and pay attention to what the teacher is telling him. So does our brain suffer from attention deficit hyperactivity disorder (ADHD)?

Let us look at the brain's hyperactivity in more detail. High resting-state activity has been observed in the brain not only at the cellular or microscopic level but also at the macroscopic level (i.e. in macroscopically identifiable regions). Early studies in humans using positron emission tomography (PET) identified high rates of oxygen and glucose consumption in the resting state in a particular set of regions, including anterior and posterior cortical midline regions such as the ventromedial prefrontal cortex (VMPFC), the dorsomedial prefrontal cortex (DMPFC), the different parts (subgenual, pregenual, and supragenual) of the anterior cingulate cortex (ACC), the posterior cingulate cortex (PCC) and the precuneus, as well as other regions, such as the lateral parietal cortex and the hippocampus (Raichle et al., 2001; Simpson et al., 2001; Greicius et al., 2004). In contrast, during presentation of external stimuli (e.g. emotional or cognitive tasks) these regions show negative changes in their neural activity, known as deactivation or negative BOLD responses (NBR), as distinct from activation or positive BOLD responses (PBR).

These regions have been subsumed under the concept of the default-mode network (DMN) (Raichle et al., 2001; Morcom and Fletcher, 2007; Buckner et al., 2008; Broyd et al., 2009). While being deactivated during stimulus- or task-induced activity, these regions show high activity and a high degree of intrinsic functional connectivity in the resting state (Greicius et al., 2004; Beckmann et al., 2005; Fox et al., 2005; Fransson, 2005; Damoiseaux et al., 2006). Interestingly, high resting-state activity as a characteristic of the DMN has been shown to be present not only in human adults, but also in human infants (Fransson et al., 2008) and monkeys (Vincent et al., 2007). Thus the concept of high resting-state activity in the DMN must be considered a characteristic feature of the brain of higher-order organisms in general, and must therefore be regarded as a truly translational concept.

However, high resting-state activity has been observed not only in the regions of the DMN (e.g. so-called task-negative regions), but also in other regions, including the sensory cortices, motor cortex, insula, and subcortical regions, mirroring so-called task-positive regions (Shulman et al., 2004, 2009; Buckner et al., 2008). Further support for spontaneous resting-state activity across the whole brain comes from electrophysiological studies that have demonstrated spontaneous neuronal oscillations and synchronizations in various parts of the brain, including the hippocampus and the visual cortex (Llinás, 1988; Arieli et al., 1996; Fries et al., 2001, 2007; Buzsaki and Draguhn, 2004; Buzsaki, 2006). This suggests that high levels of resting-state activity may be prevalent throughout the brain in both humans and other animals, and that it is not restricted to the DMN.

So far I have demonstrated that the brain itself shows intrinsic neural activity which can be described as the brain's energy. However, the question of what the brain's energy (i.e. its resting-state activity) is used for (i.e. how it is invested) now arises. The question of investment is closely linked to the precise definition of the brain's resting state. What exactly do we mean by the brain's resting-state activity? We mean by this the absence of any kind of stimulus external to the brain itself. In other words, the origin of the brain's resting-state activity is intrinsic to the brain itself, rather than being extrinsic to it (for another definition in this context, see Shulman et al., 2009).

Stimuli external to the brain can have their origin either in the outside world or within the body. The external stimuli can be described as exteroceptive if they originate in the outside world, or as interoceptive if they originate in the body. Therefore the brain's resting-state activity can be described as originating from neither exteroceptive nor interoceptive stimuli. Instead it must originate from the inside of the brain itself, and is therefore intrinsic rather than extrinsic to the brain.

Since it is considered to originate in the brain itself, I call such resting-state activity and its corresponding baseline the "neural baseline" of the brain (Northoff et al., 2010). The concept of the neural baseline may refer to truly intrinsic resting-state activity "from the inside of the brain," such as may, for example, be manifested by neuronal oscillations (Llinás, 1988) or metabolic activity (Raichle and Mintun, 2006; Shulman et al., 2009; these authors refer to such activity as a "physiological baseline") that remain completely independent of any sensory input from the outside of the brain (Shulman et al., 2004; Ben-Simon et al., 2008; Shulman et al., 2009).

The brain's neural baseline must be distinguished from other baselines that are already confounded by intero- and exteroceptive stimulus processing. Based on the results of their EEG study, Barry et al. (2007) assumed the existence of an "arousal baseline" that reflects the general state of arousal of the whole brain, as for example when the eyes are closed. Such an "arousal baseline" may reflect a mixture of the brain's intrinsic activity and interoceptively induced activity as generated by the continuous interoceptive input of the body, while ideally any kind of exteroceptively induced activity is shut off. Physiologically, the "arousal baseline," as indicated by the concept of arousal, may predispose the brain's general reactivity to any kind of exteroceptive stimuli (e.g. whether it can react strongly or weakly). As such it may be close to the concept of what Morcom and Fletcher (2007) have described as the "processing baseline."

Barry et al. (2007) distinguish such an "arousal baseline" from what they call an "activation baseline" that represents a local regionally specific threshold for the induction of neural activity (e.g. in the visual cortex when the eyes are open). In contrast to the "arousal baseline," the "activation baseline" thus includes exteroceptively induced activity, and may therefore represent a mixture of intrinsic activity and intero- and exteroceptively induced activity. However, it may be necessary to differentiate further. The "activation baseline" includes non-specific exteroceptively induced activity (e.g. the visual input when the eyes are open), thus mirroring the continuous sensory input which one may also call "background input." However, it does not include the activity elicited by specific visual stimuli. Instead, the "activation baseline" predisposes the visual cortex (and the other sensory cortices) to react to a specific visual stimulus in a certain way.

What do the brain's different baselines (i.e. the neural baseline, the arousal baseline, and the activation baseline) tell us about the brain's investment of energy? Both the arousal baseline and the activation baseline must presuppose some interaction between the brain's neural baseline, its intrinsic resting-state activity, and intero- and exteroceptive stimuli. By being exposed to both interoceptive stimuli from the body and exteroceptive stimuli from the outside world, the brain's neural baseline cannot avoid some kind of interaction with the other baselines. And such interaction between the different baselines is obviously possible only if there is some kind of investment of the brain's energy in its own neural processing of stimuli, regardless of the origin of the latter. Therefore the brain may be characterized not only by energy as mirrored in its high resting-state activity, but also by investment of its energy in its own interaction with intero- and exteroceptive stimuli, which I shall call "rest–stimulus interaction" (Northoff et al., 2010).

Taking all of the above into consideration, it is clear that the brain can be characterized by high levels of neural activity in the resting state which may be of either neural, interoceptive, or exteroceptive origin, so it is possible to distinguish between neural, arousal, and activation baselines. The brain not only has intrinsic energy as indicated by the neural baseline, but also invests this energy in the processing of intero- and exteroceptive stimuli, resulting in arousal and activation baselines. One may therefore regard the brain's resting-state activity as a specific example of the more general concept of investment of energy and thus of cathexis within the brain's neural context. By constituting the different baselines, the brain does indeed invest its own energy (i.e. its intrinsic activity), and thereby "occupies" its own neural processing of intero- and exteroceptive stimuli. It is obvious that such occupation or investment very closely resembles and corresponds to the original meaning of cathexis as envisioned by Freud himself, who in an analogous manner assumed that the psyche's energy is invested in the psychological processing of objects.

Let me be more precise. What are referred to as objects in the psychological context of the psyche's psychic apparatus are transformed into and resurface as stimuli within the neural context of the brain's neural apparatus. Meanwhile what Freud referred to as psychic energy in the psychological context resurfaces as neural energy and more specifically as the brain's intrinsic activity (i.e. its resting-state activity or neural baseline as defined in the brain's neural context). Therefore I postulate that the brain's intrinsic activity (i.e. its resting-state activity or neural baseline) may enable and predispose to what Freud called cathexis, as psychic energy that the psychic apparatus invests in the processing of objects. The brain's intrinsic activity may thus be considered a neural predisposition to rather than a neural correlate of cathexis. However, to really establish correspondence between the psychological and neural contexts, we need to investigate how the brain's intrinsic activity affects its own neural processing of extrinsic stimuli. This will be the subject of the second neuropsychodynamic hypothesis.

Rest–stimulus interaction and the investment of energy in different non-specific objects

One of the characteristic features of cathexis is the non-specific nature of the objects in which the energy is invested or cathected. A whole range of objects, be they internal or external, self- or other-related, or mental or physical, can be the target of the invested energy. Transferred to the neural context of the brain, this means that different stimuli (internal or external, self- or other-related, or mental or physical) must be able to interact with the brain's resting-state activity, thus resulting in what I call "rest–stimulus interaction." Interestingly, there is indeed empirical support for such rest–stimulus interaction with a multitude of different stimuli (for a recent review, see Northoff et al., 2010).

Let us start with rest–stimulus interaction in the case of exteroceptive stimuli (i.e. sensory stimuli). Greicius et al. (2004) investigated how the DMN network (the task-negative network) affects subsequent stimulus-induced activity in visual and auditory tasks during passive sensory tasks. They observed that the level of activity in the DMN network during stimulation predicted the neuronal activity in both the visual and auditory cortices during the auditory and visual tasks. The lower the level of activity in the DMN during auditory and visual stimulation, the higher the stimulus-induced neuronal activity was found to be in auditory and visual cortex. This clearly demonstrates that the level of resting-state activity in the DMN affects the stimulus-induced neuronal activity in sensory regions such as the auditory and visual cortex.

The assumed modulation of stimulus-induced activity in sensory regions by the resting-state activity is further supported by data from animal studies. Fiser et al. (2004) investigated juvenile and adult ferrets during rest and the presentation of natural-scene and random-noise film images. While the correspondence between stimulus-induced neuronal activity in the visual cortex and the structure of the input signal was weak in juvenile ferrets, it improved with age. This improvement could be traced back to a shift in the dynamics of spontaneous firing in the visual cortex. These results indicate the central developmental role of resting-state activity in determining stimulus-induced activity.

In a series of studies (Arieli et al., 1996; Tsodyks et al., 1999; Kenet et al., 2003) that investigated the dependence of stimulus-induced activity in the visual cortex of cats on its resting-state activity level, the variability of stimulus-induced activity in single neurons of the visual cortex could be predicted by the dynamics of the spontaneous resting-state activity. Spontaneous ongoing resting-state activity in the visual cortex even predicted the stimulus-induced activity changes in the same region during single trials in single neurons. This clearly indicates that even stimulus-induced activity at the level of the single visual cortical neuron is modulated by resting-state activity in the visual cortex.

The studies that have been discussed so far concerned the modulation of externally generated stimuli from the outside world by the brain's resting-state activity. However, there are also stimuli that do not originate from the outside world, but from the brain itself. These are internally generated stimuli such as motor and cognitive stimuli. Does the brain's resting-state activity also affect the neural processing of these motor and cognitive stimuli?

Experimental manipulation of the brain's resting state has recently been reported in an animal study. Maandag et al. (2007) used halothane and chloralose to generate high and low resting-state activity pharmacologically in rats, and subsequently measured the neural activity in fMRI during forepaw stimulation. The high resting-state activity level was associated with widespread activity across the cortex and rather weak activity in the sensorimotor cortex. This activity pattern was reversed when there was low resting-state activity, with neural activity being stronger in the sensorimotor cortex but more or less absent from the other cortical regions. These results demonstrate that the level of resting-state activity may modulate the distribution and intensity of stimulus-induced activity in the same and other regions (like the sensorimotor cortex) (see also Eijsden et al., 2009; Shulman et al., 2009).

In addition to motor stimuli, the brain's resting-state activity also affects the generation of behavioral and psychological states presupposing to cognitive stimuli. Sapir et al. (2004) investigated the impact of changes in the spatial attention cortices, as induced by a cue, on the accuracy of a spatial discrimination task. They observed that the level of resting-state activation predicted the variability in the subsequent spatial discrimination task. Similarly, Weissman et al. (2006) demonstrated that slow reaction times during a visual attention task are predicted by reduced pre-stimulus activity in the ACC and the lateral prefrontal cortex (LPFC). While these studies focus on behavioral parameters (i.e. reaction time and accuracy) as dependent variables, the aforementioned study by Boly et al. (2007) demonstrated that resting-state activity in various medial and lateral cortical regions predicts and thus influences the subsequent psychological state (i.e. somatosensory awareness, and pain).

Busch et al. (2009) presented brief flashes of light at the individual luminance (i.e. detection) threshold while EEG was recorded. Despite the fact that the stimulus was identical in all cases, subjects detected only approximately 50% of the flashes (hits), while missing the remainder (misses). When comparing phase distributions shortly before stimulus onset (i.e. –800 to 0 ms) between hits and misses, they showed different phase angles, which were strongest in alpha and theta bands. They also observed that stimuli preceded by strong alpha power are less likely to be detected. This suggests that, depending on the phase of neuronal oscillations and the power of alpha frequency in the resting state, some moments may be optimal for inducing perception while others may be suboptimal. The authors refer to the former as "perceptual moments."

What do these studies tell us? They demonstrate that the brain's resting-state activity interacts with all kinds of different stimuli, both those externally generated in the outside world as sensory stimuli and those internally generated as motor and cognitive stimuli. Although no explicit experimental support has yet been obtained, the same most probably applies to the distinction between self- and other-related stimuli. If one considers sensory stimuli as other related and motor and cognitive stimuli as self related, one may even argue (indirectly) that both self- and other-related stimuli elicit rest–stimulus interaction in the brain.

To summarize these findings, the empirical data show that the brain's resting-state activity appears to interact with a multitude of different stimuli (internal or external, sensory, motor, or cognitive, and self- or other-related). The multitude of different stimuli in rest–stimulus interaction corresponds well with the multitude of objects within the psychodynamic context of cathexis. In the same way that the psychodynamic concept of object remains non-specified in cathexis, the concept of the stimulus remains non-specified within the neural context of the brain's rest–stimulus interaction.

The non-specific nature of both stimuli within the neural context of the brain indicates that the traditional distinctions between sensory and motor, between internal and external, and between sensory/motor and cognitive may not be relevant to the brain itself and its intrinsic activity. This is because these distinctions do not appear to correspond to the main features of the brain's cathexis (i.e. its resting-state activity and its rest–stimulus interaction). The brain appears to invest its energy and to employ the same kind of rest–stimulus interaction in the different kinds of extrinsic stimuli according to the same principal mechanisms, regardless of the specific origin of the extrinsic stimuli (i.e. motor, sensory, cognitive, etc.). This means that our observer-based distinction between different contents in stimuli (i.e. sensory, motor, cognitive, affective, etc.) may not mirror the "currency" that the brain itself employs when processing them. In other words, the origin of the stimuli may not be as relevant for the brain itself as it is for us as observers of the brain.

What exactly is the "currency of the brain" according to which it structures and organizes its rest–stimulus interactions, if not according to the origin of the stimuli? Rest–stimulus interaction allows the brain to shape and influence the various incoming stimuli according to its own actual state as mirrored in its resting-state activity level. In other words, rest–stimulus interaction makes it possible for the brain to imprint itself onto its own neural processing of the stimuli that it receives from outside itself, and thus to appropriate them according to its own specific needs and purposes.[2] The "currency of the brain" therefore no longer lies in the origin of the stimuli, but rather in the degree to which the stimuli are related and linked to the brain via rest–stimulus interaction. This means that the brain's currency may rather consist of the relationship or what

2 The concept of the brain's needs and purposes is to be understood in a figurative rather than literal sense, as needs and purposes can only be attributed to people, not to the brain as observed in neuroscience (Bennett and Hacker, 2003).

I call "relatedness" of the stimulus to the brain and its intrinsic activity. The term "relatedness" describes in the current neural context of the brain how much its resting state activity impacts subsequent stimulus-induced activity: the higher the degree of rest–stimulus interaction, the more relatedness. Psychologically, however, such relatedness may correspond to what I shall later describe as "self-relatedness" (see Chapter 9).

"Relatedness" as the brain's currency may also account for the alleged purpose and exclusiveness that is the second characteristic feature of the concept of cathexis.

By relating a stimulus to its intrinsic activity, the brain may assign some purpose to that stimulus which may, for example, consist of stabilizing the brain's own intrinsic activity. The actual resting-state activity level of the brain may favor or require a specific kind of rest–stimulus interaction with one particular stimulus rather than any others, thus endowing it with some degree of exclusivity. This makes it clear that the features of purpose and exclusivity as described by Freud within the psychological context of the psyche's psychic apparatus resurface within the neural context of the brain, where they can be related to a specific mechanism (i.e. rest–stimulus interaction). Most importantly, rest–stimulus interaction and relatedness as the supposed currency of the brain appear to remain independent of the origin of the extrinsic stimuli (bodily or environmental). The non-specific nature of the objects to which the psychodynamic cathexis refers thus resurfaces in the non-specific nature of the stimuli in rest–stimulus interaction within the brain's neural context. The aforementioned conceptual ambiguity in the concept of cathexis may thus be empirically plausible, thereby lending further support to the view that it is intrinsic to the concept of cathexis.

Stimulus–rest interaction and neural organization and structure

The third characteristic feature of cathexis concerns its structural and organizational properties. By investing energy in some objects while not investing it in others, a certain structure and organization is generated. Although in his earlier work Freud considered such structure and organization to be both neural and psychological, his later work shifted in favor of the psychological context, when he associated cathexis exclusively with the psyche. Thus Freud argued that cathexis as the investment of energy in non-specified objects enables and predisposes to the constitution of a specific structure and organization of the psyche's psychic apparatus.

The question is how we can transfer the concept of structure and organization from the psychological context of the psychic apparatus to the neural context of the brain. What do the terms "structure" and "organization" mean within the neural context of the brain? Analogous to the cathected objects that organize and structure the psychic apparatus, the stimuli and more specifically their interaction with the brain's intrinsic activity may structure and organize the latter. However, in order for the stimuli to organize and structure the brain's intrinsic activity, they need to directly interact with the brain's resting-state activity, and modulate and change it. This presupposes what can be called "stimulus–rest interaction," for which there is indeed empirical evidence.

A study by Lewis et al. (2009) investigated the effects of visual perceptual learning on resting-state connectivity. The subjects in the study received training in a shape-identification task that was confined to one visual quadrant. After several days of training, the subjects underwent fMRI, which was performed first during the visual training task. This revealed an effect of training on the respective side (i.e. quadrant) in terms of visual cortical activation when compared with the untrained side. In addition, the subjects had two sets of fMRI resting-state scans with visual fixation before and after behavioral training. These revealed a difference in the resting-state connectivity between the visual cortex and other regions, including the fronto-parietal regions involved in spatial attention and typical DMN regions. These findings clearly demonstrate that visual perceptual learning has an impact on connectivity inside and outside the DMN.

Another study investigated the effects of motor learning on resting-state activity (Albert et al., 2009). Resting-state activity was investigated by performing fMRI before and after an 11-minute visuomotor training session. Neural activity in the fronto-parietal resting-state network (i.e. lateral frontal and parietal regions) and the cerebellum was significantly increased after the visuomotor training session, compared with the level of activity before the session. Interestingly, the same network was not recruited during exlusively motor performance, and was thus specific for motor learning. This suggests that resting-state activity in this network may be much more closely related to visuomotor learning than to mere visuomotor performance.

One study by Pyka et al. (2009) investigated the activation of the DMN after 1-back and 2-back blocks in a working memory task compared with the activation in the DMN after a 0-back block. As expected, the different levels of difficulty of the tasks (i.e. the 0-, 1-, and 2-back loads) yielded different degrees of deactivation (i.e. negative BOLD response) in the regions of the DMN (perigenual anterior cingulate cortex, posterior parietal cortex, and posterior cingulate cortex) during the working memory task. Interestingly, the resting-state periods following the working memory task also differed in their degree of activation (i.e. positive BOLD response). The resting-state periods following the 1- and 2-back tasks showed significantly higher levels of activation in the aforementioned regions than were observed following the 0-back task. This suggests that the subsequent resting-state period was differentially modulated by the different loads of the preceding working memory task, thus indicating stimulus–rest interaction.

A recent study by our research group (Schneider et al., 2008) investigated how the psychological features of preceding emotional pictures predicted the degree of activity during a subsequent resting-state period. Subjects perceived emotional stimuli and subjectively evaluated them according to their degree of emotional valence, emotional intensity, and self-relatedness. The emotional stimuli were followed by prolonged resting-state periods (i.e. a fixation cross). The analysis then compared all of the resting-state periods that followed a high self-related stimulus with those that followed a low self-related stimulus. This yielded significant differences in resting-state activity in various task-negative regions (i.e. VMPFC, DMPFC, and PCC). Although the effects of self-relatedness in the subsequent resting-state period involved the midline regions, the stimulus-evoked effects of self-relatedness during the stimulus period itself involved the subcortical regions (i.e. PAG/tectum, ventral striatum, and amygdale).

Moreover, we observed a parametric correlation of the resting-state activity with the degree of self-relatedness of the preceding stimulus. Since these effects were specific for self-relatedness rather than for the emotional variables, these data support the assumption that resting-state activity has a central role in a specific psychological function, namely self-related processing. Most importantly, these data show that the resting-state activity in task-negative regions can itself be modulated by the preceding stimuli.

Taken together, these findings of stimulus–rest interaction demonstrate that the brain's resting state as characterized by interregional functional connectivity and intraregional activity levels is not static but dynamic. The dynamic nature of the brain's resting state is mirrored by continuous activity changes even in the resting-state activity level itself and its continuous modulation by stimulus-induced activity via stimulus–rest interaction. Based on the observations described, one may assume that such stimulus–rest interaction may exert an organizing and structuring effect on the brain's resting-state activity level, and may thereby contribute to differentiation and adaptation between the brain's different regions.

When I described the empirical evidence that supports the modulation of the brain's resting-state activity via stimulus–rest interaction, I left open what precisely I mean by the brain's neural structure and organization. The concept of the brain's neural structure and organization may pertain first and foremost to its regional differentiation, by means of which some kind of hierarchy is established. Different regions of the brain may then be situated at different levels of the hierarchy (e.g. with sensory regions such as the visual and auditory cortex at the bottom, and the prefrontal cortex at the top). The exact nature of the brain's hierarchy mirroring its neural organization and structure will be discussed in more detail in the next section. First, however, I shall briefly relate the observation of stimulus–rest interaction within the neural context to the originally psychological context of the concept of cathexis.

I here argue that the brain's energy (i.e. its intrinsic activity) has a crucial role in enabling and predisposing to the constitution of a specific neural structure and organization. This corresponds nicely to Freud's later assumption that cathexis had a crucial role in the structure and organization of the psychic apparatus. Freud's assumption of the constitution of a (psychological) structure and organization on the basis of the investment of energy in objects (i.e. cathexis) by the psychic apparatus may correspond in the neural context to the brain's investment of its intrinsic activity in its own neural processing of extrinsic stimuli. And in the same way that the objects in which the energy is invested influence the psychic apparatus itself, the cathected stimuli exert a structuring and organizing effect on the brain's intrinsic activity across its different regions. What Freud described as shifts of energy using the terms "anticathexis," "decathexis," and "hypercathexis" may then correspond within the brain's neural context to changes (i.e. weakening or strengthening) in the brain's resting-state activity levels across the brain's different regions as enabled and predisposed to by stimulus–rest interaction.

Neural structure and organization of the brain and its hierarchical organization

I have so far described specific neural mechanisms (i.e. stimulus–rest interaction) that may be crucial in constituting the brain's neural structure and organization, while I left open how that structure and organization can be described and characterized in detail. There have been many different suggestions with regard to this, whose discussion would be beyond the scope of this book. Instead, here I want to focus on one of the most prominent examples, namely hierarchical organization, with control of lower regions by higher cortical regions, amounting to what I call "top-down control." Since the model of top-down control has recently been put into a psychodynamic context by Carhart-Harris and Friston (2010), I shall focus on their suggestions and discuss these in the context of an alternative model, namely the model of "nested hierarchies" suggested by Feinberg (2009). Moreover, it should be noted that I do not discuss the anatomical details associated with the hierarchy. This is left until Chapter 6 where, in the context of my neuropsychodynamic hypothesis of projection, I shall make some concrete suggestions in this regard and present an alternative anatomical model.

Carhart-Harris and Friston (2010) and Friston (2010) suggest a hierarchical organization of the brain in which the thalamic nuclei, the unimodal sensory regions, and the other subcortical limbic and paralimbic regions are the lowest level in a functional hierarchy. The next levels are the salience and dorsal attention systems. The salience system includes the dorsal anterior cingulate cortex, the frontoinsula cortices, the amygdala, and the ventral midbrain, while the dorsal attention system concerns the dorsolateral prefrontal cortex, the frontal eye fields, the dorsal medial prefrontal cortex, the intraparietal sulcus, and the superior parietal lobule (for further details, see Carhart-Harris and Friston, 2010, p. 6). Finally, the top level is to be found in the DMN, which shows high activity in the brain's resting state and predominant deactivation during stimulus-induced activity.

These three levels, with the DMN at the top, the attention systems at the intermediate level, and the sensory, limbic, and paralimbic regions at the bottom, form a functional hierarchy. Each level has its own energy (i.e. free energy), with which it is trying to suppress the free energy of its respective subordinate (by minimizing the prediction error). Spontaneous activity (i.e. oscillations and fluctuations) in, for example, the DMN suppresses and contains spontaneous activity in the attention systems and the limbic and paralimbic regions via top-down modulation, whereas spontaneous activity in the attention systems controls and inhibits neural activity changes induced by exogenous sensory input in the thalamic and sensory cortical regions (for further details, see Carhart-Harris and Friston, 2010, pp. 11–12).

As each system is suppressed and controlled by the one above it, one may speak here of top-down control, while the activity in the lower regions can only be controlled and suppressed. Interestingly, Carhart-Harris and Friston (2010) associate this hierarchical model with Freud's psychodynamic concepts of the id and the ego as well as with primary and secondary processes. They argue that the top-down control of the highest level, the DMN, over the subsequent lower levels corresponds to the ego, bound energy, and secondary processes, whereas the lowest level of the hierarchy corresponds to the id, free energy, and primary processes (see Chapter 9 for a discussion of the ego and the id). The levels interact with each other, with the DMN suppressing the limbic regions in very much the same way that the ego and its secondary processes contain the id and its primary processes by binding their free energy (Carhart-Harris and Friston, 2010, pp. 3, 7, 11–12) (for a detailed discussion of primary and secondary processes, see Chapter 5).

The suggestion of Carhart-Harris and Friston (2010) links a clear neuronal structure and organization, namely hierarchical organization with top-down control, to Freud's concepts of the ego and the id, and primary and secondary processes. Their hypothesis contains a number of different aspects, including anatomical structure, hierarchical organization, the ego and the id, and primary and secondary processes. At this point I want to discuss only the hierarchical organization (primary and secondary processes are discussed in Chapter 5, anatomical structure is discussed in Chapter 6, and the ego and the id are discussed in Chapter 9). I believe that the hypothesis which characterizes hierarchical organization by top-down control is problematic for both empirical and conceptual reasons. Let me start with the empirical reasons and then go on to discuss some of the conceptual reasons.

Empirically, Carhart-Harris and Friston (2010) assume first that there is clear-cut anatomo-functional segregation between the three anatomical levels, and secondly that there is inhibition and suppression of lower levels by higher levels. However, both assumptions must be considered problematic. Although there is plenty of evidence that the DMN has a special role, the concept of the DMN as being distinct from other regions (e.g. the attention system) remains an open issue. Structurally, there are many connections between the midline regions. However, there are also plenty of connections between the medial and lateral regions and thus between the DMN and the attention systems (for more empirical details, see Feinberg, 2009, as well as Chapter 5 of this book). However, this makes clear-cut segregation between the DMN and the attention systems in anatomo-structural terms problematic, and this appears to have been (implicitly) neglected by Carhart-Harris and Friston.

Instead of clear-cut segregation, there is mutual integration and dependence between the different systems, resulting in what Feinberg (2009, p. 167ff) has described as a "nested hierarchy" (see Figure 4.2). In contrast to non-nested hierarchies, where there is indeed clear-cut segregation between higher and lower levels, with the former exerting top-down control over the latter, distinction between higher and lower levels remains impossible in nested hierarchies. The latter are characterized by a combination of lower-level regions within higher-level regions, with the former resurfacing within the latter. What is described as bottom-up modulation and top-down control in non-nested hierarchies reappears as constraint of the whole system and its organization and structure (see Feinberg, 2009, p. 167ff).

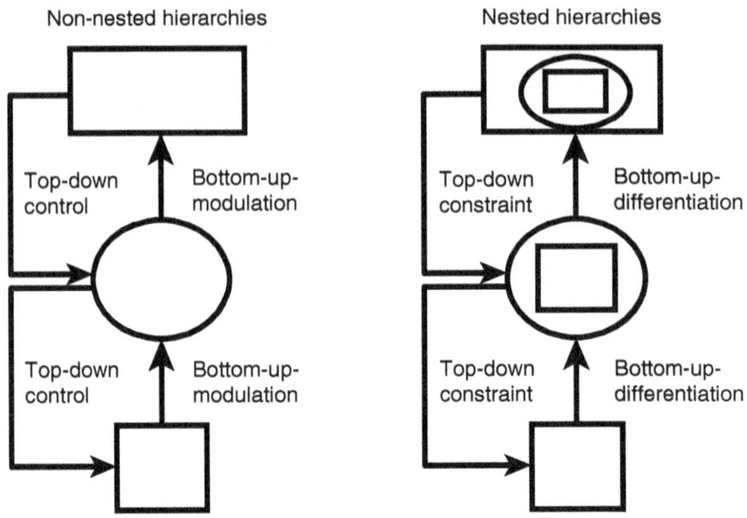

Fig. 4.2 Different forms of hierarchies.

Such a nested hierarchy can be observed not only on the anatomo-structural level of the brain, but also on the functional level. For instance, in a recent study on mind-wandering during external stimulation (e.g. a working memory task), neural activity was observed in both the DMN regions and those of the attention systems (Christoff et al., 2009). Moreover, deactivation (e.g. negative BOLD response) can be observed not only in the DMN but also in all of the other regions, including the ventral striatum and the primary sensory and motor cortices (deGreck et al., 2008). Taken together, these results shed some doubt on the first assumption, that there is clear-cut anatomo-functional segregation between DMN and the attention systems (and limbic-subcortical regions). However, if their clear-cut segregation can no longer be assumed, the assumption that there is unilateral control of the one by the other in the gestalt of top-down control may need to be questioned.

The second empirical assumption concerns the unilateral inhibitory and suppressive impact of higher regions on lower regions, which is crucial for maintaining top-down control as suggested by Carhart-Harris and Friston (2010). Although many studies have shown inhibitory and suppressive effects of the medial prefrontal cortex on the amygdala, as cited by the authors, the exact impact of other regions of the DMN, such as the posterior cingulate cortex (PCC) and the perigenual anterior cingulate cortex (PACC), on the subcortical regions remains far less clear. Also there is some evidence that the effect of the DMN on other regions may not only be suppressive but also facilitating and enabling. This is most apparent in the aforementioned results with regard to rest–stimulus interaction, where regions of the DMN may enable and predispose to stimulus-induced activity in, for example, the sensory and limbic/paralimbic regions. Therefore to assume that there is simple suppression and inhibition would be to neglect the facilitating and thus enabling and predisposing neural effects of the brain's high resting-state activity in the DMN on the neural activity in other regions during stimulus-induced activity. However, this casts some doubt on whether the hierarchical model of top-down control with suppression and inhibition as its essential ingredients is empirically plausible as an overall model of brain organization.

Let me make an important conceptual comment. Note that this is not to deny the empirical evidence of suppression and inhibition; it is only to deny that inhibition and suppression involve a certain hierarchical organization (i.e. top-down control), since that assumption does not seem to be supported clearly by the current empirical data.

Hierarchical organization, "inner–outer dichotomy," and the ego

In addition to such empirical reservations, there are also some conceptual reservations with regard to top-down control. Conceptually, to equate the DMN with the ego is to assume what I call an "inner, higher-order, and cognitive concept of the ego." The concept of the ego is here presupposed as an inner concept in that it is detached and isolated from outer stimuli from the body and the outside world, which it only suppresses and inhibits. As such the ego is supposed to be situated at the very top of the hierarchy, the DMN, which inhibits and suppresses everything else that is not ego (i.e. the id). The ego is thus a higher-order concept. Furthermore, the ego via the DMN is supposed to suppress neural activity in both the limbic and sensory regions, which have a crucial role in processing and assigning the sensory and affective features of the stimuli. This means that the ego itself must be regarded as cognitive rather than as affective and sensory.

The "inner, higher-order, and cognitive concept of the ego" can be traced back in a philosophical context to Descartes and his concept of the ego. When Carhart-Harris and Friston (2010) associate the DMN and top-down control with the Freudian concept of the ego, they project the "inner, higher-order, and cognitive concept of the ego" onto the brain. They thus project the Cartesian dichotomy of an inner mental ego and an outer physical world onto the brain itself when they distinguish between the different levels of hierarchy (i.e. the DMN and the subcortical limbic and paralimbic regions). In other words, the Cartesian dichotomy between inner mental states and outer physical states resurfaces here within the neural context as the anatomo-regional dichotomy between the DMN and the subcortical limbic and paralimbic regions. And in the same way that Descartes thought of them as opposites, Carhart-Harris and Friston declare the DMN and the subcortical limbic and paralimbic regions to be opposites when assuming top-down control of the former over the latter. Their Cartesian-inspired psychodynamic dichotomies between ego-libido and object-libido are thus supposed to correspond, and even be transferred in an almost one-to-one manner, to the anatomical and functional level, where they resurface as the anatomo-structural and functional dichotomy between the DMN and the subcortical limbic and paralimbic regions (Carhart-Harris and Friston, 2010, p. 12).

However, to assume such inner–outer dichotomy is neither conceptually nor empirically plausible. As I shall describe in detail in Chapter 5, the inner–outer dichotomy is neither empirically plausible in the neuronal context of the brain, nor conceptually plausible when considering the concepts of self and ego (see Chapter 9). Instead of assuming such dichotomy, one may instead postulate a more basic and fundamental intrinsic relationship between organism and world that first and foremost enables and predisposes to the constitution of the seemingly dichotomous entities that are so often assumed in and transferred between philosophical, psychodynamic, neuronal, and functional contexts. The seemingly almost blind transfer of the Cartesian dichotomy between mind and body into the psychodynamic and subsequently anatomo-structural context of the brain makes the inner–outer dichotomy conceptually highly implausible, especially given that its origin in Descartes is itself implausible.

Most importantly, as well as such conceptual implausibility, the inner–outer dichotomy is not empirically plausible on either anatomo-structural or functional grounds. As discussed, Feinberg's concept of nested hierarchies precludes anatomo-structural segregation between the DMN and the limbic regions, while functionally their different systems are co-activated rather than segregated in a clear-cut way and mutually inhibited. Moreover, all of the regions in the brain can be activated and recruited by both inner (i.e. mental) and outer (i.e. physical) stimuli. In the same way, physical stimuli from the outside world can recruit neural activity in both the sensory cortex and regions in the DMN, while mental stimuli from the supposedly inner world of the ego can induce activity changes not only in the DMN but also in sensory regions (see the previous sections on rest–stimulus and stimulus–rest interaction, as well as Chapter 6).

Taking all of these considerations into account, the presumed hierarchical organization of top-down control may not be plausible either empirically or conceptually. So what is the alternative? Feinberg's model of nested hierarchies provides a model of anatomo-structural hierarchy that is not based on top-down control, although it leaves open the question of what precise kind of functional hierarchy is involved. I assume that functionally nested hierarchies no longer involve top-down control, but rather what I call "neuronal contextualization."

Neuronal contextualization can be briefly explained as follows. The incoming stimulus enters the brain through the sensory regions, where it immediately encounters the brain's own activity, namely the intrinsic resting-state activity of the sensory cortex. This encounter between the stimulus and the brain's intrinsic activity means that the activity as possibly induced by the stimulus must be coded and processed in relation to the brain's intrinsic resting-state activity. This, as I shall explain below, is made possible by the stimulus directly interacting with the brain's resting-state activity (i.e. rest–stimulus and stimulus–rest interaction). By being processed and coded in relation to the brain's intrinsic resting-state activity, the stimulus and its respective stimulus-induced activity are set in the neuronal context of the brain and its intrinsic activity. This amounts to what I call "neuronal contextualization" as a functional characteristic of nested hierarchies.

In addition to neuronal contextualization, nested hierarchies may also be characterized by what I call "bottom-up differentiation" and "top-down constraint." By continuously realizing or, better still, nesting the lower system within the system immediately above it, the stimulus entering through the lower systems is continuously differentiated in a bottom-up way. Such bottom-up differentiation may be regarded as the continuation of neuronal contextualization, where the stimulus processing in the first region is contextualized by the next region, and the latter is then in turn contextualized by another region, and so on. The stimulus is thus continuously processed through the different regions, and as a result it becomes increasingly differentiated. Conversely, the more elementary regions, such as the sensory regions, may also provide the neuronal context for the stimulus processing in seemingly higher and more complex regions (e.g. the prefrontal cortex). In this case, the "higher" regions exert some constraint on (rather than control over) the "lower" regions, so that one may want to speak of "top-down constraint," which again presupposes neuronal contextualization. Accordingly, "bottom-up differentiation" and "top-down constraint" replace in nested hierarchies what are described as top-down control and bottom-up modulation in non-nested hierarchies.[3]

What are the neuropsychodynamic implications of the assumption of nested hierarchies with neuronal contextualization? As Freud had already assumed, a clear-cut segregation between the ego and the id is not empirically plausible. He demonstrated that their segregation was psychologically implausible, and we could show that it is also neuronally implausible. Looking to Freud and psychoanalysis may also give us some hints as to how such nested hierarchies with neuronal contextualization can be constituted. Realizing lower parts within higher parts and thereby contextualizing the former by the latter uses up energy, and such energy must be invested in the processing. In short, cathexis seems to be necessary to constitute a nested hierarchy with neuronal contextualization. Thus, by considering Freud's concept of cathexis within the neural context of the brain, we may be able to complement our empirical assumptions by their enabling and predisposing empirical condition, namely the investment of the brain's intrinsic activity in its own neural processing of stimuli.

[3] I am well aware that the concept of nested hierarchies suggested here requires further elaboration, but this would be beyond the scope of this book (see Feinberg, 2009, chapter 6).

Let us be more specific. The brain itself provides the energy as manifested in its intrinsic activity (i.e. its resting-state activity), which it invests in its own neural processing of stimuli, resulting in rest–stimulus interaction. The investment of the brain's intrinsic energy in its own neural processing of stimuli enables and predisposes to neuronal contextualization. This in turn, on the basis of progressive bottom-up differentiation and top-down constraint, may enable and predispose to contextualization within the psychodynamic context (i.e. contextualization of objects by the ego and, conversely, of the ego by objects).

If this is the case, we may consider neuronal contextualization as enabling and predisposing to possible contextualization within the psychodynamic context as the transformation of mere stimuli into objects, the stimulus–object transformation (for details of the latter, see Chapter 5). This raises the question of how mental states can be constituted on the basis of the brain's neuronal states. This question was left open by Carhart-Harris and Friston and has not yet been addressed here, but will be discussed in much more detail in Chapter 5. First, however, we need to shed some light on the exact role of the neuronal mechanisms of rest–stimulus and stimulus–rest interaction, with specific reference to predictive coding, which Carhart-Harris and Friston along with many present-day neuroscientists assume to be the brain's main coding strategy.

Predictive coding and cathexis

Carhart-Harris and Friston (2010) associate the brain's resting-state activity with continuous anticipation and prediction of potentially incoming external stimuli. The actual incoming stimulus is compared with the anticipated one, and in the case of a discrepancy there is an error, a so-called prediction error, which determines the amount of neural activity that is induced. This principle is called predictive coding (Frith, 1992; Friston, 2005; Stephan et al., 2009), and it provides the general framework for minimizing predictive errors when comparing the predicted or anticipated signal with the actual incoming stimulus. Following Carhart-Harris and Friston (2010), the main purpose of the brain's hierarchical organization and its top-down control is to minimize prediction error at all levels of the hierarchy, and thus to reduce the amount of free energy (for a consideration of the concept of prediction in a psychodynamic context, see Pally, 2006).

Imagine the case of an abnormal balance between anticipated and actual sensory input with, for example, too much emphasis on the anticipated input (e.g. the expectation that a particular person will speak). The actual incoming sensory input is then perceived and experienced only in the light of the anticipated input (i.e. the predicted source and thus the speaker), while leaving no room for the actual source or speaker, which is then no longer perceived as such (Friston, 2005; Stephan et al., 2009). For example, auditory verbal hallucinations in schizophrenia (and also delusions) may then be traced back to a mismatch between anticipated beliefs about a sensory input and real auditory–sensory input in the perception of auditory signals. This in turn leads to abnormal hyperactivity in the auditory cortex, as is observed in these patients (for further details, see Chapters 11 and 12).

Although a detailed empirical and conceptual critique of the concept of predictive coding is beyond the scope of this book, I here want to highlight a couple of issues. Most importantly, predictive coding is ultimately a theory about stimulus–stimulus interaction, in that it concerns the interaction between a mental stimulus (the anticipated stimulus) and a physical stimulus (the actual stimulus). However, if one presupposes stimulus–stimulus interaction it remains unclear how the anticipated stimulus can be generated and constituted as such. More specifically, both the process leading to the anticipation of a stimulus and the processes underlying the association of the anticipated stimulus with a mental state remain unclear. In the following account

I shall focus on the first question, namely the anticipation of stimuli, while deferring discussion of mental states until the next chapter.

I postulate that for anticipation of any kind of stimulus to be possible, the brain's intrinsic activity and its resting-state activity must be considered. However, this may not be sufficient. In addition to the resting-state activity itself, we need to question why the brain's resting state enables and predisposes us to generate and constitute anticipation and predictions. I here speculatively assume the following. By not only modulating stimulus-induced activity via rest–stimulus interaction but also being modulated by the latter via stimulus–rest interaction, the brain's intrinsic resting-state activity becomes tuned to the various stimuli (and, psychodynamically speaking, the objects) that it encounters. Such tuning to previously encountered stimuli (and objects) enables and predisposes the brain and its intrinsic resting-state activity to generate anticipatory and predictive stimuli. Thus what the theory of predictive coding assumes to be anticipation can be traced back ultimately to rest–stimulus and stimulus–rest interaction that first and foremost enables and predisposes the brain's resting state to generate and constitute anticipatory and predictive stimuli.

Are rest–stimulus/stimulus–rest interaction and the theory of predictive coding incompatible with each other? The answer is that in fact they are rather compatible and complementary. The consideration of rest–stimulus/stimulus–rest interaction focuses on those conditions that first and foremost enable and predispose the brain's resting-state activity to generate predictive and anticipatory stimuli. The theory of predictive coding, in contrast, looks not so much for the origin and the necessary condition of the possible generation of anticipatory and predictive stimuli via rest–stimulus/stimulus–rest interaction, but instead it focuses on the consequences and the kind of neural processing resulting from anticipation and prediction, and thus on (mental) stimulus–(physical) stimulus interaction. My focus on rest–stimulus and stimulus–rest interaction and their neuronal mechanisms, including their psychological implications, is therefore here complemented by the search for the neuronal and psychological mechanisms underlying stimulus–stimulus interaction. Thus the two accounts are highly compatible in that I am searching for those neuronal mechanisms (i.e. rest–stimulus and stimulus–rest interaction) that must precede and underlie what according to the theory of predictive coding happens neuronally during the processing of stimuli (i.e. stimulus–stimulus interaction).

In other words, the brain's resting-state activity is not to be identified with anticipation and prediction, but rather it provides the enabling and predisposing condition via rest–stimulus and stimulus–rest interaction. The brain's intrinsic resting-state activity can thus be compared with the ground upon which the legs of a table stand. If the ground is uneven, no table can stand on it. However, if the ground is even, the legs of the table can be constituted, thus corresponding to the generation of anticipatory and predictive stimuli. However, for this to be possible, the ground must be connected to the table legs in a certain way. This is provided by rest–stimulus and stimulus–rest interaction that links the brain and the anticipatory stimulus (i.e. the ground and the table legs). This in turn makes it possible to put a tabletop on the legs and thus to constitute a table, which corresponds to the interaction between the anticipatory stimulus and the real stimulus, resulting in stimulus-induced activity and predictive coding.

What are the implications of this for the neuropsychodynamic concept of cathexis? Conceptually I characterized cathexis in terms of an ambiguity, namely its characterization by seemingly opposite processes, being simultaneously static and dynamic. By allowing for anticipation of a stimulus in a mental gestalt, the free energy that the real stimulus induces may be bound and converted into bound or cathected energy (Carhart-Harris and Friston, 2010). This dynamic process of releasing and binding energy results in a static process. However, such binding and release of energy is only possible if there is a source of energy. In our case the source is the brain and the energy consists of its intrinsic activity.

By investing its own intrinsic activity in its own neural processing of extrinsic stimuli, the brain does more than merely bind the energy that the extrinsic stimuli induce in it. Most importantly, the brain opens the door to allow itself (i.e. its own intrinsic activity) to be organized and structured by the stimuli that it processes, which in turn enables and predisposes to what Carhart-Harris and Friston describe as anticipation. Thus the application of the original psychodynamic concept of cathexis within the neural context of the brain may allow us to better understand how the brain's neuronal states can be structured and organized so as to enable and predispose to the constitution of mental states as is assumed to occur in anticipation. This will be described in more detail in the next chapter.

Chapter 5

Cathexis, Neural Coding, and Mental States

Cathexis as a "neuro-mental bridge concept"

I discussed two ambiguities in the psychodynamic concept of cathexis in the previous chapter. We now need to discuss a third ambiguity in the concept of cathexis. Initially, Freud used the concept of cathexis at both levels (i.e. neural and psychological). Neurally, cathexis describes the discharge of electrochemical energy, which is especially apparent in Freud's 1895 writing, *Project for a Scientific Psychology* (see also MacIntosh, 1993, pp. 2–3). However, in the same writing the concept of cathexis is used in a different and more psychological way to describe motivation and arousal and their impact on the constitution of objects. This psychological and motivational meaning of the concept of cathexis was predominant in Freud's later writings, and the concept of cathexis referred to it exclusively, while the earlier, neural meaning was largely abandoned (see also MacIntosh, 1993, pp. 2–3).

Freud's later exclusive usage of the concept of cathexis as merely psychological–motivational may disentangle and reduce ambiguity within a conceptual–logical context. However, this raises the question of whether such exclusive usage as merely psychological does justice to the concept of cathexis itself and is thus consistent with its characteristic features. Cathexis describes the investment of energy in whatever object consecutively remains non-specified. Does the concept of cathexis involve the investment of energy in objects exclusively within the psychological context?

One may want to argue that the concept of cathexis involves an economic–structural description and usage that may be inherent in the concept of energy (see, for example, Holt, 1962, pp. 519–520; MacIntosh, 1993, p. 4). For instance, cathexis involves terms and concepts that describe economic processes of distribution, organization, and structure, whereas by itself it does not refer to psychological terms such as motivation and arousal. Economic–structural concepts such as energy and investment can be associated with and applied to the psychological context and thus to motivation and arousal. However, importantly, this is only one application to one possible context among many other possible contexts, which means that the psychological meaning is not intrinsic to or inherent in the concept of cathexis itself. Thus to exclusively associate the concept of cathexis with the psychological context is to confuse context and concept. The concept can be set in multiple contexts and can thereby acquire different meanings. In other words, the meaning of concepts such as cathexis is context dependent.

Such context dependence also holds for psychological and neural contexts. The neural context of the brain is characterized by neural states, whereas the psychological context of the psychic apparatus, as outlined by Freud, involves mental rather than neuronal states. Since the concept of cathexis is context dependent, it can be used in both neural and psychological contexts, but thereby acquires different meanings. More specifically, cathexis is by itself characterized neither by neural states nor by specific mental states. Instead, the concept of cathexis remains independent of the distinction between neural and mental states, and may therefore be a viable candidate for a concept that could bridge the gap between neural and mental states and thus between neural and psychological contexts.

I assume that the concept of cathexis can bridge the gap between neural and mental states and may therefore be regarded as what I call a "neuro-mental bridge concept." The latter concept describes features and conditions that enable and predispose to the linkage between neural and mental states, thereby allowing for what I shall later call "neuronal–mental transformation." I argue here that the concept of cathexis as a neuro-mental bridge concept describes the investment of (the brain's) energy in non-specified objects, and that such energy investment in objects allows the transformation of neural states into mental states (i.e. neuronal–mental transformation). Taking all of this into account, I characterize the concept of cathexis as a neuro-mental bridge concept by empirical and conceptual features that are believed to enable and predispose to neuronal–mental transformation. However, before considering the empirical features in detail, we need to highlight another conceptual feature, namely intentionality, as being characteristic of mental states.

Cathexis and intentionality

I have described how the concept of cathexis presupposes the concept of objects as non-specified, thus itself remaining independent of the characterization of any specific kind of object (be it psychological, economic, or neural). This raises the question of how the concept of cathexis is related to the object, if not by any specific object. It is clear that cathexis must somehow be related to objects since, as McIntosh (1993, p. 10) has stated, cathexis involves "entering into objects." However, it remains unclear how such entry of cathexis into objects is possible without the object itself being specified, thus raising the following question: How can we describe the relationship between cathexis and the objects in which it invests its energy?

McIntosh (1993, pp. 4–10) has argued that cathexis shows an intentional relationship to objects, and that cathexis is about objects and directed towards them, thus amounting to what the philosophers Franz Brentano and Edmund Husserl have described as "intentionality." However, Freud's concept of cathexis goes beyond the philosophical concept of intentionality by introducing motivational aspects that complement the usually rather cognitive notion of intentionality. This is most apparent when Freud speaks of libidinal cathexis that introduces the libido which is directed towards sensory and affective objects such as the body, rather than towards cognitive and mental objects. Since the libido is also directed towards objects, we may need to widen the scope of the concept of intentionality from its narrow philosophical focus on cognitive properties to include more bodily based sensory and affective features.

Another feature of intentionality needs to be reframed. The philosophical notion of intentionality is not only cognitive but also static, with both object and directedness remaining unchanged. This is in contrast to Freud's psychodynamic context which, adopting an economic view of the mind, is rather dynamic, involving continuous change of the object itself and the directedness towards the object, for which Freud introduces economic terms such as cathexis as "energy investment." Such reframing of intentionality in energetic terms also allowed Freud to quantify it and thus to shift it within the scientific context (see also McIntosh, 1993, p. 4). This differs from the rather qualitative notion of intentionality in the philosophical context which, as such, is supposed to remain immune to any quantification and scientific investigation.

Another distinctive feature of Freud's concept of intentionality is, according to McIntosh (1993, pp. 6–9), that it does not concern mere mental representations of objects, but rather real objects. This is well documented in psychodynamic research by, for example, Stern, who showed that infants and children are concerned with events and objects in the real outside world, rather than being restricted to their mental world. This is true even for fantasies in adults, since although fantasies and the imagination are essentially mental, they nevertheless concern objects that can be traced back to and are derived from the real outside (i.e. non-mental) world.

If cathexis concerns a form of intentionality that is directed at real objects rather than at mental representations of them, the distinction between the inside and outside world (i.e. between the mental and physical world) is undermined. If the energy of the organism (its intrapsychic energy) is invested in real objects in the outside world (the interpsychic world), the distinction between intrapsychic and interpsychic becomes blurred: "That the character of the cathexis always infuses the character of its object, especially on the unconscious level, means that it is impossible to draw any firm line between the intrapsychic and the interpersonal. The two form a single nexus. In sum, Freud's notion of a cathexis as 'entering into' its objects, which finds support in recent work in cognitive psychology, has the effect of undermining the centrality of the traditional Cartesian division between the 'inner' and the 'outer' realms of experience" (McIntosh, 1993, p. 10).

How can we describe the "single nexus" between the intrapsychic and the interpersonal in more conceptual detail? As was pointed out previously, the philosophical concept of intentionality tends to be cognitive, qualitative, mental, and static, as distinct from its use in the psychodynamic context, where it tends to be sensory–affective, quantitative, physical, and dynamic. By highlighting the sensory–affective, quantitative, physical, and dynamic features, Freud implicitly assumed a more basic and fundamental concept of intentionality that, as such, was supposed to enable and predispose to the transition from sensory–affective to cognitive functions as well as from the physical to the mental. His concept of intentionality must thus be designated as pre-perceptual, pre-cognitive, and pre-mental, which raises the following question: How can we now develop a concept of intentionality that is more appropriate for capturing the more basic, probably pre-perceptual, pre-cognitive, and pre-mental level at which Freud was operating, especially when dealing with the unconscious?

Let us return briefly to philosophy. Relying on Husserl, the French philosopher Maurice Merleau-Ponty (1962) described a special form of intentionality, which he called "operative intentionality," and distinguished it from the traditional form of intentionality described by Brentano, which he called "act intentionality": "This is why Husserl distinguishes between intentionality of act, which is that of our judgments and of those occasions when we voluntarily take up a position—the only intentionality discussed in the *Critique of Pure Reason*—and operative intentionality (fungierende Intentionalität), or that which produces the natural and antepredicative unity of the world and of our life" (Merleau-Ponty, 1958, p. xx).

Following Merleau-Ponty, operative intentionality describes a pre-reflective, pre-perceptual, and pre-cognitive "antepredicative unity of the world and of our life" which distinguishes it from "act intentionality" (for further conceptual and philosophical details, see Zahavi, 2005). What Merleau-Ponty calls "antepredicative unity of the world and of our life" may come close to what McIntosh (1993) diagnoses in Freud as "cathexis entering into the object" which is possible only when the organism (i.e. life, as Merleau-Ponty calls it) is intrinsically related to the world and its objects. The "single nexus" between the intrapsychic and the interpersonal that McIntosh pointed out may thus be traced back conceptually to operative intentionality as described by Merleau-Ponty.

Such operative intentionality may be innate or intrinsic to the organism, which means that the organism cannot avoid being in unity or a single nexus with the world. McIntosh (1993) highlights this by pointing out that infants and children cannot do other than approach the world in intentional terms that seems to be genetically predetermined: "However, recent research in cognitive psychology, some of which is summarized by Stern (1985), has overthrown the position that infantile thought is not intentional in character. Infants begin to distinguish and recognize discrete objects and events within days of birth, and within a few months they live in a world filled with clearly defined and recognizable objects and events firmly located in space and time. . . . It appears that the ability to see the world in clearly defined intentional terms is innate, i.e. genetically determined, not learned, as Freud and Piaget thought. These findings remove the main theoretical obstacle to treating all cathectic processes as intentional, i.e. as having objects" (McIntosh, 1993, p. 6).

In short, operative intentionality seems to be intrinsic rather than extrinsic to the concept of cathexis. However, this raises the question of why cathexis cannot do anything other than invest energy in objects within our world: What are the necessary empirical conditions of the apparently intrinsic nature of the operative intentional nature of cathexis? I shall later postulate that the brain's specific strategy of neural coding and of processing stimuli from the outside world enables and predisposes to the intrinsic operative intentionality of cathexis, and may thus be regarded as what philosophers might call a "necessary empirical condition."

Neuropsychodynamic hypotheses

How are these conceptual features of the psychodynamic concept of cathexis, namely the neuro-mental bridge concept and the operational intentional concept, related to the neural context of the brain? The reader may wonder why I ask this question, as in the previous chapter I developed a neuropsychodynamic hypothesis that linked the psychodynamic concept of cathexis to the corresponding features of the brain and its neural activity. However, that hypothesis only concerned neural states, not mental states.

In contrast to the features that were discussed in Chapter 4, both the neuro-mental bridge concept and the operational intentional concept assume that cathexis is related to mental states. In the case of the neuro-mental bridge concept, cathexis provides the transition from neural to mental states (with the latter understood in a purely phenomenological sense), whereas operative intentionality involves the characterization of cathexis as pre-mental. In order to understand the association of cathexis with mental states, we therefore have no choice but to reveal how the brain's neural states are transformed into mental states, and thus to account for what I call "neuronal–mental transformation."

The psychodynamic concept of cathexis is considered to be independent of the distinction between neural and psychological/mental[1] states, while at the same time cathexis is considered to be a necessary condition for bridging the gap between neural and psychological/mental states, amounting to what I called the neuro-mental bridge concept. How can we link these two apparently contradictory aspects to the neural context of the brain and its empirical functions? In the fourth chapter, I highlighted how the brain's intrinsic activity (its resting-state activity) corresponds within the neural context of the brain to Freud's use of energy investment within his psychological context of the psychic apparatus. On the basis of such correspondence between neural and psychological contexts, one would expect cathexis also to have a crucial role in enabling and predisposing to the transformation of neural states into mental states as associated with neural and psychological contexts. I therefore postulate that the brain's intrinsic activity (i.e. resting-state activity) may have a crucial role in enabling and predisposing to the transformation from the brain's neuronal states into the psyche's mental states, thus accounting for neuronal–mental transformation. This is my first specific neuropsychodynamic hypothesis that targets the empirical necessary (i.e. enabling and predisposing) conditions of neuronal–mental transformation and thus of cathexis as a neuro-mental bridge concept.

[1] Although I am well aware that psychological and mental states can be distinguished both conceptually and empirically, here for operational purposes both terms are used more or less interchangeably unless otherwise indicated.

The psychodynamic concept of cathexis was also characterized by a very basic sense of intentionality, namely operative intentionality, which described a single nexus and intrinsic linkage between the intrapsychic and interpersonal dimensions. Shifting the psychodynamic concept of cathexis and its operative intentionality into the neural context of the brain means that the brain's intrinsic activity (i.e. resting-state activity) may process and code any kind of stimulus-induced activity in such a way that it cannot do other than intrinsically link the organism (i.e. its intrapsychic and intra-biological dimension) to the world (i.e. the interpersonal and social dimension). The brain's intrinsic activity must thus code its own processing of the activity induced by extrinsic stimuli in such a way that both kinds of activities (i.e. extrinsic stimulus-induced activity and intrinsic resting-state activity) are inextricably and intrinsically linked and connected to each other. Such neural coding may in turn enable and predispose to the constitution of the aforementioned features of operative intentionality. This is my second specific neuropsychodynamic hypothesis that targets the empirical necessary (i.e. enabling and predisposing) conditions of operative intentionality as characteristic of cathexis.

Cathexis and difference-based coding

In order to demonstrate empirically that the brain's energy (its resting-state activity) has a crucial role in shaping and ultimately constituting mental states, we need to show that rest–stimulus interaction also applies to mental states. There are indeed a few studies that warrant a brief mention here.

Boly et al. (2007) investigated the effect of resting-state activity in task-negative (and task-positive) regions on subsequent behavioral performance in a somatosensory perception task. The investigators applied laser stimuli to the hand with different degrees of awareness and pain. They then investigated whether the resting-state activity (3 seconds before the stimulus onset) in task-negative and task-positive regions predicts the subsequent perception (i.e. the degree of awareness and the sensation of pain). They observed that decreased resting-state activity in task-negative regions of the DMN (i.e. the PCC/precuneus, the TPJ, and the parahippocampal region) exerted a facilitating effect upon subsequent perceptual awareness.

In addition, they observed that task-positive regions (i.e. the LPFC, the supragenual ACC, and the medial thalamus) had the reverse effect, with increased resting-state activation in these regions predicting subsequent perceptual awareness. Finally, increased resting-state activity in typical pain regions (e.g. the ACC and the bilateral anterior insula) positively predicted subsequent pain perception. This study shows that resting-state activity in various regions has a strong influence on subsequent mental states (i.e. somatosensory awareness).

Other studies demonstrated that the brain's resting-state activity level influences subsequent cognitive states. Sapir et al. (2004) investigated the effect of changes in the spatial attention cortices, as induced by a cue, on the accuracy of a spatial discrimination task. They observed that the level of resting-state activation predicted the variability in the subsequent spatial discrimination task. Similarly, Weissman et al. (2006) demonstrated that slow reaction times in a visual attention task are predicted by reduced pre-stimulus activity in the ACC and the LPFC. Kleinschmidt's research group demonstrated how resting-state activity fluctuations in the sensory cortex and the default-mode network (DMN) influence (i.e. bias) the subsequent mental states, including the behavioral reaction patterns. For example, higher resting-state activity in the DMN and the visual cortex biased the perception and subsequent perceptually based decision making (Hesselmann et al., 2008; Sadaghiani et al., 2009).

Although these studies focused on behavioral parameters (i.e. reaction time and accuracy) as dependent variables of cognitive states, the aforementioned study by Boly et al. (2007) demonstrated that resting-state activity in various regions also influences the subsequent psychological (or mental) state, namely somatosensory awareness (and pain). Taken together, all of these studies thus support the hypothesis that the brain's resting-state activity influences mental states.

However, the crucial question is what is meant by the word "influences" in the last sentence. Influence can mean mere "modulation," as distinct from "constitution," indicating that the brain's resting-state activity may not be necessary for constituting mental states as such and thus for neuronal–mental transformation. However, influence can also mean that the brain's resting-state activity enables and predisposes to the constitution of mental states, and is thus a necessary (although not sufficient) condition of possible mental states. In this case the brain's resting state is not only modulatory, but also constitutionally relevant for mental states.

How can we relate the conceptual alternative between "modulation" and "constitution" to the brain's neural function? Both options involve different kinds of neural processing of extrinsic stimuli in relation to the brain's intrinsic activity, and more specifically different methods of neural coding. However, before describing this in detail, we need to briefly consider what is meant by the term "neural coding."

What does the word "code" mean? Here it is used to refer to a metric or measure that captures and reflects purposeful and biologically or teleologically meaningful activity in a system (Descharms and Zador, 2000).[2] As such, the word "code" describes a specific processing algorithm according to which information is processed in a system. In relation to the brain, the neural code describes the metric or measure that determines and predisposes to neural activity and its possible changes during stimulation. The neural code aims to reveal a processing algorithm that organizes neural activity in the brain. As the neural code may be traced back to the brain itself, it may denote the brain's input and thus the brain as functioning rather than the brain as observed or the brain as experienced.

What kind of neural code is implied by modulation and constitution? Modulation means that the brain codes first and foremost the stimulus itself in its neural activity. This can then be modulated by the brain's resting-state activity. This form of neural coding means that the brain's resting-state activity by itself is not necessary for the neural coding and processing of the stimulus. The stimulus is thus coded (largely) independently of the brain's resting-state activity, which amountsto what I call "stimulus-based coding" (see Figure 5.1). Thus the brain's resting-state activity has here just a secondary or modulatory effect on the stimulus-induced activity and its associated mental states.

The case of constitution is different. Here the constitution of mental states as such is necessarily dependent on the brain's resting-state activity and its interaction with the stimulus involving rest–stimulus interaction. In contrast to the situation in modulation, the relationship between the brain's resting-state activity and the stimulus is supposed to enable and predispose to the transformation of neural states into mental states. If this is so, the brain's neural coding and processing of the stimuli can no longer be based upon the stimulus alone, as in stimulus-based coding, but must involve coding the interaction between the brain's intrinsic activity and its neural processing of extrinsic stimulus-induced activity. This interaction may be coded by coding the difference between the actual resting-state level of the brain and the stimulus-induced activity (as ideally induced independently of the brain's resting-state activity level). This amounts to what I call "difference-based coding." Rather than relying on the stimulus itself, such neural coding relies on the *difference* between the intrinsic and extrinsic activities within the brain itself. As I shall argue in the next section, such difference-based coding may enable and predispose to and

---

[2] See also Friston (2000, p. 219), who however only determines the meaning of the term "code" with regard to the brain.

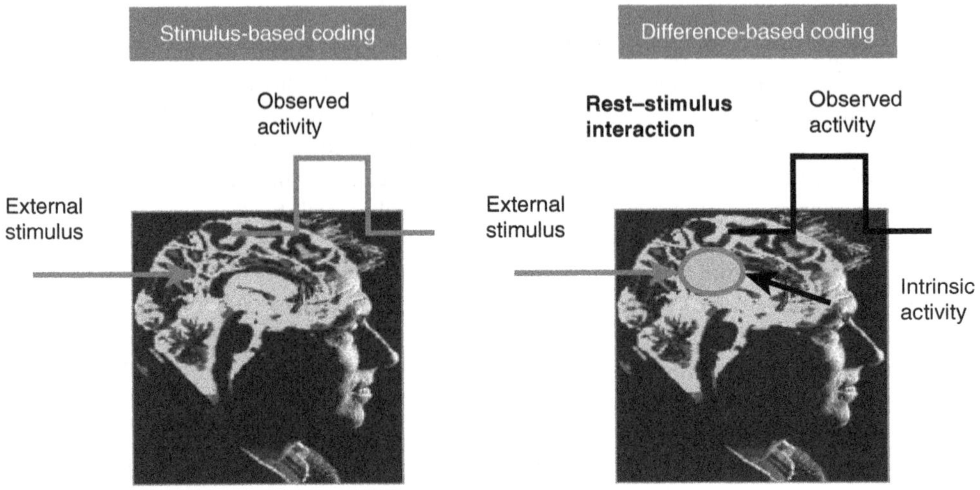

Fig. 5.1 Stimulus-based coding versus difference-based coding as neural code of the brain.

thus be a necessary empirical condition of possible neuronal–mental transformation. First, however, I want to present some empirical support for my hypothesis of difference-based coding.

Let us consider the motor cortex. Georgopoulus et al. (1986) demonstrated that the activity of a given motor cortical neuron is changed as a function of the activity of the other neurons that together with the neuron in question constitute a population or, better still, a neuronal assembly. More specifically, each individual neuron from the primary motor cortex codes for a given direction of a movement, but the level of precision is rather low because, at the same time, that neuron also participates in the coding of all other movements in every direction. How can a single neuron code such a multitude of different directions at the same time? The best and most economic way would be for it to code the difference between the different directions, rather than each direction independently and separately, thus involving what I describe as difference-based coding.

Thus one single neuron does not code one particular direction which would make it possible for us to infer the actual direction of the movement from the neural activity of this neuron. Instead, individual neurons code the difference between different directions in most probably different ways (maybe in part because of their different spatial location), so that in order to infer the actual direction of the movement, we need to calculate the neural activity of many neurons in the gestalt of a so-called "population vector" (Georgopoulos et al., 1989). This example demonstrates that even at the level of the individual neuron, activity changes are determined and coded by the difference between different stimuli, rather than by the stimulus and its characteristics.[3]

[3] Fiorillo (2008) too considers the single neuron and characterizes them by difference-based coding. Based on the neuron's biophysical properties like its K+ and Cl- channels, Fiorillo (2008, 3–4) argues that single neurons in general "integrate current information about its stimulus from one pool of ion channels and synapses, and prior information from another pool. Its membrane potential signals prediction error." The single cells membrane potential, i.e. its activity, is predetermined and disposed by the difference between current and prior information as signalled by the differences in activity levels between different ion channels and synapses. Fiorillo thus applies and extends the concept of prediction error beyond the behavioral level and the domain of reward to the level of the single neuron. Similar to the behavioral level of the whole organism, the single neuron's goal is to minimize prediction error and thus to maintain the difference between anticipated and actual activity levels as low as possible. Even the selection of both prior and current information sources is very much oriented on keeping this difference low. Only those inputs from the

I have demonstrated that difference-based coding is empirically plausible at the microscopic level of the individual neuron. Is this also true at the mesoscopic level of several neurons (i.e. a population of neurons)? Let us consider the motor cortex again. Grammont and Riehle (2003) demonstrated that each motor cortical neuron within a neuronal assembly depends on its relationship to its neighboring neuron. Whether, for example, the individual neuron might synchronize its activity with the activity of the other neurons depends on its relationship to its neighboring neurons, while the synchronizing activity of the latter depends in turn on the activity of the former, and so on. The individual neuron's activity cannot be considered independent of and in isolation from the activity of other neurons, and it is this relationship to other neurons' activity level that seems to be encoded in the individual neuron's activity. In other words, the individual neuron's activity level is constitutively dependent upon the activity level of the other neurons, and it is this difference that is coded in the actual activity changes in neuronal assemblies.

The activity level of the other neurons provides what may be called the neuronal context. More specifically, the input of the other neurons is regarded as an important source of prior information for the neuron in question, and it is the difference between its own activity level and that of the other neuron that predetermines and predisposes the neuron's anticipated activity and ultimately its reactivity to actual stimuli (i.e. whether it synchronizes with the other neuron or not). Synchronization can then no longer be considered to be caused by the stimulus itself (as might be assumed by an external observer), but rather by the neuron's need to maintain and adapt its activity level. This links synchronization much more to the neuron itself than to the stimulus, and outs it as a rather narcissistic or self-related mechanism.

Examples from Georgopoulos et al. (1989) and Grammont and Riehl (2003) are cited by Legrand and Grammont (2005). Taken together, these examples provide empirical support for the hypothesis that neural activity at both microscopic and macroscopic levels is coded in terms of differences between the activities that stem from different activities and stimuli. The neural coding of neural activity in terms of differences between different activities and stimuli (i.e. difference-based coding) must therefore be distinguished from neural coding of the stimuli themselves (i.e. stimulus-based coding). Based on the data described in Chapter 4, I postulate that rest–stimulus and stimulus–rest interaction do also presuppose difference-based coding rather than stimulus-based coding.

What are the implications of difference-based coding for the concept of cathexis and its characterization as a neuro-mental bridge concept? I postulate that the brain's resting-state activity in conjunction with difference-based coding can be shown to enable and predispose to and thus be a necessary empirical condition of possible neuronal–mental transformation. In this case, cathexis referring to the resting-state activity as the brain's energy may indeed be regarded as a concept that predisposes and enables to bridging the gap between neuronal and mental states, thus being a necessary empirical condition of possible neuronal–mental transformation. Since I demonstrated empirical support for both difference-based coding and the effect of the brain's resting-state activity on mental states, the characterization of the concept of cathexis as a neuro-mental bridge may be not only conceptually but also empirically plausible. However, in

various stimuli are selected that can contribute to minimize the difference between anticipated and actual activity levels. This implies that the neuron produces output signals only when there is a difference between anticipated and actual activity level; if the actual activity level is higher exceeding its anticipation, the output signal is positive while in the reverse case the output signal may be negative.

order to obtain stronger empirical support, we need to demonstrate the exact mechanisms whereby difference-based coding enables and predisposes to the transformation of neuronal states into mental states (i.e. neuronal–mental transformation). This will be the focus of the next section.

Difference-based coding and neuronal–mental transformation

I now want to shed some light on how difference-based coding enables and predisposes to neuronal–mental transformation and thus the constitution of mental states. Conceptually, I here consider the concept of mental states in a very phenomenological sense, as the subjective experience of phenomenal or mental states from first-person perspective as distinct from observation of physical states from third-person perspective. As such, mental states are supposed to be characterized by individuality and privatization (which will be explained in more detail in Chapter 7). It is important to note that such a definition of mental states does not involve any cognitive or representational features, but a purely phenomenal or phenomenological approach. However, this does not exclude the possibility that the latter enable and predispose to the former (although an explanation of this would be beyond the scope of this book). Having clarifying some conceptual details, let me now describe some empirical data.

Sommerfield et al. (2006) applied two different stimuli (houses and faces), matched their physical characteristics, and matched the stimuli to the individual subjects' thresholds for perception. When comparing face stimuli with non-face stimuli, irrespective of their perceptual sets they could tap into those regions that were related to the physical characteristics of the stimuli, yielding neural activity changes in the fusiform face area, the TPJ and the inferior occipital cortex. Their design also allowed them to compare the perceptual sets irrespective of the physical characteristics (face sets versus house sets), which yielded neural activity changes in the ventro- and dorsomedial prefrontal cortex that were dependent on the perceptual characteristics but not the physical characteristics of the stimuli.

The authors argued that the activity in the frontal regions reflects the generation of some template, a prediction or anticipation that top-down modulates neural activity changes in the visual cortex induced by the stimuli themselves and their physical characteristics. The frontal cortex generates some prediction of possible sensory changes, a template, against which the sensory changes that actually occur are matched and compared. Such matching and comparison between frontal top-down signals (i.e. the predicted sensory input) and visual cortical bottom-up signals (the actual sensory input) allows us to shape and constitute a percept (a visual percept). Since the percept is based upon matching and comparison between the anticipated and actual input, it can no longer be traced back to the actual visual stimulus itself. The constitution of the percept as distinct from the stimulus itself thus presupposes difference-based coding rather than stimulus-based coding.

Due to such difference-based coding, the percept (the mental state) "goes beyond" the actual stimulus and its physical information (Edelman and Tononi, 2000, p. 173; Prinz, 2000b, pp. 18–19), which is nicely expressed and formulated in the following quote by Prinz: "For instance, we cannot look at a written word without hearing, as it were, how it sounds, and understanding what it means. In these cases, the stimulus information on which perception is grounded is purely visual, but the information that is perceived goes far beyond—capturing sound, meaning, and so on. The same applies to the perception of physical events. When we observe the billiard ball that hits another, the information our percept is grounded on is again purely visual, but the information perceived includes the perception of a causal relationship between the movements of the two balls. In the same vein, when we watch other people's action, what we see goes far beyond what is contained in the underlying stimulus configuration. We see, for example, what objects people are reaching for, what goals they strive for, and whether they fail or succeed. . . . We do it by ordinary perception—which by definition implies that we go beyond the information given" (Prinz, 2000b, pp. 18–19).

I demonstrated that our percept may depend on a difference, namely the difference between anticipated and actual stimulus input, which presupposes difference-based coding rather than stimulus-based coding. Most importantly, difference-based coding allowed for the constitution of a mental state, the percept, that is to be distinguished from the stimulus itself and its physical features. On a more general level, I therefore postulate that difference-based coding enables and predisposes to the possible constitution of mental states and their transformation from neuronal states as physical states. By "going beyond" the stimulus itself and its physical features, the constitution of mental states as distinct from neuronal (i.e. physical) states becomes possible. Thus difference-based coding must be regarded as a necessary empirical condition of the possible constitution of mental states and their distinction from physical states and thus what we can observe as neuronal states in the brain.

This is different in stimulus-based coding. Stimulus-based coding may be incompatible with the constitution of mental states and their distinction from physical states. In the case of stimulus-based coding, our visual information would no longer "go beyond" the actual visual stimulus, which means that the distinction between stimulus and percept (with perception in the current sense becoming impossible, too) would remain impossible. However, this precludes stimulus-based coding from enabling and predisposing to the transformation of neuronal states into mental states (i.e. neuronal–mental transformation). Accordingly, unlike difference-based coding, stimulus-based coding leaves the neuronal mechanisms underlying neuronal–mental transformation open.

Why is there difference-based coding rather than stimulus-based coding? I postulate that the brain's intrinsic activity and its constitutional impact on stimulus-induced activity enable and predispose the brain's neural coding to difference-based rather than stimulus-based coding. If the brain's activity is solely determined by the stimulus itself, there would be no need for the brain to code and process stimuli in terms of differences, but rather in terms of stimuli involving stimulus-based coding. However, as the brain shows intrinsic activity, any stimulus that is coded and processed by the brain, unavoidably, and then necessarily, encounters the brain's resting-state activity. Thus the only way for the brain and its intrinsic activity to deal with the extrinsic stimuli entering the brain is to code and process it in relation to itself (i.e. its actual resting-state activity level). This relationship between the brain's resting-state activity and the stimulus means that it is neurally coded in terms of a difference (i.e. difference-based coding). I therefore postulate that the brain's intrinsic activity enables and predisposes to difference-based coding rather than stimulus-based coding.

What are the implications of this for the concept of cathexis? I argued that cathexis corresponds to the brain's intrinsic activity within the neural context. If now the brain's intrinsic activity enables and predisposes to difference-based coding, which in turn enables and predisposes to neuronal–mental transformation, cathexis must be considered a necessary (i.e. enabling and predisposing) condition of possible neuronal–mental transformation. In other words, by enabling and predisposing to neuronal–mental transformation, cathexis must be considered to have a crucial part in providing the transition from the neural context of the brain to the psychological context of the psychic apparatus. This means, within the conceptual context, that cathexis must be regarded as a neuro-mental bridge concept (in the literal sense of the term), with this characterization being both empirically and conceptually plausible.

What are the implications of this for the psychodynamic context? This involves not only the transition from neuronal to mental states and from neuronal to psychological contexts, but also the transformation of stimuli into objects, which I call "stimulus–object transformation." Stimulus–object transformation describes how stimuli within the neuronal context of the brain are transformed into what are called objects within the psychodynamic context of psychoanalysis.[4] Thus the transition from the neuronal to the psychological context (i.e. from neuronal to mental states) is accompanied by the transformation of mere stimuli into what are called objects within the psychodynamic context. As I shall argue later, cathexis (i.e. the brain's intrinsic activity and its difference-based coding) may enable and predispose to stimulus–object transformation as well. First, however, I want to shed some more light on the psychodynamic implications of difference-based coding and neuronal–mental transformation.

Neuronal–mental transformation and primary and secondary processes

Carhart-Harris and Friston (2010) associate primary processes and the id with neural activity in the subcortical and limbic regions, while they assume secondary processes and the ego to be related to neural activity in the DMN, which are supposed to suppress and control the former (for further details, see also Chapter 4). How is this compatible with the account of neuronal–mental transformation suggested here?

One might want to argue that what Freud called primary processes remain essentially unconscious and may therefore not be relevant to my account of neuronal–mental transformation. However, this is to presuppose that mental states have a certain meaning which is, very much in the traditional philosophical sense, equated with conscious states as distinct from unconscious states. Thus, although neuronal–mental transformation may hold for secondary processes that become conscious, it does not apply to primary processes as they remain unconscious.

However, I shall here argue that neuronal–mental transformation applies to both primary and secondary processes. This presupposes a different concept of mental states that is no longer exclusively restricted to conscious states but also includes unconscious states (as distinct from non-conscious states; see Chapter 8 for details). Thus I consider unconscious states to be as mental as conscious states. This is, for instance, supported by the fact that dreams that are usually classified as unconscious are as mental as conscious states (see also Chapter 7). However, this runs counter to the traditional philosophical characterization of mental states as conscious states as opposed to unconscious states. More specifically, the characterization of unconscious states as mental states shifts the distinction between mental and non-mental states from the distinction between conscious and unconscious states to that between conscious/unconscious and non-conscious states. This means that, contrary to the traditional philosophical context, the concept of mental states presupposed here is extended to include both conscious and unconscious states.

What are the implications of this for the relationship between primary and secondary processes? Primary processes surface in dreams, hallucinations, changes in sense of self, déjà vu, etc. as described by Carhart-Harris and Friston (2010, pp. 8–9), which they trace back to the lack of suppressive and inhibitory control by secondary processes and the ego (see also Chapter 4). This explains the differentiation between primary and secondary processes, while leaving open how both primary and secondary processes are constituted on the basis of the brain's neuronal states. In other words, the question that is left unanswered by Carhart-Harris and Friston refers to how the brain must process and code its own neuronal processing of intero- and exteroceptive stimuli

[4] I am fully aware that the concept of the object is rather complex, although this issue cannot be discussed in detail here. I refer the reader to the description of the concept of the object at the beginning of this and the previous chapter, as well as to footnote 3 in chapter 6, where I make more detailed conceptual determinations.

so that primary and secondary processes possible at all. Since both primary and secondary processes as unconscious and conscious states are supposed to involve mental states (see previous paragraph), this question touches upon the issue of neuronal–mental transformation, which may be reformulated in the current context in the following way. Why do we access and experience both primary and secondary processes in a mental gestalt (i.e. in terms of certain contents, events, and objects) rather than in a neuronal gestalt (i.e. as the activity of our brain's neurons)?

How does this compare with the approach adopted by Carhart-Harris and Friston? Neuronal–mental transformation targets the question of how our brain's neuronal states must be transformed in order for them to be accessible to us as mental states, whether in a conscious (i.e. as secondary process) or unconscious (i.e. as primary process) way. In contrast, Carhart-Harris and Friston (2010) focus only on the differentiation between conscious or unconscious access, while they do not target the transformation of neuronal states into mental states. My question about neuronal–mental transformation thus targets a more basic and fundamental issue than Carhart-Harris and Friston (2010) addressed in their focus on the differentiation between the unconscious/primary process and conscious/secondary process. They thus leave the very grounds upon which they stand open, and it is these "grounds" that enable and predispose to the transformation of neuronal states into mental states, and their subsequent differentiation into unconscious and conscious states and thus primary and secondary processes.

So far I have determined such grounds empirically by rest–stimulus and stimulus–rest interaction. I have also shown that difference-based coding building upon rest–stimulus and stimulus–rest interaction enables and predisposes to neuronal–mental transformation, thereby providing the basis upon which mental states can be constituted and subsequently differentiated into unconscious and conscious states (i.e. primary and secondary processes). Carhart-Harris and Friston may now want to argue that this explains secondary processes but not primary processes, which (like dreams or hallucinations) remain independent of any external stimulus. This means that rest–stimulus interaction and difference-based coding may not be appropriate for accounting for primary processes such as dreams and hallucinations.

I postulate that primary processes such as dreams and hallucinations are highly compatible with both the brain's resting-state activity and difference-based coding, as will be demonstrated in detail in Chapters 8, 11, and 12. Let us consider exactly what happens in the brain's resting state itself. This is not a static state, but rather it is dynamic, as it covers different regions with different resting-state activity levels that may interact with each other, amounting to what can be called "rest–rest interaction" (for an overview of recent results, see Northoff et al., 2010). Rest–rest interaction describes the neuronal interactions during the absence of (specific) exteroceptive stimuli. These neuronal interactions include spontaneous oscillations within regions as well as interactions across different regions by means of, for example, synchronization (for a recent review, see Uhlhass et al., 2009).

How is the neural activity associated with these neuronal interactions in the resting state coded and processed? I postulate that the brain does in the resting state exactly the same as it does when it processes and codes intero- and exteroceptive stimuli. That is, it codes its different resting-state activity levels in terms of their differences (rather than by themselves in an isolated way in terms of stimulus-based coding). One may thus speak analogously of "rest–rest interaction." And in the same way that difference-based coding enables and predisposes to the transformation of neuronal states into mental states, also does this in the resting state when there is rest–rest interaction. This means that the coding and processing of different resting-state activity levels in terms of difference-based coding involves the transformation of the brain's neuronal state into mental states even in the resting state. Thus difference-based coding may also account for neuronal–mental transformation in the resting state during rest–rest interaction, and thus for the occurrence of mental states (as defined previously) such as dreams, hallucinations, etc., even in the absence of exteroceptive stimuli (see Chapters 8, 11, and 12).

"Stimulus–object transformation" and primary and secondary processes

At this point the hardcore psychoanalyst may want to object. All I have shown is that difference-based coding enables and predisposes to the occurrence of mental states in the resting state, while I did not demonstrate why we experience mental objects rather than simple stimuli. So long as stimulus–object transformation remains unclear, difference-based coding cannot adequately account for the unconscious and thus for primary processes. How can we account for stimulus–object transformation? As described in Chapter 4, there is not only rest–stimulus interaction but also stimulus–rest interaction that describes that the brain's resting-state activity, and its subsequent structure and organization are shaped by the stimuli. Metaphorically speaking, due to stimulus–rest interaction, the stimuli are ingrained in the brain's resting-state activity (as for example is psychologically most apparent in autobiographical memories).

Coding resting-state activity differences in terms of differences, the different stimuli (e.g. neural and also previous intero- and exteroceptive stimuli; see Part IV of this book) associated with the brain's resting-state activity are related and linked to each other, thereby transforming them from mere stimuli to objects (i.e. stimulus–object transformation). Therefore difference-based coding of resting-state activity differences and thus of rest–rest interaction may enable and predispose not only to neuronal–mental transformation but also to stimulus–object transformation. Together, neuronal–mental transformation and stimulus–object transformation can easily account for the occurrence of objects in mental states such as dreams and hallucinations that occur in the absence of any exteroceptive stimuli.

What are the implications of this for the distinction between primary and secondary processes? The only empirical (i.e. neuronal) difference between primary and secondary processes is the presence or absence of exteroceptive stimuli.[5] In the case of primary processes, difference-based coding applies to the intrinsic resting state itself, and it is what I later call "rest–rest interaction," while exteroceptive stimuli remain absent. In secondary processes, difference-based coding applies to the interaction between intrinsic resting-state activity and stimulus-induced activity as related to intero- and exteroceptive stimuli. In both cases mental states as well as objects in those mental states are constituted, although they may be conceptualized in different ways. In the absence of exteroceptive stimuli, objects may be characterized as internally generated, whereas in the presence of exteroceptive stimuli, objects may be deemed to be externally generated.

Carhart-Harris and Friston (2010) may want to object that their distinction between primary and secondary processes resurfaces in my distinction between internally and externally generated objects in mental states. They are correct in asserting this. However, they are wrong to associate a primary difference between internally and externally generated objects. As I have demonstrated, the same principle (i.e. difference-based coding) underlies the constitution of both internally and externally generated objects in our mental states. My focus here is the constitution of objects in mental states as primarily distinguished from mere physical stimuli, involving what

5 Empirically speaking this is not entirely true. Even in the resting state, extrinsic stimuli still penetrate the brain (e.g. interoceptive stimuli from the body and exteroceptive stimuli from the sensory organs that cannot be completely shut down). This pertains to the definition of the resting state, which is beyond the scope of this book (for details, see Northoff et al., 2010). I here assume the resting state in an operational definition to be the absence of any specifically employed extrinsic stimuli (as, for instance, during sleep).

I described as neuronal–mental and stimulus–object transformation. Both neuronal–mental and stimulus–object transformation are assumed to enable and predispose to the constitution of objects and mental states including their differentiation from both single stimuli and neuronal states.

Thus it is clear that Carhart-Harris and Friston are answering a different question to the one posed by me. They ask whether it is possible to differentiate between primary and secondary processes, and thus between internally and externally generated objects in our mental states. Thus they take for granted the constitution of both mental states and objects on the basis of the brain's neuronal processing, whereas I in contrast address exactly what they take for granted, namely the constitution of mental states and objects, and relate it to specific neuronal mechanisms in the brain when discussing neuronal–mental and stimulus–object transformation.

Although the questions are complementary, their answers are not. As soon as Carhart-Harris and Friston start to project the difference between primary and secondary processes upon the brain itself and its different regions, they implicitly deny the common ground underlying both processes, including internally and externally generated objects in our mental states. By denying that common ground and assuming that different mechanisms underlie primary and secondary processes, they implicitly reject my answer, namely that both primary and secondary processes are traced back to difference-based coding as the common neuronal mechanism that underlies both internally and externally generated objects in our mental states. However, this makes their answer incompatible with the one suggested here.

Unlike Carhart-Harris and Friston, I here refrain from projecting the difference between primary and secondary processes (i.e. internally and externally generated objects in mental states) onto the brain itself. Instead, I look for those neuronal mechanisms that may enable and predispose to neuronal–mental transformation and stimulus–object transformation. I delineated a neural mechanism (i.e. difference-based coding) that is supposed to enable and predispose to both neuronal–mental and stimulus–object transformation. And I showed that the same principles (i.e. difference-based coding and rest–stimulus/stimulus–rest interaction) that enable the constitution of externally generated objects (i.e. secondary processes) also apply to internally generated objects in our mental states, which Freud associated with primary processes. This makes any projection of the difference between primary and secondary processes onto the brain and its different regions superfluous. In contrast to Carhart-Harris and Friston (2010), this allows us to explain the generation and constitution of both primary and secondary mental processes, including their differentiation by internally generated objects (i.e. primary processes) and externally generated objects (i.e. secondary processes), by referring to one and the same underlying neuronal mechanism.

What are the implications of this for the concept of cathexis? As described in the previous chapter, Macdiarmid (1989) distinguishes between two basic objects of cathexis, namely the self (amounting to what he calls "self-cathexis") and the other person (leading to what he describes as "other-cathexis"). In the case of "self-cathexis," energy is predominantly invested in the self and the ego, whereas in the case of "other-cathexis," investment of energy in the other person predominates (see Macdiarmid, 1989, p. 850). As conceptually indicated by the term "cathexis," both must be regarded as specific instances of the more general process of cathecting energy which may be invested in either the self or the other. This is mirrored in the energy that the brain invests via its resting-state activity in rest–stimulus interaction as corresponding to other-cathexis. Alternatively, the brain may invest its energy in stimulus–rest and rest–rest interaction which, in the psychological context, may correspond to self-cathexis.

The distinction between self- and other-cathexis is here traced back to the same underlying process, namely the investment of energy as the brain's resting-state activity in its interaction with either resting or stimulus-induced activity. This prevents us from making the same mistake as

Carhart-Harris and Friston (and nowadays many others), who project the difference between self- and other-cathexis onto the brain itself. This allowed us to address the question that Carhart-Harris and Friston (2010) left open, namely how mental objects in both primary and secondary processes can be constituted on the basis of the brain's neuronal stimuli (i.e. neuronal–mental and stimulus–object transformation).

Let me conclude this section with a brief comment about methodology. So long as we project the observer-related conceptual distinction between internally and externally generated (whether in its psychological version as internal and external attention, its psychodynamic version as self- and other-cathexis or as ego- and object-libido, or its neural version as internally and externally operating networks; Boly et al., 2008a,b) onto the brain itself, we shall not be able to state what kind of neuronal mechanism enables and predisposes to neuronal–mental and stimulus–object transformation. This is because by forcing and imposing our observer-related internal–external framework upon the brain itself (as, for example, is very clear in Carhart-Harris and Friston, 2010), we do not give the brain any space in which to, metaphorically speaking, breathe and reveal itself independent of the concepts that we as observers employ in order to understand it. To put this in more philosophical terms, so long as we give in to the tendency to project our observer-related conceptual distinctions onto the brain itself, we shall be unable to differentiate between the brain as functioning and the brain as observed (for a description of this distinction, see Part I) both empirically and conceptually.

Constitutive context dependence and operative intentionality

As described in the initial sections of this chapter, the concept of operative intentionality describes the intrinsic linkage and connection—the single nexus—between the organism and the world, and between the intrapsychic and the interpersonal dimension. Transferred into the neural context of the brain, this raises the question of how the brain's input can be related to the world's input in such a way that both become intrinsically connected and linked.

How can the world's input be related to the brain's input? The world's input includes not only the actual stimulus itself, but also its respective environmental context. This is supported by empirical examples, such as that of the visual cortex. It can be shown (Shulman et al., 1997a,b; Somers et al., 1999; Gilbert et al., 2000; Hopfinger et al., 2000; Macaluso et al., 2000; Smith et al., 2000) that activation in the primary visual cortex (i.e. V1) is context dependent and thus dependent on the respective environmental situation. Activation in the so-called receptive fields depends not only on local stimuli inside the respective receptive field, but also on stimuli that are outside the receptive field.

The "contextual influence" is neurally modulated by long-range horizontal (i.e. lateral) connections that are formed by the axons of cortical pyramidal cells, which link cells with widely separated receptive fields. These horizontal connections allow for global response properties of local receptive fields. Events outside the receptive fields are thus indirectly represented within the receptive field itself. These horizontal connections intrinsic to V1 are modulated by feedback connections from the prefrontal cortical areas, which may either facilitate or block the generation of neural activity. Consequently, there is feedback or "top-down modulation" of neural activity in V1 (i.e. in particular receptive fields).

However, this modulation of neural activity in V1 is indirect—that is, via the modulation of horizontal connections (i.e. of "contextual influence"). This implies that perception is intrinsic to the respective context: "Perception always occurs within a context. Kenet and colleagues' findings show that, to fully understand how our brains allow us to perceive the world in real time, we must investigate the context in which incoming visual signals are received and that context interacts with the signals to produce a behavioral response" (Ringach, 2003, p. 913).

The recently emerged discipline of neuroeconomics (Fehr and Camerer, 2007; Montague, 2007; Krueger et al., 2008; Rangel et al., 2008) presents particularly striking examples of how social contexts determine and constitute neural activity during reward. A recent fMRI study (Fliessbach et al., 2007) demonstrated that the activity in reward circuitry was highest when the person in the scanner received $30 in a gambling task and knew that another fictive player got less (i.e. $10). However, the activity in reward circuitry decreased when the fictive player received $60, although the person in the scanner still received the same amount (i.e. $30). Thus neural activity in reward circuitry is determined not so much by the actual stimulus (the $30) as by its relationship to the stimulus that another person receives at the same time (either $10 or $60). What is coded in the neural activity of reward circuitry is not so much the absolute amount of the actual stimulus that the person in the scanner receives, but rather its difference and thus its relationship to the other person's stimulus. Thus it is the social relationship, the difference between the self and the other person, that determines and codes neural activity changes in reward circuitry.

Another example is the so-called "endowment effect." This describes the tendency to attach greater value to items that one owns than to those that one does not. Knutson et al. (2008a,b) showed that the medial prefrontal cortex exhibits greater changes in neural activity in response to low prices during buying than during selling. This indicates that this region's neural activity changes are sensitive to the difference between selling and buying with regard to the same item, thus mirroring the endowment effect. Moreover, it was found that neural activity changes in the insula predicted differences in the individual subjects with regard to the endowment effect during selling (i.e. whether or not they owned the item). Taking these findings together, the endowment effect demonstrates that the value of a product is not determined by the absolute value of the product, independent of whether one owns it or not, but rather by the relationship of that product to the person concerned (i.e. whether they own it or not). Thus there is what is called "reference-dependent neural activity" that constitutes the value and salience of a stimulus depending on its relationship to the person (i.e. ownership) (for similar findings of reference-dependent neural activity in the amygdala and the orbitofrontal cortex in the endowment effect, see Fliessbach et al., 2007).

Although the contents differ in the various examples, being either financial–economic or visual–perceptual, they all share the context dependence. More specifically, what these examples have in common is that the relative difference between different stimuli and their respective contents determines and codes the respective neural activity changes rather than the particular stimulus or content by itself independent of the others. All of these examples thus show that neural activity changes are organized and coded by differences between different stimuli, be they mental, social, or personal stimuli. What seems to be crucial for determining and organizing neural activity changes is not so much the content of a particular stimulus as its relationship to other stimuli (i.e. the difference).

By coding the difference between the actual stimulus and other stimuli (be they social, personal, or mental), the brain can account for the respective context of the actual stimulus, which involves intrinsic linkage between stimulus and context (i.e. stimulus–context relationship). This intrinsic relationship between context and stimulus is only possible if the stimulus is coded and processed in relation to its respective context (e.g. social, personal, or neuronal context). This amounts to what I call "constitutive context dependence." Constitutive context dependence describes the necessary inclusion of the context in the neural processing and coding of stimuli, which means that they cannot be coded and processed independent of their respective context by the brain's neural activity.

What is meant and denoted by the two concepts, "context" and "dependence"? Let us start with "dependence," while the concept of "context" will be discussed in the next section. I take the concept of "dependence" to mean that neural activity is not merely modulated but also constituted and determined, and it is this that I define as constitutive dependence, as distinct from mere modulatory dependence. In the case of modulatory dependence, the fruit juice, the $30, or the product would induce some activity in reward circuitry by itself independent of its mental, social, and personal context. This would be compatible with what I call stimulus-based coding, where the single stimulus itself (i.e. its absolute value) determines and codes neural activity. In this case, the context could have at best some modulatory impact, thereby only modulating but not generating and constituting the stimulus-induced changes in neural activity. Neural activity changes in, for example, reward circuitry must then be assumed to consist of two distinct parts, one of which is context independent while the other is context dependent.

Such doubling of neural activity in two distinct parts is denied in the case of constitutive dependence. In this case, neural activity changes in reward circuitry are supposed to be adequately and exclusively determined by the relationship of the actual stimulus to other stimuli in the respective context (i.e. the social, mental, or personal context). What is of primary importance here is the relationship (i.e. the relative difference) between the different stimuli, thereby including the stimulus–context relationship as the prime determinant of constituting (not only modulating) neural activity. This precludes that neural activity is completely and exclusively determined by the single stimulus, itself independent of its respective context and thus its relation to other stimuli. Thus neural activity in, for example, reward circuitry no longer includes two parts (i.e. context-independent and context-dependent parts), but only one (i.e. a context-dependent part as denoted by the stimulus–context relationship).

How is such constitutive context dependence of the brain related to the psychodynamic concept of cathexis and its characterization by operative intentionality? Operative intentionality describes the intrinsic linkage and connection between intrapsychic and interpersonal dimensions and thus between the organism and the world. Analogous to intentionality, the brain's neural activity during the processing and coding of extrinsic stimuli appears to be determined by the relationship of the stimuli to their respective context (i.e. the stimulus–context relationship). This intrinsic stimulus–context relationship within the neural context resurfaces as the nexus between the intrapsychic and interpersonal dimensions within the psychological context. One may consequently assume that the constitutive context dependence within the brain's neural context may enable and predispose to what Freud, within the psychological context, described as the intrinsic linkage between the intrapsychic and interpersonal dimensions.

Since Freud implicitly assumed the concept of operative intentionality in the aforementioned way to describe the nexus between the intrapsychic and interpersonal dimensions, the constitutive context dependence of the brain's difference-based coding may be considered an enabling and predisposing and thus necessary empirical condition of possible operative intentionality. However, it should be mentioned that this concerns only the concept of operative intentionality in the sensory–affective, dynamic, pre-mental, and pre-cognitive sense. It may be necessary to discuss separately how much of that also applies to the traditional philosophical concept of intentionality as cognitive, static, and mental. However, since Freud and other psychoanalysts have assumed such a revised concept of intentionality (i.e. operative intentionality), rather than the traditional one, I shall not provide an account of the latter here (see Figure 5.2).

Constitutive context dependence and embeddedness as silent presuppositions in Freud

What does the concept of "context" mean? I have already demonstrated different contexts, namely mental context in the anticipation of reward, social context in the case of the fictive

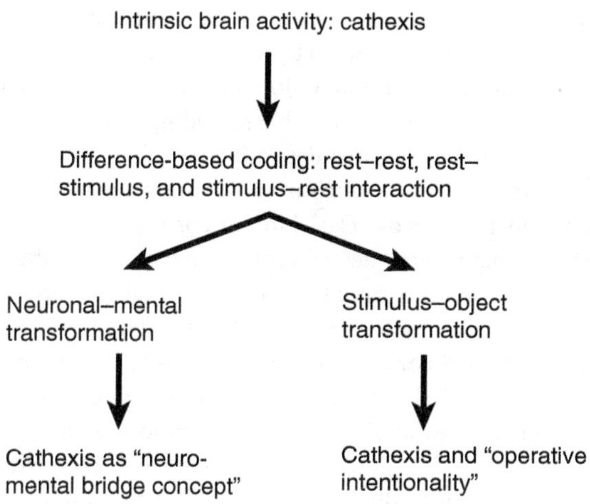

Fig. 5.2 Cathexis, mental states, and intentionality in the neuronal context of the brain.

player, and personal context in the case of the endowment effect. Thus I use the concept of context in a wide and non-specific way that remains independent of a specific domain (genetic, neural, mental, personal, social, or otherwise). Such a wide domain-independent concept of "context" is also assumed by Mareschal et al. (2007, p. 91ff) who, in a similar way, consider the concept of context in a wide sense, including genetic, cellular, regional–neural, bodily, and social contexts. This serves as the starting point of my argument for what I call the "domain independence" of constitutive context dependence.

Mareschal et al. (2007) showed impressively that the context is crucial even at the cellular level in that the single cell's activity is determined and coded in relation to other cells, whether they are close or distant (Mareschal et al., 2007, pp. 29–47). By coding neural activity in relation to the respective context, whether it is cellular, genetic, regional, bodily, social, or mental, the respective contextual stimuli are constitutively integrated in the neural activity changes that are induced by the actual stimulus. This is what Mareschal et al. (2007) call "encellment" (i.e. the integration of other cells in the activity of one particular cell) (Mareschal et al., 2007, pp. 29–47), "embrainment" (i.e. the integration of particular brain regions in the activity of another region) (Mareschal et al., 2007, pp. 49–68), "embodiment" (i.e. the integration of one's own and other bodies in the brain's neural activity) (Mareschal et al., 2007, pp. 69–90), and "ensocialment" (i.e. the integration of the social environment in the activity of the brain and the body (Mareschal et al., 2007, chapter 10).

In other words, it is the cellular, genetic, neuronal, bodily, and social contexts that exclusively and entirely determine neural activity in the brain, and this is what I mean by constitutional context dependence as distinct from mere modulatory context dependence. As Mareschal et al. (2007) have also suggested, constitutional context dependence presupposes a wide and domain-independent concept of "context," which I call "domain-independent context dependence," that is closely linked to constitutional context dependence. This must be distinguished from a narrow and domain-dependent concept of context where it is restricted to one particular domain, such as the bodily and/or social domain, which I call "domain-dependent context dependence," that often accompanies the assumption of modulatory context dependence.

How can such domain-independent context dependence be neurally realized? This is only possible if, metaphorically speaking, the different stimuli (visual, social, personal, etc.) "speak the same language or use the same currency," thus presupposing the same format or code. Despite their differences in content, the different stimuli and their respective contexts should be coded and organized in orientation on the same measure or standard using the same format. I assume that this format is the relative difference between the different stimuli (i.e. difference-based coding) that allows the coding of stimuli across different domains (personal, social, mental, physical, etc.) in relation to each other. If this is so, difference-based coding may enable and predispose to the domain independence of constitutive context dependence.

What does the domain independence of constitutive context dependence imply at a conceptual level? It means that we are always already related to our context, regardless of the exact gestalt of that context. This relates to what is often described as the concept of embeddedness. The concept of embeddedness describes the necessary and thus constitutional dependence of our mental states on our relationship to the environment (Northoff, 2004a,b; Thompson, 2007). The concept of "embeddedness" must be specified conceptually with regard to its relationship to embodiment. In the current literature (Clark, 2003; Thomspon, 2007), the concept of embeddedness is most often considered to be a mere extension of the concept of embodiment, with the latter supposedly involving the former (see Northoff, 2004a,b as well as Northoff, 2008 with regard to emotional feelings). By linking the brain to the body and its sensorimotor action, the brain is granted access to the environment and may be able to interact with it. Embodiment then involves embeddedness (which is sometimes also called "enactment") (Thompson, 2007).

However, this is different in the concept of embeddedness that is assumed here. Difference-based coding is considered to be a necessary neural condition of the possible predisposition of embeddedness and thus, to use phenomenological terms, of our interaction with the world. This predisposes us to interact with the world, which may occur via the body (e.g. via the libido or our drives). If so, the concept of embeddedness, or at least the predisposition to it, may be considered primary, and embodiment, being an extension of the former, as secondary. Embeddedness would then involve embodiment, not vice versa as in current phenomenological approaches (see, for example, Thompson, 2007).

Although one may argue that this remains a purely conceptual detail, it pertains to different presuppositions (i.e. different problems or questions) that are addressed by the respective concepts of embeddedness. The concept of embeddedness as a mere extension of embodiment addresses the question of the tools or means of our interaction with the environment, which must necessarily be bodily and thus sensorimotor. As such it targets the contents of that interaction and may thus be designated as "content problem." In contrast, the (here presupposed) concept of embeddedness as primary predisposition when compared with embodiment focuses on the issue of how the brain's input can interact with and relate to the world's input and thus what I call "relation problem." The relation problem describes the problem of how the brain's input and the world's input can be linked and related to each other in such a way that a mental state as distinct from a physical state can be constituted. This in turn paves the way for the subsequent question, which concerns the tools or means of such relationship or interaction and thus the "content problem."[6] Therefore the "relation problem" may be considered more fundamental and basic than the "content problem," which as such can be raised only on the grounds and basis of our intrinsic relationship to the world (i.e. embeddedness).

What are the implications of the concept of embeddedness for the neuropsychodynamic context? Freud must have silently presupposed embeddedness when focusing on development,

[6] Since these are more conceptual issues, I shall not pursue them here. However, I am fully aware that the two different concepts of embeddedness that have been suggested here require more detailed discussion at conceptual, epistemic, and empirical levels. This can only be done within a (neuro)philosophical framework, not the neuropsychodynamic one that is the focus of this book.

including his assumption of a specific set of mechanisms, often referred to as defense mechanisms (for details, see Chapter 6). Defense mechanisms describe different forms of interaction of the psychic apparatus with the world in order to cope with the intrusions of the latter into the former. Such mechanisms of action as well as his developmental considerations strongly suggest that Freud must have silently presupposed embeddedness in the sense described here. Since such embeddedness builds empirically on what I have described as domain-independent constitutive context dependence, Freud must have silently and implicitly presupposed both assumptions. Only on the basis of both embeddedness and constitutive context dependence can his developmental considerations and the concept of defense mechanisms be accounted for conceptually and empirically. This is of not only conceptual but also empirical importance, because it can help us to find the respective enabling and predisposing neuronal mechanisms within the brain's neural context, and thus to establish the link between neuroscientific mechanisms and psychodynamic concepts.

Difference-based coding and Freud's *Project for a Scientific Psychology*

Freud's 1895 writing, *Project for a Scientific Psychology*, has aroused much controversy and subsequently been interpreted in various ways (for recent overviews, see Brook, 1998; Levin, 2003; Peled, 2008). It is not my intention to follow any of these interpretations here, nor do I aim to provide a point-by-point translation of the concepts used by Freud in the *Project* into neuroscientifically amenable concepts, or to discuss the reasons for the change in Freud's focus from neuroscience to psychology and metapsychology after 1895. I leave open the question of whether the move from the brain to the psychic apparatus and thus from neuroscience to metapsychology is a major step from brain to mind involving major and primary differences, or whether Freud's development after 1895 must be considered a rather small step, since he remained within the initial mechanistic framework, which he now only applied to the psychic apparatus rather than to the brain (Brook, 1998).

My aim is much more modest. I shall focus on some concepts and issues that were raised by Freud, and put them into the neuroscientific context presupposed here, thereby linking them to my hypothesis of difference-based coding and the brain's resting-state activity. I shall therefore focus on three issues that Freud raised in the *Project for a Scientific Psychology*, namely the idea of endogenous activity, the necessity of object relations, and the crucial role of inhibition.

Freud raises the issue of endogenous activity when he distinguishes a class of neurons that receive endogenous input from neurons that receive either exogenous input or qualitative information. How can we relate this assumption of the prevalence of endogenous activity to the brain? As already illustrated, the brain itself can be characterized by intrinsic activity and thus by endogenous activity. Even more specifically, there does indeed appear to be a core system, namely the default-mode network (DMN), which seems to have a special although as yet unclear role in providing resting-state activity when compared with other regions.

This is not to say that the other regions of the brain (e.g. sensory regions) do not show any endogenous activity. In fact regions such as the auditory cortex also show high resting-state activity with neuronal oscillations and fluctuations (Woodruff et al., 2006), which may, for example, be especially relevant in the generation of auditory hallucinations in schizophrenia (see Part IV of this book). One may draw a parallel between Freud's distinction between neurons with exogenous and endogenous inputs and the anatomical distinction between sensory regions and the DMN. However, such one-to-one mapping between the different concepts may run into difficulties because of the different contexts, as Freud presupposed a dynamic and functional context that has to be distinguished from the current anatomical–regional context, which is rather static.

Another issue that Freud raises in the *Project for a Scientific Psychology* is the necessity for object relationships such as the mother–infant relationship in the constitution and development of the psychic apparatus. He argues that the mother–infant relationship constitutes and provides a template for structuring and organizing the psychic apparatus, which he associates with different neurons in the different systems (this aspect has been emphasized by Schmidt-Hellerau, 1995, pp. 1180–1182). How can we put such need for object relationships in the current neuroscientific context?

I demonstrated that the brain's neural structure and organization are strongly dependent neuronally on stimulus–rest interaction, with the stimuli shaping the brain's resting-state activity, which may then be carried forward even into rest–rest interaction. This indicates that the neural processing of extrinsic stimuli is essential for the brain to develop (i.e. structure and organize its own intrinsic activity), and it is such structuring and organizing of the brain's neural activity that enables and predisposes to difference-based coding and thus neuronal–mental and stimulus–object transformation. Therefore what Freud described as the need for object relations within the psychological context may correspond within the neural context to the dependence of the brain's neural structure and organization on the brain's stimulus processing in the gestalt of stimulus–rest interaction.

However, the relationship between neural and psychological contexts goes beyond mere correspondence. I postulated that the brain's resting-state activity in conjunction with difference-based coding enables and predisposes to and is thus a necessary empirical condition of possible neuronal–mental and stimulus–object transformation. Thus the brain's need to process extrinsic stimuli in order to structure and organize its own intrinsic activity may be crucial for enabling and predisposing to what Freud described so nicely as the necessity of object relations within the psychodynamic context. This would need to be filled by empirical details from neurodevelopmental research within a psychodynamic context. However, such details are beyond the scope of this book.

Neural inhibition and Freud's *Project for a Scientific Psychology*

Another issue that Freud emphasizes in his *Project for a Scientific Psychology* is the crucial role of inhibition, as illustrated by his famous drawings. I do not intend to go into detail about Freud's concept of inhibition in this early writing, but I do want to make a conceptual comment and then show the possibly crucial relevance of neuronal inhibition to rest–stimulus interaction and thus to difference-based coding.

Let me start with the conceptual comment. One difficulty that one encounters when reading Freud's *Project for a Scientific Psychology* is the constant switching between different contexts, from the neuronal to the psychological to the behavioral context, and back. This makes it difficult to determine the precise meaning of concepts such as inhibition that are used in several different contexts. The first context is the neuronal one, with Freud describing the inhibition of neurons as being crucial for neuronal processing and transmission of stimuli from one region to another. This presupposes the use of the concept of inhibition in a neuronal sense.

However, one may want to make a further distinction here. Neuronal inhibition can be understood in a cellular context as the inhibition of one neuron by another, as was apparently presupposed by Freud himself. Alternatively, neuronal inhibition can be understood in a regional context (e.g. inhibition of a whole region which contains a multitude of interacting neurons), and if one wanted to one could distinguish many more additional contexts of neural inhibition. Such a neuronal context of inhibition must be distinguished from a psychological context, where inhibition pertains to the suppression of specific psychological functions and their respective contents. Such a psychological context may, for example, be most apparent when Freud characterizes the defense mechanism of repression by suppression of specific psychological contents (for details, see Chapter 6). Finally, one may also consider the concept of inhibition in a behavioral context, as for instance Freud does when he assumes that moral behavior is associated with the superego, to suppress sexual behavior as related to the libidinal drives, the id.

I here want to focus on inhibition in the neuronal context, and more specifically in the regional (rather than cellular) sense. If there is indeed inhibition in a regional context, one would expect inhibition to play a central part in rest–stimulus interaction, as the latter involves different regions and their interaction, as has been demonstrated previously. A couple of recent findings suggest that the degree of resting-state neuronal inhibition as mediated by specific neurons (i.e. GABA-ergic neurons) may be crucial at a regional level in determining the neuronal activity of that particular region induced by specific external stimuli. Let me describe briefly some recent empirical findings.

Using combined magnetic resonance spectroscopy (MRS) and fMRI, a recent study by our research group (Northoff et al., 2007b) investigated the level of gamma-aminobutyric acid (GABA) in a typical DMN region, namely the perigenual anterior cingulate cortex (PACC), which as part of the DMN shows a predominant negative blood oxygen level dependent (BOLD) response (NBR). The resting-state level of GABA in the PACC correlated with the degree of NBR as induced by an emotional judgment task in exactly the same region. It was found that the higher the resting-state concentration of GABA in the PACC, the higher the degree of NBR during stimulus-induced activity. This study demonstrated that the resting-state concentration of GABA in the PACC may indeed influence stimulus-induced activity changes in the PACC. This suggests that the resting-state activity level (i.e. GABA) in a DMN region appears to influence the degree of stimulus-induced activity in the same region.

Another study in healthy subjects investigated the resting-state concentration of GABA in the visual cortex and its effects on subsequent stimulus-induced activity in the visual cortex itself and gamma frequency bands. Muthukumaraswamy et al. (2009) measured resting-state levels of GABA in the visual cortex with MRS, and employed fMRI and magnetoencephalography (MEG) to measure stimulus-induced activity changes in the visual cortex. The resting-state concentration of GABA in the visual cortex predicted the degree of stimulus-induced activation (i.e. positive BOLD response) and the gamma frequency in the same region.

Despite focusing on different regions (i.e. the PACC and the visual cortex), both studies observed prediction of stimulus-induced activity by the resting-state level of GABA. In contrast, resting-state levels of glutamate (an excitatory transmitter) were not related to rest–stimulus interaction. This suggests that the resting-state concentration of GABA as the major inhibitory transmitter may have a crucial role in linking resting-state activity to stimulus-induced activity, thereby making possible what I here describe as rest–stimulus interaction (for empirical details, see Northoff et al., 2010).

Taken together, these findings suggest that GABA-ergic mediated neuronal inhibition may be crucial for enabling and predisposing to the linkage of the resting-state activity of a particular region to its activity changes during stimulation by an external stimulus. However, although these findings do suggest some relationship between resting-state GABA-ergic mediated neuronal inhibition and stimulus-induced activity, their exact relationship remains far from clear. Moreover, it is even more unclear exactly how neuronal inhibition and GABA mediate the relationship between resting-state and stimulus-induced activity within the same region (i.e. rest–stimulus interaction). For instance, if one assumes that difference-based coding may be crucial for mediating such rest–stimulus interaction, the coding of stimuli in terms of differences (i.e. difference-based coding) must be crucially linked to GABA and thus to neuronal inhibition. However, this has yet to be demonstrated.

How is neural inhibition in a regional sense related to the concept of inhibition within the psychological context? Freud himself aimed to address the transition from neural to psychological contexts when he associated specific classes of neurons with inhibition of specific psychological functions. However, he himself considered his attempt to have failed, with the transition from the neural to the psychological context remaining rather blurred. Are we more advanced in our knowledge these days? While there is certainly much more detailed knowledge about both neuronal and psychological contexts than in Freud's time, the mechanisms underlying the transition from the neuronal to the psychological context still elude us. We still do not know why and how neuronal states are transformed into psychological and ultimately mental states.

My account in this and the previous chapter has focused on exactly those neuronal mechanisms that may prove essential for enabling and predisposing to such a transition from the neuronal to the psychological context. More specifically, I postulated that rest–stimulus interaction and difference-based coding are necessary empirical conditions of possible neuronal–mental and stimulus–object transformation. If inhibition now plays a major role in this transition, one would expect it to be centrally involved in both rest–stimulus interaction and difference-based coding. Although I could gather some empirical support for the view that inhibition is involved in rest–stimulus interaction, no such data were available in the case of difference-based coding. However, even with regard to rest–stimulus interaction, caution is needed. The data support the involvement of GABA and possibly neural inhibition in the neuronal mechanisms of rest–stimulus interaction, whereas they do not show the involvement of GABA in the transition from neuronal to psychological and thus mental states. For that we would need empirical data to demonstrate that GABA mediates not only rest–stimulus interaction but also the respective associated mental states. In other words, we would need to show that GABA is a necessary empirical condition of difference-based coding and thus of possible neuronal–mental transformation. Even then the careful neuroscientist may want to argue that GABA by itself does not necessarily indicate inhibition. Although GABA is clearly inhibitory in the cellular context of individual neurons, the net effect of GABA-ergic neurons at a regional level, as measured in brain imaging, may correspond to excitation rather than inhibition, whether this is understood in a neuronal–regional or even psychological sense.

This brief discussion has shown that we are still a long way from demonstrating a possible central role of inhibition in mediating the transition from neuronal to psychological and thus mental states. However, it is also clear that we need to address this transition in both conceptual and empirical terms in order to avoid the haziness that Freud himself diagnosed in his *Project for a Scientific Psychology*, which led him to abandon the whole pursuit of a brain-based psychoanalysis. This makes it clear what is at stake for our enterprise of neuropsychoanalysis. So long as we do not carefully address the transitions themselves (i.e. the transition from neuronal to psychological and mental states), our attempts to establish neuropsychoanalysis as a discipline that stands on its own two feet may be futile in the same way that Freud's attempts in his *Project for a Scientific Psychology* turned out to be highly problematic. Furthermore, I believe that we cannot afford to fail and give up our current neuropsychoanalytic endeavors as Freud did, and then wait for another hundred years before again tackling the issue of a brain-based psychoanalysis.

Difference-based coding and Solms' concept of "dynamic localization"

Mark Solms considered Freud's suggestion in his *Project* (Freud, 1895) to be merely metaphorical, which, according to him, "does not provide a bridge between psychoanalysis and neuroscience: It is therefore not the correct place to seek an understanding of the nature of the relationship between psychoanalysis and neuroscience" (Solms and Saling, 1986, p. 401). Instead, Solms looks to an earlier writing of Freud, on aphasia, entitled *Zur Auffassung der Aphasien* (Freud, 1891): "'On aphasia,'" and not the Project, is the 'missing link' between Freud's neurological and his psychoanalytic years. It is to the monograph on aphasia that one must return if one wishes to define the relationship between neuroscience and psychoanalysis. And it is from there that one must begin if one wishes to understand how psychoanalysis and neuroscience might collaborate in the future" (Solms and Saling, 1986, p. 407).

In his early writing on aphasia, Freud (1891) rejects the predominant localization of mental and psychological functions in specific brain regions (i.e. circumscribed cortical regions), as was the predominant view of his time. He argues that mental phenomena must be properly analyzed and that their complexity makes their static localization in a discrete cortical region impossible. More specifically, he claims that complex phenomena such as speech functions require proper psychological analysis and a dynamic rather than static psychological theory. Solms subsequently argues that Freud's early writing on aphasia shows that the neuroscience at his time "had little to offer to Freud," and that this writing "laid the dynamic foundations for his later psychopathology and the whole of psychoanalysis" (Solms and Saling, 1986, p. 407).

Although Freud rejected any kind of static localization of psychological and mental functions in discrete regions of the brain, he did not offer any kind of dynamic account of the brain and thus a theory of dynamic (rather than static) localization. Following Solms, this was up to the Russian neuropsychologist A. R. Lurija (1973) who, on the basis of work with lesion patients, developed a theory of how complex psychological functions must be linked to various regions of the brain and thus be localized in a dynamic rather than static way (Solms and Saling, 1986; Solms, 1998).

Instead of localizing a psychological function in one specific brain region, he associated the former with so-called functional systems which include various regions and their interconnections that are recruited according to the needs of the respective psychological function. The complexity and inner structure of the psychological and mental functions are supposed to correspond to the functional system recruited and the linkages (i.e. connections) between the different regions within that particular functional system. Mental functions are then supposed to be "localized" in the brain in a distributed and thus dynamic rather than regional and static way.

How does the concept of dynamic localization stand up to the hypothesis of difference-based coding as the brain's neural code, as described here? Let us consider the questions to which both dynamic localization and difference-based coding are supposed to provide the answers. We start with the question that Solms presupposes in his reading of Freud and his concept of dynamic localization. The latter concept presupposes the question of how the complex contents of psychological and mental states can be related to the brain and its different regions. This was, following Solms, the question that guided Freud in his 1891 writing on aphasia.

While rejecting a static and regional localization, Freud did not consider that there was any possibility of answer this question in his time, and he subsequently abandoned neuroscience and reverted to psychoanalysis. This is the point at which the concept of dynamic localization comes in. Following Solms' interpretation of Freud, it addresses the hitherto unanswerable question of how the complex contents of psychological and mental states can be related to the brain and its various regions. The "concept of dynamic localization" provides an account of how complex psychological contents can be related to and correspond to the brain and its various regions. Solms opts for a distributed and dynamic (as opposed to regional and static) localization of the complex psychological contents in the brain. One may thus characterize the concept of dynamic localization as content based. Solms argues that Freud was not able to link psychodynamic concepts to the brain because he was not yet exposed to Lurija's concept of dynamic localization at this time.

Dynamic localization answers the question of how different contents can be related to the brain and its different regions. This distinguishes dynamic localization from the question underlying and guiding my hypothesis of difference-based coding as the brain's neural code. Rather than explaining and accounting for the complex contents of psychological and mental states, it aims to address and answer the question of how the brain must organize and structure its neural activity in order to make possible the constitution of complex contents and their dynamic localization in the brain. Thus the question underlying my hypothesis of difference-based coding as the neural code focuses on a specific neural structure and organization, rather than the complexity of contents, and must therefore be considered organization based rather than content based.

Let me describe the relationship between Solms' concept of dynamic localization and my hypothesis of difference-based coding. The concept of dynamic localization raises the question of how we can (and cannot) relate the complex contents of psychological and mental states to the brain so that we can find corresponding contents (i.e. neural contents). Solms argues that we cannot find the corresponding neural contents in specific regions, which would involve static and regional localization, but rather across various regions, which would involve dynamic and distributed localization. Thus his aim is to search for the neural correlates of the complex psychological contents.

The hypothesis of difference-based coding that I have proposed here has a different focus. I raise the question of how the brain can dynamically distribute what we describe as complex psychological contents across various regions, rather than localizing (and representing) them in a specific region. Thus the hypothesis of difference-based coding raises the question of what enables and predisposes the brain to process psychological contents in a dynamic and distributed rather than static and localized way. The guiding question does not concern the relationship between contents and regions, but rather the brain's neural structure and organization. How must the brain be structured and organized in order to enable and predispose to the constitution of contents and their subsequent dynamic and distributed localization in the brain? As such, my account of difference-based coding targets the enabling and predisposing and thus the necessary conditions of possible dynamic and distributed (or some other form of) localization, rather than addressing the issue of localization itself, as for instance Solms has done.

Since the hypothesis of difference-based coding can be regarded as addressing the enabling and predisposing (i.e. necessary) empirical conditions of possible dynamic and distributed localization, both difference-based coding and dynamic localization may be considered compatible and complementary rather than mutually exclusive (as, for instance, dynamic/distributed and static/regional localization). One may even go one step further and argue that difference-based coding involves dynamic and distributed localization rather than static and regional localization. By coding the differences between different stimuli, difference-based coding makes the precise, distinct, and mutually exclusive localization of specific stimuli and their respective associated contents in one specific region impossible.

Such static and regional localization becomes even more problematic when considering the crucial role of the brain's resting-state activity in organizing and structuring stimulus-induced activity. Since in the resting state itself there are already various interactions between different brain regions (e.g. rest–rest interaction), the rest–stimulus interaction cannot be confined to one particular region. Even if the direct rest–stimulus interaction is limited to one particular region, there is always already some indirect impact on that region's resting-state activity level by that of another region, and so on. Thus rest–rest interaction and its direct or indirect impact on rest–stimulus interaction in either the same or another region precludes strict localization in a static and regional sense. One may therefore postulate that rest–rest interaction and rest–stimulus interaction enable and predispose to dynamic and distributed rather than static and regional localization.

This makes it even clearer that Solms' concept of dynamic localization does not contradict but in fact tends to support my hypothesis of rest–rest and rest–stimulus interaction and difference-based coding. Solms focuses on the neural processing of stimuli and their respective associated contents, whereas I focus on what the brain's neural structure and organization must be like in order to enable and predispose to the kind of neural processing and localization that Solms describes. Thus his stimulus-driven approach can be regarded as presupposing the kind of resting-state-driven approach that I have outlined here.

By outlining the effect of the resting state, I am also able to account for how the brain constitutes an intrinsic relationship to the world which in turn must be presupposed by Solms, since it is only on the basis of such a brain–world relationship that contents can be constituted. I consequently focus on the necessary conditions that enable and predispose to the kind of dynamic and distributed localization Solms assumes on the basis of Lurija. Going way beyond his 1891 writing on aphasia, Freud aimed to tap into this very foundation itself in his 1895 writing in both conceptual and empirical terms, writing without being able to provide proper empirical support and an appropriate conceptual framework. As I have argued in this and the previous chapter, we are now ready to do exactly that Freud had in mind 1895, namely to tap into the very foundation (i.e. the brain and its relationship to the world), and to show how it enables and predisposes to the constitution of those mental contents and objects which Freud later described so well in psychodynamic terms.

Chapter 6

Defense Mechanisms and Brain–Object and Brain–Self Differentiation

Defense mechanisms and internalization

The concept of defense mechanisms was introduced in Sigmund Freud's early work, where he regarded them as mental operations that keep painful thoughts and affects out of awareness. Freud considered repression to be the most significant defense mechanism that describes how painful thoughts and affects can be suppressed in order to prevent them from entering consciousness, and thus to remain in the unconscious. Anna Freud (1946) broadened the concept of defense mechanisms and described them as psychological mechanisms (e.g. the unconscious part of the ego) for coping with intrapsychic conflict in order to protect the ego and the self from psychological damage. More recently, the original psychodynamic concept of defense mechanisms has aroused the interest of empirical researchers who aim to associate them with specific psychological and/or neuronal mechanisms (Vaillant, 1992; Cramer, 2000; Northoff et al., 2007c; Feinberg, 2010).

What does the concept of a defense mechanism represent? This concept was originally intended to refer to psychological mechanisms that adapt and defend the structure and organization of the ego, or the self (using the two terms synonymously for the sake of operational simplicity; see Chapter 9 for details of their exact definition and differentiation), in the face of conflict arising from threat and danger. This means that defense mechanisms have what may be called an "adaptive function." At the same time, however, the concept of defense mechanisms may go beyond mere defense by contributing to the building and constructing, or re-building and re-constructing, of the structure and organization of the ego (the self). The adaptive function of defense mechanisms may then go along with what I call "constructive function." Thus the overall concept of defense mechanisms may be characterized by a certain ambiguity as it includes both adaptive and constructive functions. This ambiguity is especially apparent in what are known as the early and immature defense mechanisms, as will be discussed below.

Several authors distinguish early and immature defense mechanisms from later and more mature ones. Early and immature defense mechanisms include splitting, projective identification, dissociation, projection, denial, and fragmentation, while later and more mature defense mechanisms include intellectualization, rationalization, repression, displacement, identification, ego objectivization, isolation, and idealization (Vaillant, 1993; Cramer, 2000; Gabbard, 2005a,b; Northoff et al., 2007c; Feinberg, 2010). Both mature and immature defense mechanisms aim to preserve and maintain the boundaries of the self (i.e. the ego boundaries), thereby allowing for some kind of protection of the self from potentially damaging objects. This reflects what I earlier described as adaptive function. Early and immature defense mechanisms in particular are also involved in constructing and building the structure and organization of the ego (or self) in, for example, the early (and more immature) developmental stages in infancy and childhood. This pertains to the constitutive function of defense mechanisms, whereby they are crucial in constructing and constituting the developing ego (or self). More specifically, they help to construct the ego or self of the infant or child as distinct from the various objects that he or she encounters. As such, early and immature defense mechanisms may enable and predispose to what I called "self–object differentiation" in the Introduction. However, to better understand their role in self–object differentiation we need to characterize early and immature defense mechanisms in more detail, in terms of the more specific mechanisms of internalization and externalization.

Let us start with internalization. The concept of internalization describes how objects external to the self are related to the self and, in the most extreme scenario, become part of it. As such, internalization may contribute to constituting and constructing the self. Mentzos (1991) has characterized internalization as the "constitution of the self," thus nicely mirroring what I earlier called constitutive function. He assumes that such constitutive function is realized by three more specific mechanisms, namely incorporation, introjection, and identification, all of which can be subsumed under the umbrella of "internalization." These three mechanisms will be described in more detail below.

The first and most immature mechanism of internalization is "incorporation." Here objects are completely integrated into the self by becoming part of it, which means that the distinction between self and object is blurred, with the latter being transformed into part of the self and thus the self itself. This is, for example, the case in early infancy when the infant regards other people as part of its own self rather than as objects distinct from the self (i.e. "The world is me and I am the world," which mirrors primary narcissism; see Chapter 7 for further details) . Since objects become part of the self, there is as yet no distinction between self and objects, which implies the absence of what I earlier called "self–object differentiation."

The second and more mature mechanism of internalization is introjection. This describes the internalization of an object that is taken from the outside of the world to the inside of the self. However, in contrast to incorporation, the object remains distinct from the self, thus presupposing some degree of differentiation between self and object, although this remains blurred. Due to such partial integration of the object into the self, the introjected object is highly personally loaded with affects and ambivalence (see next section for details). More specifically, the introject (i.e. the result of introjection) is loaded with aggressive, libidinal, and narcissistic conflicts, which makes the associated object (i.e. the introjected object) affectively highly ambivalent, leading to high degrees of anxiety (especially in the context of rather "bad" and problematic objects).

Although the separate reality of the existence of the object as distinct from the self is acknowledged in the case of introjection, this acknowledgment remains highly ambivalent, and helps to transform the originally "objective object" into a "subjective object" as "transformation of something objective into something subjective" (Mentzos, 1991, p. 45). For instance, introjection in adulthood is typical in patients with depression, who often perceive other people's behavior as personal and subjective, and relate it to their own self even if there is no connection (i.e. resemblance) at all (see Chapter 10 for more details). For instance, the secretary who usually makes good coffee for her boss in the morning feels guilty and worthless if her boss does not like the coffee on one particular morning, although this is not related to the coffee at all (on the other mornings he likes it), but rather to his bad mood after having had a difficult weekend (for further details, see Chapter 10). Thus the secretary is unable to completely detach herself from her boss, who (due to previous experiences in her life) becomes the object of her introjection and thus an introjected object for her, including all of its affective ambiguity.

One characteristic of introjection is low tolerance of ambiguities of the object, which oscillates between the poles of the self and the object, thus being somehow distinct but not yet completely different from the self. This leads to decreased tolerance of conflicts and ambiguities, as is apparent in the case of the secretary in the example just described. The tolerance of ambivalence and conflict is much higher in the third form of internalization, namely identification, which describes a good and stable relationship of the self to the objects. The object is specifically related to the self, while its existence and reality as distinct from the self are acknowledged and much better tolerated when compared with introjection. A typical example here is the soccer fan who identifies himself with his team or a particular player while usually being able to acknowledge the separate existence of the object (i.e. the player), as distinct from his own self.

These descriptions of the three mechanisms, namely incorporation, introjection, and identification, clearly point to their specific role in constituting and establishing the distinction between self and objects. In the case of incorporation, objects are integrated or incorporated into the self without yet acknowledging their separate existence and reality. In introjection, however, objects are considered distinct from the self, although in an ambivalent and unstable way, with the objects remaining halfway between the extremes of total integration into the self (as in incorporation) and total detachment from the self (as in identification). Taking all three together, there is thus progressive differentiation of the self from mere objects, with this differentiation being lowest in incorporation and highest in identification. In other words, all three mechanisms of internalization enable and predispose to what I call self–object differentiation by differentiating the self from mere objects.

Externalization and the "co-occurrence and co-constitution of self and objects"

The concept of internalization targets the constitution of the self and its differentiation from objects. It thus presupposes the point of view of the self. However, one may also take the opposite point of view, the point of view of the object, and focus on how objects are differentiated from the self, thus approaching self–object differentiation from the point of view of the object. The point of view of the object targets the mechanisms that allow objects to be constituted and differentiated from the self. These mechanisms are subsumed under the concept of externalization. Externalization describes the constitution of objects and their differentiation from the self while at the same time maintaining and preserving their relation to the self. Thus externalization may be regarded the "mirror image" of internalization on the side of the objects. Following Mentzos (1991), externalization may therefore be characterized by three mechanisms, namely excorporation, projection, and self-objectivation, which must be considered analogous to those on the side of internalization (i.e. incorporation, introjection, and identification).

Excorporation describes how parts of the self are outsourced; they are alienated and detached completely from the self and are considered as pure object rather than as self. These outsourced and detached parts are no longer perceived and experienced as part of the self and its inner mental, but are rather attributed to objects in the environment, as for instance to other individuals. Excorporation can thus be regarded as the counterpart of incorporation. Whereas incorporation adds objects to the self in order to constitute the self and its inner mental world, excorporation detaches parts from the self and its inner mental world by transforming them into mere objects when shifting them from the self to, for example, other individuals in the respective environmental context. Thus excorporation may describe a process on the side of the objects that seems to be more or less analogous to that occurring on the side of the self (i.e. incorporation). Excorporation can, for instance, be observed in young infants when they deny any relationship of specific objects to their own self. For example, this may be the case if the father is violent, with the subsequent excorporation of him by the infant being an adaptive mechanism.

Projection describes a second form of externalization. Unlike excorporation, the separate existence and reality of the self and the objects are acknowledged. The self or parts of it are imposed upon and ingrained in objects that thus become related to the self, thereby constituting and constructing what I call "self–object relations" (or what in psychoanalysis is described as "object relations"; despite some subtle differences, for operational purposes I shall use both terms "object relations" and "self–object relations" in an identical way in the following account). Analogous to introjection, but in the opposite way, something subjective (i.e. the self or parts of it) is transformed into something objective (i.e. an object) (Mentzos, 1991, p. 45). Projection can, for example, be observed in psychotic patients when they attribute their own anxieties and conflicts to other individuals whom, in their paranoid delusions, they perceive to be and experience as dangerous (see Chapters 11 and 12 for details).

Analogous to introjection, as its counterpart on the side of internalization, projection can be characterized by a rather unstable self–object relation, since the self still remains part of the object and is thus not yet completely detached from it. In contrast to excorporation, the differentiation between self and objects is not complete in projection, with the object as the former part of the self oscillating between complete detachment from the self and complete integration into the self. As in the mirror case of introjection, there is strong affective ambiguity and conflict in the relation between self and object (the self–object relation) in projection. This is different in the third form of externalization, namely self-objectification, where the self (or parts of it) is converted into an object, thereby being completely detached from the self. Such detachment of the self from itself, with its subsequent change in role to an object, is for instance manifested in self-reflection, when one reflects about the self as if it is just another object in exactly the same way that one reflects about other objects (e.g. another person).

This short (and incomplete) description of the different mechanisms of internalization and externalization makes it clear that they seem to be almost mirror images of each other. The mechanisms whereby the self is constituted and constructed by relating to objects as described by internalization finds its counterpart on the side of the object in the different mechanisms of externalization that allow for the constitution and construction of objects by implicating the self. Thus both self and objects seem to be constituted at the same time in mutual dependence on each other. Such mutual dependence in their respective constitution is only possible if there is some kind of prior relationship between self and objects e.g. a commonly underlying ground or foundation. I assume that this prior relationship between self and object enables and predisposes to the mutual dependence in the constitution of self and objects, which I shall describe below in terms of the concept of "co-constitution and co-construction" of the self and its objects.

What does such "co-constitution and co-construction" of self and object involve both psychodynamically and empirically? Psychodynamically, "co-constitution and co-construction" of self and object describes a dynamic and continuously changing balance between internalization and externalization. This balance between internalization and externalization involves a balance between the inner world of the self and the outer world of objects or, as Melanie Klein describes it, a "balance between giving out and taking in": ". . . the ego can then also feel that it is able to re-introject the love it has given out, as well as take in goodness from other sources, and thus be enriched by the whole process. In other words, in such cases there is a balance between giving out and taking in between introjection and projection" (Klein, 1955, pp. 312–313).

What are the implications of the psychodynamic balance between "giving in" and "taking out" in internalization and externalization for the empirical context and more specifically for the underlying psychological processes? One may assume two different psychological processes, one for underlying internalization and one for externalization. However, such different psychological processes raise the question of how the mutual dependence in the constitution between self and objects, their co-occurrence and co-constitution, is made possible, thus raising the question of the enabling and predisposing empirical conditions. If there are different underlying psychological processes, one would assume externalization and internalization to be independent rather than mutually dependent and co-constitutive.

Alternatively, one may assume a psychological process that commonly underlies both internalization and externalization (i.e. to which both can be traced back). Such commonly underlying psychological processes must enable and predispose to the co-constitution and co-construction of self and objects, thereby preventing them from being constituted independently of each other. I postulate that what I shall call "self-related processing" is the psychological process that may commonly underlie the co-constitution and co-construction of self and objects and thus, more generally, of self–object differentiation. Self-related processing will be the subject of Part III of this book, where I shall discuss the mental equipment.

We also need to determine the neuronal mechanisms associated with such a psychological process, in order to formulate a neuropsychodynamic hypothesis. I therefore focus here more on the neural equipment and thus the neuronal mechanisms that may correspond to what I here described as the self–object relation which is assumed to form the very foundations of the co-constitution and co-construction of self and objects. However, before going into neuronal detail, I want to elaborate the psychodynamic side further. I shall provide more details specifically about introjection and projection that, as intermediates forms of internalization and externalization, may be crucial in enabling and predisposing to self–object differentiation.

Projection, perception, and object relation

Analogous to internalization and externalization, I here focus on the constitutive and constructive side of introjection and projection in constituting and constructing self and objects. While projection and introjection may be constitutive (and also adaptive) in the early stages of children's development, they are adaptive in the later stages and in adulthood, where they defend the adult self or they alternatively become disruptive for the self when used in an abnormal context (e.g. in pathological behavior such as that in psychosis or depression). I here focus in particular on the constitutive aspect of introjection and projection and thus their essential role in constituting and constructing self and objects in infancy and childhood. I shall not be considering in detail their adaptive role when defending the self in adulthood, but I will discuss how introjection and projection can become disruptive in adults, specifically in relation to their rather disastrous effects in depression and psychosis (see Part IV).

Let us start with the concept of projection. Malancharuvil (2004) considers projection as a perceptual process that tests and evaluates specific objects in terms of their acceptability to the self. More specifically, the object of the outer environment is matched and compared with unconscious fantasies, cognitive schemata, and past experiences of the self and its inner mental world. The perception of the object is consequently not only about the object itself but also about the relation of the self to the object. Malancharuvil (2004, p. 375) says that "the perception of the object goes along with insertion of part of the self into the object," which he also describes as "penetration of the self into the object" (Malancharuvil, 2004, p. 376).

By inserting and penetrating itself into the perceived object, the self and its inner mental world become related to the object and the respective environmental context. In the absence of such insertion and penetration, the self becomes isolated and autistic, thereby "remaining unnourished and out of touch with the world" (Malancharuvil, 2004, p. 376). Projection in this sense must be regarded as constitutive and constructive, because it relates the self to objects in the respective environment by enabling and predisposing self–object relation. This in turn allows the self to constitute an image of itself in interaction with the objects. By relating self and object in perception, projection may be regarded as a way of continuously matching, comparing, and testing the self against the various objects that one encounters. More specifically, Malancharuvil (2004, p. 379) argues that projection can be regarded as the method by which all reality is tested and postulated in perception, which he describes as the "assimilation or verification process of anticipations/expectations." This matching, comparison, and testing make it possible to organize and structure our perception of the environment.

Thus the subject can impose itself on the perception of its environment, thus going along with a "subjectivization of reality" (Malancharuvil, 2004, p. 379) and thereby transforming reality from something that is merely objective to something rather subjective (which therefore "goes beyond" the actual stimulus and becomes mental rather than physical; see Chapter 5). What we perceive in our environment is subsequently influenced at least as much by our own self as by the respective object itself, which in turn allows us to constitute and construct a relationship between self and object: "Projection therefore is the process of contacting and evaluating the stimulus in terms of one's experiential structure (i.e. the self). It is not necessarily a defensive process, while defensiveness could profoundly affect the perceptual process. Projection understood as a foundational and commonplace perceptual process enriches the psychoanalytic theory of cognition" (Malancharuvil, 2004, pp. 381–382).

Projection concerns the imprinting and ingraining of the self and its inner mental world onto the objects as they are located in the outer environment. Thus projection establishes some kind of relationship between self and object, the self–object relation, which means that it is not merely defensive and thus adaptive, but also constructive and thus constitutive. Projection "is the primary phase of forming an image of self in interaction with the object. Projection is not merely a mechanism of ejecting persecutory objects, because projection precedes the formation of an image of the object as persecutory. Therefore the traditional notion of projection as an ego-defense requires reconsideration" (Malancharuvil, 2004, p. 376).

This quote demonstrates that projection in the constitutive sense as used here describes the very first or initial contact or encounter between self and object. This also enables and predisposes the self to internalize the object by means of introjection, and thus to relate to it in an ambivalent and affectively loaded way. Thus projection in the constitutive sense that is understood here precedes introjection. By constituting and constructing a basic relation to the object, projection enables and predisposes the self to introject objects. Projection may thus enable and predispose to the constitution of objects by relating the subject to events or individuals in the environment which are thereby transformed into what are called objects within the psychodynamic context. This allows the constitution of what is usually described as object relation.

Introjection, affect, and self–object relation

The constitution of objects and thus of object relation by projection represents only half of what I here call self–object relation. The other half, namely the self, is supposed to be constituted by subsequent introjection where the objects, as constituted on the basis of perception and thus projection, are related to the perceiving subject that is thereby transformed into what psychodynamically is described by the term "self" (or ego). Introjection accompanying projection may therefore allow complementing of the object by enabling and predisposing to the subsequent constitution of the self, whereby it transforms the initial object relation into a self–object relation. The concept of introjection must then be regarded as a way to constitute and thus complement the second half of the self–object relation, namely the self, whose first half, namely the object relation, is constituted and constructed on the basis of projection. Whereas projection describes the reaching out of the subject (i.e. what later becomes the self) to the object, introjection concerns the assimilation of the object by the subject, which is thereby constituted as self (or ego).

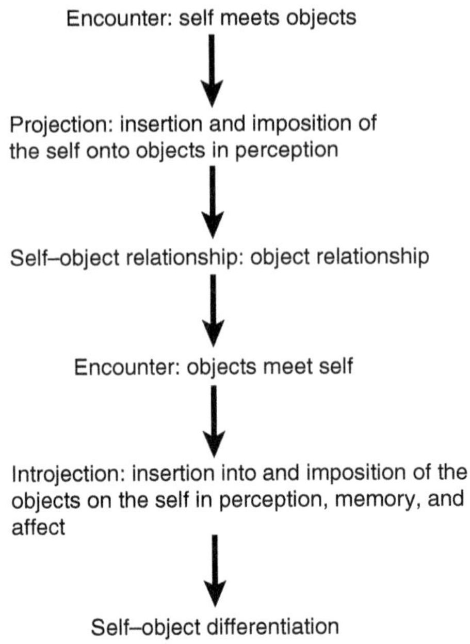

Fig. 6.1 Projection, introjection, and self–object differentiation.

Considering their rather complex relationship, the concepts of projection and introjection may be characterized by both symmetry and asymmetry. They may be characterized as symmetrical with regard to their direction in the relationship between self and object. Projection is directed from the (future) self towards the object, whereas introjection occurs in the opposite direction, from objects to the (future) self. This has already been pointed out in the original descriptions by Ferenczi and Freud. Although both concepts must be regarded as asymmetrical with regard to their dependence on each other, introjection is unilaterally and necessarily dependent upon projection and its constitution and construction of objects and object relation, without which there would be no objects that could be introjected (see Figure 6.1). In contrast, projection and more specifically its constitution and construction of objects are not necessarily dependent upon introjection, although in the later stages of development the two are reciprocally linked to each other in an almost circular manner.

Analogous to projection with regard to perception, introjection plays a crucial part in constituting and constructing the structure and organization of psychological functions. Projection structures and organizes perception, whereas introjection affects not only the structure and organization of perception but also that of memories "as they are linked with early object relationships" (Kernberg, 1966, p. 359). By means of memories, certain interactions with the environment (or relationships between self and object, as I would say) become reproduced and fixated, which makes possible the internalization and thus mentalization (as used in the sense of representation, as distinct from the more phenomenal sense of mental states in Chapter 5) of object relationships. This is well reflected in Kernberg's definition of introjection: "I will consider introjections as independent psychic structures, mainly growing out of primary autonomous functions (perception and memory) as they are linked with early object relationships, and although introjections will be seen as strongly influenced by oral conflicts, they will not be seen as growing out of them" (Kernberg, 1966, pp. 358–359).

In addition to perception and memories, introjection is also closely associated with another psychological function, namely emotions or affects (for the sake of operational simplicity I here use both terms synonymously). Kernberg assumes that the "external perception (e.g. perception of the external world and its outer objects)" is linked to the "perception of primitive affect states representing drive derivatives" (Kernberg, 1966, p. 360). All introjections that are assigned a positive affect may then be grouped together and distinguished from the grouping of those with a rather negative affective valence, thereby constituting and constructing what I here call "object–affect structure": "The affective colouring of the introjection is an essential part of it and represents the active valence of the introjection, which determines the fusion and organization of introjections of similar valences. Thus, introjections taking place under the positive valence of libidinal instinctual gratification, as in loving mother–child contact, tend to fuse and become organized in what has been called somewhat loosely, but pregnantly, 'the good internal object.' Introjections taking place under the negative valence of aggressive drive derivatives tend to fuse with similar negative valence introjections and become organized in the 'bad internal objects' (Kernberg, 1966, pp. 360–361).

By constituting and constructing the object–affect structure according to positive and negative valences, both self and object, or more precisely self-image and object-image, become structured, organized, and further differentiated. This in turn may provide the ground for the development of more complex and higher forms of internalization and externalization, such as identification and self-objectivization. What I here call "object–affect structure" therefore denotes and specifies self and objects as such, and thus the self–object relationship itself: The more complex and differentiated the object–affect structure that is developed, the more sophisticated the self–object relationship will be, with increasing differentiation between self and objects. The constitution and construction of the object–affect structure on the basis of initial projection and subsequent introjection may therefore be considered a necessary (i.e. enabling and predisposing) condition of possible self–object differentiation (and also, as we shall see later, of brain–self differentiation) (see Figure 6.1).

Let us return to Kernberg, who has nicely described the close relationship of the object–affect structure to self–object differentiation and, more specifically, how the former leads to increasing sophistication of the latter: "In the process of the fusions of introjections with similar valence, homologous components of introjection tend to fuse, self-images with other self-images and object-image with other object-images. Since by this fusion more elaborate self-images and object-images are being 'mapped out', this process contributes to the differentiation of self and object and to the delimitation of ego boundaries. This in turn further organizes and integrates the apparatuses of perception and memory; thus later introjections contain an ever growing complexity of information about both the object and the self in any particular interaction" (Kernberg, 1966, p. 361).

Psychodynamically this results in an ever changing "world of inner objects" that is continuously matched and compared with both the experience of the inner mental world of the self and the perception of the objects in the external environment. The inner objects are continuously modified and reshaped in order to adapt them as well as possible to the perception of the external environment of objects while at the same time maintaining their connection and linkage to the experience of an inner mental world of the self: "The world of inner objects, then, gradually changes and comes closer to the 'external' perceptions of the reality of significant objects throughout childhood and later life, without ever becoming an actual copy of the environmental world. 'Confirmation', intrapsychically speaking, is the ongoing process of reshaping the world of inner objects under the influence of the reality principle, of ego maturation and development, and through cycles of projection and introjection" (Kernberg, 1966, pp. 364–365).

If these cycles of projection and introjection are interrupted by, for example, traumatic events, the development of object–affect structure and self–object differentiation comes to a halt. Instead of developing object–affect structure and self–object differentiation in an ever more complex and sophisticated way, all energy is invested in preserving the status achieved so far by defending early and rather immature introjections. This results in the pathological fixation on severely disturbed early object relationships or, as Kernberg (1966, p. 365) has described it, in the "persistence of 'non-metabolized' early introjections." To prevent these early introjects from being metabolized, their negative valences are not integrated and fused with positive valences, so that the traumatic event becomes isolated from other more positive introjects and thus from the rest of the self in order to protect the latter (see Chapter 10 for further details). This may be different if the traumatic event occurs later after the fusion and integration of earlier positive and negative valences in object relationships. Splitting is then no longer needed to defend this more mature self. If traumatic events occur, the defense of the more mature self can then focus on suppressing the supposedly damaging objects and their associated affective states, resulting in what Kernberg describes as repression (Kernberg, 1966, p. 376).

Rest–stimulus interaction and projection

So far I have discussed the psychodynamic mechanisms underlying the constitution and construction of self and objects (i.e. the self–object differentiation). I thereby assumed that internalization and externalization in general and introjection and projection in particular are crucial for the construction and constitution of both self and objects, thereby predisposing to and enabling self–object differentiation. The concept of self–object differentiation is a psychodynamic one that, as previously demonstrated, can be linked to the psychological context (e.g. the psychological functions of perception, memory, and affect/emotion).

How can we now relate and link the psychodynamic concept of self–object differentiation and its psychological characterization in the neuronal context of the brain? I postulate that self–object differentiation is enabled and predisposed to by the differentiation of both self and objects from the brain. For that we have to consider what I described in the Introduction as "brain–self differentiation" and "brain–object differentiation." These two concepts refer to the mechanisms that enable and predispose to the constitution and construction of self and objects as distinct from the brain. I postulate that these mechanisms are related to the way in which the brain processes and codes stimuli (as discussed in Chapters 4 and 5). Both brain–self and brain–object differentiation may thus be related to specific neuronal mechanisms as their necessary empirical (i.e. enabling and predisposing) condition. My first specific neuropsychodynamic hypothesis is that the structure and organization of brain–stimulus interaction enables and predisposes to projection and thereby brain–object differentiation (see Figure 6.2).

Let me start with the first aspect of this neuropsychodynamic hypothesis, namely the relationship between projection and rest–stimulus interaction, while I discuss its second aspect, namely the relationship to brain–object differentiation, later on. In Chapters 4 and 5 about cathexis I distinguished between rest–stimulus and stimulus–rest interaction. In the case of rest–stimulus interaction, the brain's resting-state activity level modulates and shapes the neuronal activity that can possibly be induced by stimuli from outside the brain, be they interoceptive (originating in the body) or exteroceptive (originating in the outside world). Rest–stimulus interaction in this sense means that the brain and its actual resting-state activity level (as defined in an operational sense as the absence of external stimuli) is imposed and ingrained upon the stimulus-induced activity, thereby shaping and modulating the stimulus itself.

I demonstrated various empirical examples of rest–stimulus interaction, especially in the sensory domain and thus in perception. What are the implications of rest–stimulus interaction for

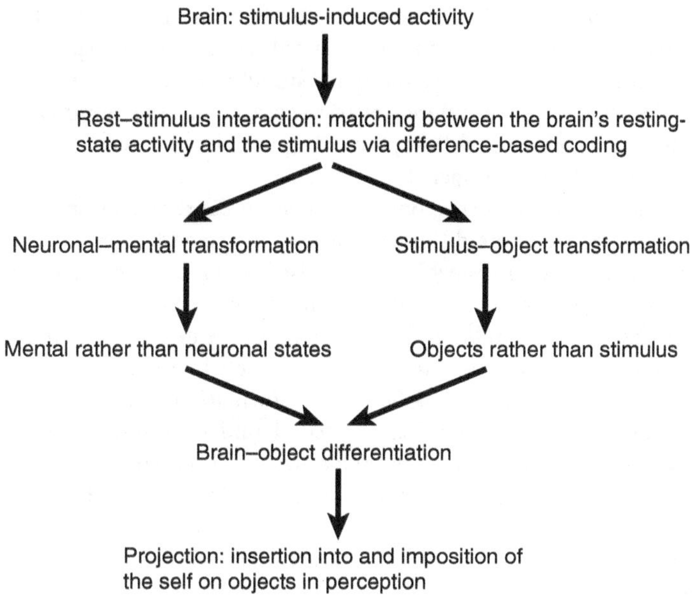

Fig. 6.2 Projection, rest–stimulus interaction, and brain–object differentiation.

perception? Rest–stimulus interaction means that the brain inserts itself and penetrates into the stimulus-induced activity and thus onto the subject's perception of the respective stimulus. Therefore what we perceive is not so much the stimulus by itself, isolated from the brain itself, but rather its relationship to our own brain and its respective resting-state activity level. What are the implications of this for the psychological context? As described in Chapter 4, according to the concept of predictive coding we perceive not so much the stimulus itself as the difference between the anticipated or expected stimulus and the actual stimulus. Predictive coding thus describes the process that allows the comparing, matching, and testing of anticipated and actual stimuli with each other.

Both neuronal and psychological descriptions correspond well to what I described with regard to projection within the psychodynamic context. In the same way that the self inserts and penetrates into the perception of objects in projection, the brain and its resting-state activity appear to insert and penetrate into the brain's own neural processing and coding of stimuli (i.e. the stimulus-induced activity). And in the same way that Kernberg describes projection as matching and testing between self and objects (i.e. as a reality test of the self), the brain's resting-state activity, including its anticipation and expectations of a specific stimulus, are matched and tested against the actual stimulus. Thus there appears to be correspondence between neuronal, psychological, and psychodynamic contexts with regard to insertion and testing, matching, and comparison.

Let me describe this correspondence in more detail. There is correspondence between the different concepts in that there are analogous concepts (i.e. resting-state activity/anticipation/self and stimulus-induced activity/real stimuli/objects) as well as analogous processes (i.e. insertion/penetration and testing/matching). One may therefore want to speak of what can be called "conceptual correspondence." However, such "conceptual correspondence" must be distinguished from "empirical correspondence." Rather than concerning concepts and their analogies as in "conceptual correspondence," "empirical correspondence" refers to analogous or similar contents. However, these differ in all three contexts—neuronal, psychological, and psychodynamic—as they can all be characterized by different contents. In the neuronal context, the brain's resting-state activity inserts itself into its own stimulus-induced activity. Psychologically, the anticipation of the possible stimulus inserts itself into the actual stimulus. Psychodynamically, the self inserts itself into the object. Due to such differences in their contents, the empirical correspondence between the neuronal, psychological, and psychodynamic contexts remains at best intuitive (as based on their conceptual correspondence).

To go beyond mere intuitive correspondence in empirical regard, we need to better understand the transitions between the different contexts and to address the question of how the neuronal mechanisms of insertion enable and predispose to the psychological mechanisms and ultimately the psychodynamic mechanisms. For instance, we need to better understand how stimuli and their respective stimulus-induced activity in the neuronal context of the brain are related to and more specifically transformed into objects in a psychodynamic sense, which involves what I earlier called "stimulus–object transformation" (see Chapter 5). Insight into stimulus–object transformation will also give us some understanding of how the brain and its resting-state and stimulus-induced activity enable and predispose to the constitution of objects as distinct from itself (i.e. brain–object differentiation). This leads us to the second aspect of my first neuropsychodynamic hypothesis, namely the relationship between rest–stimulus interaction and brain–object differentiation. For this we first need to understand stimulus–object transformation.

Rest–stimulus interaction and brain–object differentiation

How can we provide the linkage between stimuli and objects? The psychodynamic concept of objects refers to whole people, events, or objects that are perceived and experienced as unity, while the concept of stimuli, as used in the neuronal context, refers to a multitude of single units that are supposed to induce neuronal activity (i.e. stimulus-induced activity). How are stimuli and objects related to each other? A multitude of stimuli composed in a certain way are supposed to constitute and construct people, events, or objects as associated with the psychodynamic concept of objects. Thus in order to provide the link between stimuli and objects, we need to understand how the multitude of different stimuli are transformed into what we perceive as an object. I call this process of transformation "stimulus–object transformation."

Let me go back to rest–stimulus interaction and its precise neural coding mechanisms. I demonstrated in Chapter 5 that rest–stimulus interaction is made possible by coding of the difference between the actual resting-state activity of the brain and the ideal stimulus-induced activity. However, since a multitude of different stimuli enter the brain at the same time as the specific stimulus in question, the latter is coded not only as the difference from the brain's resting-state activity, but also in relation to the other stimuli (be they anticipated or actually occurring). This means that difference-based coding provides linkage not only between brain and stimulus but also between different stimuli. By providing the linkage between different stimuli, the various single stimuli are composed into an object, thus enabling and predisposing to stimulus–object transformation (for details, see Chapter 5).

The psychoanalyst may still want to raise objections here. Even if the various single stimuli are transformed into an object on the basis of difference-based coding, the objects as understood in the psychodynamic concept are mental (often taken as representational in this context) and related to the self, something which is not captured by mere stimulus–object transformation as described here. How can we account for the mentalization of the object as transformed from the multitudes of stimuli? Remember that difference-based coding was at the bottom and was therefore a necessary (i.e. enabling and predisposing) empirical condition of the transformation of neuronal states into mental states, namely neuronal–mental transformation (see Chapter 5).[1] Thus stimulus–object transformation and neuronal–mental transformation are supposed to be enabled and predisposed to by the same neuronal mechanisms (i.e. difference-based coding).

What are the implications of this for the psychodynamic context? Difference-based coding allows the transformation of single stimuli into whole objects, with the underlying neuronal states being transformed into mental states. This co-occurrence between stimulus–object transformation and neuronal–mental transformation allows the constitution of objects in our mental states, which Kernberg called "inner objects." And most importantly, these objects are distinct not only from the single stimuli that entered the brain but also from the brain itself, thus enabling and predisposing to brain–object differentiation. Therefore I postulate that rest–stimulus interaction in conjunction with difference-based coding enables and predisposes to and is thus a necessary empirical condition of possible brain–object differentiation.

Radial–concentric organization and subcortical–cortical systems

My second neuropsychodynamic hypothesis argues that the psychodynamic concept of introjection as previously outlined corresponds within the brain's neuronal context to stimulus–rest interaction, which in turn may enable and predispose to brain–self differentiation. Let me go into some more detail as I first discuss the relationship between stimulus–rest interaction and introjection and then consider their relationship to brain–self differentiation.

In addition to rest–stimulus interaction, I also gathered some empirical evidence for the reverse interaction, namely stimulus–rest interaction (see Chapter 4). In the case of stimulus–rest interaction, the stimulus-induced activity affects and ingrains the brain's resting-state activity level. This means that the brain's resting-state activity is shaped by the incoming stimulus. In other words, the stimulus is ingrained and encoded in the brain's structure and organization of its own resting-state activity. This corresponds nicely to the way that Kernberg characterized introjection within the psychodynamic context. In the same way that the stimulus ingrains and encodes itself into the brain's resting-state activity, the object ingrains and encodes itself into the self in introjection.

Kernberg went even further and argued that the object ingrains and encodes itself into the perception, memories, and affects of the self. Beyond the merely conceptual and thus intuitive correspondence between stimulus–rest interaction and introjection, this provides a more content-based linkage with empirical correspondence between neuronal and psychodynamic contexts. By gaining some insight into how perception, memories, and affects are mediated and processed in

[1] The psychoanalyst may still want to object. The overlap between what he calls "mentalization of objects" and what I describe as "neuronal–mental transformation" is at best only conceptual (i.e. terminological), with both referring to the term "mental," while it is not substantial and thus empirical with regard to specific contents. The term "mental" in "mentalization of objects" as used in the psychodynamic context usually refers to the representation of objects, with such representation being possibly understood in different ways as well. However, this is very different to what I associate with the term "mental" in "neuronal–mental transformation," which has strong phenomenological (i.e. subjective–experiential, rather than representational implications (see Chapter 5). Although I fully concur about the different usages of the term "mental" in the two contexts, I do not agree that they are mutually exclusive. I would argue and postulate that the phenomenological meaning of the term "mental" in neuronal–mental transformation is a necessary (i.e. enabling and predisposing) condition of the rather representational processes associated with the concept of "mentalization of objects" within the psychodynamic context. However, to demonstrate this in detail would be beyond the scope of this book, and may be left for future endeavors.

the brain, we may establish tighter and thus empirical correspondence between stimulus–rest interaction and introjection that is based on (empirical) contents rather than mere concepts. However, for this we need to go back to the brain itself and its neural organization and structure with regard to its subcortical–cortical anatomy (see also Chapter 4, where I discuss an alternative anatomical model that is based less on anatomical features and more on cognitive processes).

Nieuwenhuys (1996) and Nieuwenhuys et al. (1988) assumed a medio-lateral trend in subcortical regions that are located concentrically or radially around the ventricle (i.e. the aqueduct), with progressive extension from median to lateral locations. Based on various distinct features, he classified the subcortical regions into three distinct territories—core, median and lateral paracore, and lateral regions—which despite being closely interconnected can be distinguished from each other. The core subcortical regions are those regions that are located in direct proximity to the ventricle and may thus be described as paraventricular or periaqueductal. They include the PAG, the pontine central gray, the medial hypothalamus, the septum, the parabrachial nuclei, and the dorsal vagal complex. The subcortical median paracore regions are located directly adjacent to the core regions. They include the series of raphe nuclei, the lateral hypothalamus, and the bed nucleus of the stria terminalis. These are closely connected to the bilateral paracore regions, which include the ventral tegmental area (VTA), the locus coeruleus, the substantia nigra, and the nucleus reticularis.

The two inner rings (i.e. the core and median and lateral paracore regions) can be distinguished from the outermost ring (i.e. the lateral regions) with regard to their fibers (myelinated or unmyelinated), biogenic amines (e.g. serotonin, noradrenaline/adrenaline, dopamine, histamine), circumventricular organs, gonadal-steroid receptors, and coherent behavior (e.g. as induced by localized electrical stimulation) (for details, see Nieuwenhuys, 1996, pp. 560–567; Panksepp, 1998). This can lead us to consider the core and median and lateral paracore regions together as one entity which can be functionally characterized by its predominant involvement in processing interoceptive stimuli, thereby regulating the body's homeostatic milieu, vegetative–autonomic functions, and a range of specific emotional and motivational processes. The core regions (i.e. the innermost ring) in particular are strongly implicated in homeostatic and basic emotional regulation. This distinguishes them from the most lateral regions at the subcortical level, such as the crus and the colliculi which, by containing and mediating the respective fibers to the spinal cord, are implicated in the processing of exteroceptive and sensorimotor stimuli.

Based on Nauta and McLeans' concept of the limbic system, Nieuwenhuys (1996) proposes extension of the core–paracore system into the forebrain. Mesencephalic core–paracore regions are closely connected to the hypothalamus and various regions in the forebrain, including the amygdala, septum, hippocampus, and parahippocampal gyrus. This led to the concept of the "greater, distributed or extended limbic system" (Niewenhuys, 1996, pp. 567–574, Heimer, 2003; Morgane et al., 2005, 2006). This was even further developed by Mesulam (2000) (see also Morgane et al. 2005, 2006), who extended the originally mesencephalic core regions over the forebrain (i.e. the limbic system) into the cortex itself, and referred to what he called "paralimbic areas." Paralimbic areas are those regions in the cortex that are anatomically linked to ancient emotional and motivational networks, located right around and thus adjacent to the ventricle, so those heavily paleocortical regions can be regarded as extensions of the originally mesencephalic grouping of the core regions. These paralimbic areas include the lower parts of the orbitofrontal cortex, the perigenual anterior cingulate cortex (PACC) and supragenual anterior cingulate cortex (SACC), the posterior cingulate cortex (PCC), the retrosplenial cortex (RSC), the temporal pole and the insula.

Taking as a starting point the radial–concentric organization into rings at the subcortical mesencephalic level and its extension into the forebrain, Feinberg (2009) argues that the three postulated rings are also manifested and visible at the level of the cortex. The outer or most peripheral ring, which is furthest away from the ventricle, includes regions like the sensory cortex, the motor cortex, and the lateral prefrontal, parietal, and occipital regions. These regions may be considered as the cortical extension of the most lateral regions at the subcortical level (i.e. the mesencephalic crus and colliculi). Analogous to the lateral regions at the subcortical level, the lateral cortical regions (the outer ring) are predominantly involved in processing external stimuli as distinct from the interoceptive stimulus processing that takes place in the innermost ring. This provides an anatomical distinction between exogenously driven cognitive processes and endogenously driven primary homeostatic, motivational, and core emotional processes (Panksepp, 1998).

Feinberg (2009) also assumes the existence of a middle ring at the cortical level that is located in between the inner and outer rings and thus between the paralimbic and lateral cortical regions. This middle ring at the cortical level includes regions like the medial orbitofrontal cortex, the ventromedial prefrontal cortex (VMPFC) and dorsomedial prefrontal cortex (DMPFC) and the medial parietal cortex (MPC), which have recently been subsumed under the concept of cortical midline structures (CMS) (Northoff and Bermpohl, 2004; Northoff et al., 2006). Since it is sandwiched in between the inner and outer rings and their involvement in intero- and exteroceptive processing, respectively, Feinberg assumes that this middle ring is responsible for the integration and linking of both kinds of stimuli (i.e. intero-exteroceptive integration).[2] The CMS do grossly overlap with what particularly in the imaging domain is often described as the default-mode network (DMN), which is supposed to be characterized by particularly high resting-state activity (i.e. intrinsic activity) (see Chapter 4 and also Raichle et al., 2001; Buckner et al., 2008). However, how such high intrinsic or resting-state activity in the CMS is related to intero-exteroceptive integration, as postulated by Feinberg, remains unclear.

Let us summarize this subcortical–cortical system that is based on various anatomical (i.e. cytoarchitectonic, chemoarchitectonic, and connectional) features. The subcortical core regions adjacent to the aqueduct extend over the forebrain and the limbic regions into the cortex, where they surface as paralimbic regions. One may consequently want to speak of a "core–paralimbic system." The next concentric ring consists of the median and lateral paracore regions at the subcortical level that extend over the forebrain and its limbic system into the midline regions at the cortical level. One may consequently want to speak of a "paracore–midline system." Finally, the very lateral subcortical regions, where sensory and motor fibers connect to the spinal cord, extend cortically into lateral cortical regions as the outermost or peripheral ring, which can be envisioned as a "lateral–lateral system" (see Figure 6.3). Such lateral subcortical extensions at lateral cortical level suggest that the information processing of these subcortical perceptual regions are refined at the cortical level. This is highly compatible with the hierarchical organization of a "nested hierarchy" as discussed in

---

[2] However, it remains unclear whether such intero-exteroceptive integration at the cortical level corresponds to analogous processes at the level of the forebrain and the mesencephalon. One could for instance imagine that what is described as the core system at the mesencephalic level may extend into the paralimbic areas, since both are located directly adjacent to the aquaeduct/ventricle. However, the median and lateral paracore regions at the mesencephalic level may correspond to the middle ring at the cortical level and thus the CMS. Here support comes from the connectivity pattern. Cortical regions such as the anterior cingulate (PACC, SACC, and PCC), the caudal orbitofrontal cortex, the temporal poles and the insula are characterized by strong inputs especially from the subcortical core regions such as the PAG (Nieuwenhuys, 1996, p. 573). In contrast, the VMPFC and the DMPFC receive strong input from in particular the raphe nuclei as median paracore regions and the locus coeruleus as lateral paracore region (Nieuwenhuys, 1996; Morgane et al., 2005).

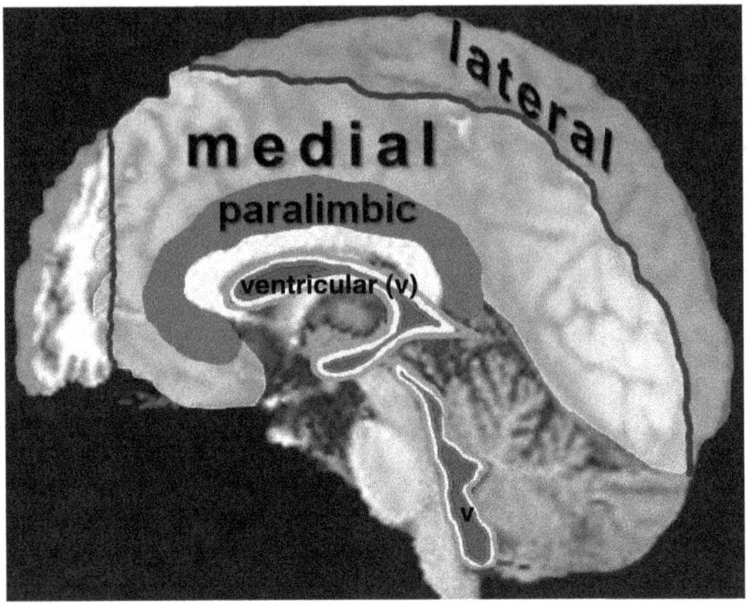

Fig. 6.3 Radial–concentric organization and subcortical–cortical systems in the healthy brain.

Chapter 4. Further empirical support for the integration of subcortical and cortical regions and thus for such hierarchical organization will be described later in this chapter.

Stimulus–rest interaction and introjection

How can we look at the three different subcortical–cortical anatomical systems (core–paralimbic, paracore–midline, and lateral–lateral) in more detail and put them into the psychodynamic context? The distinction made by Freud (1895) between different neurons (i.e. those that receive endogenous input, exogenous input, or qualitative information), may spring to mind for the versed neuropsychoanalyst. However, exactly how Freud's different classes of neurons correspond to the neuroanatomical distinction suggested here remains unclear. It is true that the lateral–lateral system receives predominantly exogenous input (i.e. exteroceptive stimuli), while the core–paralimbic system is certainly characterized by predominantly endogenous input (i.e. interoceptive stimuli). And one may associate the paracore–midline system with Freud's third class of neurons, namely those that receive qualitative information.

Alternatively, one may also argue that what Freud called endogenous input may in fact correspond to the paracore–midline system and its characterization by intrinsic activity as suggested by its strong overlap with the DMN. Then, however, the question arises as to the exact anatomical location of what Freud described as the neurons receiving qualitative information. Are these neurons then related to the core–paralimbic system? Does this mean that what Freud called "qualitative information" corresponds to interoceptive stimulus processing as associated with the core–paralimbic system? This demonstrates that direct one-to-one translation of the concepts of Freud (1895) into the current neuroanatomical framework may be difficult if not impossible. I shall therefore refrain from making any such attempt.

However, what may be relevant in the present context is how the three systems thath Kernberg assumes to be central in introjection (i.e. perception, memories (especially autobiographical memories), and affects) are related to the neuroanatomical distinctions described here. Perception is the easy one. It is clearly related to the processing of exteroceptive stimuli and thus to the lateral–lateral system. Affects, in contrast, are supposed to be strongly based on interoceptive stimulus processing (see Part III for details), and may consequently be associated with the core–paralimbic system. However, memories have been shown to be related to many of the regions designated as paracore–midline (e.g. hippocampus, precuneus, and ventromedial and dorsomedial prefrontal cortex. Taken together, this suggests that the three functions (i.e. perception, memories and affects) that Kernberg assumes enable and predispose to introjection may be related to different subcortical–cortical neural networks (lateral–lateral, core–paralimbic, and paracore–midline).

What are the implications of this for the psychodynamic context of introjection and, even more important, for brain–self differentiation? Kernberg assumes that the encounter between self and object in introjection is mediated (and processed, if one wants to describe it as such) by perception, memories, and affects. Put within the neuroanatomical context, this means that the stimuli must be processed in all three systems (lateral–lateral, core–paralimbic, and paracore–midline) in order for them to be transformed into an object. This means that, functionally, the stimuli must influence the resting-state activity level in all three systems, thus presupposing stimulus–rest interaction to enable and predispose to stimulus–rest interaction. However, in addition to the already well characterized neuronal mechanisms of stimulus–rest interaction, we also have to consider the interaction between the three different anatomical systems (i.e. lateral–lateral, core–paralimbic, and paracore–midline), which I shall discuss below (see Figure 6.4).

How can we specify the interaction between the three networks? Due to difference-based coding, the neural activity in one network provides the neuronal context for the neural activity in the respective other, including both its resting-state and its stimulus-induced activity. This amounts at the anatomical level to what I called (see Chapter 4) "neuronal contextualization." Neuronal contextualization, as used here, describes the effects of difference-based coding at a regional network level (i.e. across the different networks, including lateral–lateral, core–paralimbic, and paracore–midline). Neuronal contextualization in this sense has major implications for the processing of stimuli. Stimuli are processed not only directly within one specific network (e.g. interoceptive stimuli in the core–paralimbic network) but also, via neuronal contextualization, indirectly in the respective other networks (e.g. lateral–lateral and core–paralimbic systems), even though they do not enter these networks directly themselves.

Put within the psychodynamic context, such direct and indirect neural processing of stimuli across all networks may account for what Kernberg described as the mediation of introjection by perception, affect, and memories. Due to neuronal contextualization, the stimulus cannot avoid influencing the resting-state activity in all three networks (lateral–lateral, core–paralimbic, and paracore–midline), thus involving stimulus–rest interaction. This in turn may enable and predispose to the transition from the neuronal to the psychological context, with the stimulus being transformed in all three systems (perception, affect, and memories). Following Kernberg, the processing of the stimulus (i.e. the potential object) in the systems of perception, affect, and memory enables and predisposes not only to its transformation into an object (i.e. stimulus–object transformation), but also, even more important, to its internalization by the self by means of introjection. However, it remains unclear how, as being not only adaptive but also constitutive, internalization enables and predisposes to the constitution of the self and its differentiation from the brain (i.e. brain–self differentiation), both of which are necessary for the object to be introjected by the self. If there is no self, objects cannot be introjected at all, as there would be nothing that (or nobody who) could introject the objects.

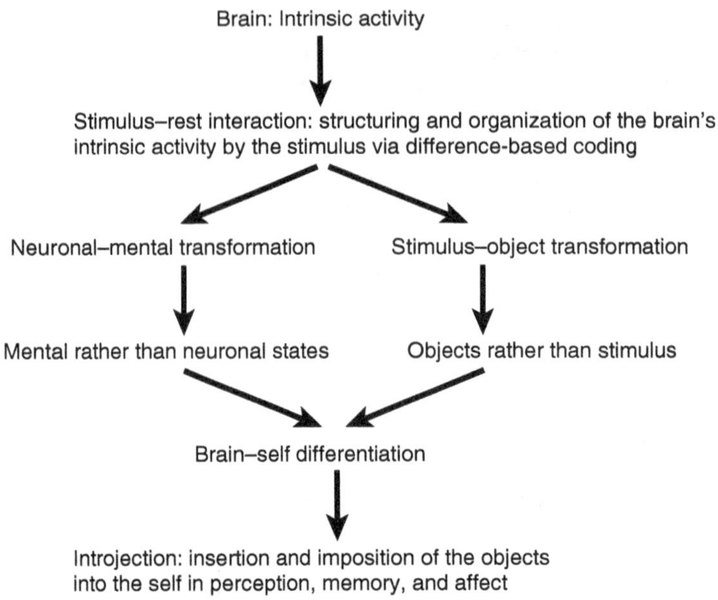

Fig. 6.4 Introjection, stimulus–rest interaction, and brain–self differentiation.

Rest–rest interaction, neuronal contextualization, and brain–self differentiation

One may now want to ask how stimulus–rest interaction enables and predisposes not only to introjection but also to brain–self differentiation. The latter refers to the mechanisms that make possible subjective perception and experience of one's self as distinct from one's own brain. Thus we need to understand how the brain itself and its own mechanisms enable and predispose us to experience and perceive ourselves as self rather than as brain (or body or mere organism). I discussed the epistemic implications of brain–self differentiation in Part I when introducing the concept of autoepistemic limitation. Moving now to the empirical context, I intend to link neuronal and psychodynamic contexts by showing how the brain's neuronal mechanism of neuronal contextualization enables and predisposes to the differentiation of the self from the brain (i.e. brain–self differentiation).

Let me summarize this briefly. I demonstrated that there may be different neural networks in the brain (lateral–lateral, core–paralimbic, and paracore–midline). Even if they are not encountering stimuli directly themselves, all three networks are considered to be involved directly or indirectly in the neural processing of the stimulus via stimulus–rest interaction, neuronal contextualization, and difference-based coding. Moreover, I described how the neuronal states associated with stimuli are transformed into mental states that refer to objects rather than stimuli. I referred to this as neuronal–mental and stimulus–object transformation (see also Chapter 5). More specifically, I postulated that neuronal–mental transformation and stimulus–object transformation were enabled and presupposed to by stimulus–rest interaction and rest–stimulus interaction by means of difference-based coding and neuronal contextualization.

What are the implications of this for the relationship between the three networks in the resting state? If the aforementioned neuronal mechanisms (i.e. difference-based coding and neuronal contextualization) enable and predispose to neuronal–mental and stimulus–object transformation in rest–stimulus and stimulus–rest interaction, they should also do so when they are employed in the interaction between the three anatomical networks in the resting state. The interaction between the three networks in the resting state may be described as rest–rest interaction (as distinct from rest–stimulus and stimulus–rest interaction), and it is by such rest–rest interaction that the one system is neuronally contextualized by the other, and vice versa. I now postulate that the same mechanisms (i.e. difference-based coding and neuronal contextualization) that are active during rest–stimulus and stimulus–rest interaction are also employed during rest–rest interaction. If so, one would expect rest–rest interaction to enable and predispose to neuronal–mental and stimulus–object transformation, and thus mental states and objects, in just the same way as in rest–stimulus and stimulus–rest interaction.

Let me be more specific in an empirical regard. In the same way that the neural activity induced by intero- or exteroceptive stimuli is coded as the difference compared with the brain's resting-state activity, with the latter providing the neuronal context for the former, the resting-state activity of the one network is coded as the difference compared with the other networks (and so on). Since difference-based coding is supposed to enable and predispose to neuronal–mental and stimulus–object transformation regardless of the origin of the neuronal difference (see Chapter 5), it may also do so in the case of rest–rest interaction across the different networks. If this is the case, the neural activity differences associated with rest–rest interaction across the different networks undergo the same kind of transformations (i.e. neuronal–mental and stimulus–object transformation), as in rest–stimulus and stimulus–rest interaction. In short, I postulate that not only rest–stimulus and stimulus–rest interaction but also rest–rest interaction is associated with neuronal–mental and stimulus–object transformation, which means that analogous mechanisms are at work during both the resting-state and stimulus-induced states.

What are the implications of this for the psychodynamic context of brain–self differentiation? In the same way that the intero- or exteroceptive stimuli from the body and the environment are transformed into an object (stimulus–object differentiation) on the basis of their differentiation from the brain and its stimulus-induced activity (i.e. brain–object differentiation) (see the previous section as well as footnote 3 below about the concept of the object as presupposed here), the brain's neural stimuli associated with its resting-state activity are transformed into an object by themselves. And in the same way that intero- and exteroceptive stimuli are associated with mental states, the brain's neural stimuli and their respective neuronal states are also transformed into a mental state as distinct from the neuronal state (such mental states based on rest–rest interaction may, for instance, surface in what is called mind-wandering, random thoughts, or day dreaming or night dreaming; see Chapter 8 for details). I consequently postulate that analogous to any stimulus-induced activity, the brain's resting-state activity itself and its rest–rest interaction enable and predispose to the constitution of objects and mental states.

Psychodynamically, this means that the constitution of objects on the basis of extero-interoceptive stimuli and their respective stimulus-induced activities goes hand in hand with the constitution of objects on the basis of the brain's neural stimuli (i.e. its resting-state activity). The first pertains to what I called brain–object differentiation, while the second is here conceptualized as brain–self differentiation. Therefore, analogous to the case of the brain's stimulus-induced activity and subsequent brain–object differentiation, I here postulate that the application of difference-based coding and neuronal contextualization to the brain's rest–rest interactions across its three different networks enables and predisposes to brain–self differentiation. The shift from the concept of the object to that of the self means that now the brain itself rather than the intero- or exteroceptive stimulus is transformed into an object and associated with a mental state. Due to their different origins in either intero-exteroceptive or neural stimuli, both objects can be distinguished from each other, with this empirical difference resurfacing in the conceptual distinction between self and object.[3] I therefore assume that the brain's resting-state activity and its rest–rest interaction enable and predispose not only to brain–self differentiation but also to the differentiation of the self from objects (i.e. self–object differentiation).

One may now want to argue that this only argues in favor of the transformation of one's own brain into an object, while it does not make a case for the transformation of one's own brain into a self as distinct from objects. Thus I may have shown another instance of brain–object differentiation, while I appeared to have confused that with brain–self differentiation. How can we counter such an argument? For this we need to investigate the differentiation between object and self (i.e. self–object differentiation), and how that relates to the neuronal mechanisms previously outlined.

Trilateral interaction and the balance between introjection and projection

Let us go back to the empirical details and the three different neural networks. I argued so far that the three neural networks interact with each other via difference-based coding and neuronal contextualization. What are the implications of this for the neural processing of stimuli? The three different subcortical–cortical networks (core–paralimbic, paracore–midline, and lateral–lateral) are associated with the neural processing of different stimuli (interoceptive, intrinsic/neural

[3] The conceptually versed psychoanalyst may want to argue that I here use the concept of objects in different ways. There is the use of object in the concept of stimulus–object transformation which, taken exclusively within the neuronal context, may refer to persons and events as distinct from mere stimuli, but not to objects as used in the psychodynamic sense. However, this changes once I associate stimulus–object transformation with neuronal–mental transformation, as I claim is the case. This means that those persons and events as objects in a purely empirical (i.e. physical) sense are then related to what the psychoanalyst may want to call their "representation in mental states," where the same hitherto physical objects resurface as mental objects as coming close to the meaning of object in the psychodynamic context. However, further conceptual confusion may be diagnosed when I claim stimulus–object transformation to occur in both cases (extero-interoceptive stimuli and the brain's neural stimuli), while later associating the stimulus–object transformation of the latter as resulting in the self as distinct from objects. This is correct. I here do indeed use the concept of the object in different ways (a narrow and a wider sense). The concept of stimulus–object transformation does presuppose a wider conception of the term "object," that is not necessarily yet associated with mental states, as is the case in the psychodynamic context, and one that remains independent of the origin of the stimuli (i.e. intero-exteroceptive or neural). However, this changes when I use the term "object" outside the concept of stimulus–object transformation and thus in an isolated way standing by itself independent of the term "stimuli," or in the concept of brain–object differentiation. Then the term object is used in a more narrow sense as associated with mental states and referring to objects only as constituted on the basis of intero- and exteroceptive stimuli, thus referring to the meaning that is often presupposed in the psychodynamic context. However, this more narrow meaning of the term "object" is admittedly violated when I argue that the brain's rest–rest interaction across its different networks enables and predisposes to stimulus–object transformation and subsequent constitution of the brain as object. The latter use of the term "object" should, if keeping to the narrow meaning, be replaced by a different term, such as the self. Although I agree that this may look like conceptual confusion, I nevertheless use the term here to emphasize the analogy with regard to the processes and the results between stimulus–object transformation in both extero-interoceptive and neural stimuli. Thus I may be forgiven such intentional conceptual confusion for the sake of pointing out neuronal and psychodynamic analogies between the brain's resting states and stimulus-induced states (which in turn may be helpful for better understanding our concepts).

and exteroceptive). Since, however, the neural activity in one neural network interacts with that in the others, there is also interaction between the different stimuli within each of the networks.

Such interaction may occur either directly within the same region or network (e.g. between neural and intero-exteroceptive stimuli, as reflected in rest–stimulus and stimulus–rest interaction on the basis of difference-based coding), or it may occur indirectly across different regions via neuronal contextualization (i.e. between intero- and exteroceptive stimuli). On the basis of this one would expect that, for instance, interoceptive stimuli would induce neural activity not only in the core–paralimbic network but also in the regions of the paracore–midline and even the lateral–lateral network, which are responsible for processing neural and exteroceptive stimuli. Conversely, one would also expect exteroceptive stimuli, as associated with the lateral–lateral network, to induce neural activity in core–paralimbic and paracore–midline networks. I shall now present some empirical evidence for these assumptions.

Let me start with the case of exteroceptive stimuli inducing neural activity in interoceptive regions. A recent study by our research group (Wiebking et al., 2011) investigated the neural processing of the heartbeat as interoceptive stimulus and compared it with the processing of an external tone as exteroceptive stimulus. We observed that exteroceptive stimuli induced even stronger activity changes in seemingly typical interoceptive regions like the bilateral anterior insula, the dorsomedial thalamus, and the dorsal anterior cingulate cortex. This demonstrates that exteroceptive stimuli do indeed induce neural activity in core–paralimbic and also paracore–midline regions that, following Feinberg's model, should be associated only with the neural processing of interoceptive stimuli. These data may be considered to support the assumption of neuronal contextualization at an anatomical level.

Furthermore, subjective–experiential variables like stress, stress perception, and bodily awareness did not correlate in healthy subjects with the purely interoceptive signal changes in the core–paralimbic regions, as one would have expected. Instead, bodily awareness and stress perception correlated with the difference between intero- and exteroceptive signal changes within the seemingly purely interoceptive regions (i.e. the core–paralimbic network). This lends support to my assumption that difference-based coding enables and predisposes to neuronal–mental transformation, with the mental states of one's own body corresponding to a neuronal difference (i.e. that between intero- and exteroceptive stimuli) rather than to stimulus-induced activity as related to the single and isolated stimulus itself.

These data clearly demonstrate that the neural activity in the distinct neural networks (i.e. core–paralimbic, paracore–midline, and lateral–lateral) does not appear to be confined to the neural processing of one particular stimulus type (interoceptive, exteroceptive, or neural). Instead, all three stimulus types seem to be processed (either directly or indirectly) in all three different neural networks. This means that all three types of stimulus do seem to interact with each other in all three neural networks. Since such interaction involves three different types of stimuli, I here speak of "trilateral interaction." The concept of trilateral interaction thus describes the interaction between intero- and exteroceptive and neural stimuli in and across the different regions and networks of the brain.

How does the trilateral interaction between intero-exteroceptive and neural stimuli correspond to the psychodynamic concepts of introjection and projection? Intuitively one may postulate that the processing of extero- and interoceptive stimuli within the paracore–midline network, as strongly overlapping with the DMN and its high intrinsic activity mirroring neural stimuli, may correspond to the concept of introjection. By inducing activity changes within the paracore–midline system, and thus the DMN via stimulus–rest interaction, intero- and exteroceptive stimuli penetrate into the brain's resting-state level within its typical resting-state network, which psychodynamically corresponds well to the penetration of objects into the self, and hence introjection.

In contrast, the penetration of the DMN's resting-state activity level into the core–paralimbic and lateral–lateral networks of intero- and exteroceptive stimulus processing via rest–stimulus interaction may correspond psychodynamically to the penetration of the self into the objects, and hence projection.

If the different kinds and directions of interactions between the three different neural networks may enable and predispose to different psychodynamic mechanisms, namely introjection and projection, way may need to ask what is the relationship between them. How can (and cannot) introjection and projection be related to each other if they are enabled and predisposed to by interactions between the three neural networks? I postulate that in the same way that the three different neural networks are mutually and reciprocally dependent on each other due to their various interactions, introjection and projection are also mutually and reciprocally dependent on each other. I consequently postulate that the balance in the neural activity between the three distinct neural networks may enable and predispose to the balance between introjection and projection, and more generally between internalization and externalization. This will be discussed in detail in the next section.

"Hybrid neural activity" and self–object differentiation

What are the implications of the trilateral interaction for the neural processing of stimuli? Due to difference-based coding and neuronal contextualization, the trilateral interaction seems to be unavoidable, which means that no one stimulus can escape interaction with the others. However, this means that the neural activity observed in association with specific stimuli (e.g. stimulus-induced activity) cannot be associated exclusively and completely with a specific and isolated stimulus (i.e. intero- or exteroceptive or neural), and thus be what one may want to call "pure." Instead, rather than being pure, the neural activity resulting from the trilateral interaction must be considered a composite or hybrid of all three different stimulus types (i.e. intero- and exteroceptive and neural). Thus trilateral interaction involves what I describe as "hybrid neural activity" (see Figure 6.5).

Does the "hybrid neural activity" mean that the neural activity is identical across the different regions and neural networks? This is to neglect the possibility that the hybrid or composite nature of neural activity may well allow for and make possible different balances between the different types of stimuli. For instance, neural stimuli may predominate over intero- and exteroceptive stimuli in the para–midline network, while interoceptive stimuli may predominate in the core–paralimbic network, and exteroceptive stimuli may predominate in the lateral–lateral network. The neural activity in the different regions and neural networks may thus reflect different balances or constellations between different stimulus types.

In addition to the balance or constellation between different stimuli, the "hybrid neural activity" may also be characterized by the balance between different temporal constellations (e.g. time dimensions). Specific exteroceptive stimuli, for instance, do not occur continuously but periodically, whereas interoceptive stimuli from one's own body (e.g. from one's own heart) must be processed more continuously, although some periodic changes may still occur. Finally, the brain's intrinsic activity may show certain continuous oscillations or fluctuations (see Chapters 4 and 5 for details), and thus a rhythm that may be different from that presupposed by intero- and exteroceptive stimuli.

The trilateral interaction between all three types of stimuli means that their different temporal dimensions interact and must be integrated with each other (for more empirical results and details, see Chapter 12). Such integration between different temporal dimensions must thus be mirrored in the hybrid neural activity. Depending on which stimulus type predominates, one

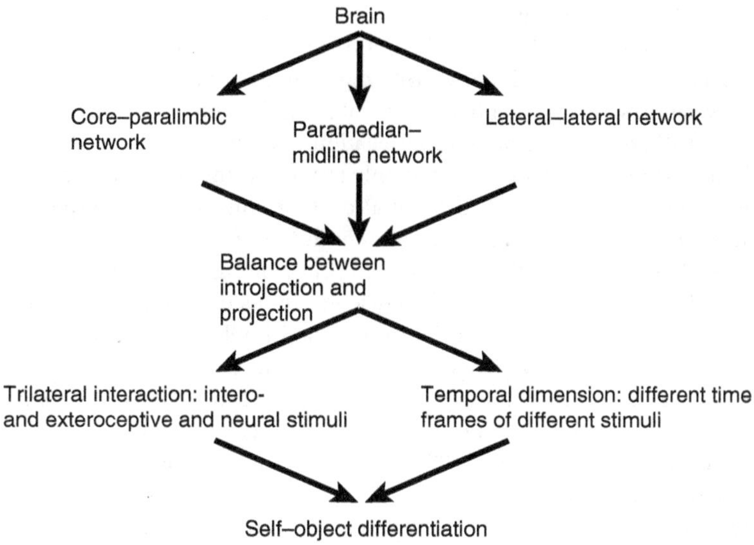

Fig. 6.5 Hybrid neural activity, different networks, and self–object differentiation.

may assume the predominance of one particular temporal dimension over the others in the trilateral interaction between the different stimuli. For instance, neural activity in the paracore–midline network that is dominated by neural stimuli and their continuous occurrence may show a different temporal pattern (e.g. a more continuous neural activity) than for instance lateral regions as related to the periodic occurrence of exteroceptive stimuli. Although this seems to be a viable hypothesis, it urgently needs further experimental investigation.

In summary, I characterized the "hybrid neural activity" by two main features, namely a specific stimulus constellation and a specific temporal constellation. Depending on which stimulus type (intero- or exteroceptive or neural) predominates, neuronal–mental and stimulus–object transformation will yield the corresponding mental state and object. If, for instance, the interoceptive stimuli predominate, our own body as object will dominate our mental states as "bodily object." If exteroceptive stimuli predominate, the respectively associated environmental event, person, or object will become the dominating object, an "environmental object." In the case of predominant neural stimuli, the respectively associated objects will outbalance intero- and exteroceptive inputs in our mental states, resulting in mind-wandering, daydreaming, or dreams that have what one might call "mental objects"[4] (see Part III for details).

The psychoanalyst may now argue that the hybrid neural activity may well explain the different kinds of objects, such as environmental, bodily, or mental objects, and thus the differentiation between different objects, including their differentiation from the brain. However, there hs not yet been any mention of the self, which means that that the hybrid neural activity falls short of explaining the differentiation of self from objects (i.e. self–object differentiation). This also means that there is no proper distinction between brain–self differentiation and brain–object

[4] I am aware that here again I use the term object in a rather blurred way that does not fully correspond to the narrow meaning of the term object outline in footnote 3; but again this conceptual confusion is intentional for the sake of showing neuronal and psychodynamic analogies. Such conceptual confusion may especially apply to my term "mental object" as used here but as I said it is intentional for the aforementioned reasons.

differentiation, with the former just being a specific instance of the latter. Therefore I provided further arguments and empirical support for brain–object differentiation but not for brain–self differentiation as distinct from and specific with regard to the self.

However, this ignores the fact that hybrid neural activity involves not only different stimulus constellations but also different temporal constellations. For instance, the predominance of exteroceptive stimuli enables and predisposes not only to the constitution of a specific type of object ("environmental objects," as I call them), but also to the constitution of a specific time frame (a discontinuous or periodic one). This is different in the case of neural stimuli that originate neither from the body nor from the environment, but from the brain itself. Their continuous occurrence implies a continuous rather than discontinuous time frame. Put in the psychodynamic context, such a continuous time frame means that stimulus–object transformation and the resulting object and its associated mental state (i.e. the "mental object" as I call it) are continuously present, not just periodically present like the environmental objects. Meanwhile the interoceptive stimuli from the body do range between the extremes of the mental objects and the environmental objects, so that temporally "bodily objects," as I call them, may be situated between the extremes of the other two.

I postulate that the temporal difference between mental and environmental objects corresponds to the distinction between brain–self and brain–object differentiation and thus to the difference between self and object (self–object differentiation). Due to the continuous presence of mental objects,[5] we distinguish them from environmental objects and associate the former with our own organism and person (i.e. what we conceptualize by the term "self"), and the latter with everything that is not self (i.e. objects). I therefore postulate that the different time frames of the different forms of hybrid neural activity may enable and predispose to brain–self differentiation as distinct from brain–object differentiation.

The empirical (i.e. neuronal) distinction between brain–self and brain–object differentiation by different time frames involves differentiation between self and objects (self–object differentiation). Based on this account, I assume that the main difference between self and objects is temporal, with both presupposing a different temporal framework, the self being associated with a more continuous and less periodic framework than objects.[6] Thus it is the temporal dimension which may enable and predispose to the self–object differentiation which, in the neuronal context, may resurface as different forms of trilateral interaction and hybrid neural activity (i.e. as different stimulus and temporal constellations).[7]

---

[5] For the concept of the object presupposed in the term "mental object," please refer to footnote 3.

[6] This might be interesting to consider for the specific case of self-objects as intermediate between mere objects and the self (see Chapter 7). Self-objects are supposed to exert a stabilizing, preserving, and maintaining function for the self, and it may be argued that this could be enabled and predisposed to by a temporal pattern that may come close to that supposed to underlie the self itself. One would consequently assume that the continuous presence of objects may enable and predispose to their transformation from mere objects into self-objects. Although this is certainly an interesting hypothesis, it cannot be pursued further here.

[7] One may finally raise the question for the body (see also the end of Chapter 7 for further discussion of this issue). Some authors consider the body to be part of the self when they assume what they call the "proto-self" as distinct from the non-bodily mental self (Panksepp, 1998; Damasio, 1999; Northoff and Panksepp, 2008). This ambiguity of the body as self and non-self is mirrored in the phenomenological characterization of the body as both "lived body" and "objective body." While the concept of the "objective body" describes the body as mere environmental object, the phenomenological concept of the "lived body" refers to it as mental object, thus oscillating between the two extremes. I tentatively postulate that this ambiguity

Methodological issues

How to experimentally investigate introjection directly

I have so far developed neuropsychodynamic hypotheses about the relationship between introjection/projection and the corresponding neuronal mechanisms. This yielded novel insights into the psychodynamic concepts that may then, for instance, be specified by value and reward, intero-exteroceptive linkage, and temporal extension. Although I indicated novel conceptual and empirical insights, the methodological strategy (i.e. how introjection can be investigated empirically using the techniques and methods of current neuroscience) remains unclear. I therefore want to briefly discuss some possible methodological strategies for future empirical investigation, focusing on introjection.

How should a psychodynamic process such as introjection be investigated? Should it be investigated directly or indirectly? Direct investigation means that the neurophysiological data obtained are supposed to be related to introjection itself, without any mediating function. I shall suggest two strategies for such direct investigation, namely simulation and high versus low comparison. Indirect investigation assumes that the neurophysiological data obtained do not mirror introjection itself but rather some associated processes, functions, or mechanisms that are supposed to be closely related to introjection. I shall suggest two psychological indirect strategies, namely affective–cognitive functions and comparison with basic functions, and one indirect neuronal strategy.

The first direct strategy is simulation. This strategy assumes that the process of introjection can be simulated in the experiment. Subjects may thus perform tasks that require the process of introjecting, including its different ingredients. The validity of the data depends strongly on the task. One of the main challenges here is the specificity of the task for introjection. Introjection cannot be considered in isolation from other defense mechanisms such as projection, rationalization, etc. It is not that we employ only one specific defense mechanisms in specific situations. Rather we employ different defense mechanisms at the same time as well as in different situations, and even within the same situation we usually recruit several different defense mechanisms. Introjection thus cannot be considered in isolation either from other defense mechanisms or from the respective environmental and personal context.

Experimentally, we are therefore limited to investigating constellations or configurations of different defense mechanisms, with some standing out and others remaining in the background. This means that the relationship of introjection to other defense mechanisms needs to be included as an experimental measure, rather than introjection being considered to be isolated and independent of other defense mechanisms.

Furthermore, the fact that the recruitment and use of defense mechanisms are strongly dependent on the specific personal and environmental context indicates that the respective context should also be included as one experimental variable. The task may then be to focus on investigating the relationship between these two variables (i.e. the variation of the relationship between introjection and other defense mechanisms in dependence on the personal and environmental context, and vice versa). For such an experimental set-up to be created, the following steps may be necessary. First, we need to develop a task that has a strong requirement for introjection.

---

may, within the neuronal context, correspond to the specific time frame of the interoceptive input, which seems to be intermediate between those of the periodic exteroceptive input and the continuous neural input. However, this has yet to be demonstrated.

Secondly, we need to define the personal and environmental context (i.e. the situation) in which predominant involvement of introjection in the task becomes possible. Thirdly, we need to define subjective–phenomenal and behavioral (and possibly vegetative) parameters that independently validate the presupposed psychodynamic process (i.e. introjection). Once these requirements are fulfilled in prior phenomenological and behavioral investigation of introjection, one may move on to functional brain imaging, develop an appropriate and suitable design, and implement the task in, for instance, fMRI.

The other strategy of direct investigation of introjection is to compare high versus low introjection. One could, for instance, investigate subjects with a high degree of introjection in functional brain imaging and compare them with subjects who show a low degree of introjection. This strategy is often employed in brain imaging of personality traits and other task performers (high versus low performers) (Northoff et al., 2006; see also Fan et al., 2010 for details of the application of such a strategy with regard to narcissism; see Chapter 7). However, the problem here is how to obtain valid and reliable measures of introjection that can distinguish subjects specifically with regard to their degree of introjection, while other variables remain fixed. If there is insufficient specificity in the experimental measures of high and low introjection, one may measure other factors that either dependently or independently vary between high and low introjecting subjects.

This implies that the experimental data may not really correspond to the difference between high and low introjection, but rather to other variables which may be dependent on or independent of introjection. In addition to such an inter-subject design, which compares subjects with high introjection with those who show low introjection, one may also employ an intra-subject design in which high and low degrees of introjection may be induced in the same subject by two variations of a specific task. Taken together, although viable, both direct strategies present some methodological problems. I shall therefore now focus on indirect strategies for investigating the neural activity underlying introjection as mediated by a third variable.

How to experimentally investigate introjection indirectly

A rather indirect strategy of investigation is to focus on those affective and cognitive functions that are supposed to be implicated in introjection. This presupposes an exact analysis of which affective and cognitive functions are involved in introjection, which may need to be even more specific and detailed, thus going beyond what Kernberg described so nicely. For instance, emotional perception and judgment may be strongly recruited in introjection, whereas emotional reappraisal appears to be less predominant (as it is, for instance, in intellectualization and rationalization). Working memory and especially episodic and autobiographical memory may be strongly implicated in introjections, though only in their implicit mode, not their explicit mode, otherwise introject cannot be constituted. Spatial cognition may also be involved, which allows representation of an object as introject in a mental form, which also requires visuo-spatial attention. Creating a specific constellation of these functions in an affective–cognitive task may then be considered to come close to and at best mirror introjection.

However, the question remains whether such an affective–cognitive set-up really captures the psychodynamic process of introjecting, and particularly its subjective–experiential aspects. Furthermore, it may be suggested that it still leaves out a more basic function which seems to be crucial, especially in introjection (i.e. the ability to relate and consequently to internalize objects from the environment to/within one's self). Panksepp (1998) has referred to seeking in this context a specific function that describes the ability to reach out and relate to the environment. Since relating and internalization are crucial in introjection, one may assume that seeking plays a crucial part in introjection. This example demonstrates that it may be difficult (if not largely impossible, because it leaves out the subjective aspects) to quantify and experimentally set up the affective–cognitive constellation that may be specific for introjection.

Finally, the comparison of introjection with a more basic function may be another indirect strategy. Even if introjection itself remains unclear, one may still be able to assume some related processes, such as reward seeking, etc., that are better defined and have already been successfully investigated experimentally. Although it does not concern introjection itself, one such example has recently been published in an imaging study conducted by our group. Self-relatedness is a rather difficult concept that remains unclear due to its essentially subjective nature, among other reasons. However, what is clear is that self-relatedness implicates in some way the concept of value (i.e. that a certain object possesses specific personal significance and meaning of a specific stimulus to a person or organism) (Montague et al., 2002, 2006). Value in turn is implicated in reward, which has been thoroughly investigated in both animals and humans.

Using fMRI, de Greck et al. (2008) compared self-relatedness with reward by using exactly the same stimuli for both tasks (see also Enzi et al., 2009; Ersner-Hershfield et al., 2009). They observed increased neural activity in reward circuitry during both reward and self-relatedness, although the latter showed prolonged neural activity when compared with the former. This indicates that reward circuitry and value are implicated in self-relatedness. Is our self thus nothing but reward? The study by de Greck et al. (2008) points out some differences in the duration of the BOLD response associated with self and reward. Based on raw data analysis (rather than model-based analysis), they observed sustained BOLD signals during high self-related, i.e. personally relevant stimuli in the aforementioned reward circuitry, while reward induced shorter and more phasic signals in the same regions. Although, due to the sluggishness of the BOLD response, this finding must be regarded as preliminary, these researchers further substantiated the assumption of possible dissociation between self and reward by investigating psychiatric patients. They observed that both detoxified alcoholic patients and pathological gamblers showed "normal" neural activity changes in VMPFC, VTA, and ventral striatum (VS) during reward, but that the self-task (judgment of the stimuli as self- or non-self-related) did not induce any activity changes in the same regions in both patient groups (de Greck et al., 2010).

Although preliminary, this study can be regarded as an example of how unknown seemingly higher-order processes, such as self-relatedness, may be investigated by comparing them with better known but rather lower-order processes, such as reward, which provide both psychological and neurophysiological information about the former. While this example concerns self-relatedness and reward, it may in an analogous way also apply to introjection when, for instance, comparing a task that requires introjection with one for self-relatedness, reward, perception, memory, or emotions (in order to support the hypothesis previously stated). Finally, one may also investigate specific neuronal mechanisms, such as stimulus–rest interaction, or different stimulus and temporal constellations which vary in degree and include introjection as the dependent (rather than independent) variable in the experimental design. This would be an indirect neuronal strategy, and it could also be used to test for the postulated neuropsychodynamic hypotheses.

Given our hypothesis that different stimulus and temporal constellations enable and predispose to the balance between introjection and projection, future research may also want to focus on the various kinds of interactions postulated here. More specifically, one could envisage studies that investigate the interaction between interoceptive processing and self-relatedness, and see how this affects the balance between introjection and projection. Another possibility here is the timing of the stimuli in such interaction designs, which is probably better investigated with EEG than with fMRI. This would allow us to test in particular the temporal hypothesis of the interaction and predominance between the different temporal frameworks (i.e. periodic, semi-continuous, and continuous) that are associated with the distinct stimuli (i.e. exteroceptive, interoceptive, and neural) and the respective objects (i.e. environmental, bodily, and mental).

Methodologically, these interaction studies may target the proclaimed neuronal mechanism and tests for its empirical plausibility. The psychoanalyst may now want to object that this is all fine but it misses the point because it does not link neuronal to psychodynamic mechanisms. He would be both right and wrong here. He is right in that we do not test the psychodynamic mechanism directly by itself as an independent variable in our neuronal investigation, and thus remain unable to reveal its sufficient conditions (i.e. its neural correlates) (see Introduction). However, what I suggest here as an indirect strategy is to test first the presumed neuronal mechanisms and then the psychodynamic mechanisms. This means that the experimental design (i.e. its independent and dependent variables) is neuronal, with no consideration of any psychodynamic variables at this stage. However, these can be included later as dependent variables once the neuronal mechanism itself is clear and can be taken by itself as an independent variable. Rather than revealing the neural correlates (i.e. the sufficient but non-necessary conditions), such a methodological strategy targets those neural conditions that are necessary but non-sufficient, and thus what I called in the Introduction "neural predisposition."

Pursuing such an indirect methodological strategy that targets neural predispositions rather than neural correlates will tell us not only about how our brain enables and predisposes to specific psychodynamic mechanisms, but also how it functions as such. Thus it will provide us with plenty of insights, such as why there are mental states at all, why we experience and perceive our own organism as self and, most importantly, how that is enabled and predisposed to by our brain's specific functioning. And, as I understand him, this is exactly what Freud had in mind when he embarked on the adventure of psychoanalysis, namely to understand human nature and more specifically its mind or "psychic apparatus" as he calls it.

Acknowledgments

I am grateful to Marina Farinelli and Stephan Doering for reading and providing helpful comments on previous drafts of this chapter.

Part III

Mental Equipment

I have so far discussed the conceptual and neural equipment necessary to understand brain–self and brain–object differentiation. The discussion of the conceptual equipment discussed how to approach the brain as well as different concepts of the brain, while the discussion of the neural equipment focused on the brain's own contribution (i.e. its intrinsic activity) and how it affects the brain's own neural processing of intero- and exteroceptive stimuli from the body and the environment. Thus I assumed that the brain's intrinsic activity may correspond to what Freud called "cathexis," namely the investment of energy.

I demonstrated that the brain's intrinsic activity involved a specific kind of neural coding of the brain's neural processing of bodily and environmental stimuli in terms of the difference between brain and stimuli. This amounted to what I described as difference-based coding (see Chapter 4). Difference-based coding in turn was shown to be crucial in transforming the brain's neuronal states and its neural processing of intero- and exteroceptive stimuli into mental states and objects (i.e. neuronal–mental and stimulus–object transformation) (see Chapter 5). This provided the neural basis or neural equipment for the subsequent investigation of how self and objects differentiate from the brain (i.e. brain–self and brain–object differentiation) (see Chapter 6). I here shifted my focus to early defense mechanisms like internalization and externalization, and more specifically to projection and introjection, which are supposed to be central in constituting (and later defending) self and objects as distinct from brain, body, and environment (see Chapter 3).

The part about the neural equipment provided us with insight into how the brain's neural mechanisms and neural coding enable and predispose to brain–self and brain–object differentiation. This let us understand the transition from the brain's neural apparatus to the psyche's psychic apparatus. There is though more to the psychic apparatus than mere self and object. The psychic apparatus shows certain mental features or mental equipment, as I here call it, which was the target of Freud's and his follower's psychodynamic investigations. I here want to pick out some specific features of the mental equipment while neglecting others. Freud assumed the psychic apparatus to be characterized by narcissism, with followers like Kohut assuming a specific kind of object (i.e. self-objects). And Freud assumed the psychic apparatus to function in a mostly unconscious rather than conscious mode and, most importantly, he characterized it by his famous topological distinction between the ego, the id, and the superego.

The complex mental equipment of the psyche's psychic apparatus indicates that we need to go beyond mere brain–self and brain–object differentiation. More specifically, we need to understand how self and object themselves can further differentiate into the kind of mental equipment Freud and others attributed to them, as for instance is manifested in narcissism and self-objects, unconscious mode, and the threefold distinction between ego, id, and superego. I here postulate that the further differentiation of self and objects into more specific mental equipment is enabled and predisposed to by the brain's neural structure and organization. This means that the kind of neural equipment that I discussed in Part II has laid the groundwork or, to resume my comparison with the Scottish castle from the Introduction, the neural platform for the mental equipment. By being shaped in a certain way,as for instance by difference-based coding, our brain's neural platform enables and predisposes to the development and constitution of a certain kind of possible mental equipment (while thereby at the same time excluding other kinds of mental equipment that hitherto remain impossible).

Before venturing into the mental equipment, one final comment is needed about the methodological approach presupposed here. When discussing conceptual equipment, I took a stance in the conceptual (and epistemic) domain and henceforth considered the brain in predominantly a conceptual and epistemic context. This stance (or perspective as one may also want to call it) was shifted in the discussion of neural equipment, where I switched from the conceptual–epistemic to the empirical domain, and more specifically to the neural states of the brain itself. The brain was consequently no longer considered in a conceptual–epistemic context, but rather in a neural context. This allowed me to investigate what the brain itself contributes (i.e. its neural equipment) to its own neural processing of intero- and exteroceptive (i.e. bodily and environmental) stimuli.

In Part III about mental equipment I now change the context again. Rather than considering the brain in its neural context, I move to the mental context and investigate the brain in the context of the kind of mental equipment that psychoanalysis has analyzed so well. My methodological direction is consequently different now. Rather than proceeding forward from the brain's neural states to the constitution of mental states, as in Part II about neural equipment, I now take the opposite direction (i.e. a move backwards) when inferring from the mental equipment to the brain's neural structure and organization and its specific features that allow it to enable and predispose to the former.

Again, as already mentioned, I must emphasize that I am not able here to recount all the empirical and psychodynamic details that any specialist (be they a neuroscientist or a psychoanalyst) expects and is used to from his intradisciplinary background. Both empirical data and psychodynamic concepts are only detailed in so far as it seems useful and relevant for linking them with each other. In contrast, I have ignored those data and concepts that seem to be of purely neuroscientific or psychoanalytic relevance, but which are not relevant to the neuropsychodynamic context presupposed here (see also the Epilogue for an account of the tension between a transdisciplinary approach and intradisciplinary expectations).

Chapter 7

Narcissism, Self-Objects, and the Brain

Psychodynamic concepts

Concept of narcissism in Freud

Freud first mentioned the concept of narcissism in a footnote that he inserted in his essay on a theory of sexuality, where he associated it with self-love in homosexuality (Freud, 1905). This sets the stage for Freud's determination of narcissism in a predominantly sexual–libidinal context in a later writing on Leonardo da Vinci, whom he characterizes by a repressed form of homosexuality which Freud associates with narcissism (Freud, 1910).

However, Freud's concept of narcissism changed over the years, with Mancia (2004), for instance, assuming five different stages in Freud's usage of the term. The first stage is clearly sexual, pertaining to autoeroticism and homosexuality, as is apparent in his writing on Leonardo da Vinci. This is followed by a more complex concept of narcissism, which is related to withdrawal of the cathexis from objects and its shift back to the ego (this is the second stage). In the third stage he again associates narcissism with instincts, thus returning somehow to his first meaning. This is followed by a fourth stage where, on the basis of his topological distinction, narcissism no longer describes the investment of energy in the ego but in the id. Finally, the fifth stage puts narcissism in the context of masochism and the death instinct. I here follow mainly the first and second meanings, thereby also implicating the third and fourth contexts while ignoring the fifth one.

What does the term narcissism describe in Freud's work? Following Freud (1914), the concept of narcissism describes a specific stage in the development from the primitive autoeroticism of the infant to the mature object love of others. The narcissistic person remains in the stage of self-love, loving themselves and their own body, and consequently becomes homosexual because they remain unable to direct their love towards other people of the opposite sex (i.e. objects) to constitute object-love: "(Narcissism) is a necessary step in the transition from autoeroticism to object love. The infatuation with the self (= genitalia) is a necessary developmental stage. From there, one moves on to similar object" (Freud's 1909 speech, cited in Nunberg and Federn, 1977).

Let us be more specific. Freud (1914) distinguishes between primary and secondary narcissism. Primary narcissism describes the initial love of the own body and its genitals. The libido is thus applied to the person's own body, resulting in self-love and autoeroticism. The libidinuous cathexis, the libidinal energy, is released into their own body, resulting in what Freud described as "a libidinous extension of the egoism of the self-preservation instinct" (Freud, 1914, p. 138). Following Freud (1914), we are born into a state of primary narcissism in which the libido is cathected to one's own ego and its body. However, development takes place, which allows us to withdraw the libido from our own ego, including its body, and to redirect it towards objects which enable us to outgrow our narcissism and become object-related rather than body-related, as one might want to say. Love of other objects (i.e. people) replaces the love of one's own ego and its body, with the consequent replacement of homosexuality by heterosexuality. Primary narcissism is thus nothing but a stage that is to be overcome by directing the libidinous cathexis towards other people and their bodies rather than one's own ego and its body.

However, if some disruption takes place in the respective developmental stages, such as a traumatic event, one may fall back into the earlier stages of development. More specifically, the libido may then withdraw from the objects outside one's own ego (i.e. other people) to one's own ego and its body, thus falling back into a narcissistic stage which Freud calls secondary narcissism (Freud, 1914). In addition to such developmental disruptions, narcissism may also occur in our daily life, as for instance during sleep and dreams in which the libido is withdrawn from the objects of the external world and narcissistically invested in the ego.[1] Other instances of secondary narcissism include pathological conditions such as depression, where the objects are lost so the cathexis has no choice but to be invested in its own ego (which thus identifies itself with the lost object; see Chapter 10 for further details).

Freud's concept of narcissism has a predominant negative implication in that it describes an immature developmental stage that has to be overcome on the way towards the more mature stage of object-love (i.e. love of other persons as distinct from love of one's own person). Such negative connotation is well mirrored in Freud's consideration of narcissism as a regressive phenomenon that can be described as "object avoidance." This is the case in, for instance, psychosis which Freud describes as fixation in a stage of primary narcissism (Freud, 1914) (see Chapters 11 and 12).

Taken together, Freud's concept of narcissism can be characterized as developmental, sexual, and negative. His concept of narcissism is developmental in that it describes one early and immature stage of development, which has to be overcome by subsequent developments into more mature stages. And it is sexual in that it concerns the self-love of one's own body, including its genitals, resulting in homosexuality as distinct from homosexuality. Finally, narcissism has a rather negative connotation in that it is just a developmental and rather immature stage or regressive phenomenon that hinders the more mature stage of object-love, and therefore needs to be overcome. Freud's threefold characterization of the concept of narcissism as developmental, sexual and negative was challenged by Kohut and his development of self psychology, to which we shall return now.

Concept of narcissism in Kohut

Following Freud, Melanie Klein (1935) rejected Freud's presupposition of narcissism of an essentially object-less state. Already the infant narcissistically invests energy in its relationship with the mother, as for instance during breastfeeding. Narcissism can thus not be considered an object-less state but must rather be regarded as a specific way of relating to objects, with the latter having primary importance even for the narcissistic self. Instead of the instincts and the libido, we thus have to investigate the objects themselves and more specifically our relationship to them in order to understand narcissism. This sets the stage for a relational concept of narcissism on the basis of specific object relations, which was further developed by Kohut as the founder of self psychology.

Kohut (1971, 1977, 1984) considers all three characteristics of Freud's concept of narcissism in a broader context, which leads him to revise it (see Figure 7.1 for a comparison of Freud

[1] One may, however, also take a different perspective. Cathexis is no longer invested in real and physical objects in sleep and dreams. Instead it is invested in imaginary and mental objects which may though be ultimately traced back to the former ones. If so, sleep and dreams cannot be characterized by mere withdrawal of cathexis from objects and retreating back to the ego, which means that sleep and dreams may be as (object-)relational and thus as narcissistic or non-narcissistic as the awake state. This is the position I shall take later, as it will surface in Chapter 8 on the unconscious.

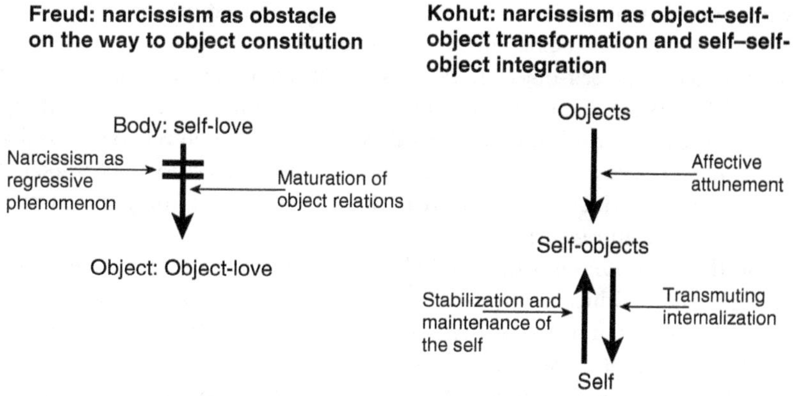

Fig. 7.1 Concepts of narcissism in Freud and Kohut.

and Kohut). First and foremost, he no longer exclusively considers narcissism within the libidinal–sexual line of development. Instead, in his earlier writings he assumes a second line of narcissistic development parallel to the libidinal one. However, he later replaces the two developmental lines (libidinal–sexual and narcissistic) with one supraordinate one with narcissism as its characteristic feature (Fosshage, 2009, pp. 4–5; Hartman, 2009, p. 86).

Kohut's later assumption (Kohut, 1977) that narcissistic development is supraordinate means that narcissism is crucial in preserving and maintaining the self (see also Fosshage, 2009, pp. 4–5). This, though, implies that narcissistic development is no longer a mere developmental stage that simply needs to be overcome to achieve the real goal (i.e. object-love). Instead, narcissism itself must then be considered crucial in allowing and enabling object-love. This means that narcissism is no longer assumed to hinder object-love, as in Freud, but rather to enable and predispose to it and thus to make it possible (or in more philosophical terms, it is a necessary condition). Thus narcissism is considered to be the ground or basis upon which any kind of self and its object stands, and it is therefore self- and object-maintaining and sustaining.

The shift of the concept of narcissism from Freud's narrow developmental–sexual context to Kohut's broader context of (self-)objects[2] that maintain and sustain the self which has a positive rather than negative connotation. And most importantly, it shifts the question of the self and how it achieves a relationship to objects to the centre of attention, the primary self–self-object relation as one may call it (Hartman, 2009, p. 91). This may be divided into two questions. First, how should the self be constituted to enable and predispose it to narcissistically reach out and direct its energy towards objects? Secondly, how must the objects be shaped and constituted by the self in order for them to be related to the self in such a way that they can serve the kind of self-maintaining and self-sustaining function that Kohut considers central to narcissism?

Let me start with the first question. How should the self be constituted to enable and predispose it to narcissistically reach out and direct its energy towards objects? Kohut (1977, 1984) assumes that there is development of the self from a "virtual self" through an "archaic self" to a "mature self." The infantile omnipotence and grandiosity of the initial virtual self is gradually replaced by the archaic self, which is characterized by merging needs that gradually recede into the background.

[2] Kohut introduces the term self-object to characterize the special relationship of the self to specific objects that serve to maintain and sustain the self. This will be discussed later on. Thus when speaking of objects in this sense, I shall denote this by putting parentheses around the term "self" before the term object.

Most importantly, the archaic self is in need of mirroring, idealization, and admiration by other objects (i.e. persons), such as the parents, in order to develop further to a "mature self" that is related to specific objects (i.e. self-objects), and thus become embedded in what can be called a "self–object matrix" (Wolf, 1980). Such development of the self leads to what is called "self-coherence," namely the structure and organization of the self as it is related to objects (i.e. self-objects). The degree of self-coherence determines, for instance, how a self can react to and compensate for loss or failures of self-objects, which in turn, as we shall see later on, determines the degree to which narcissistic personality features surface.

The view on the development of the self already indicates the central role of what Kohut described as "self-object," which leads us to the second question. How must the objects be shaped and constituted by the self in order for them to be related to the self in such a way that they can serve the kind of self-maintaining and self-sustaining function that Kohut considers central to narcissism? Kohut describes those objects that are important for constituting, maintaining and preserving the self as "self-objects," and distinguishes them from "mere objects," which are those objects that are not relevant to self-constitution, self-maintenance and self-preservation.

How can an object that is irrelevant to the self be transformed into a self-object that is relevant to the self? Kohut assumes that there are three central needs and processes, namely admiration, mirroring and idealization. The self wants to be admired and the person who admires the self becomes a self-object (e.g. mother or father). The self wants some affective resonance or mirroring as Kohut describes it, and the objects who provide that become self-objects. Finally, there is a need for idealization of, for example, the parents, resulting in an idealized parent image. These three needs (admiration, mirroring and idealization) do thus allow the self to distinguish relevant from irrelevant objects and thereby to constitute what Kohut calls self-objects which, unlike mere objects, are important for constituting, maintaining and preserving one's own self.

Let me be more specific. In the early developmental stages, self-objects take over the function of self-constitution, self-maintenance, and self-preservation from the external caregiver. By a process of what Kohut (1984) calls "transmuting internalization," the developing self becomes able to gradually take over the self-maintaining and self-preserving functions provided by the external caregiver. This allows the regulation and its control to be exerted by the self itself rather than by the external caregiver.

However, the need for self-objects does not stop once the self is constituted. Instead, self-objects need to be constituted continuously throughout life in order to maintain and preserve the self. In other words, even as adults we are still in need of some form of admiration, idealization, and mirroring to constitute self-objects that are relevant and important for maintaining and preserving our self and its self-coherence, including self-regulation. Self-objects are thus important and relevant throughout life, so that narcissism is considered a basic dimension of human existence and its self. Accordingly, in contrast to Freud's conception, narcissism is no longer a mere developmental stage that needs to be overcome, but rather it is a continuous and existential need throughout life.

In summary, Kohut converts Freud's developmental, sexual, and negative connotation of the concept of narcissism into an existential, affective, and positive context. Narcissism is no longer a mere developmental stage, but a need that persists throughout life, thus being existential rather than developmental. The purpose of narcissism is not merely to clear the way to object-love, but to allow for the constitution of self-objects that are central in self-constitution, self-maintenance and self-preservation. This detaches the concept of narcissism from the predominantly sexual context in Freud to a more affective (and cognitive) context, also manifested in the three needs of admiration, idealization, and mirroring. Finally, the concept of narcissism assumes a positive rather than negative connotation in that it is central to constituting, maintaining, and preserving the self. In short, the developmental, sexual, and negative characterization of Freud is here replaced by an existential, affective, and positive concept of narcissism.

Such redesignation of the concept of narcissism as existential, affective, and positive entails a broader context when compared with Freud's developmental, sexual, and negative characterization. Let us return briefly to Freud himself. Pulver (1970), for instance, distinguishes between four different usages or contexts of the concept of narcissism in Freud. He uses narcissism in a genetic context to describe a specific stage in development, he refers to it in a clinical sense to describe specific forms of sexual perversion (i.e. homosexuality), he describes specific forms of object relations (i.e. in relation to one's own body) in narcissism), and he considers the concept and the associate narcissism with the regulation of self-esteem.

These different contexts (genetic, clinical, bodily, and psychological)[3] resurface in Kohut, although in a different and much broader way. Here the concept of narcissism is no longer used exclusively in a genetic–developmental sense. Instead, the genetic–developmental sense describes just one aspect of the broader existential function of narcissism persisting throughout life. This means that narcissism is shifted from the clinical–pathological context into a rather healthy or existential context. In a similar way to Freud, narcissism describes specific forms of object relation, although in a different sense. Kohut no longer distinguishes between "self relation" and "object relation" as Freud does, but rather, on the basis of his concept of self-objects, between "self–object relation" and "object relation." The relation to the body may be one such self–object relation, which may though be complemented by others (i.e. non-bodily self–object relations). Finally, Kohut also regards narcissism as essential for regulating self-esteem, as in Freud, but its importance goes beyond that, as it is relevant to sustaining and maintaining the self as such, and not just its self-esteem.

Pathological narcissism and narcissistic personality disorder

Clinically, Freud regarded narcissism as a sexual disorder and more specifically as self-love or homosexuality. This changed in Kohut's concept, where clinically narcissism is considered to be a "self-object disorder" with symptoms that arise from a temporary loss of self-coherence. Self-coherence describes the self-state which can be characterized by intrapsychic structure and organization of the self. In the case of low coherence or even incoherence of the self-state (i.e. self-incoherence), self-objects cannot be constituted, resulting in a lack of self–object experience. Lack of self–object experience triggers various processes to resolve and compensate for this deficit, which Kohut calls "autoplastic adaptation." The concept of "autoplastic adaptation" describes essentially compensatory mechanisms aimed at restructuring and reorganizing the psychological structure and organization of the psychic apparatus to cope with the lack of self-objects. The recruitment of such compensatory mechanisms may, however, lead to symptoms like depression, hypochondria, hypersensitivity, and lack of vitality (Kohut and Wolf, 1978; Wolf, 1988; Hartman, 2009). Although not spelled out by Kohut himself, such autoplastic adaptation may also be crucial for generating symptoms in other disorders such as depression or psychosis, which will be discussed in Part IV of this book. Most importantly, the concept of autoplastic adaptation means that the symptoms are not the effect or the cause of the respective disorder, but rather the consequence or effect of the compensatory and rather healthy mechanisms that are set in motion by the cause, as is apparent in pathological narcissism.

[3] As well as these different contexts, one may also assume a sociological context for the concept of narcissism when it is used to describe the behavior of whole groups of individuals or even of society as a whole (Hartmann, 2009, pp. 87–88).

The concept of autoplastic adaptation means that the clinical symptoms of narcissism arise from the compensatory mechanisms of self-object failure, rather than being identical to self-object failure itself. Since it is essential for the kind of compensatory mechanisms that can be set in motion when encountering self-object failure, the degree of self-coherence must be considered crucial in providing the transition from healthy to clinical (i.e. pathological) narcissism. The higher the degree of self-coherence, the better compensatory mechanisms can be recruited to resolve the self-object failure, and the better the latter can be compensated, resulting in less clinical symptoms than in the case of lower self-coherence (Hartman, 2009, p. 91). Clinical symptoms of narcissism arise therefore as a result of an inability to cope with self-object failure, which in turn may be closely related to the degree of self-coherence. For instance, in the case of high self-coherence, sexual and aggressive behavior by another person may be integrated into the self by transforming the respective person into a self-object that as such can be perceived and experienced as sexually attractive. While in the opposite case, namely low self-coherence (i.e. self-incoherence), going along with self-object failure, the other person may instead be perceived and experienced in terms of mere sexualization and aggressiveness, and thus as a threat that induces (narcissistic) frustration-related rage (this is a slight modification of the basic idea described by Hartmann, 2009, p. 92).

In addition to the degree of self-incoherence, the respective biological and social contexts also determine whether a specific behavior remains within the normal range of narcissism or must be considered to transgress the border to pathological narcissism. Hartmann (2009, p. 92) cites the example of a 16-year-old boy who blow dries his hair for an hour in front of the mirror, which may be regarded as normal healthy age-related narcissism, whereas it would be regarded as pathological in a 45-year-old father. Complementing the biological context, the social and cultural contexts and their respective habits may also need to be considered when making the distinction between normal (i.e. healthy) and pathological narcissism.

Following Kohut, the inability to cope with self-object failure in the case of self-incoherence may lead to various clinical symptoms, including emptiness, futility, hypochondria, depression, labile self-esteem, lack of vitality, hypersensitivity to perceived affronts, and addictive or delinquent behavior (Kohut and Wolf, 1978). These clinical symptoms have been subsumed under the concept of a hypervigilant, covert, and thin-skinned type of narcissism (Gabbard, 2005a,b; Hartmann, 2009, p. 92) or vulnerability–sensitivity–depletion (Wink, 1991; Ritter and Lammers, 2007; Cain et al., 2008). Kohut's concept of pathological narcissism needs to be distinguished from that described by Kernberg (1984, 1988), which is characterized by fantasies of grandeur and exhibitionistic and aggressive behavior. Symptoms such as aggressive and arrogant behavior, self-absorption, inability to receive from others, insensitivity to feelings being hurt by others, and the need always to be the centre of attention are typical of this type of narcissism (Gabbard, 2005a,b). These symptoms have been subsumed under the description of the oblivious, overt, and thick-skinned type of narcissism, or grandiosity–exhibitionism, which as narcissistic personality disorder (NPD) entered the current classification of psychiatric disorders in *DSM-IV* (Wink, 1991; Cain et al., 2008; Hartmann, 2009; Thomaes et al., 2009).

In contrast to Kohut, Kernberg (1984, 1998) considers that frustration-related narcissistic rage results not from the inability to compensate for self-object failure, but rather from the fixation on drive-related oral aggression. This may be traced back to early emotional deprivation of the infant by emotionally cool, absent, or even latent aggressive parents to which the infant may react with rage and frustration, which in turn may reinforce the parent's behavior. To escape this cycle and to receive some emotional involvement from the parents, the child will focus on those aspects that may be appreciated and emotionally gratified by the parents, such as performance, good grades, etc. At the same time, fantasies of the self as being grandiose and fantastic develop (see also Raskin and Novacek, 1991) and are dissociated from the real self, thereby predisposing the individual to develop the aforementioned symptoms of grandiosity–exhibitionism, amounting to what nowadays is described as NPD.

What are the implications of these clinical considerations for the concept of narcissism? The relevance of the degree of self-coherence and the kind of context (i.e. biological, social, or cultural) in distinguishing between healthy and clinical narcissism implies a continuum rather than a dichotomy between normal and pathological narcissism (see also Kohut, 1984; Hartmann, 2009). This means that the concept of narcissism may apply to both healthy and clinical subjects. While in healthy subjects it describes a specific personality trait, it concerns two different types of symptoms in the clinical context as described by Kohut and Kernberg. The common roots of both healthy and pathological narcissism are to be found in the constitution of self-objects and the ability to compensate or adapt to self-object failure. Therefore, in order to better understand pathological narcissism and thus NPD, we need to go back to the basics (i.e. what self-objects are and how they are constituted, amounting to what I call "self-object constitution"). Due to their primary importance in narcissism, in the remainder of this chapter I shall focus on self-objects in both conceptual and empirical contexts.

Concept of self-objects and the "self-object relation"

What exactly are self-objects? They are those objects that constitute, maintain, and preserve the self of a person. As such they must be distinguished from mere objects that, following Kohut, do not have such a maintaining, preserving, and constitutive function for the self. Although self-objects often refer to other individuals (e.g. mother, father, spouse, friends, etc.), they do not necessarily have to concern people. Music, arts, religion, or other objects can also become self-objects so long as they contribute to the homeostasis and regulation of the self, thus being central to what I referred to as self-constitution, self-maintenance, and self-preservation (Gehrie, 2009, p. 35; see also Kohut, 1984).

The possible reference of self-objects to both persons and non-persons immediately raises what Gehrie (2009, p. 35) has called the central problem in defining self-objects. Self-objects may either refer to a specific way of perceiving and experiencing objects, in which case they can be characterized as a specific intrapsychic experience, or they may concern specific objects outside the self, which means that they are intersubjective rather than intrasubjective or intrapsychic. This is well expressed in the following quote by Gehrie: "The central difficulty of the self-object concept centered on the implications of its definition; did it refer to an internal intrapsychic experience, or to an external 'object,' or both?"(Gehrie, 2009, p. 35). While Kohut himself seemed to oscillate between an intrapsychic and an intersubjective conception of self-objects, some of his successors elaborated especially the intersubjective view when for instance developing a relational self psychology and a therapy of personal relationships (for an excellent overview, see Milch, 2001, pp. 30–45). This emphasis on the intersubjective nature of self-objects is consistent with the concept of the object presupposed in object relation theory as developed by Ferency, Balint, Klein, and Winnicott (Milch, 2001, pp. 277–279; Hartmann, 2009, pp. 90–91).

Are self-objects intrapsychic (i.e. intrasubjective) or intersubjective? How can we resolve this ambiguity in the definition of self-objects? Let me return to the constitution of self-objects, or "self-object constitution" as I shall call it in the following account. Self-object constitution must allow for the transformation of a mere object into a self-object. This is only possible when experiencing a particular object in a subjective way that is specific to that particular person. Self-objects can thus be characterized by a specific individual subjective experience: "I suggest a return to the idea that the self-object per se is not a person in external reality but an internal experience that may occur with a person or a context in the external world and that may be used by the self for various functional purposes, including self-regulation. And this experience is the functionally important piece, even in the presence of an actual person" (Gehrie, 2009, p. 48; see also Milch, 2001, pp. 65–66).

How is this intrasubjective experience of self-objects possible? In other words, what enables and predisposes self-objects to be experienced intrasubjectively? In order for self-objects to be experienced intrapsychically (or intrasubjectively), they must first and foremost be related to the self. More specifically, the object must be related to the self, amounting to what I called in Part II of this book "self–object relation" (see Chapter 6). Such self–object relation enables and predisposes to the conversion of mere objects into self-objects, including their intrapsychic (or intrasubjective) experience by the self itself. Intrapsychic (or intrasubjective) experience may thus be regarded as the placeholder of the self-object relation on the side of the self, while the intersubjective concept of self-objects may be considered the placeholder of the self-object relation on the side of the object. The concepts of intrapsychic (or intrasubjective) experience and the intersubjective nature of self-objects may consequently describe the self-object relation from different perspectives, namely from the perspective of the self as in intrapsychic experience, and from the perspective of the object in the case of the intersubjective nature, while (and this is crucial) both refer to the self-object relation, although admittedly from different perspectives. Thus the difference between intrapsychic (or intrasubjective) and intersubjective conceptions of self-objects may be more perspectival than substantial.

Let me be more specific. Milch (2001, p. 278) has nicely shown that we can approach self-objects either from the viewpoint of object relation theory, thus taking the perspective of the objects themselves, or alternatively from the viewpoint of Kohut's self-object theory, thereby taking the perspective of the self. Following and extending Milch, I here make an epistemic distinction and distinguish between the self-object relation itself and our perception and cognition of it. This epistemic distinction is relevant to the way in which and what we can (and cannot) investigate with regard to the self-object relation, as will be briefly demonstrated below.

I assume that the self-object relation can be approached and thus considered from two different perspectives, namely the perspective of the self and the perspective of objects, with the former being presupposed in intrapsychic (or intrasubjective) and the latter in intersubjective conceptions of self-objects. Due to such a double perspective on self-objects, there is an (epistemic) ambiguity in the way that we can approach and investigate the latter. However, this (epistemic) ambiguity concerns only the perspectives from which we can perceive and cognize the self-object relation, and it does not apply to the latter itself, as it remains independent of our perception and cognition of it. The ambiguity is consequently more related to us as observers and investigators of self-objects than to self-objects themselves (i.e. as independent of our investigation and observation of them). In other words, the ambiguity is intrinsic to our perception and cognition of self-objects and thus purely epistemic, while it remains extrinsic to self-objects themselves, including their underlying self–object relation.

What are the implications of this for our characterization of self-objects as either intrasubjective or intersubjective? It means that the characterization of self-objects by self-object relation remains itself immune to the distinction between intrasubjective and intersubjective. Rather than being related to self-objects themselves and thus the self-object relation, the distinction between intrasubjective and intersubjective is associated with ourselves as observer or investigator and thus the way that we are (or are not) able to perceive and cognize the self-object relation. To characterize self-objects as either intrasubjective or intersubjective would thus be to confuse the self-object relation itself with our perception and cognition of it, and thus ultimately with our epistemic abilities and inabilities.

This shifts the focus from the intrasubjective–intersubjective dichotomy in the determination of self-objects to the self-object relation itself. What must the self-object relation be like in order to enable and predispose to the constitution of self-objects as described by Kohut? As pointed out, we can access the self-object relation only indirectly through our perception and cognition of self-objects. We may thus consider the features that we attribute to self-objects and may infer from them what the self-object relation itself must be like in order to enable and predispose to those features as we observe and perceive them.

Self-objects are considered to be specific for the particular individual self, which distinguishes that individual self from other individual selves. Self-objects may thus be characterized as individual and therefore be distinguished from mere objects that may not be so individual but rather more general. Moreover, as pointed out, self-objects are intrasubjectively experienced, which means that they cannot be shared with other subjects. This means that our access to the individualized objects as being constitutive of our own self is private, rather than public. What are the implications of this for the characterization of the self-object relation that is supposed to underlie self-objects? The self-object relation must be such that it enables and predisposes to the individualization of objects as well as the privatization of the individualized objects and the self itself. I shall therefore characterize the self-object relation in the following account by what I describe as "individualization of objects" and "privatization of the self."

Individualization of objects and privatization of the self

How can we characterize the individualization of objects and the privatization of the self in more detail? The guiding question here is as follows. How can the initially general and publicly accessible self-object relation be transformed such that it can enable and predispose to the individualization of objects (i.e. individual objects) and the privatization of the self (i.e. a publicly non-accessible self as stabilized by very private self-objects)? Let us return to Kohut himself and how he conceived the constitution of self-objects.

Kohut (1977, 1984) emphasized the central importance of mirroring and admiration, with both being strongly related to affective function. In particular he pointed out the central role of empathy or affective exchange and attunement with objects in the environment. It is the mirroring of one's own affect in the other (the mother, the father, or other objects) that allows the constitution, maintenance, and preservation of one's own self. For instance, the mutual affective attunement between infant and mother allows the former to constitute the mother as self-object. This means that, to use my phraseology, the affective attunement enables and predisposes to transforming the initially publicly accessible relationship to the mother or father, where they are merely some non-specific object, into a very private and individual relationship, where they are transformed from mere objects into self-objects. In other words, affective attunement allows the merely biological role of father or mother to be transformed into a psychological one. However, it should be noted that the affective engagement must be mutual and thus occur on both sides (infant and father or mother, or self and object), to enable and predispose to self-object constitution. If, for instance, the affective engagement is unilateral on the side of the self (i.e. the infant), but not reciprocated on the side of the objects (i.e. the mother and father), the transformation of the latter into self-objects for the former is blocked.

The relevance of mutual affect and empathy between self and objects has been emphasized not only by Kohut highlighting the crucial relevance of empathy, but also by recent psychological accounts of narcissistic personality disorder (NPD). Gehrie (2009, p. 36) assumes a negative affect in NPD which results in what he calls "negative self-objects" as "consequences of early self-object relations that were profoundly negative in quality." Assignment of predominantly negative emotions and affect to early self-objects may later result in difficulties in accessing (i.e. perceiving and identifying) one's own positive emotions and the other person's positive emotions, as has also been emphasized by Dimaggio et al. (2007a,b, 2008a,b, 2010) in patients with NPD.

Kohut and self psychology pointed out that emotions and affect (I use the two terms interchangeably here for the sake of operational simplicity) are central to (although not sufficient for) the transformation of an object into a self-object. Once an affect or emotion has been assigned to an object, this object can become relevant and important for the self, and can thereby be transformed into a self-object. Affective assignment may thus enable and predispose to what can be called "object–self-object transformation." However, since the relational concept of self-object as self-object relation considers both sides, namely that of the object and that of the self, there must be more to self-object constitution than mere affective assignment with consecutive object–self-object transformation. This means that object–self-object transformation is by itself not a sufficient but only a necessary condition of self-object constitution. If so, object–self-object transformation only enables and predisposes objects to become possible self-objects, while not by itself constituting the actual self-object. Strictly speaking, one may therefore want to speak of object–possible self-object transformation. In the following account, for the sake of brevity, I refer to this as object–self-object transformation.

What is the additional process besides object–self-object transformation that is necessary for the constitution of actual self-objects? The affect as reciprocated by the object has to be internalized by the self in order for it to become true affective (bilateral) attunement which, following Kohut, allows the transformation of the object into a self-object. Kohut described this process of internalization by the self as "transmuting internalization," since it is by the internalization of the other's affect that the mere object, or possible self-object as I would say, is transformed or transmuted into a self-object (thereby also transforming the self itself) (see also Schaefer, 1972; Kohut, 1984; Gehrie, 2009, p. 48; Hartmann, 2009, p. 94). Thus internalization in general and introjection in particular (see Chapter 6) may be crucial in transforming and transmuting the possible self-object, as provided by object–self-object transformation, into an actual self-object for the self.

The concept of transmuting internalization describes how the internalization of the object goes along with the transformation of the possible self-object into an actual one. What exactly happens in internalization? Internalization describes how the possible self-object as distinct from mere objects is linked to and integrated within the self, and it is such integration that enables and predisposes to the transformation of the possible self-object into an actual self-object. I call such integration "self–self-object integration," which again, by analogy with object–self-object transformation, refers to possible self-objects rather than actual self-objects in order to avoid logical circularity within the conceptual context (where actual self-objects would precede the process, i.e. transformation or integration, by means of which they are constituted).

Taken together, based on and extending Kohut and his school of self psychology, I here characterize self-object constitution by two processes, namely object–self-object transformation and self–self-object integration.[4] Moreover, again based on Kohut himself, I assume both object–self-object transformation and self–self-object integration to be mediated psychologically by affect and emotions (i.e. affective attunement as Kohut himself described it). How are object–self-object transformation and self–self-object integration related to what I described as the individualization of objects and the privatization of the self? I here postulate that object–self-object transformation is

[4] The reader may be surprised that I introduce new psychodynamic concepts here, where the primary aim is neuropsychodynamic. I have done so because these novel concepts will serve as bridge concepts that mediate between the psychodynamic and neuroscientific concepts and their respective contexts (see Chapter 3 for methodological details of such mediating or bridge concepts). Thus their function and purpose will become clearer once we put them into the neuropsychodynamic context, as described later on.

the process that leads to the individualization of objects, whereas self–self-object integration results in the privatization of the self.[5]

I have now determined the processes (object–self-object transformation and self–self-object integration) and their results (individualization of objects and privatization of self). This provides the starting point for developing empirical hypotheses and more specifically neuropsychodynamic hypotheses. As we discussed, object–self-object transformation could be characterized by affects and emotions, which may thus be crucial in enabling and predisposing to the individualization of objects. If affect and emotions are indeed central in object–self-object transformation, they must be assigned to the object in order for the latter to be transformed into a possible self-object. I refer to this as "affective assignment." I now postulate that the brain's neural processes underlying affective assignment (i.e. the generation of affect and emotions, including their assignment to specific stimuli) may enable and predispose to the individualization of objects and their consecutive transformation into possible self-objects (i.e. object–self-object transformation). This is my first neuropsychodynamic hypothesis (see Figure 7.2).

Self–self-object integration was psychodynamically associated with "transmuting internalization" as the specification of the more general process of internalization. As I discussed in Chapter 6, the processes of internalization in general and introjection in particular are made possible by linking the extero- and interoceptive stimuli associated with environmental or bodily objects to the brain itself and its intrinsic activity and thus its neural stimuli. Such linkage between intero- and exteroceptive and neural stimuli results in what I described as "trilateral interaction" and composite or "hybrid neural activity." This in turn was considered essential for differentiating the self from the brain, thus enabling and predisposing to brain–self differentiation (see Chapter 6). Based on these assumptions, I postulate that the trilateral interaction and the hybrid neural activity are crucial for enabling and predisposing to the privatization of objects (i.e. possible self-objects), with their subsequent integration into the self (i.e. self–self-object integration). This is my second neuropsychodynamic hypothesis (see Figure 7.2).

Neuropsychodynamic hypotheses

Psychological and neural findings in narcissism and affective assignment

My first hypothesis is that the brain's neural processes underlying affective assignment (i.e. the generation of affect and emotions, including their assignment to stimuli, and ultimately objects), may enable and predispose to what I here describe as the individualization of objects and consecutive object–self-object transformation. If so, one would expect that deficient individualization of objects and thus self-object failure, as in extreme or pathological forms of narcissism, should go along with deficits in affective assignment in such individuals. This is an experimentally testable hypothesis which has indeed been investigated, although only in a few studies.

Based on observation of single cases, Dimaggio et al. (2007a,b, 2008a,b, 2009a,b) have argued that pathological narcissism may be related to deficits in identifying and recognizing one's own and others' emotions. More specifically, NPD patients show enduring difficulty in accessing and identifying their own emotions and inner states. This also applies to others' emotions, leading to reduced empathy. A detailed study of metacognition abilities has shown that NPD patients have

[5] Thus, conceptually, object–self-object transformation and self–self-object integration refer to processes, whereas the terms "privatization" and "individualization" are supposed to describe the results of those processes.

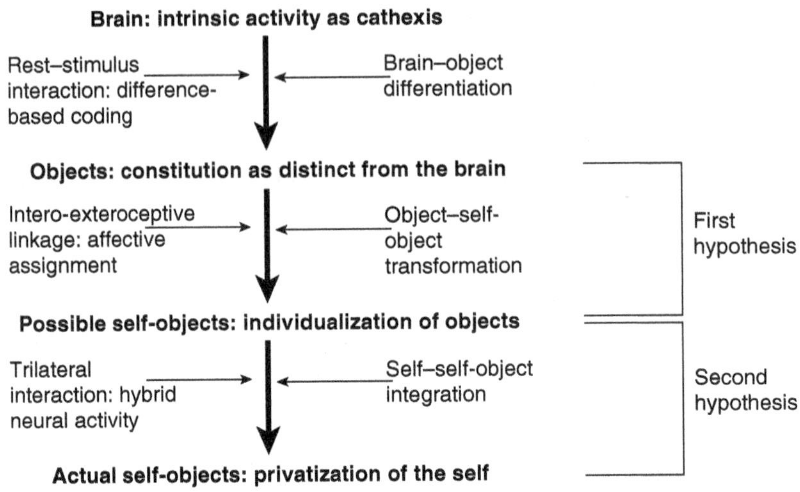

Fig. 7.2 Neuropsychodynamic hypotheses.

particular difficulty in recognizing their emotional and inner states and linking these to their respective environmental and psychological causes. These patients thus suffer from what can be described as "difficulty in identifying and relating emotions and inner states" (Dimaggio et al., 2007a,b, pp. 11–12).

Based on these observations, Dimaggio et al. (2008a,b) assume what they call an "encoding problem of affect" in NPD, with the patients experiencing and perceiving an intense but vague and non-specific sensation, rather than detailed emotional feelings that focus on specific objects. In other words, the assignment of affect to specific objects appears to be deficient in these patients. The combined difficulty in identifying and relating also seems to lead to lack of affect and emotional feeling when the patients reflect themselves. They cannot identify and relate their own emotions, and remain unable to attribute any emotions to themselves when reflecting on their own self. Their self thus remains devoid of emotions in both subjective experience and self-reflection.

The difficulty of identifying, relating, and introspecting one's own emotional feeling has been described as alexithymia, which may therefore be a crucial feature of NPD (Dimaggio et al., 2007a, pp. 11–12). One would consequently expect higher scores on scales that measure alexithymia, such as the Toronto Alexithymia Scale (TAS). In a combined psychological and imaging study by our research group, 30 healthy subjects were investigated and grouped according to low or high scores on the Narcissistic Personality Inventory (NPI) (Fan et al., 2010). They were then subjected to different psychological tests, including measures of alexithymia (the TAS) and general symptoms (the Symptom Checklist-90, or SCL-90). High narcissistic subjects had significantly higher alexithymia scores than low narcissistic subjects. Interestingly, they also had significantly higher scores on the various symptom dimensions of the SCL-90, including anxiety, paranoia, depression, and neuroticism. Therefore these results may be considered to lend some empirical (i.e. psychological) support to the assumption that an emotional deficit (i.e. alexithymia) is crucial in high narcissism.

However, caution is needed. Alexithymia is most probably non-specific to narcissism, as it occurs in various psychiatric disorders (e.g. eating disorders, depression, etc.), thus raising some questions about its psychological specificity for narcissism. This could be empirically tested by investigating and comparing narcissists with other groups of subjects suffering from other disorders that are also associated with alexithymia. Unfortunately, no such studies have yet been reported. Moreover, one may need to distinguish between emotions and self-related emotions. Narcissists may not suffer from deficits in processing emotions in general, but rather they may suffer from a deficit in relating emotions to the self (or to other people), so that specifically the self-relational aspect of emotions appears to be altered in these subjects. Future studies may therefore need to distinguish between self-related and non-self-related emotions, with deficits in the former in particular supposedly characterizing high narcissism, while other groups of patients may tend to have more general deficits in the processing of emotions (i.e. non-self-related emotions).

How about neural data? We also conducted an imaging study using fMRI in the aforementioned group, and employed an empathy task. Subjects were instructed to empathize with faces that they saw on a screen while merely perceiving them (which, for behavioral control, was followed by a judgment or evaluation period), while the non-empathic task consisted of a smoothed picture of a face (Fan et al., 2010). High narcissistic subjects showed significant differences in neural activity changes during empathy, especially in the right anterior insula and other regions such as the dorsolateral prefrontal cortex and the posterior cingulate cortex, when compared with low narcissistic subjects. This difference could be traced back neither to emotional or rewarding effects of the stimuli nor to non-specific task-related effects as revealed in careful control tasks and analyses.

High narcissistic subjects were not able to properly lower or deactivate the activity in these regions during non-empathic stimuli. Unlike low narcissistic subjects, they could no longer distinguish between empathic and non-empathic stimuli. This was also visible in a subjective–experiential context when high narcissistic subjects did not show proper differentiation between empathic and non-empathic stimuli in post-scanning subjective evaluation on the basis of a visual analog scale. Taken together, our findings indicate that high narcissistic subjects could neither neurally nor subjectively distinguish properly between empathic and non-empathic stimuli. Put into the psychodynamic context of Kohut, this means that high narcissistic subjects were not able to properly distinguish between affective and non-affective attunement. If, however, they cannot do this, they remain unable to generate and assign affect to specific objects (affective assignment), and thus to transform objects into possible self-objects (i.e. object–self-object transformation). In short, neural and subjective alterations in empathy may involve deficient affective assignment to objects, which in turn may make object–self-object transformation impossible. Although these considerations give us some psychological (and neural) insight into object–self-object transformation, the exact neural processes and, even more importantly, the individualization of objects by such affective assignment remain unclear. This will be discussed in the next section.

Affective assignment, intero-exteroceptive linkage, and object–self-object transformation

In order to better understand affective assignment, we have to venture into the territory of affects and emotions and their neural basis in our brain (see Panksepp, 1998; Panksepp was one of the main founders of the field of affective neuroscience). Current neuroscience often presupposes the James–Lange theory (for an overview, see Northoff, 2008). This well-known theory determined emotional feelings as perceptions of physiological body changes in the autonomic, hormonal, and motor systems. Once we become aware of physiological bodily changes induced by danger, we feel fear and subjectively experience emotional feelings. James (1884, p. 190) therefore considered bodily changes to be central to emotional feelings: "We feel sorry because we cry, angry because we strike, afraid because we tremble, and not that we cry, strike, or tremble, because we are sorry, angry, or fearful, as the case may be" (James, 1884, p. 190).

Modern empirical versions of this theory resurface in current neuroscientific models of emotion as, for instance, in those of Damasio and others (Craig, 2003, 2004, 2005; Bechara, 2004;

Niedenthal, 2007).[6] Based on the James–Lange theory, these accounts assume that the neural networks underlying emotions involve those regions that process interoceptive stimuli coming from one's own body. And most importantly, one would expect clear-cut distinction and segregation from the processing of exteroceptive stimuli. Critchley et al. (2004, 2005, 2005) and Pollatos et al. (2007a,b,c,d, 2008, 2009) conducted a series of studies in which they demonstrated the neural association between interoceptive awareness and emotional feelings in regions such as the insula, subcortical regions such as the PAG, tectum and colliculi, and anterior midline regions such as the DMPFC, VMPFC and SACC (for details, see Northoff, 2008). Affective assignment is then based solely and exclusively on interoceptive stimulus processing and must, even more importantly, be clearly distinguished from exteroceptive processing.

Is affective assignment indeed based solely and exclusively on interoceptive processing and the respective neural processes in the brain? Let us consider the data again. The aforementioned regions, the insula, many subcortical regions, and the anterior cortical midline system not only process interoceptive stimuli, but are also involved in exteroceptive stimulus processing. This is highly consistent with their connectivity pattern receiving input (i.e. afferences) from both intero- and exteroceptive regions that converge into common neurons (Panksepp, 1998, 2005; Rolls, 1999). Due to their connectivity pattern, these regions cannot avoid relating intero- and exteroceptive stimuli and thus what I here call "intero-exteroceptive linkage" in their neural processing (for details, see Northoff, 2008). Since the very same regions are apparently also involved in processing emotions and more specifically emotional feelings, affective assignment seems to be closely linked to intero-exteroceptive linkage. More specifically, as I argued recently (Northoff, 2008), emotional feeling and thus affective assignment may be considered the result of intero-exteroceptive linkage rather than isolated interoceptive processing.

Is there any empirical evidence in favor of intero-exteroceptive linkage underlying affective assignment? Critchley (2005, p. 162), one of the main investigators of interoceptive processing in imaging, states that the "insula maps bodily arousal states" and "it does so contextually," which therefore "represents an integration of external emotional information with peripheral states of arousal" (Critchley et al., 2005, p. 759). What seems to be coded in the brain is not so much the interoceptive stimulus itself as its relationship to the respective exteroceptive stimulus (i.e. the actual balance between intero- and exteroceptive stimuli). We tested this further by investigating whether intero- or exteroceptively induced signal changes in the insula were related to perception of one's own body and stress (Wiebking et al., 2010). Surprisingly, it was only the difference (i.e. the relationship) between intero- and exteroceptively induced signal changes that predicted body and stress perception, while the latter were not related to either intero- or exteroceptive signal changes themselves. This clearly provides some evidence in favor of intero-exteroceptive linkage rather than isolated interoceptive stimulus processing being crucial in determining neural activity in the insula and its relationship to affective assignment. The same also holds for the other regions (i.e. subcortical and anterior cortical midline regions) implicated in the neural processing of emotional feelings (for details, see Northoff, 2008; Wiebking et al., 2010).

In summary, I here provide structural–connectional and functional–physiological evidence that affective assignment presupposes intero-exteroceptive linkage rather than mere isolated

[6] Conceptually, the embodied approach to emotion emphasizes the crucial role of the body in emotional feeling. If the body and its vegetative and sensorimotor function play a crucial role in constituting emotional feelings, the body can no longer be regarded in a merely objective way, but rather should be considered as subjective and experienced. The mere Koerper as objective body must be distinguished from the lived body as subjectively experienced body in emotional feeling (Colombetti and Thompson, 2005, 2007; Colombetti, 2008).

interoceptive stimulus processing (as assumed in the James–Lange theory). This not only presupposes a relational concept of emotional feeling (for details, see Northoff, 2008), but also has several implications for the psychodynamic context. More specifically, I postulate that intero–exteroceptive linkage in apparently purely interoceptive regions such as the insula and other subcortical–cortical midline regions (see Chapter 6 for anatomical details) may be crucial in assigning affect and emotions to objects, thereby enabling and predisposing to the individualization of objects and their subsequent transformation into possible self-objects (i.e. object–self-object transformation).

First, though, we need to describe the relationship between intero- and exteroceptive stimulus processing in a little more detail. The exteroceptive stimuli associated with objects are processed not only in exteroceptive regions (e.g. sensory cortex), but also in regions of the interoceptive network (e.g. subcortical and cortical medial regions) (see Chapter 6 for anatomical details). There they encounter interoceptive stimuli, so the neural activity of these regions and networks may be a composite or hybrid of both intero- and exteroceptive stimuli. Rather than being coded in isolation from each other, intero- and exteroceptive stimuli are coded in relation to each other, in particular with regard to the difference between them, which amounts to what I described in Part II as difference-based coding. In other words, difference-based coding enables and predisposes to intero-exteroceptive linkage rather than isolated processing of both intero- and exteroceptive stimuli.

What are the implications of this for the psychodynamic context? Intero-exteroceptive linkage means that stimuli from the body and the environment are not processed separately and isolated in the brain. In other words, intero-exteroceptive linkage enables and predisposes the objects originating from the environment to be related to the individual body of the respective person. This means that by linking objects to the body, the former are not only assigned a specific affect, but are also at the same time individualized. Thus intero-exteroceptive linkage may enable and predispose not only to affective assignment, but also to the individualization of objects, both of which enable objects to be transformed into possible self-objects (object–self-object transformation).

One may finally want to ask why affective assignment individualizes the object and thus why distinct and individual emotional feelings are assigned to the same objects by different individuals. One could, on a purely logical basis, also imagine that the same affect and emotional feeling are assigned to all objects, as the body is the same across all individuals. Let me briefly speculate about this. The physiological state of the body and its interoceptive stimuli are continuously changing, which means that the different objects do not encounter the body in exactly the same physiological state in all individuals. By being related to different physiological and thus individual states of the body, the respective intero-exteroceptive constellations change and may thereby be individualized, too. This in turn may enable and predispose to the assignment of distinct and thus individual affects and emotional feelings (I use the terms "affects" and "emotional feelings" interchangeably for the sake of operational simplicity) to objects which thereby become individualized. If so, one would postulate that, neuropsychodynamically, the individual affective assignment to objects may be related to specific and individual constellations between intero- and exteroceptive stimuli (i.e. individually specific intero-exteroceptive linkages).

Finally, we need to put these hypotheses back into the context of narcissism. If the interoceptive processing and more specifically intero-exteroceptive linkage are indeed crucial in affective assignment, one would expect alterations in vegetative (i.e. interoceptively based) reactions to exteroceptive stimuli associated with emotions in subjects with pathological narcissism (i.e. NPD). One psychophysiological study has indeed focused on various autonomous parameters indicating interoceptive processing. Sylers et al. (2008) investigated skin conductance (SC), pre-ejection period of the heart (PEP) and respiratory sinus arrhythmia (RSA) during happy and threatening (i.e. involving anticipation of a countdown) emotional scenes in patients with NPD and antisocial personality disorder (APD). Changes in SC and PEP did not correlate with and thus predict NPD and were, in contrast to those in APD patients, where a negative correlation was observed (higher APD features were associated with decreased SC and PEP), unrelated to the threatening emotional scenes. This led the authors to conclude that NPD patients have difficulty in coping with distress, and mobilize and activate themselves during threat.

In contrast, RSA decreases and PEP shortening predicted NPD during positive (i.e. happy) emotional scenes. The authors interpret this finding as showing that NPD patients react negatively towards the happiness of others, rather than positively as one would normally expect. They thus show the autonomic changes that are usually associated with negative emotions in others. While reacting negatively towards others' positive emotions, they at the same time appear to remain unresponsive to their own negative emotions, as indicated by the absence of any correlation in the countdown, as previously described. Taken together, these results thus demonstrate vegetative alterations in patients with NPD in reaction to exteroceptive emotional stimuli. This lends some initial although indirect empirical support to the here assumed crucial role of interoceptive processing and consecutive intero-exteroceptive linkage being crucial in self-object constitution and hence in object–self-object transformation. I am aware, though, that empirical support for my first neuropsychodynamic hypothesis is rather sparse and must therefore be considered preliminary and tentative, awaiting further experimental investigation.

Brain regions and self–self-object integration

I characterized self-objects and their supposedly underlying self–object relation by object–self-object transformation and individualization of objects, and provided empirical evidence in favor of such a hypothesis. In addition, however, self-objects were supposed to also require self–self-object integration with the subsequent privatization of the self, which leads me to my second neuropsychodynamic hypothesis. I postulate that the trilateral interaction between intero- and exteroceptive and neural stimuli may be considered crucial in enabling and predisposing to self–self-object integration privatization of the self. This is my second neuropsychodynamic hypothesis (see Figure 7.2). In this first section, I want to consider the research findings that assume specific regions to be crucial in what I here call self–self-object integration. For this I shall focus on the right hemisphere and the ventromedial prefrontal cortex. In the next section I shall discuss the processes and thus trilateral interaction in further detail.

Todd Feinberg (2010) investigated a series of patients with brain lesions with regard to changes in their self, or "neuropathologies of the self" as he calls them. He assumes that different neuropathologies of the self, such as delusional anosognosia, asomatognosia, and delusional misidentification, are strongly associated with the thought and defense style of the normal child. He argues that immature and early defense mechanisms such as denial, splitting, projection, and wishful fantasy are as crucial in the adult's neuropathologies of the self as they are in the normal development of the child's self.

Neuroanatomically, he observes the occurrence of these early defense mechanisms in particular in patients with right frontal cortical lesions. He therefore concludes that proper function of the right hemisphere, especially in the frontal cortex, may be crucial in overcoming immature early defenses, since the presence of lesions appears to trigger their recurrence in adulthood. Feinberg (2010) therefore assumes that there must be a "left to right brain defensive shift" between the ages of 3 to 8 years, where there is a developmental shift away from immature defense functions and fantasies towards more mature defenses and the inhibition of fantasies, with the latter being mediated by the maturation of the right hemisphere. A lesion of the right hemisphere allows the left hemisphere, which is associated strongly with verbal functions, to take over. The fact that early immature verbal defenses such as projection, splitting, verbal denial, and fantasy are still functioning in the case of right hemispheric damage indicates that their left hemispheric mediation can be assumed. In contrast, the occurrence of more mature defense mechanisms, such as isolation, reaction formation, rationalization, etc., appears to be linked to normal right hemispheric functioning.

A recent brain imaging study of the relationship between early and late defense mechanisms and the rate of glucose metabolism in different regions has confirmed the left–right hemispheric difference (Reznikova et al., 2004). Investigating patients with multiple sclerosis, those researchers observed that almost all defense mechanisms were negatively correlated with the glucose metabolism rate, especially in the limbic, frontal and temporal cortical regions. Moreover, they observed that immature defenses such as denial, projection, and regression were associated with left hemispheric glucose metabolism, whereas all other defenses (i.e. more mature ones such as intellectualization, reaction formation, compensation, and repression) were correlated with right hemispheric glucose metabolism.

Besides the more general account of the right hemisphere as a whole, more specific proposals with regard to specific right hemispheric regions have been put forward. Solms (1999) has discussed several patients with lesions in the right perisylvian convexity, and observed a high degree of narcissism in these individuals. Neuropsychologically, these patients suffer from unawareness or anosognosia of their plegic body parts, with active denial of their paralysis, neglect of the left side of space, and disorders of spatial perception and cognition. Treating them psychoanalytically, Solms observes that these patients cannot tolerate the loss of the functions of their bodily parts, and that they therefore deny their loss by introjecting the lost object. By taking the lost object into themselves via internalization/introjection, they retain the lost body part inside themselves, which consecutively leads to the denial of its loss (i.e. anosognosia). Solms characterizes such an object relationship where a lost object is preserved by introjection as a narcissistic one, which in the context of Kohut would probably be described as "autoplastic adaptation" to self-object failure (with the lost body part being a lost or failed self-object).

Based on these observations, Solms characterizes patients with lesions in the right perisylvian convexity by narcissism. From this he infers that a functioning right perisylvian convexity "is a crucial component of the neuroanatomical substrate of object representation and therefore a vehicle for whole object cathexis and the capacity for mature object love" (Solms, 1999, p. 28). One should, though, consider that Solms, being very much in the Freudian tradition, regards narcissism as primarily a regressive and thus pathological phenomenon, rather than as existential as is presupposed here. Framing it within the current context of Kohut and self psychology, one would assume that the right perisylvian convexity may enable and predispose to the representation of an object as self-object, and may therefore be central to object–self-object transformation. By linking the spatial and emotional aspects (Solms, 1999, p. 29), the right perisylvian convexity may be crucial in transforming objects into self-objects. Lesions in the right perisylvian convexity may then lead to self-object failure, which has to be compensated for by introjection with the subsequent denial of the loss (i.e. anosognosia).

But what about self–self-object integration? While the right perisylvian regions may be involved in "narcissistic object representation" and what I have here called object–self-object transformation, Solms assumes that the ventromedial prefrontal cortex, especially the right one, is a "crucial neuroanatomical substrate of the more primitive, narcissistic object representations that form the nucleus of the self regulatory functions of the ego and superego, which remained largely intact in the right perisylvian cases but which were almost completely destructuralized in the ventromesial frontal cases" (Solms, 1999, p. 28; also see Solms, 1998 for the ventromesial cases). Despite the different terminology, Solms seems to suggest that what I call self–self-object integration may be mediated by the ventromedial prefrontal cortex, with lesions in these regions leading to self-object failure (although he would never describe it in this way, because he presupposes a Freudian rather than Kohutian context).

What makes the ventromedial prefrontal cortex so special that it may enable and predispose to self–self-object integration, including the privatization of the self? Let us return to the anatomy as discussed in the context of defense mechanisms in Chapter 6. I demonstrated that the medial–lateral dichotomy from subcortical regions may extend to the cortex. More specifically, based on Feinberg and others, I suggested that the subcortical distinction between core and median/lateral paracore regions extends over the forebrain (i.e. the extended or greater limbic system) into the cortex, where it resurfaces in the distinction between the paralimbic and midline regions (see Chapter 6). The paralimbic regions include the anterior cingulate and its distinct subregions, such as the PACC, the SACC, and the PCC, as well as other related regions such as the insula and the RSC, while the midline regions include the VMPFC, the DMPFC, the MPC, and the precuneus. Most importantly, the anatomical subcortical–cortical extension may be accompanied by a functional extension. The predominant interoceptive processing from the subcortical core region may be continued in the paralimbic regions, while the cortical midline regions, due to their sandwiched position as the median and lateral subcortical paracore, may be predominantly involved in integrating intero- and exteroceptive stimuli (Feinberg, 2009).

This anatomical characterization gives the VMPFC a prominent role. The VMPFC is involved in integrating intero- and exteroceptive stimuli, which distinguishes it from both paralimbic and lateral cortical regions that are implicated in processing the intero- and exteroceptive stimuli themselves. At the same time, the VMPFC is very close to most of the paralimbic regions, such as the PACC, the SACC, and the insula, which distinguishes it from other midline regions such as the DMPFC, the MPC and the precuneus. The role of the VMPFC as mediating and integrating between intero- and exteroceptive input is also mirrored in its connectivity pattern. It receives afferent connections from all of the lateral cortical sensory regions and their exteroceptive inputs, from the subcortical reward network, including the ventral striatum and the ventral tegmental area, and from the limbic system, including the amygdala and the insula and subcortical interoceptive regions. This allows them to integrate both intero- and exteroceptive inputs. If this is the case, one would expect the VMPFC to be recruited during emotional processing and especially during emotional feeling that, as previously shown, are supposed to be characterized by the linkage between intero- and exteroceptive stimuli. This is indeed supported by many imaging studies that show strong neural activity in the VMPFC and other midline regions during presentation of emotional stimuli (see Phan et al., 2002 for a review and Northoff, 2008 for a discussion with regard to emotional feelings).

What are the implications of this for the psychodynamic context? According to Shore this connectivity pattern makes the right VMPFC/orbitofrontal cortex ideally suited for representing both self- and object-images. The convergence of connections from the other systems allows the right VMPFC/orbitofrontal cortex to control them and, as Shore says (1996, 2003), to gate them (see also Feinberg, 2010, p. 55).[7] Let me put these observations in the terms and context presupposed here. The connectivity pattern of the VMPFC and its anatomical location in the cortical paracore region between the medial (core) and lateral cortical regions makes this region a perfect candidate for enabling and predisposing to the integration of possible self-objects into the self (self–self-object integration). More specifically, its peculiar anatomical and connectional characterization may give the VMPFC a unique role not only in integrating intero- and exteroceptive stimuli, but also in integrating the intero-exteroceptive linkage, with all other stimulus processing going on

[7] One may speculatively assume that such control or gating may possibly be mediated by inhibition. However, in contrast to Solms (1998, 1999), this does not target the inhibition of primary processes in favor of secondary processes, but rather the inhibition of specific self-objects while others that are not being inhibited may then be integrated into the self.

elsewhere in the brain. This in turn may enable and predispose to the integration of the possible self-objects into the self, whereby the former is transformed into an actual self-object (self–self-object integration).

We have to be careful, though. Although this sheds some light on the involved regions and the anatomy, the exact neural processes and mechanisms remain unclear. However, we need to consider the neural processes and mechanisms in order to fully understand why self–self-object integration is associated with what I described as the privatization of the self. This leads me to the trilateral interaction as the second aspect of the second neuropsychodynamic hypothesis.

Trilateral interaction and the privatization of the self in self–self-object integration

I characterized the paracore regions at the cortical level as midline regions or cortical midline structures (CMS) that include the VMPFC, the DMPFC, the MPC and the precuneus. These regions may be crucial in linking and integrating intero- and exteroceptive stimuli. In addition, however, I assumed that integration of such intero-exteroceptive linkage with the stimulus processing in other regions of the brain enables and predisposes to self–self-object integration. How is such integration of intero-exteroceptive linkage with the stimulus processing in other regions of the brain mediated neuronally? Let me return to the physiological properties of the midline regions. The midline regions and thus the CMS show a specific physiological feature, namely high resting-state activity, which, in conjunction with the lateral parietal cortex, distinguishes them from other regions. The CMS and the lateral parietal cortex have thus been subsumed under the concept of the DMN that describes high resting-state activity, a specific resting-state connectivity pattern among these regions, and a specific activation/deactivation pattern during task-induced stimulation (Raichle et al., 2001; Buckner et al, 2008; see Chapter 4 for details).

What are the implications of this for the neural processes and mechanisms that are presumably at work in the VMPFC and other midline regions? The observation of a specific type of resting-state activity means that there is not only bilateral interaction, as in intero-exteroceptive linkage, but also trilateral interaction between intero- and exteroceptive stimuli from the body and the environment and neural stimuli from the brain's intrinsic activity (i.e. its resting-state activity). All exteroceptive stimuli as linked to interoceptive stimuli encounter here the brain itself (i.e. its intrinsic activity or resting-state activity). The neural activity in the VMPFC and other midline regions must thus be considered to be a composite or hybrid of all three stimuli (intero- and exteroceptive and neural), which predisposes the VMPFC and other midline regions to trilateral interaction and hybrid neural activity (see Chapter 6 for an introduction to and definition of both concepts).

What are the implications of such trilateral interaction for the psychodynamic context of self–self-object integration? I consider intero-exteroceptive linkage to enable and predispose to object–self-object transformation. Moreover, I postulate trilateral interaction that has been shown to enable and predispose to brain–self differentiation (see Chapter 6). If now the intero-exteroceptive linkage and the consecutive constitution of possible self-objects (via object–self-object transformation) are subjected to trilateral interaction, they have, metaphorically speaking, no other option but to be related to the brain itself (i.e. its resting-state activity), and, due to prior brain–self differentiation, to the self. The trilateral interaction thus enables and predisposes to the integration of intero-exteroceptive linkage in the brain and its resting-state activity, which psychodynamically may correspond to the integration of the possible self-object into the self, thus enabling and predisposing to self–self-object integration.

How about what I earlier called the privatization of the self? Privatization of the self refers to how the self-object becomes exclusively and solely associated with the self of the respective person, which in turn makes the self of that person, including its self-objects, unique and inaccessible to others. As in the case of the body and its individual interoceptive processing, the brain's resting-state activity level and how it integrates the intero-exteroceptive linkage may also be individual. However, it may go beyond that by also being unique to the respective person, due to the different kinds of rest–stimulus and stimulus–rest interactions in their life (see Chapter 4 for more details about both kinds of interaction, as well as Part IV for more details about the impact of life experiences on the resting state). This means that the trilateral interaction is both individual and unique to the respective person and the self. And it is this individual and unique character of the trilateral interaction that may enable and predispose to the integration of the possible self-object into the self in a unique way such that the resulting actual self-object remains inaccessible to any other person (i.e. self). Unlike the possible self-object, the actual self-object is therefore no longer publicly accessible, but only accessible by the self itself thus being private. The actual self-object can thus be accessed only by the self itself in the gestalt of intrapsychic experience as previously described. And by privatizing the actual self-object, the self itself becomes privatized via its relationship to the former and its stabilizing, maintaining, and preserving function for the latter.

What are the implications of this for the psychodynamic context of emotions and affects and their relationship to self-objects? Based on neuroscientific evidence, I considered intero-exteroceptive linkage to enable and predispose to the generation and assignment of affect and emotional feelings to objects, resulting in object–self-object transformation. However, these self-objects are not yet integrated within the self which would convert them from possible into actual self-objects via self–self-object integration. How is the psychodynamic difference between possible and actual self-objects and thus between object–self-object transformation and self–self-object integration manifested within the psychological context? I postulate that possible self-objects are manifested psychologically in emotions that are not yet associated with and linked to the self, amounting to what I earlier described as non-self-related emotions that may not be as private as self-related emotions. As soon as object–self-object transformation is complemented by self–self-object integration, the non-self-related emotions and thus the possible self-objects are converted into self-related emotions and thus actual self-objects, which then also show a much higher degree of privacy.

Finally, our hypothesis also has important clinical implications. As pointed out already, this difference between non-self- and self-related emotions may be particularly relevant in pathological forms of narcissism (e.g. NPD), where subjects seem to remain unable to relate emotions and affects to themselves in a self-related way. This inability to constitute and generate self-relatedness of emotions (self-related emotions), and thus to integrate possible self-objects into the self, is well reflected in the already reported findings of emotional abnormalities in subjects with high narcissism and NPD. I postulate that subjects with NPD suffer from changes in the trilateral interaction between intero- and exteroceptive and neural stimuli.

More specifically, I assume that although intero- and exteroceptive stimuli can be well linked to each other, they can apparently not interact accordingly with the brain's resting-state activity in pathological forms of narcissism. This in turn may impede psychologically the linkage of affect and emotion to the self (self-related emotions), and psychodynamically the self–self-object integration. Since they remain apparently unable to integrate possible self-objects as based on intero-exteroceptive linkage into their own self, narcissists may compensate for their lack of self–self-object integration by taking their own self as self-object. This may be necessary to experience some degree of self-related emotions and thus some kind of self-object in order to stabilize the self. The original self–self-object integration is then transformed into what may be called "self-self as self-object integration." Depending on the focus (and the respective balance between the three different stimuli), this may result clinically in either narcissistic fantasies of grandeur about the self as self-object, as described by Kernberg, or narcissistic depression about the failure of self-objects, as emphasized by Kohut.

Conceptual implication: body, brain, and the existential necessity of narcissism

Let us briefly return to the opposite characterization of narcissism by Freud and Kohut. Freud considered narcissism to be developmental and negative when he characterized it by the direction of cathexis towards one's own body rather than towards external objects. Kohut, in contrast, gave a positive account of narcissism by describing self-objects as existential and reaching out beyond one's own body. This raises the question of how the relational concept of self-objects as based on self-object relation relates to the body. I assume that the body is the first self-object in very much the sense that Solms described it: "The ego is first and foremost a bodily ego and the first object it discovers in the outside world is its own physical body. Initially, this body—which is attached to us wherever we go—is experienced as something alien, a source of peculiar sensations of various types, arising from the superficial and deep sensory receptors (including from passive awareness of movement) and, more importantly, from the major somatic needs which constantly demand satisfaction, and press themselves on our attention in the form of acutely distressing states of colicky, hungry unpleasure" (Solms, 1999, p. 21).

Although he does not use the term "self-object" (due to his rather Freudian background), Solms nicely describes here the role of the body as the first self-object in the infant's life.[8] One would thus assume that the body as mere object also undergoes object–self-object transformation and self–self-object integration in the very same way as I described for objects from the environment (e.g. persons, objects, etc.). This means that the body may not be essentially different from other self-objects (e.g. persons or events in the outside world), although arguably it may be special in that, unlike other self-objects, the own body is continuously present from the very beginning and may therefore be regarded a particularly intimate and early self-object.

The reader may wonder that I consider the body to be a first self-object rather than being part of the self itself. Taking the perspective of the self, one may consider the body to be part of the self itself rather than an object that is to be transformed into a self-object. If, in contrast, one takes the perspective of the brain,[9] the body is to be distinguished from it and may then be regarded as the origin of the first stimuli (i.e. interoceptive stimuli) that the brain and its intrinsic activity encounter. This is nicely expressed by Mark Solms (1996, p. 509; my translation): "From the perspective of the externally oriented cortex (e.g. of the brain), the body itself is nothing but a physical object of the outside world."

One would consequently assume that the brain's specific coding of any kind of stimuli (i.e. difference-based coding) may also apply to interoceptive stimuli and their linkage to the brain's intrinsic activity. This means that the brain's neural stimuli and the body's interoceptive stimuli are linked and coded in terms of difference-based coding, entailing what I call "neural–interoceptive linkage" (see also Figure 7.3). By allowing for neural–interceptive linkage via rest–stimulus interaction, the brain from early on does exactly the same with interoceptive stimuli from the body as it does later with regard to exteroceptive stimuli from the environment, linking them to itself and

[8] I am aware that I am contradicting Kohut and self psychology here. I would strongly argue that the body, especially the mastering of the own body, is associated with self-esteem; failure of the mastering of the own body can induce lack of self-esteem as manifest in frustration and strong emotional feelings as it can for instance be observed in lesion patients. Thus the body may well serve as the first self-object thereby fulfilling the criteria postulated in self psychology. Although this would obviously require much more conceptual elaboration and empirical support. I owe this remark about the body to Moritz de Greck.

[9] This is meant in a figurative way, as the brain itself does not possess any perspective by itself: brains do not see and perceive, only persons do so (see Bennett and Hacker 2003).

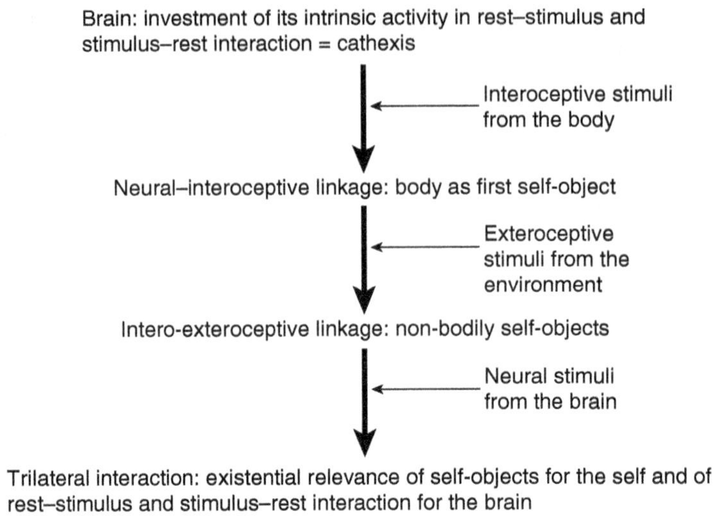

Fig. 7.3 Existential relevance of self-objects for self and brain.

its own intrinsic activity (neural–exteroceptive linkage). Hence, despite the different stimuli (intero- and exteroceptive) and their different objects (bodily and non-bodily), the brain applies the same mechanisms and coding to them (rest–stimulus and stimulus–rest interaction and difference-based coding). The application of the same neural mechanisms and coding to different stimuli (i.e. intero- and exteroceptive stimuli) enables and predisposes to the transformation of both bodily and non-bodily objects into possible self-objects (object–self-object transformation), and their integration into the self (self–self-object integration).

One may finally raise the question of why the brain employs neural mechanisms and a type of neural coding that enable and predispose to what within the psychodynamic context are described as self-objects. Kohut (1984, p. 61) assumed that "a self can never exist outside a matrix of self-objects," indicating that independence from self-objects remains impossible. Throughout life we need self-objects. Although the contents of our self-objects may change with changing persons, events, and aging, we can never survive without self-objects, as they provide the matrix (the self–object matrix) for our self and our very existence: "The self-object concept illuminates a central dimension of human experience: the basic enduring needs for emotional sustenance from others as they emerge and develop through the life cycle" (Gehrie, 2009, p. 48). In the same way that we need to breathe, since without oxygen we and our bodies could not survive physically, we and our selves need self-objects in order to survive mentally (Kohut, 1971, cited by Milch, 2001, p. 66). Thus there is a lifelong dependence on self-objects (Gehrie, 2009, pp. 35–36, 48). In short, self-objects are existential rather than being merely developmental (for empirical support for a healthy narcissism, see Sedikides et al., 2004).

Why are self-objects so existential? Self-objects, as demonstrated previously, are existential because they serve the purpose of self-constitution, self-maintenance, and self-preservation. Plenty of energy is invested (i.e. cathected) in self-object constitution. I neuropsychodynamically postulated that the brain invests its own energy (i.e. its intrinsic activity or resting-state activity) in intero-exteroceptive linkage and trilateral interaction that in turn enable and predispose to the constitution of self-objects via object–self-object transformation and self–self-object integration. Why does the brain invest its own intrinsic activity (i.e. its resting-state activity) in its own neural processing of intero- and exteroceptive stimuli by linking them to itself via difference-based coding? One could for instance imagine purely logically (and thus not empirically) that the brain's resting-state activity does not interfere at all with the neural activity induced by intero- and exteroceptive stimuli, so that there would be neither rest–stimulus and stimulus–rest interaction nor difference-based coding. This scenario is, however, as we demonstrated in Part II, not empirically plausible since there is plenty of evidence to the contrary (i.e. for rest–stimulus and stimulus–rest interaction and difference-based coding).

Why then does the brain use its own energy for its neural processing of intero- and exteroceptive stimuli? Let us again take the perspective of the brain (in a figurative sense as already described). By encountering stimuli from outside itself (i.e. from the body and the environment), the brain's intrinsic activity, the pattern of its own resting-state activity across its different regions is disturbed or, in systems theory parlance, perturbed. Allowing for rest–stimulus and stimulus–rest interaction via difference-based coding, the brain may be able to neutralize the foreign intruders (i.e. intero- and exteroceptive stimuli) by simply integrating them via modulation and adaptation of its own intrinsic activity. One could, however, also envisage other forms of neutralization (e.g. by ignoring or eliminating the impact of intero- and exteroceptive stimuli, which would amount to the strategy that I call "neutralization by ignorance or elimination"). However, there is no empirical evidence for any such ignorance or elimination of intero- and exteroceptive stimuli by the brain and its intrinsic activity. Instead there is, as discussed in Part II of this book (and also in Part IV), plenty of evidence for modulation and adaptation and thus for integration, so that I here speak of "neutralization by integration" as distinct from "neutralization by ignorance or elimination."

One may further ask why the brain employs the strategy of "neutralization by integration." This is of course a speculative issue, but let me at least make an attempt. Integration of intero- and exteroceptive stimuli into its own intrinsic activity may serve the purpose of adapting itself and thus the whole organism to the environmental demands by allowing itself (i.e. its resting-state activity) to be structured and organized by the intero- and exteroceptive stimuli.

Analogous to the existential importance of self-objects for the self, the adaptation to the body and the environment via the structuring and organizing effects of its intero- and exteroceptive stimuli may be existential for the brain. Only if the organism and its body are preserved is the rather high energy input to the brain, that represents about 20% of the total body energy, secured. And the organism and its body can best be maintained and preserved by adapting to the environment. The brain's strategy of neutralization by integration may thus serve the purpose of maintaining the brain's high energy demand and thus its very existence as such.

Metaphorically speaking, the brain's strategy of "neutralization by integration" serves the rather narcissistic interest of the brain of maintaining and preserving its existence (see also Atkins, 1996, who also speaks of narcissism in the context of the brain, although in a slightly different way that is more concerned with the processing and thus the rest–stimulus interaction than with the here presupposed existential meaning). One may therefore want to speak of narcissism within the neural context of the brain, too, and thus consider the brain to be narcissistic in the very same way that Kohut described it with regard to our self. As generous and non-selfish as it appears to be, the brain's rest–stimulus and stimulus–rest interaction as realized by its employment of difference-based coding may ultimately serve the rather selfish purpose of maintaining and preserving its own existence. In short, Kohut's characterization of narcissism as existential may apply not only to the psychodynamic context of the self but also to the neural context of the brain.

Acknowledgments

I am grateful to Moritz de Greck who, as a specialist in Kohut and self psychology, read an earlier draft of this chapter and offered some helpful suggestions.

Chapter 8

Unconsciousness and the Brain

Psychodynamic concepts

Freud and the concept of the unconscious

What is the unconscious? Notions of the unconscious can be traced back as far as Plato and Aristotle, and have been elaborated since in philosophical and later in psychological thought. Whatever frameworks have been presupposed, unconscious states have been characterized by hidden characteristics of a person's self (fate, temperament, soul, or character) that need to be inferred and cannot be accessed directly. Such hidden characteristics were distinguished from those that were believed to be transparent, experienced directly, open to introspection, and thus accessible to consciousness (Uleman, 2005).

Building on descriptions of such hidden and inaccessible features in philosophy (Schopenhauer, Nietzsche) and literature (Dostoyevsky), Freud was probably the first to attempt to offer a systematic empirically based psychological account of the unconscious. According to Freud, the unconscious in a psychoanalytic sense, the "dynamic or repressed unconscious" (or "dynamic unconscious" as I shall call it), not only operates outside awareness but is also extremely complex, including distinct aspects of the person's self. For instance, the dynamic or repressed unconscious includes innate and inherently sexual and aggressive blind drives (i.e. the id), most of its conscience and ego ideals (i.e. the superego), and processes (perception, action, etc.) that deal with reality (i.e. the ego).

The dynamic or repressed unconscious mediates wishful, associative, instinctual primary processes and their respective contents which must be repressed because they are threatening the person and its self. As such the dynamic or repressed unconscious must be distinguished from what Freud called the "systems preconscious."[1] The systems preconscious, or "preconscious" as I shall call it, concerns contents that are no longer threatening or offensive to the person's self, but which have not yet entered consciousness. In other words, these contents are ready to become conscious but are not yet so. Unlike the contents of the dynamic unconscious, those of the preconscious are no longer as instinctually driven and irrational, but are more rational, disciplined, reality oriented and energetically "bound."

Originally, Freud described the contents of the unconscious in terms of predominantly primary processes such as wishes, drives, instincts, and associative "blind" non-rational processes which he subsumed under the concept of primary processes (as distinct from secondary processes). This focus on non-rational contents has been complemented by more recent research that includes both rational and non-rational contents in the unconscious,[2] the "new unconscious"

---

1 One may also distinguish between a repressed unconscious and an unrepressed unconscious, with the latter being free and non-suppressed (Mancia, 2006c, pp. 49–50). However, since my focus here is not on repression, I shall not pursue this distinction.

2 However, in his later work he refrained, at least in part, from exhaustive and exclusive definition of the dynamic unconscious by primary and primitive processes by including more rational processes associated with the ego and superego in the unconscious (Westen, 1999).

(Uleman, 2005, p. 6). The contents of the unconscious may concern cognition, emotions, movements, perceptions, behavior, etc. This includes the "cognitive unconscious" (Kihlstroem, 1987), "emotional unconscious" (Kihlstroem, 1987; Kihlstroem et al., 1992; Berridge and Winkielman, 2003), "behavioral unconscious" (Uleman, 2005), and "procedural unconscious" (Schuessler, 2002).[3] The association of the unconscious with different contents shifted the concept of the unconscious from its initially exclusively psychodynamic context and its association with instincts and libidinal drives to a broader psychological context. One may consequently speak of the unconscious as the "psychological unconscious" (Kihlstroem et al., 1992).

How can we put the psychological unconscious into a neural context and ultimately develop neuropsychodynamic hypotheses? Two strategies are possible. Either focuses on one particular manifestation of the psychological unconscious and its respective contents, for which one may then search for correlating neural contents. This is the strategy that I designated as "neural correlates" (see Introduction). Alternatively, one may refrain from the focus on the different contents of the psychological unconscious and instead search for the necessary empirical conditions that enable and predispose to the unconscious as distinct from the conscious. In other words, we then search for what I called "neural predispositions" rather than neural correlates (see Introduction). These neural predispositions must serve a double function. They must enable and predispose to the primary distinction and differentiation between the "dynamic unconscious" and consciousness (i.e. the "unconscious–conscious differentiation," as I shall it). And they must, at the same time, also account for the transition from unconscious to conscious states, which may (more or less) correspond to what Freud described as the "systems preconscious."

However, before developing specific neuropsychodynamic hypotheses, I would like to venture briefly into the conceptual context and more specifically to consider the philosopher J. R. Searle and his characterization of the unconscious. Searle (2004, pp. 165–172) distinguishes between four different types of unconsciousness, which he describes as "preconscious," "dynamic unconscious," "deep unconscious," and "non-conscious." He speaks of the "preconscious" which corresponds more or less to Freud's notion of the "systems preconscious." Then there is the "dynamic unconscious," which describes cases where the "unconscious mental states function causally, even when unconscious" (Searle, 2004, p. 167). This mirrors more or less what Freud described as the "dynamic or repressed unconscious."

The third type of unconscious is what Searle describes as the "deep unconscious." Here the unconscious mental state[4] is neither conscious nor accessible at all in consciousness; since in principle it cannot be accessed by consciousness, one may characterize the deep unconscious as necessary unconscious, as the philosophers would say, which means that it remains largely impossible to bring such an unconscious mental state into consciousness, as it is not "the sort of thing that can form the content of a conscious intentional state" (Searle, 2004, p. 168). Such a "deep unconscious" must be distinguished from what he refers to as the "dynamic unconscious," where the mental state remains factually or actually unconscious, too, but can in principle be brought into consciousness. He highlights this by citing the example of the computational rules that we follow unconsciously when acquiring language. Although we can be preconscious or dynamically

[3] The variety of contents associated with the unconscious suggests that one can no longer speak of "the unconscious" as though it were a unified and isolated sector of the mind. The concept of "the unconscious" has thus outlived its usefulness (Westen, 1999, p. 1064).

[4] This also makes it clear that I here use the concept of mental states in a broader sense. Philosophically, the concept of mental states is traditionally used to describe the conscious state, thus being linked to consciousness as distinct from unconsciousness. This rather narrow use of the term is here replaced by a wider usage that includes both conscious and unconscious states.

unconscious about the language and its letters, we remain deeply unconscious about the rules and principles of its universal grammar that guide our learning of the language. Thus the rules that guide the acquisition of language are simply not the sort of things of which we can become conscious at all. This may be extended to other rules, too, such as those that determine the computations and processing in the retina and the visual cortex during visual perception.

Finally, there is a fourth type of unconscious, namely the non-conscious, that, following Searle, concerns neurobiological phenomena that remain non-conscious and cannot become cases of consciousness at all: "There are all sorts of things going on in the brain, many of which function crucially in controlling our mental lives but that are not cases of mental phenomena at all. So, for example, the secretion of serotonin at the synaptic cleft is simply not a mental phenomenon. Serotonin is important for several kinds of mental phenomena, and indeed some important drugs, such as Prozac, are used specifically to influence serotonin, but there is no mental reality to the behavior of serotonin as such. Let us call these sorts of cases the 'non-conscious.' There are other examples of the non-conscious that are more problematic. So, for example, when I am totally unconscious, the medulla will still control my breathing. This is why I do not die when I am unconscious or in a sound sleep. But there is no mental reality to the events in the medulla that keep me breathing even when unconscious. I am not unconsciously following the rule 'Keep breathing'; rather, the medulla is just functioning in a non-mental fashion, in the same way that the stomach functions in a non-mental fashion when I am digesting food" (Searle, 2004, p. 168).

Unconscious–conscious differentiation and the brain's neural code

Searle assumes that neither the deep unconscious nor the non-conscious can in principle be brought into consciousness. Why can they not be brought into consciousness like, for instance, the contents of the dynamic unconscious and the preconscious? What is the main difference between the deep unconscious and the non-conscious on the one hand and the preconscious and the dynamic unconscious on the other?

Presupposing a purely conceptual–logical context, Searle assumes that there is a logical connection between unconscious and conscious states. Unconscious states in both the dynamic unconscious and the preconscious are logically connected to consciousness in that they can become principally conscious. This principal possibility of becoming conscious remains independent of whether the respective contents actually become conscious or not. If there is such a primary possibility of becoming conscious, one must assume a conceptual–logical connection between unconsciousness and consciousness. Searle calls this the "connection principle," which states that the notion of unconsciousness is logically connected to the notion of consciousness, with an unconscious mental state being "the kind of thing that could be a conscious mental state" (Searle, 2004, p. 171; see also Strawson, 1994; Revensuo, 2005, p. 63).

Searle's "connection principle" holds for both the dynamic unconscious and the preconscious, as the contents of both can in principle become conscious in very much the same way as Freud described. In other words, the dynamic unconscious and the preconscious have the potential to become conscious. In contrast, the connection principle does not apply to the deep unconscious and the non-conscious, which thus do not have the potential to become conscious. They can even in principle not become conscious. Based on Searle's connection principle, I therefore want to subsume the dynamic unconscious and the preconscious under what I call the "principal consciousness.'[5] The concept of the "principal consciousness" describes the principal possibility

[5] The concept of the "principal consciousness" may come close to what Panksepp has described as "core consciousness," which he regards as being closely related to primary processing and affects (Panksepp, 1998).

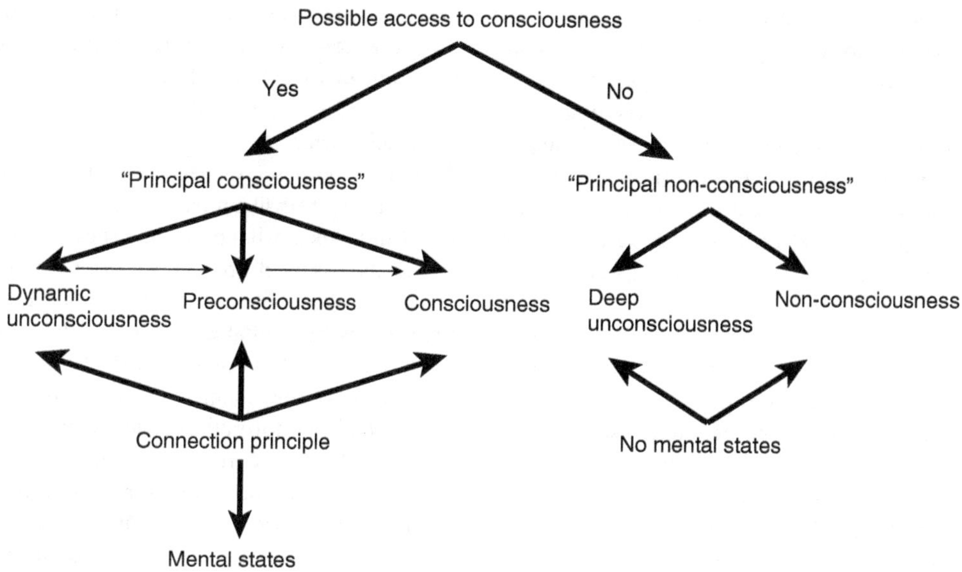

Fig. 8.1 Conceptual characterization of consciousness and non-consciousness.

of a specific state or content becoming conscious, independent of whether it is actually or de facto realized or not. As such, the principal consciousness must be distinguished from what I call "principal non-consciousness," which describes the principal impossibility of a state or content entering consciousness. The principal non-consciousness subsumes, for instance, what Searle described as the deep unconscious and the non-conscious. In other words, the difference between "principal consciousness" and "principal unconsciousness" concerns the principal access to consciousness, which remains open in the former and closed in the latter.

In the reminder of this chapter I shall focus on the principal consciousness and its empirical realization. I shall not discuss the principal non-consciousness that touches upon the epistemic domain, as I dealt with this in Part I of this book when discussing the autoepistemic limitation (see Chapter 2). The focus on the principal consciousness raises the question of what connects unconscious states to consciousness, thus accounting for what Searle described as the "connection principle" as the characteristic feature of what I call "principal consciousness" (see Figure 8.1).

The connection principle describes the necessary linkage between unconscious and conscious states within a conceptual–logical context. However, I here presuppose an empirical (i.e. psychological and neuronal) context rather than a conceptual–logical one as Searle did. This means that I aim to target those psychological and neuronal mechanisms that primarily link and connect unconscious states (i.e. the dynamic unconscious and the preconscious) to consciousness. In other words, I shall here search for those neural mechanisms that enable the unconscious state to become conscious, amounting to what I call "neural predispositions of consciousness" (i.e. neural predispositions). These neural predispositions are thus supposed to enable and predispose to what I described as "unconscious–conscious differentiation."

What then is the difference between my description and that of Searle? The difference between Searle's connection principle and my concept of unconscious–conscious differentiation lies in the context. As indicated, the connection principle presupposes a purely conceptual–logical context, whereas my concept of unconscious–conscious differentiation refers to an empirical (i.e. neuronal and psychological) context. This makes it possible to develop neuropsychodynamic hypotheses about the neuronal mechanisms that enable and predispose to the distinction and transition between unconscious and conscious states (i.e. unconscious–conscious differentiation), and thus to account for principal consciousness. In contrast to neural correlates and more specifically the neural correlates of consciousness (NCC) (see, for instance,Crick, 1994), I here focus not on actual consciousness and its correlating neural states, but rather on possible or principal consciousness and its enabling and predisposing neuronal mechanisms.

However, before developing specific neuropsychodynamic hypotheses, we need to briefly discuss the kind of possible feature that enables and predisposes to the transition from unconscious to conscious states. For this we may start at the opposite end and look at what prevents a specific state from becoming conscious, as for instance what Searle described as the deep unconscious and non-conscious. Despite the differences with regard to their states and contents (i.e. mental versus non-mental), both can in principle not be brought into consciousness. What exactly prevents them from entering and being brought into consciousness? Revensuo (2006) argues that the format or code presupposed by a specific state or content may be crucial in either enabling or preventing the entry of that state or content to consciousness:" There is biological information coded in the DNA of our brain cells, but that type of information is in a totally non-conscious format and we will never be able to read it out just by reaching into our own minds and trying to retrieve it into consciousness. It is in a format unreadable at the phenomenal level" (Revensuo, 2006, p. 63).

What does the concept of "format" denote? It refers to the rules, principles, or algorithms according to which states or contents are processed, independent of whether they are for instance mental or non-mental. For example, Searle refers to such rules and principles in the context of language and its grammar in his description of the deep unconscious. Although I here neglect the logical paradox that the rules and principles which guide the principal consciousness remain by themselves principally unconscious, this example demonstrates what the concept of format refers to. As such the concept of format is very close to that of the "code" that I introduced in Chapter 4 in the context of the brain. Ignoring subtle differences between the concepts of "format" and "code," I shall use both synonymously in the following account. Based on these considerations, one may thus assume that the principal consciousness may be characterized by a specific format or code which distinguishes it from the principal unconsciousness.

What are the implications of the characterization of the principal consciousness by a specific format or code for the unconscious–conscious differentiation within our empirical (i.e. psychological and neuronal) context? Let us return briefly to Freud. Freud aimed to reveal the principles and rules according to which the psychic apparatus organizes and structures the distinction and transition between the unconscious and the conscious (i.e. from the dynamic or repressed unconscious through the systems preconscious to the conscious). One such principle was, for instance, repression or suppression, which was complemented later by the broader concept of defense mechanisms (see also Chapter 6). Taken within the current context of consciousness, defense mechanisms such as repression may be taken to describe the rules and principles that regulate the traffic (e.g. distinction and transition) between the unconscious and the conscious. Thus Freud aimed to reveal the format or code of the psychic apparatus that allowed it to differentiate between the unconscious and conscious (unconscious–conscious differentiation).

I here intend to move from Freud's psychological context of the psychic apparatus to the brain's neuronal context. What are the implications for the brain and its neuronal context of the characterization of the principal consciousness by a specific format or code? One may assume that the brain must employ a specific format or coding with specific rules and principles that regulate the border traffic between unconscious and conscious states, thereby enabling and predisposing to unconscious–conscious differentiation. I have already demonstrated the brain's neural code and defined it as difference-based coding. I have also demonstrated some of the rules and principles that the brain applies to its own neural processing of stimuli (e.g. rest–rest, rest–stimulus, and stimulus–rest interaction) (see Chapter 4). What I need to show now is how the very same neural code (difference-based coding) and rules (rest–rest, rest–stimulus, and stimulus–rest interaction) enable and predispose to both differentiation and the transition between unconscious and conscious states.

On the basis of these considerations, I develop two neuropsychodynamic hypotheses about the principal consciousness and unconscious–conscious differentiation. First, I postulate that the same neuronal mechanisms (i.e. difference-based coding, and rest–stimulus and stimulus–rest interaction) are operational in both unconscious and conscious states, although in a slightly modified way. Secondly, I postulate that these same neuronal mechanisms enable and predispose to the transition from unconscious to conscious states. For both neuropsychodynamic hypotheses I shall use dreams[6] to illustrate the transition of unconscious material and contents into consciousness, but will not address other forms of the unconscious.

Why was such a detour into the complicated conceptual territory of consciousness and unconsciousness necessary in order to arrive at this point? It was important to clarify the conceptual ground upon which my neuropsychodynamic hypotheses stand, and this is of particular significance given the complexity of a concept such as consciousness that needs to be sharpened and clearly defined within the conceptual context in order to develop specific hypotheses within the empirical context. Such conceptual clarification allowed me to clearly point out what I mean by neural predispositions of the principal consciousness (NPPC), and to distinguish these from the concept of the neural correlates of consciousness (NCC) (see below for details).

Neuropsychodynamic hypotheses

Rest–stimulus interaction and perception in dreams

As indicated, I shall use dreams to illustrate the transition of unconscious content and material to consciousness, which will serve to illustrate my neuropsychodynamic hypotheses. Thus I split my first hypothesis, concerning the operation of similar neuronal mechanisms in unconscious and conscious states, although in a slightly modified way, into four parts, namely rest–stimulus interaction, intero-exteroceptive linkage, stimulus–rest interaction, and trilateral interaction. I aim to demonstrate how each mechanism enables and predisposes to one particular feature of dreams (these features will include perception, emotions, proto-self, and personal relevance) (see Nir and Tononi, 2010, for an excellent overview and the relevant references). I therefore start with that part of the first hypothesis that concerns the relationship between rest–stimulus interaction and perception in dreams (for a summary and overview, see Figure 8.2).

Dreams can be characterized by vivid perception of scenario-like structures as the simulacrum of the world, thereby integrating highly disparate images and themes into a seamless scenario in the absence of any external stimulation (see, for instance, Hobson, 2009, p. 804 for a nice description of dream perception in a painting by Salvador Dali). There is often intense visual and/or auditory imagery despite the absence of exteroceptive input from the environment. Although asleep, the dreamer perceives and experiences the images and his dream hallucinations as if he is awake, and is thus deluded about his own state and the origin of his images. While pertaining somehow to reality, these images and the dream hallucinations often include bizarre distortions

[6] The reader may also want to ask what precise concept of dreams I here presuppose (e.g. whether dreams are considered in a wide sense as any kind of mental activity during sleep, or in a narrow sense as the recollection of vivid and bizarre perceptions and emotions. I here presuppose the latter rather than the former (see also Mancia, 2006c, p. 91).

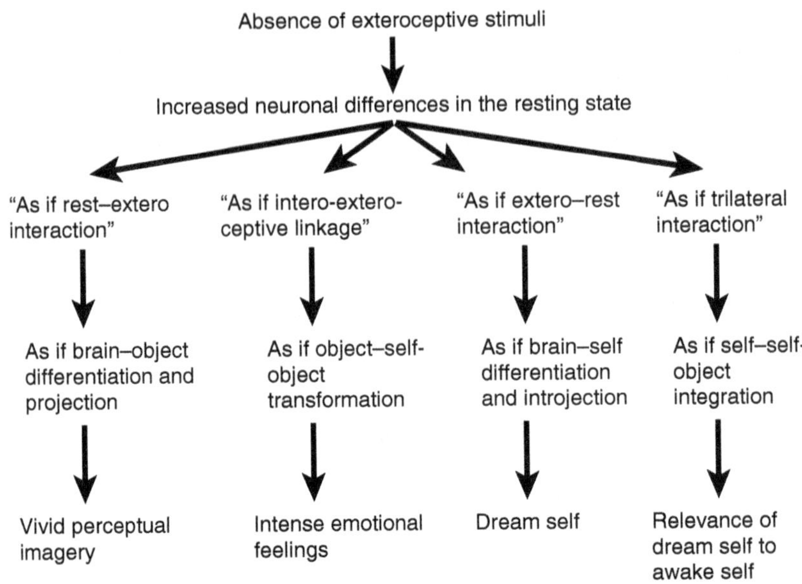

Fig 8.2 Neuronal, psychological, and psychodynamic mechanisms in dreams.

of objects, persons, and events which Freud suggested were highly meaningful and personally relevant to the self of the dreamer.

How can we account for the vivid perceptions in the gestalt of imagery and hallucinations in dreams in the neuronal context of the brain? I postulate that the same neuronal mechanisms are at work in dreams and thus in the unconscious as in the awake and conscious state, namely rest–stimulus interaction. We characterized perception by the interaction of an exteroceptive stimulus with the brain's intrinsic activity (its resting-state activity), and designated this as rest–stimulus interaction (see Chapter 4). This led to the assumption that difference-based coding takes place, with the brain's neural activity coding the difference between the exteroceptive stimulus and the brain's resting-state activity, rather than the stimulus being coded by itself independent of the brain's intrinsic activity (see Chapter 4). Most importantly, such difference-based coding was supposed to enable and predispose to the transformation of mere stimuli into objects of perception, (i.e. stimulus–object transformation). In addition, difference-based coding was considered crucial in predisposing to the transformation of neuronal into mental states (i.e. neuronal–mental transformation) (see Chapter 5). By that means the mental states of, for instance, perception could be constituted and characterized by "going beyond" the actual exteroceptive stimulus.

I now postulate that the very same neuronal mechanism is also at work in the perceptions in dreams, and thus in the unconscious state. How though is this possible? There is one essential difference between the awake state and the dreaming state. The awake state can be characterized by exteroceptive input, whereas the latter is absent in the dreaming state. One must therefore assume that rest–stimulus interaction is absent in the dreaming state. However, we need to outline the scenario of dreams in more detail.

Let us start with the brain itself and its intrinsic activity. The brain's intrinsic activity is still present in the dreaming state, as denoted by a specific resting-state activity level. Empirically, there is indeed abundant support that the brain's resting-state activity is high both in the awake state and especially during those stages of sleep when dreaming occurs frequently (i.e. REM sleep). In contrast, the brain's resting-state activity is rather low during the non-dreaming stages of sleep (i.e. NREM sleep).[7] This similarity in resting-state activity level has allowed Hobson (2009, pp. 808–809) to assume that dreams result from self-activation of the brain and more specifically the brainstem, with the pons as the generator that extends and spreads its activation to the forebrain (see also Solms 1995, 1997, 2000 as well as below for further discussion of the relationship between the brainstem/pons and forebrain in dreams).

Can the similarity in resting-state activity account for the vivid perceptions in the REM stages of sleep? It cannot do so by itself, because for perception we need some other stimulus from outside the brain. In the awake state these are exteroceptive stimuli, but these stimuli remain absent during REM sleep. However, in addition to the brain's intrinsic activity, there is also continuous interoceptive input from the body in dreams. Even while we are sleeping our body is still active, sending interoceptive signals to the brain and thus recruiting those networks that are implicated in interoceptive processing (see the next section for further details). This has the following implications for rest–stimulus interaction. Rest–stimulus interaction can still occur, although no longer as "rest–extero interaction" but rather as "rest–intero interaction."

However, there is another form of interaction that we have not yet considered, namely the interaction between the different resting-state activity levels across the different brain regions, amounting to what I earlier called "rest–rest interaction." Is there any empirical evidence for such rest–rest interaction in dreams? Let me briefly describe some of the findings with regard to dreams and then return to rest–rest interaction. Resting-state activity in the pontine tegmentum, the amygdala, the hippocampus, the occipital cortex, the mediobasal prefrontal cortex, the anterior commissure, the parietal operculum, the midline thalamus, the deep frontal white matter and the anterior cingulate cortex has been found to be increased in REM sleep when compared with the waking state (see Hobson, 2009, p. 810 as well as Wehrle et al., 2007; Walker, 2009). In contrast, other regions such as the dorsolateral prefrontal cortex and the posterior cingulate show relatively decreased resting-state activity in REM sleep/dreams compared with the waking state, whereas NREM sleep activity levels range between the two extremes.

The differences in distribution of resting-state activity levels across the different brain regions suggest that some rest–rest interaction must occur during dreams. Most importantly, due to the different resting-state activity levels, such rest–rest interaction across the different brain regions must be different in dreams from that which occurs during the waking state. If there is indeed such rest–rest interaction across different regions, one would expect different functional connectivity among the different regions in the dreaming and the waking state. This is indeed supported by empirical evidence. Kaufmann et al. (2006) observed increased connectivity of several cortical and subcortical core (i.e. median) regions with the hypothalamus in the dreaming state when compared with the waking state. Such differential connectivity patterns could be indicative of specific forms of rest–rest interactions in the dreaming state. Thus there seems to be empirical support for different connectivity patterns and possibly rest–rest interaction in dreams (see Walker, 2009 for a recent review, as well as Larson-Prior et al., 2009).

Can rest–rest interaction account for the occurrence of perceptions in dreams? How can the vivid perceptions, including their perceptual distortions, be generated in dreams despite the absence of exteroceptive stimuli? Let us return to the sensory cortex. Hobson (2009, p. 809) assumes that there is what he calls "input–output gating" in REM sleep/dreams. More specifically, he assumes that during dreams the sensory cortex (as well as the motor cortex) is actively suppressed and thus shut down. This makes it impossible for exteroceptive stimuli to enter the brain and

7 I here ignore the fact that dreams can also occur in NREM sleep stages, where they seem to be less accessible for subsequent remembrance and report than dreams associated with REM sleep (Mancia, 2006c, pp. 90–91).

interact with its intrinsic activity—in his words, the brain is kept "off-line" (Hobson, 2009, p. 809). The inhibition of the motor cortex in dream/sleep has been well documented. However, the data concerning the sensory cortex are less clear, but they do seem to indicate decreased reactivity to external stimuli in the dreaming state.

For instance, Wehrle et al. (2005, 2007) investigated auditory stimulation during phasic and tonic REM sleep and NREM sleep, with the former in particular being associated with dreams. They observed significantly decreased activity (i.e. increased deactivation) in the auditory cortex during auditory stimulation in phasic REM sleep as associated with dreams. A similar finding was reported in an earlier study by the same group that observed increased deactivation in visual and auditory cortex in NREM sleep, indicating decreased activity (Czisch et al., 2004; Kaufmann et al., 2006; Walker, 2009). Taken together, these findings indicate that there is an altered (i.e. decreased) reactivity to exteroceptive stimuli in the sensory cortex (i.e. visual and auditory cortex).

In contrast to reduced stimulus-induced activity, the resting-state activity level appears to be preserved in the sensory cortex during dreams. One would thus assume that there is reduced rest–extero interaction in the sensory cortex, although this cannot be traced back to the resting-state activity level itself. With the capacity for rest–extero interaction being apparently reduced in the dreaming state, the following question arises. What happens in the sensory cortex during dreams such that the vivid perceptions can be generated despite reduced rest–extero interaction?

As it shows decreased capacity for rest–extero interaction, I postulate that the dreaming state may be characterized by increased rest–rest interaction. However, this needs to be specified. As demonstrated, the relationship between different resting-state activity levels appears to change between the different regions in dreams. This may lead to increased neuronal differences between the different resting-state activity levels within each region across different time points. This may, for instance, apply to the sensory cortex, and would then lead to increased intraregional rest–rest interaction. In addition, there may be increased neuronal differences between the sensory cortex and other regions, such as the DMN or subcortical regions at one particular time point. This may amount to increased interregional rest–rest interaction. Such increased neuronal differences may, for instance, be generated and indicated by decreased cortical synchronization (i.e. increased cortical desynchronization) in EEG. This is indeed empirically supported, as increased cortical desynchronization is an electrophysiological characteristic feature of REM sleep. Moreover, increased spontaneous activity as observed in the visual system in the gestalt of ponto-geniculo-occipital (PGO) waves during dreams could be another indicator of increased neuronal differences in the resting state. Low K-complexes and delta waves are other measures of increased cortical desynchronization (Czisch et al., 2004) that could indicate increased neuronal differences and thus increased rest–rest interaction. Taken together, there is solid evidence for endogenously generated activity with increased neuronal differences and rest–rest interaction in the dreaming state.

How, though, can the apparently increased endogenously generated activity with its increased neuronal differences and rest–rest interaction lead to perceptions that are vivid and distorted in the dreaming state? The endogenously generated activity may not only occlude sensory input, so that we remain asleep, but may also at the same time enable and predispose to the constitution of perceptions in our dreams. Hobson (2009, p. 809) assumes that "the brain, isolated from the outside world, treats this endogenous stimulation as if it were exogenous." How can the brain confuse its own resting-state activity and its rest–rest interaction with rest–extero interaction? I postulate that the increased neuronal differences in the resting state in dreams lead to confusion of rest–rest interaction with rest–extero interaction. The neuronal differences generated in the resting state (i.e. during rest–rest interaction) appear to be of the same degree as is usually associated with rest–extero interaction. This may make subsequent neural processing prone to confuse rest–rest interaction with rest–extero interaction, and to treat the former as if it were the latter. I therefore speak of "as if rest–extero interaction" in the following account, to indicate the increased rest–rest interaction and neuronal differences in the dreaming state.

What happens in the case of such virtual "as if rest–extero interaction"? Exactly the same thing happens as in the case of real rest–extero interaction. The brain codes its increased neuronal differences in terms of difference-based coding which enables and predisposes to stimulus–object transformation and neuronal–mental transformation. This results in mental states and objects as characteristics of perception. And this is exactly what we observe in dreams—vivid perception of objects in mental states. Corresponding to "as if rest–extero interaction," one may therefore want to speak of "as if perception" in dreams. The occurrence of perceptions in dreams is thus assumed to be enabled and predisposed to by such "as if rest–extero interaction." One would therefore assume that different resting-state activity levels would lead to different "as if rest–extero interactions" and thus to different "as if perceptions" in the dreaming states. However, to my knowledge this has yet to be tested.

Why then are the "as if perceptions" in dreams so real and vivid, being as much in time and space as real perception, and thus having what Hobson (2009, p. 808) calls "built-in predictions of external space and time"? By employing rest–rest interactions across its different brain regions, the brain enables and predisposes to the constitution of space and time in a phenomenal sense. Such phenomenal space and time are experienced and perceived in the respective perceptions in both awake and dreaming states, and it is this that makes the internal "as if perceptions" in dreams so real and vivid (i.e. a remarkable simulacrum of the external world). However, this needs to be tested by, for instance, comparing phenomenal space and time between dreaming and awake states.

Why are there objects in the vivid and almost real perceptions in our dreams? To answer this question we may need to move from the purely neuronal context back to the psychodynamic context. I associated rest–stimulus interaction with brain–object differentiation and projection in the psychodynamic context (see Chapter 6). How do both mechanisms apply to the present context of dreams? The answer is very simple. They apply in exactly the same way, although in a slightly different sense (i.e. in an "as if sense"). More specifically, based on "as if rest–extero interaction," there is as much brain–object differentiation taking place in dreams as in the awake state. However, due to the absence of real objects from the outside world, these objects must be characterized as either "internal objects" in the sense of Melanie Klein (1935), or as I would call them "as if objects," involving "as if brain–object differentiation" in the dreaming state. Since they do not draw on exteroceptive stimuli but resting-state stimuli (and their formerly associated objects), the objects in dreams become distorted and different from the objects in the awake state. One may now also question why many of the "as if objects" are so real and close to the real and external objects in our perception of the external world. I shall defer answering this question until I consider the third part of the first neuropsychodynamic hypothesis.

The psychodynamic characterization of the dreaming state by "as if objects" and "as if brain–object differentiation" does not, however, explain why the "as if perceptions" and their "as if objects" are often so distorted in our dreams. For this we have to go even deeper into the psychodynamic context. We remember that brain–object differentiation was associated with the defense mechanism of projection (see Chapter 6). Projection describes the imposition of the self on the objects of its perception, by means of which the objects become altered in our perception when compared with the ones in the real world that remain independent of us and our perception. If dreams can be characterized by "as if brain–object differentiation" and "as if perception," one would also assume that there is projection taking place in the dreaming state, although again not real projection as in the awake state but "as if projection" as I call it. (This is because brain–object differentiation is invariably associated with projection, as previously demonstrated). The occurrence of "as if projection" means that the "as if objects" as internal objects in the "as if perception" of the dreaming state become as distorted as the external objects in the perceptions of the awake state. Thus, as in the dreaming state, projection in the gestalt of "as if projection" may enable and predispose to the distortions of the "as if objects" in the "as if perceptions" of our dreams. Neuronally, this may correspond to increased rest–rest interaction which though taking the perspective of the brain must rather be characterized as "as if rest–extero interaction."

Intero-exteroceptive linkage and emotions in dreams

Dreams are characterized by vivid emotions. We do not only perceive vivid objects in our perceptions in the dreaming state, but these "as if objects" are associated and closely tied to emotions in our "as if perceptions" in the dreaming state. How is this possible? There are not real external objects in perceptions. There are only "as if objects," and despite the fact that they are not real and external, they are nevertheless associated with strong emotions that in part account for the intensity (and personal relevance) of dreams (see Figure 8.2).

How are objects related to affect and emotions? Let me return briefly to the previous chapter on narcissism. There I discussed this question under the heading of "affective assignment" in the context of self-objects. More specially, I postulated that affective assignment corresponds within the neuronal context to the linkage between intero- and exteroceptive stimuli (i.e. intero-exteroceptive linkage). Such intero-exteroceptive linkage involves convergence between intero- and exteroceptive stimuli, particularly within the interoceptive network, the core subcortical and cortical regions around the ventricle (see Chapters 6 and 7 for details). These core regions include subcortical regions such as the amygdala and the brainstem (e.g. the pons and the tegmentum), and cortical regions predominantly in the forebrain (e.g. the subgenual and pregenual anterior cingulate cortex).

I assumed that these core subcortical and cortical regions are crucial in intero-exteroceptive linkage, thereby enabling and predisposing to affective assignment to objects. Interestingly, exactly these regions and especially the amygdala show increased resting-state activity in the dreaming state when compared with the waking state (see Czisch et al., 2004; Wehrle et al., 2005, 2007; Hobson, 2009; Walker, 2009). Involvement of the amygdala in REM sleep is often considered to account for the presence of strong emotions in dreams (see, for instance, Solms, 1995, 1997, 2000; Hobson, 2009). Yoo et al. (2007) observed a stronger response to negative stimuli in the amygdala and decreased amygdala–medial prefrontal connectivity in sleep-deprived subjects when compared with non-sleep-deprived ones. The exact mechanisms whereby emotions are generated in dreams, including their attribution to the "as if objects" in the "as if perception," remain unclear.

I postulate that, analogous to "as if rest–extero interaction," there is also "as if intero-exteroceptive linkage" in dreams. Once a stimulus is designated as exteroceptive, be it a real one or an "as if exteroceptive stimulus," it interacts with the interoceptive stimuli that originate continuously from the body even in the dreaming state. In other words, the "as if exteroceptive stimulus" has no choice (metaphorically speaking) but to undergo intero-exteroceptive linkage although in an "as if" sense (i.e. as an "as if intero-exteroceptive linkage"). Analogous to the intero-exteroceptive linkage in the case of real exteroceptive stimuli, the "as if intero-exteroceptive linkage" in the dreaming state may be mediated by those regions where there is strong convergence between the intero- and exteroceptive inputs. These regions include subcortical and cortical core or median regions such as the amygdala, the brainstem, the pons, the tegmentum, and the perigenual and subgenual anterior cingulate cortex, and it is precisely here where the aforementioned resting-state hyperactivity during dreaming has been observed. This therefore lends some indirect support to my hypothesis of "as if intero-exteroceptive linkage" in the dreaming state.

How can we now account for the occurrence of strong emotions in the dreaming state? In the same way that intero-exteroceptive linkage leads to affective assignment of objects in the awake state, the "as if intero-exteroceptive linkage" enables and predisposes to the assignment of affects and emotions to the "as if objects" in our dreams. Based on "as if intero-exteroceptive linkage," the "as if objects" are assigned affect and emotion, which may be experienced as strong emotional feelings in the dreaming state. If the occurrence of emotions in dreams is indeed enabled and predisposed to by "as if intero-exteroceptive linkage," one would assume that different bodily states providing different interoceptive input would lead to different "as if intero-exteroceptive linkage," and thus to different emotions in the dreaming state (this is because one variable, namely the interoceptive input, is changed). However, to my knowledge this has yet to be tested.

One may want to argue that, analogous to the case of perception, the emotions that occur in dreams can only be considered in an "as if" sense. Thus, analogous to "as if perception," one may want to speak of "as if emotion." This certainly holds if one considers the fact that the objects to which the emotions are assigned are not real objects but only "as if objects." While the exteroceptive input in the "as if intero-exteroceptive linkage" is indeed only an "as if exteroceptive stimulus," the interoceptive input is real and not in an "as if" sense, since even in dreams there is continuous interoceptive input from the body. The concept of "as if intero-exteroceptive linkage" would thus need to be refined in order to be more precise. Rather than speaking of "as if intero-exteroceptive linkage," one might do better to refer to "intero-as if exteroceptive linkage." However, to keep things simple, I do not use that term here, and subsume it under the former.

The assumption of an "as if intero-exteroceptive linkage" as "intero-as if exteroceptive linkage" may be very compatible with Walker's model of sleep-dependent emotional memory processing (Walker, 2009; Walker and van der Helm, 2009). He assumes that while emotions are initially coupled to a specific event when the latter is encoded, emotions are decoupled from that event when retrieved in memory during sleep. Continuous reactivation and retrieval of the event in question are supposed to be accompanied by decreased assignment of affect to the retrieved event. Throughout several cycles this ultimately results in strong memories of the event, although these are detached from the originally assigned affects and emotions. Such decoupling of emotions and affects from retrieved events may be mediated by increased activity and theta oscillations in regions such as the hippocampus, the amygdala, and the medial prefrontal cortex. Most importantly, at a neurochemical level there is decreased aminergic tone, whereas the cholinergic tone is maintained. This aminergic–cholinergic imbalance promotes the downgrading of the emotions while at the same time supporting the storage and retrieval of the particular event. Through many reiterations in REM sleep this will ultimately result, during the awake state, in recall of the event as detached from its original affects and emotions.

Is the Walker model of sleep-dependent emotional memory processing compatible with the assumption of "as if intero-exteroceptive linkage" as "intero-as if exteroceptive linkage"? One may assume that the decoupling of the emotions from the event may be mediated by the linkage between the interoceptive stimuli and the "as if" exteroceptive stimuli. Due to the "as if" character of the retrieved exteroceptive stimuli, the neural difference between the two may no longer be as strong and large as in the original experience of the event and its associated exteroceptive stimuli. Furthermore, the interaction between the two stimuli (the interoceptive stimulus and the as-if exteroceptive stimulus) may not be further impeded by the decreased aminergic tone that may further weaken the intero-as-if exteroceptive interaction by assigning decreased value to it. This will ultimately result in the kind of downgrading and detachment of affect and emotions that Walker assumes in his model. Thus his assumption and my own may be highly compatible and complementary in that one describes the processes and the other describes the results.

However, let us return to the occurrence of emotions in dreams. Even if they are downgraded, how is it possible for emotions to be experienced as real in dreams? One may postulate that the real origin and nature of the interoceptive input in "as if intero-exteroceptive linkage" may make the subsequent assignment of affect and emotions so real in dreams, and almost indistinguishable from those in the case of real objects in the awake state. Thus it may be due to the interoceptive input that the dreaming person regards their experience of emotional feeling as real. Therefore the associated "as if objects" are experienced as emotionally real and thus as real as the objects in the awake state. Due to the real character and reality of the emotional feelings in the dreaming state, I shall refrain from speaking of "as if affective assignment." Instead I consider the affective assignment in dreams to be as real as that in the awake state.

Why are the objects and the emotional feelings so individual in the dreaming state, thus being specific for a particular dreaming person as distinct from others? To answer this question we may have to turn back to the psychodynamic context and investigate what happens psychodynamically in the dreaming state. In Chapter 7 on narcissism I demonstrated that intero-exteroceptive linkage and affective assignment enable and predispose to the transformation of objects into self-objects (object–self-object transformation), including the individualization of the latter. I postulate that exactly the same thing happens in the dreaming state. Due to affective assignment, the "as if objects" are transformed into self-objects (i.e. "as if self-objects") in the dreaming state, so that one may want to speak of "as if object–self-object transformation." In the same way that object–self-object transformation accompanies the individualization of the self-object in the awake state, the "as if object–self-object transformation" in the dreaming state enables and predisposes to the individualization of the "as if self-object." (This may be why the "as if objects" and their respectively assigned emotions are constituted in a highly individual way and thus experienced as such in the dreaming state.)

Stimulus–rest interaction and the occurrence of an often bizarre self in dreams

As well as featuring perceptions and emotions, dreams can be characterized by the experience of an often bizarre self. The perceptions and emotions are attributed to a self (a dreaming self) who experiences them. Based on these observations Hobson (2009) argues for the "instantiation of a protoself" in dreams. However, the dreaming self is quite different and often rather bizarre when compared with the awake self, although each is closely related to the other, as Freud in particular pointed out. In this section I want to focus on the constitution of a self in dreams and why it is often rather bizarre, while in the next section I shall discuss the personal relevance of the dreaming self.

Let me recap briefly. I assumed that there is "as if rest–extero interaction" in dreams which enables and predisposes to the constitution of objects as "as if objects" and thus "as if perceptions" in dreams. As demonstrated in Chapter 4, rest–stimulus interaction is accompanied by stimulus–rest interaction, which is considered crucial for constituting a sense of self (see Chapter 6). Based on the occurrence of "as if rest–extero interaction" as already discussed, one would also assume stimulus–rest interaction in the dreaming state, although again in an "as if" sense (i.e. "as if extero–rest interaction").

What could such "as if extero–rest interaction" look like and how could we demonstrate it empirically? Since "as if extero–rest interaction" occurs in the absence of any real external input, it must be manifested in spontaneous activity changes in, for example, the visual cortex. This is indeed the case, as such spontaneous activity changes have been observed and described as the aforementioned PGO waves that occur predominantly in the visual system, including the occipital cortex (see also Wehrle et al., 2007). Other spontaneous activity changes are reflected in slow-wave oscillations such as the K-complexes and the delta waves that occur both spontaneously and when induced by a stimulus (see, for instance, Czisch et al., 2004 for an investigation of these measures in both resting and stimulus-induced states). Although there is clear evidence for such spontaneous activity changes in the dreaming state, the kinds of processes that are associated with them remain unclear. Therefore we can only speculate at this point that these spontaneous activity changes may reflect the impact of the neuronal activity changes associated with the "as if object" on the resting-state activity (i.e. "as if stimulus–rest interaction") in the dreaming state.

Although we are currently unable to provide direct neuronal evidence in favor of "as if extero–rest interaction," there may however be at least some empirical support (albeit indirect) for the alleged personal relevance of the "as if objects" in our dreams. Let me start with the neuroanatomical evidence. Based on extensive lesions studies, Solms (1995, 1997, 2000) considers the forebrain regions to be crucial in enabling and predisposing to dreaming. These forebrain regions include, for example, the hypothalamus, the amygdala, the subventral striatal areas, the subgenual and pregenual anterior cingulate cortex, the insula, and the ventromedial prefrontal cortex. Patients with lesions in these regions report cessation of dreams, whereas this is not the case in patients with lesions in the pons and the brainstem. Thus the pons and the brainstem may be important for generating REM sleep patterns, but they may not be necessary for constituting dreams, including their vivid "as if objects" and "as if perceptions" for which the forebrain may be crucial. Accordingly, Solms presupposes that there is neuronal dissociation between REM sleep and dreaming, which sets him apart from Hobson (2009), who considers the pons/brainstem rather than the forebrain to be a necessary condition of dreams.

How can we characterize Solms' forebrain regions in more detail? They (more or less) correspond to the subcortical and cortical paralimbic and midline regions that I described (see Chapter 6) as distinct from more lateral regions. These regions are predominantly involved in interoceptive processing and its linkage to exteroceptive processing, thus accounting for intero-exteroceptive linkage. Moreover, these regions have been shown to be strongly implicated in self-related processing, which is the processing of intero- and exteroceptive stimuli in relation to the self of the respective person according to the meaning, significance, and relevance of the stimuli for the self (see Northoff et al., 2006 as well as Chapter 9 of this book for further details).

As previously described, these forebrain regions that are situated in the paralimbic and midline location have been observed to show increased resting-state activity in the dreaming state. Moreover, I assumed that increased resting-state activity is related to increased "as if rest–extero interaction" and "as if intero-exteroceptive linkage." Based on the recruitment of the same regions during self-related processing in the awake state, one would postulate increased "as if rest–extero interaction" and "as if intero-exteroceptive linkage" to accompany increased self-related processing in the dreaming state. More specifically, one would assume increased self-related processing in the dreaming state in the gestalt of "as if self-related processing." This in turn may enable and predispose to the attribution of increased personal significance, relevance, and meaning to the "as if objects" and their respectively associated emotions in the dreaming state.

Why, however, is the personal significance, relevance, and meaning of the "as if objects" and their relationship to the self so bizarre in dreams, such that it can barely be deciphered by the dreaming person him- or herself? I speculatively postulate that altered "neuronal contextualization" is crucial here. Neuronal contextualization describes how the constitutional dependence of neural activity in, for instance, one region or network is dependent on that in the other (see Chapter 6). Increased resting-state activity in the forebrain and thus in the paralimbic/midline regions of the brain during dreaming appears to accompany decreased activity in more lateral regions, including the dorsolateral prefrontal cortex and the sensory cortex (Solms, 2000; Kaufmann et al., 2006; Hobson, 2009; Walker, 2009). This means that the neuronal balance between the forebrain regions and the lateral network changes, and as a result the neuronal contextualization is different in the dreaming state to that in the waking state.

Different neuronal contextualization may be traced back to different neuronal balances in the resting-state activity levels of the forebrain when compared with the lateral network. These different neuronal balances may lead to increased rest–rest differences between paralimbic/midline and lateral–lateral networks. Increased inter-regional neuronal differences may lead to increased difference-based coding, which in turn may stipulate increased self-related processing. One may now speculatively assume that the different neuronal contextualization of the forebrain regions in relation to the lateral network may account for the often bizarre relationship of the "as if objects" to the self in dreams, including increased "as if self-related processing."

However, it is not only the "as if objects" and their relationship to the self that are bizarre in dreams, but also the dreaming person's self, or the "proto-self" as Hobson calls it. The "proto-self" surfaces in a rather bizarre gestalt in our dreams such that we cannot even link it with our awake self. How is such change of the self possible? To answer this question we have to go back to the neuropsychodynamic context, where stimulus–rest interaction enabled and predisposed to brain–self differentiation and introjection (see Chapter 6). In the same way that stimulus–rest interaction in the awake brain enables and predisposes to brain–self differentiation, the "as if extero–rest interaction" in the dreaming state may make possible differentiation between brain and self, which might amount to what I call "as if brain–self differentiation."

In the same way that brain–self differentiation in the awake brain is psychodynamically based on introjection, the "as if brain–self differentiation" of the dreaming self may also rely on introjection. More specifically, it may rely on introjection of the "as if objects" from the "as if perception," thus resulting in what I call "as if introjection." However, this "as if introjection" concerns objects that have already been altered and distorted (i.e. "as if objects"), and is possible only on the basis of apparently abnormally increased "as if self-related processing." This may thus enable and predispose to the constitution of a self, a dreaming self, which however is altered and changed significantly and thus experienced as bizarre and strange when compared with the awake self.

Trilateral interaction and the personal relevance of dreams

One of the main features of dreams that Freud pointed out is their personal relevance. The material and contents that are processed and experienced in dreams have something to do with the self of the dreaming person, their past experiences, and their life as a whole. More specifically, the dreamed objects, the "as if objects," and the dreaming self are closely related to the awake self and its specific self-objects. I characterized the constitution of self-objects in the awake state by self–self-object integration which, in the neuronal context, was supposed to correspond to trilateral interaction between neural and intero- and exteroceptive stimuli leading to composite or hybrid neural activity (see Chapter 7). If the objects and the self in the dreaming state are closely related to the awake self and its specific self-objects, one may need to investigate what happens to self–self-object integration and thus to trilateral interaction in the dreaming state.

Let us return to the neuronal context of the brain. I assumed "as if rest–extero interaction," "as if intero-exteroceptive linkage," and "as if extero–rest interaction" in the dreaming state. Based on "as if exteroceptive stimuli" and the simultaneous presence of the brain's neural stimuli and the body's interoceptive stimuli, one may assume "as if trilateral interaction" in the dreaming state. Thus there is composite or "as if hybrid neural activity" that can be traced back to "as if trilateral interaction" between "as if exteroceptive stimuli," interoceptive stimuli, and neural stimuli. However, this "as if trilateral interaction" needs to be identified in more detail. Let us start with the brain's neural stimuli and thus its resting-state activity. Although the resting-state activity level has been shown to be altered in the different networks in the dreaming state, it may nevertheless be structured and organized according to previous stimulus–rest interactions that occurred during the preceding day(s) and indeed the entire life of that particular person. Thus the neural structure and organization specific for that particular self enter its dreaming state in the gestalt of its resting-state activity configuration as shaped by previous stimulus–rest interactions in the awake state.

What are the implications of this for the trilateral interaction? It means that the bodily interoceptive stimuli and the "as if exteroceptive stimuli" from the "as if object" encounter, at least in part, the neural structure and organization associated with the awake self. The resulting composite or "as if hybrid neural activity" is thus hybrid not only with regard to the origin of the different stimuli (i.e. neural and intero- and "as if exteroceptive"), but also with regard to the state (i.e. awake or dreaming). There is thus a mixture of the neural structure and organization associated with the awake state and the actual resting-state activity level in the dreaming state, with each influencing the other (i.e. restructuring and reorganizing) via "as if stimulus–rest interaction" and "as if rest–stimulus interaction."

How, though, does the double hybrid nature of the "as if hybrid neural activity" enable and predispose to personal relevance or, as I would say, self-relatedness in our dreams? Why do the dreams and their content reveal something about our self (i.e. the awake self)? To answer this question we have to turn to the psychodynamic mechanisms already discussed. I demonstrated in Chapter 7 that the self-objects provided by object–self-object transformation are not yet actual self-objects but only possible ones still awaiting integration into the self via self–self-object integration. This integration of self-objects converts the possible self-objects into actual ones, and as a result the latter are able to stabilize and maintain the self (i.e. the awake self).

Is there an analogous self–self-object integration in the dreaming state? There is certainly some integration of the "as if self-objects" into the dreaming self, so that one may want to speak of "as if self–self-object integration." How though is such "as if self–self-object integration" affected by the double hybrid nature of the "as if hybrid neural activity"? There is "as if trilateral interaction" between the different stimuli (neural, interoceptive, and "as if exteroceptive") in "as if self–self-object integration," accounting for half of the double hybrid nature. However, the "as if self–self-object integration" is most severely affected by the other half of the double hybrid nature of the "as if trilateral interaction," namely the mixture between the awake and dreaming states.

Let us be more specific. The integration of the "as if possible self-object" into the self encounters a highly hybrid self that is based on both the awake and the dreaming self. The self is on the one hand based upon the awake self and its self-objects as ingrained in the neural structure and organization, the latter mirroring the personal life and thus the biography of the awake self. On the other hand, the "as if possible self-objects" encounter the dream self that is constituted on the basis of the "as if brain–self differentiation" in the dreaming state. This means that the "as if possible self-objects" in the dreaming state, as provided by "as if object–self-object transformation," converge here with the actual self-objects of the awake self and the dreaming self. The "as if self–self-object integration" in the dreaming state can thus not do other than integrate the possible "as if possible self-objects" of the dreaming state with the actual self-objects of the awake self into the dreaming self. This results in the constitution of highly hybrid actual "as if actual self-objects" in the dreaming state as an integration of actual self-objects from the awake state and "as if possible self-objects" from the dreaming state. "As if self–self-object integration" may thus be more properly described as "dream self/awake self with actual self-objects–as if possible self-object integration." Since this is rather complex, I here subsume its meaning under the concept of "as if self–self-object integration" that therefore must be considered intrinsically hybrid as composite or a mixture between the dreaming and waking states.

The intrinsically hybrid nature of "as if self–self-object integration" as composite between both dreaming and waking states may also account for the observation that dreams reach beyond the dreaming state itself into the waking state. More specifically, dreams in particular and sleep in general have been shown to enhance cognitive performance in the awake state (e.g. in learning) (see Hobson, 2009, p. 807 and Walker, 2009 for recent reviews). Most importantly, it may be due to the intrinsically hybrid nature of "as if self–self-object integration" as composite between dreaming and waking self-objects that our dreams show personal relevance or self-relatedness with regard to the awake self. The "as if self–self-object integration" in the dream state cannot do other than rely on the awake self and its personal life history and biography as manifested in its actual self-objects and their mixture and integration with the self-objects constituted in the dreaming self, the possible "as if self-objects." This is why the past personal biography as manifested in the specific actual self-objects may be mirrored in the dreaming state, as was so nicely pointed out by Freud.

In addition to the linkage of the past awake self to the present dreaming self, the latter also reaches out into the future. Analogous to the awake state, the "as if self–self-object integration" serves the purpose of stabilizing and maintaining the self, be it the awake or the dreaming self, by converting "possible into actual as if self-objects." Thus the "as if self–self-object integration" may stabilize the dreaming self which, due to its intrinsically hybrid nature, may also extend to the awake self. In other words, the awake self (i.e. the self that wakes up in the morning) may be stabilized by the "as if actual self-objects" as constituted on the basis of the "as if self–self-object integration" during its dreams. Following my account of the brain in the last section of Chapter 7, the stabilizing function of dreams with regard to the self may also apply to the brain itself. By allowing for the hybrid nature of "as if self–self-object integration," the brain may stabilize itself by reorganizing and restructuring its own resting-state activity during dreaming and sleeping. This in turn may enable and predispose the brain to better integrate and process future stimulation in the awake state.

In summary, the process of "as if self–self-object integration" provides the bridge between the awake self before the night and the awake self after the night, and thus between the past and future awake self. Freud provided us with the ingenious insight that the former's actual self-objects (those of the awake self before the night) enter the night, are modified by our dreams, and then exit the dreams in the morning in a modified gestalt that allows them to stabilize the awake self after the night. Although I have here described this modification of the actual self-objects by the concept of "as if self–self-object integration" and its hybrid character, this still falls short of the detailed mechanisms such as censorship and repression that Freud demonstrated so nicely. This would for instance include the exact role of memories in the dreaming state (see, for instance, Mancia, 2006a,b; Walker, 2009), which may be crucial in constituting what I here called "as if self–self-object integration." However, such an account would be beyond the scope of this book. Thus the task for the future would be to link Freud's psychological mechanisms to the concept advanced here of the "as if self–self-object integration" and its corresponding neuronal mechanisms.

Transition from the unconscious to the conscious state and the "process-based hypothesis of dreams"

I demonstrated here that the same neuronal mechanisms that determine our mental states in the conscious state are also at work in the unconscious state, relying on the dreaming state to illustrate the latter. More specifically, I pointed out that rest–stimulus and stimulus–rest interaction may be operational in the dreaming state, too, in the gestalt of "as if rest–extero interaction" and "as if extero–rest interaction." Moreover, I showed that intero-exteroceptive linkage and trilateral interaction from the awake state also resurface in the dreaming state as "as if intero-exteroceptive linkage" and "as if trilateral interaction," with the latter in particular being intrinsically hybrid as a composite of the awake and dreaming states. These mechanisms can only be exemplified if there is a specific format or code applied to the processing of the respective stimuli (i.e. "as if stimuli"), and I supposed that this format, as in the awake state, is difference-based coding. I therefore postulate that difference-based coding is as active and utilized in the dreaming state as it is in the waking state.

Why is there any difference at all between the waking state and the dreaming state if the same neuronal mechanisms (i.e. rest–stimulus and stimulus–rest interaction as well as intero-exteroceptive linkage and trilateral interaction) and the same neural code (i.e. difference-based coding) are employed in both states? How can the same neuronal mechanisms enable and predispose to such divergent states? Although the same neuronal mechanisms are indeed recruited, they are employed in a slightly modified way. This is due to the absence of exteroceptive stimuli in the dreaming state. I postulate that the brain may compensate for the absence of exteroceptive stimuli by processing its own neuronal differences in rest–rest interactions. The brain in the dreaming state may thus show two specific features when compared with the brain in the awake state, namely absent exteroceptive stimuli and increased neuronal differences in rest–rest interaction (see Figure 8.3). This results in the conversion of rest–rest interaction into "as if rest–extero interaction" as previously described.

This difference at the very beginning (i.e. the difference between "as if rest–extero interaction" and "rest–extero interaction") involves all the other differences described already. However, this is only possible on the basis that the same mechanisms are applied to "as if rest–extero interaction" in the dreaming state as to "rest–extero interaction" in the waking state. Thus it is not the absence of external stimuli itself, but rather the application of the very same mechanisms in the dreaming state as in the awake state that enables and predisposes to dreams. If, for instance, the increased neuronal differences in rest–rest interaction are not treated analogously to exteroceptive stimuli (i.e. as if exteroceptive stimuli), the subsequent neuronal mechanisms would not be exemplified at all and dreams would then, as I postulate, remain impossible. One may therefore postulate that increased neuronal differences with rest–rest interaction during sleep may predict the occurrence

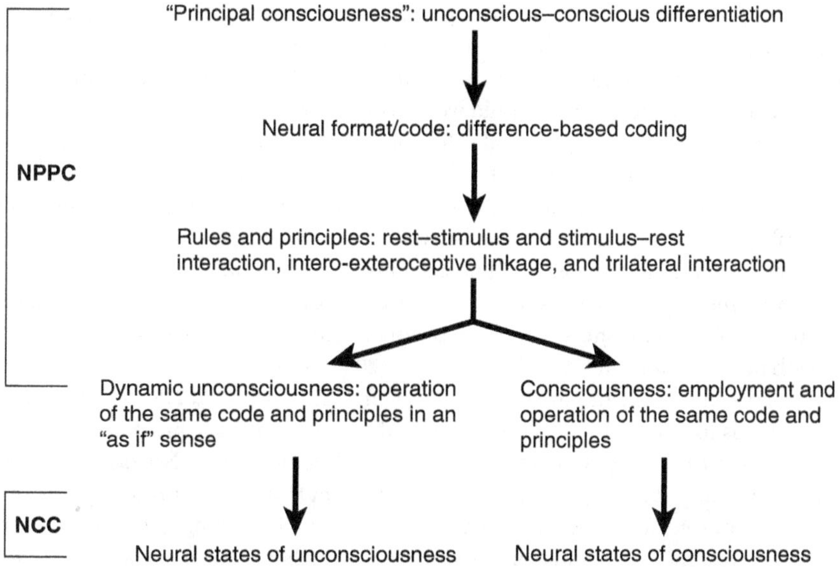

Fig. 8.3 Neural predisposition of principal consciousness (NPPC) versus neural correlates of consciousness (NCC).

of "as if stimulus–rest interaction," "as if intero-exteroceptive linkage," and "as if trilateral interaction," and consequently the manifestation and higher intensity of dreams.

Where does this leave us? So far I have demonstrated that the same neuronal mechanisms are employed in both the dreaming and awake states. The same holds for both the psychological and psychodynamic context. Psychologically, perception, affect and emotions, self and personal relevance exist in the awake state as in the dreaming state. This makes it highly likely that in the dreaming state the same cognitive and affective mechanisms are recruited as in the awake state, as has for instance been postulated by Solms (2000, p. 848). Furthermore, I demonstrated that the same psychodynamic mechanisms hold in both states (e.g. brain–object differentiation with projection, brain–self differentiation with projection, object–self-object transformation, and self–self-object integration). However, these psychodynamic mechanisms are employed in a slightly different sense (i.e. an "as if" sense). Taking all of these considerations into account, I here postulate that the same mechanisms (neuronal, psychological, and psychodynamic) are operational and thus at work in both the dreaming and the waking state. Since I assume that specific mechanisms or processes rather than specific regions enable and predispose to dreams, I call the hypothesis suggested here the "process-based hypothesis of dreams."

What are the implications of the process-based hypothesis of dreams for our second neuropsychodynamic hypothesis? Employment of the same mechanisms provides the transition and linkage between the two different states. The dreaming and waking states can consequently not be regarded as primarily different states that occupy different non-overlapping and mutually exclusive spaces (i.e. unconscious and conscious space). Instead, the dreaming and waking states may rather be regarded as a continuum within a commonly underlying space. Such a commonly underlying space includes both unconsciousness and consciousness, thus allowing for their differentiation (i.e. "unconscious–conscious differentiation").

I described this commonly underlying space by the term "principal consciousness" within the conceptual context, while within the empirical context I assumed the commonly underlying space to be characterized by the aforementioned neuronal, psychological, and psychodynamic mechanisms. I consequently postulate that the various empirical mechanisms (i.e. neuronal, psychological, and psychodynamic) enable and predispose to the space I here described as the concept of "principal consciousness." If this is so, the very same mechanisms should also allow for the differentiation between conscious and unconscious states, which I have here illustrated using the example of dreams. The mechanisms could well account for some of the core features of dreams, perception, emotions, self and personal relevance, with the respective underlying neuronal changes making the assumption of the same mechanisms empirically plausible. This in turn lends support to my second neuropsychodynamic hypothesis, namely that the very same neuronal, psychological, and psychodynamic mechanisms enable and predispose to the transition from the unconscious to the conscious state (i.e. "unconscious–conscious differentiation").

Before comparing my hypothesis about principal consciousness with other current neuroscientific hypotheses of consciousness, I want to briefly compare my "process-based hypothesis of dreams" with those by Hobson and Solms. Hobson (2009) also presupposes the existence of some continuum between waking and dreaming states (e.g. REM sleep, and NREM sleep) in his AIM model. "A" stands for "activation." This describes the level of intrinsic brain activity, which may be either high as in REM sleep and the awake state, or rather low in the NREM sleep periods. "I" stands for "input–output gating," which describes different levels of closure and gating of sensory input and motor output. The sensorimotor system is open in the waking state and closed and shut off during sleep. Finally, "M" stands for "modulation," which Hobson associates with cholinergic "REM-on" cells in the pedunculopontine nucleus and with noradrenergic/serotonergic "REM-off" cells in the locus coeruleus/raphe nucleus in the brain stem/midbrain. As indicated by the terms "on" and "off," these cells in the brainstem/midbrain may be either activated or inactivated during sleep, dreaming, and waking states, and may thus may have a crucial role in, for instance, causing dreams.

Solms (1995, 1997, 2000) rejects Hobson's assumption that dreams are causally associated with REM sleep and brainstem/midbrain activity. Based on lesion studies, he argues that rather than the brainstem, the forebrain structures are crucial in enabling and predisposing to dreams. Only patients with lesions in these regions show a cessation of dreams, while patients with lesions in the brainstem continue to dream. Unlike Hobson, who assumes the forebrain structure to be only secondarily modulated by primary changes in the brainstem, Solms regards the forebrain regions as primarily (and causally) involved in dreaming. This means that dreams can no longer be associated exclusively with REM sleep as generated by the brainstem, as Hobson assumes. Instead, there seems to be double dissociation between REM sleep and dreams, with the latter occurring in the absence of the former, and vice versa. Moreover, Solms links the various psychological features of dreams (e.g. vivid perceptions, emotions, etc.) with the cognitive processes associated with the forebrain regions rather than with their spontaneous generation by brainstem activity as suggested by Hobson. Finally, rather than regarding the cholinergic–noradrenergic/serotonergic balance in the brainstem/midbrain as central, Solms postulates that the dopaminergic system, and specifically the mesolimbic/mesocortical system that is situated in the forebrain regions, enables and predisposes to dreaming, including its various psychological features.

How do these hypotheses stand against the hypothesis suggested here? First and foremost there is a difference in focus. Although disagreeing with regard to the kind of region, both Solms and Hobson focus on specific regions, be it the brainstem or the forebrain. They thus pursue what I call a "region-based approach," which must be distinguished from my focus on mechanisms and processes rather than on regions, involving what I called a "process-based approach" (see also Introduction). However, this is not to say that I do not include reference to specific regions or networks in my hypothesis. It only means that I consider them within the context of specific mechanisms, as for instance neuronal contextualization, rather than in an isolated way by themselves. I therefore name my hypothesis the "process-based hypothesis of dreams" which, in a wider and more general sense, may also be described as a "process-based hypothesis of principal consciousness."

Secondly, I focus on the similarity of the processes and mechanisms involved in the dreaming and waking states, as indicated by the employment of the same neuronal, psychological, and psychodynamic mechanisms in both states. This is at least in part different from Solms and Hobson, who both seem to focus more on the differences between the two states. However, Solms (2000, p. 848) appears to make the same presupposition (i.e. usage of the same mechanisms in waking and dreaming states) as I do here. This is apparent when he argues that an intact forebrain is necessary for dreaming because the latter requires the kind of cognitive processes that are associated with these regions in the waking state.

Thirdly, the focus on the similar mechanisms makes it possible for me to describe the transition between unconscious and conscious states (i.e. unconscious–conscious differentiation) in more empirical detail and to treat both states as a continuum within a commonly underlying space (i.e. principal consciousness), rather than as two largely different states.

Fourthly, unlike Hobson but in a similar way to Solms (1995, 1997), I consider the reference to psychodynamic mechanisms necessary to account for the psychological features of dreams. However, unlike Solms I do not apply a predominantly cognitive framework to explain the psychodynamic mechanisms.

Fifthly, unlike either Hobson or Solms, my hypothesis does not include any reference to specific biochemical mechanisms in dreams in particular and in principal consciousness in general, leaving this for future investigation.

Another issue that has not been touched upon by my hypothesis is its relationship to specific psychological functions (e.g. cognitive functions, such as memories) (see, for instance, Mancia, 2006a,b; Walker, 2009; Nir and Tononi, 2010, with regard to neurocognitive theories of dreams). I consider the implication of specific neuropsychological functions such as memories to be secondary to the here described processes of difference-based coding and neuronal contextualization. Thus there is much research to be done in the future. One may need to link the various processes suggested here, such as "as if rest–extero interaction," "as if intero-exteroceptive linkage," "as if trilateral linkage," and "as if stimulus–extero interaction," to specific neuropsychological functions such as memories. I also left unaddressed why there is such shutdown of exteroceptive input in the dreaming state, which I took for granted and thus as a starting point while not elaborating its own underlying mechanisms (see, for instance, Nir and Tononi, 2010 for a discussion of this issue in the gestalt of "disconnection from the external environment"). One may want to assume that the increased neuronal differences with subsequent confusion of rest–rest interaction with rest–extero interaction may be central here. However, that remains speculative at this point.

I also left open the biological function of dreams. One possible speculation is that the brain needs dreams in order to restructure and reorganize itself after having being exposed to so many exteroceptive stimuli during the day. For this it shuts itself off from exteroceptive sensory input and restructures and reorganizes itself by "turning inwards," as one might say metaphorically. Let us compare the dream to digestion. In the same way that the stomach continues its digestive work even during the night without any actual food intake, the brain continues to structure and organize itself during the night in the absence of exteroceptive stimuli. And in the same way that the nightly activity of the stomach serves to account for the digestion of heavy food that is difficult to digest, the dreaming activity of the brain aims to integrate or isolate difficult stimuli and objects from the day. Analogous to the stomach's nightly activity that produces more or less the same kind of output as during the day, the brain constitutes similar mental phenomena during the night and during the day. In other words one may, metaphorically speaking, regard dreaming as the brain's way of digesting the events and emotions of the day. Dreaming may thus be nothing but the brain's way of digesting the inputs that it received during the day.

Neural correlates of consciousness (NCC) versus neural predispositions of principal consciousness (NPPC)

Several neuroscientific hypotheses of consciousness have been suggested recently. For instance, Edelman and Tononi (2000) and Edelman (2003) consider cyclic processing and thus circularity crucial for the brain's neural organization, which may allow the constitution of consciousness. Cyclic processing describes the re-entrance of neural activity in the same region after looping and circulating in so-called re-entrant (or feedback) circuits. Baars (2005) assumes that there is a global distribution of neural activity across many brain regions in a so-called "global workspace" as characteristic of neural organization that is necessary for consciousness to emerge. Another candidate is neural synchronization. This describes the temporal coordination and integration of activity changes across different brain regions, as for instance by gamma oscillations that are regarded as necessary for consciousness to occur (Fries et al., 2001, 2007; Varela et al., 2001; Baars, 2005; Koch, 2005; Tononi, 2008; Tononi and Koch, 2008; Uhlhaas et al., 2009a,b).

These and other neuroscientific hypotheses target the sufficient conditions, thus focusing on those neuronal mechanisms that constitute consciousness. Since they refer to the sufficient conditions of consciousness, they are subsumed under what are called the "neural correlates of consciousness" (NCC) (Koch, 2004; Tononi and Koch, 2008). The NCC focus on consciousness and how it is constituted in its various manifestations (e.g. perceptions, emotion, cognition, etc.) during stimulation by external stimuli. In other words, the NCC refer to hypotheses about the neuronal mechanisms that make and bring external stimuli into consciousness.

The approach taken here is different from that presupposed in the NCC. Rather than focusing on the sufficient conditions, I have here restricted my attention to those neuronal mechanisms that are necessary but not sufficient by themselves. Therefore I used the terms "enabling and predisposing" rather than "causing," and I referred to "neural predispositions" rather than "neural correlates." Moreover, I did not limit my focus to consciousness as distinct from unconsciousness as in the NCC, but rather I claim, as illustrated by the example of dreams, that my hypothesis applies to both conscious and unconscious states, including their transition. To make this difference clear, I introduced the concept of principal consciousness to describe the common space within which the transition between conscious and unconscious states as what I called "unconscious–conscious differentiation" can be accounted for. Taking the two differences together, I shall now characterize my hypothesis about consciousness as neural predispositions of principal consciousness (NPPC), and distinguish it from the neural correlates of consciousness (NCC) (see Figure 8.3).'

The focus on neural predispositions and the principal consciousness rather than neural correlates and consciousness also involves a shift from stimulus-induced activity to the brain's intrinsic activity (its resting-state activity). I consider the brain's intrinsic activity and its specific neural organization and structure as enabling and predisposing and thus as a necessary although non-sufficient condition of possible consciousness (i.e. principal consciousness). One recent proposal in this direction is that slow cortical potentials that describe neural activity in the frequency range 0.1–4 Hz are postulated to play a central part in consciousness (He et al., 2008; He and Raichle, 2009; Raichle, 2009; see also Shulman et al., 2009 for a step in this direction). This, however, leaves unclear how exactly stimulus-induced activity must be coded in neural activity in order to enable and predispose to the kind of rest–stimulus and stimulus–rest interaction that can be observed and that is here considered essential or crucial in accounting for mental states in both consciousness and unconsciousness (i.e. neuronal–mental transformation and stimulus–object transformation as I conceptualized it here).

Are the NPPC as suggested here contradictory to the NCC that dominate the current neuroscientific research? They are not conflicting because the NPPC target different conditions (necessary rather than sufficient), different phenomena (principal consciousness rather than consciousness), and different neuronal mechanisms (resting-state activity rather than stimulus-induced activity) by comparison with the NCC. Thus the NCC and NPPC may not be as mutually exclusive and incompatible as they appear at first glance. For instance, the kind of rest–stimulus and stimulus–rest interaction that I described here may enable and predispose to the kind of neuronal synchronization or global workspace that is focused on so much in the NCC. Thus future research may focus on integrating the kind of neuronal, psychological, and psychodynamic mechanisms that are associated with the NPPC with those described in the NCC.

The shift in focus that the NPPC involves may best be illustrated by a metaphorical comparison. Let us imagine an old Scottish castle. The Scottish Highlands are hilly and being located prominently and well visible from everywhere the castle stands on a high hill where a special platform had to be constructed to build the walls and thus the castle. Coming from the distance one does not see this platform, and may therefore ask how the castle could be built on such hilly terrain. While it is certainly true that the platform should not be confused with the castle itself, the construction of the latter may nevertheless have been impossible without the former. Hence, the platform is not a sufficient and executing condition but rather an enabling and necessary (and non-sufficient) condition of the possible construction of the castle and its walls. More specifically, it is the design, the shape, and the configuration of the platform that enables and predisposes the castle's wall to be organized, shaped, and structured in a certain way, while at the same time excluding other ways. If for instance the platform is rectangular, it may make the construction of the castle as round impossible.

What I describe here as a platform corresponds well to the brain's resting-state activity and the subsequent specific mode, code, or format that it applies to the neural processing of any incoming stimulus. The brain's specific method of neural coding (difference-based coding) provides the platform for subsequent stimulus-induced activity in the same way that the platform in the Scottish Highlands provides the ground for the construction of the walls of the castle. Stimulus-induced activity may correspond to the walls of the castle and the respectively associated rooms rather than to the platform, while consciousness may be analogous to the castle as a whole, as it is visible from the outside when coming from the Highlands.

When shifting my focus from the brain's neural correlates of consciousness to the brain's resting-state activity as the neural predisposition, I shift my focus from the castle itself, including its various rooms, to its underlying platform, without which the castle's construction and its specific form and shape would have been impossible to achieve. In the same way that the specific shape of the castle, including its main room, the ballroom, would have been impossible to achieve without the specific configuration of the underlying platform, I assume that the brain's stimulus-induced activity during consciousness would be impossible to achieve without a specific structure and organization of its underlying resting-state activity. Thus, in order to fully understand the latter and especially why it is like that and not otherwise, we need to go back to the "neural platform," our brain's resting-state activity and the specific neural code that it applies to its own neural processing of intero- and exteroceptive stimuli.

Conceptual implication: conceptual specification of consciousness

Mark Solms claims that Freud had a special concept of consciousness and unconsciousness which has often been neglected (Solms, 1997). For instance, Solms argues that, following Freud, mental states and ultimately consciousness must be regarded as a construction or interpretation of reality as it appears to us through inner and outer sense (Solms, 1997, p. 687). Meanwhile he rejected the concept of consciousness as mere representation of the inner and outer world itself that may be traced back to the comparison of different stimuli and their processing according to a "certain algorithm" (Solms, 1997, pp. 693, 694). The assumption that consciousness is "construction and interpretation" of reality fits well with the assumptions that have been made here.

Let me be more specific. I considered that, neuronally, rest–stimulus and stimulus–rest interaction, trilateral interaction, and intero-exteroceptive linkage are central mechanisms in structuring and organizing the brain's resting-state activity. Thus different stimuli (neural, intero- and exteroceptive) are compared, matched, and integrated, which makes the process of structuring and organizing within the brain's neuronal context akin to what Solms described as "construction and interpretation" on the basis of a "certain algorithm" (Solms, 1997, pp. 693, 694) within the psychological context of consciousness. What Solms here calls a "certain algorithm" may thus correspond within the neuronal context of the brain to the kind of neural coding that it employs (i.e. difference-based coding), and the respectively involved neuronal mechanisms of rest–stimulus and stimulus–rest interaction, intero-exteroceptive linkage, and trilateral interaction as already sketched. Psychologically, the "certain algorithm" may correspond to what I described as neuronal–mental transformation, stimulus–object transformation, and intero-exteroceptive assignment. Within the psychodynamic context, the concept of the "certain algorithm" may refer to the aforementioned mechanisms of brain–object differentiation, brain–self differentiation, self–self-object integration, and object–self-object transformation.

As demonstrated, the very same mechanisms (neuronal, psychological, and psychodynamic) and thus the same "certain algorithm" are apparently employed in both unconscious and conscious states. Based on these considerations, I assume that the "principal consciousness" is enabled and predisposed to by a "certain algorithm," namely difference-based coding, with subsequent "(neural) construction and interpretation," namely rest–stimulus and stimulus–rest interaction, trilateral interaction, and intero-exteroceptive linkage, and their corresponding psychological and psychodynamic mechanisms.

How can the "principal consciousness" (including consciousness, dynamic unconsciousness, and preconsciousness) be distinguished from the "principal non-consciousness" (encompassing the deep unconsciousness and the non-consciousness) within the empirical context? One may speculate that the main distinguishing feature may consist of exactly what Solms described as the "certain algorithm" and "construction and interpretation." I speculatively assume that the "certain algorithm" must be different in principle in the case of the principal non-consciousness that may for instance not involve subsequent "construction and interpretation." This means that in such a case the brain would employ a neural code other than difference-based coding that enables and predisposes to "principal consciousness." Such an assumption amounts, however, to pure speculation, as the principal consciousness also defines the borders of our possible (brain-based) knowledge, beyond which we remain largely unable to know anything (e.g. the mechanisms underlying the principal non-consciousness) (see Northoff, 2004a,b for details). In other words, the principal consciousness also delineates the border between the empirical and epistemic domains, as well as that between possible and impossible knowledge (see also Part I of this book).

In addition to "construction and interpretation," Solms, claiming to follow Freud's lead, assumes that there is a dichotomy between internal subjective and external objective states: "The perception of internal states (of the person) results in subjective consciousness analogous to the way the perception of external states (of the world) leads to objective consciousness" (Solms, 1997, pp. 685–686, 693–694). Corresponding to the internal–external distinction, Solms distinguishes between internal and external sensory modalities, with affect being the internal modality corresponding to the five well-known external ones (Solms, 1997, pp. 692–693).

How does Solms' claim of the internal–external dichotomy relate to the here presupposed framework of "principal consciousness"? In contrast to Solms, I here do not consider affect and emotion to be purely internal. If this was so, affect and emotion would need to be based on interoceptive processing alone, rather than on intero-exteroceptive linkage as I postulated here. In the same way, I do not consider the external senses to be purely external, as due to rest–stimulus interaction they are linked to the internal resting-state activity of the brain. This provides me with the empirical basis for undermining the alleged internal–external dichotomy by showing that it does not correspond to analogous distinctions within the neuronal, psychological, and psychodynamic contexts. I therefore assume that the internal–external dichotomy within the conceptual context is not empirically plausible when considering the neuronal, psychological, and psychodynamic mechanisms.

The empirical implausibility of the internal–external dichotomy postulated here implies that neither the brain itself nor the psychic apparatus is structured and organized in such a way. However, if neither the brain nor the psychic apparatus (and the mind) can be characterized by such an internal–external dichotomy in an empirically plausible way, the question of the origin of the idea of such dichotomy arises. Rather than being traced back to the brain or the psychic apparatus itself, the internal–external dichotomy may be related to us as observer and the way we can and cannot perceive and cognize ourselves, our brains, and our world. I therefore assume that the internal–external dichotomy is intrinsic to the observer and their specific methods of possible (and impossible) observation. Thus I assume the internal–external dichotomy to be observer based and thereby intrinsic to the observer himself while remaining extrinsic to the brain itself, thus not being brain based.[8]

Why, though, does Solms insist so strongly on the internal–external dichotomy by determining both the psychic apparatus and the brain along its lines? I assume that the reason is conceptual. As demonstrated (see Part I), I distinguish between the concept of the brain as the brain as functioning and that of the brain as observed. When Solms claims that the brain is related to the external dimension and associates consciousness with the internal dimension, he must presuppose the concept of the brain as observed. Meanwhile he appears to neglect the brain as functioning that must be presupposed to be distinct from the brain as observed, at least within the conceptual context.

Let me be more specific. The neuronal, psychological, and psychodynamic mechanisms I here demonstrated target the brain as functioning rather than the brain as observed, thereby leaving it open whether or not both concepts of the brain may be identical within the empirical context. Even if ultimately I may remain unable to determine whether or not the brain as functioning and the brain as observed are identical empirically, the conceptual distinction demonstrated that the internal–external dichotomy may be extrinsic rather than intrinsic to the brain, and thus may not be empirically plausible. However, this remains impossible if, like Solms, one employs the internal–external dichotomy from the very beginning in the presupposed concept of the brain (i.e. the brain as observed), since one can then observe only what oneself as observer has put in into one's own investigation of the brain (i.e. the internal–external dichotomy).

Solms, however, points us in the right direction in the following quote, which makes it even more clear that he lacks the concept of the brain as functioning as distinct from that of the brain as observed: "If we wish to understand anything about mental life as such, that is about the essential processes that lie behind both the brain and subjective experience, then we need to investigate the causal processes underlying both. And we will know that we are approaching a satisfactory scientific understanding when a unified theoretical account begins to emerge that explains both of these perceptual manifestations. That theoretical picture will represent neither the brain [*i.e. the brain as we can observe it*], nor subjectivity, but rather the abstract, natural thing [*i.e. the brain as functioning*] that generates both of them, and that can never be known directly [*due to autoepistemic limitation*]" (Solms, 1998, p. 701; my comments in italics).

Let us go back from the rather slippery conceptual and epistemic terrain to the empirical context of the principal consciousness. What our investigation demonstrated is that the space of the principal consciousness includes the self. Neither in the conscious awake state nor in the dreaming and unconscious state can we avoid some kind of self, be it the dreaming or the awake self. The space of the principal consciousness thus corresponds or maps to the space of the self. This means that we cannot eliminate the self, as that would mean eliminating principal consciousness itself, including dynamic unconsciousness, preconsciousness, and consciousness. There is no consciousness or unconsciousness outside the self, and it was the genius of Freud that he clearly recognized exactly this, namely that nothing in the psychic apparatus including its various conscious and unconscious states is processed outside the self. The various modern scientific attempts to

8 The avid philosopher may of course immediately see a contradiction at work. If our cognition and knowledge and thus observation are considered to be brain based, the internal–external dichotomy cannot only be related to the observer himself as distinct from the brain, but must also be related to the brain itself. In other words, the observer-based nature of the internal–external dichotomy presupposes the brain and may therefore be regarded as brain-based, unless one assumes that knowledge and observation are based not on the brain but rather on the mind as distinct from the brain.

eliminate the self from both consciousness and unconsciousness for methodological–scientific reasons must therefore be doomed to failure (see also Mancia, 2006c, pp. 49–50).

Extending Freud from the psychic apparatus to the neuronal context of the brain, I here aimed to demonstrate that the self cannot be left out of the brain either (i.e. the brain as functioning). Only if we include the self and consider it (i.e. with a basic sense of subjectivity) central to the neuronal mechanisms and neural coding that our brain employs can we understand how the very same brain enables and predisposes to the various conscious and unconscious states that Freud showed to be intrinsically related to the self. I have thus demonstrated the neuronal mechanisms that enable and predispose to what I have here called "brain–self differentiation" (see Chapter 6) and followed it through its manifestations in the conscious state, such as narcissism (see Chapter 7), and in the unconscious state, such as dreams (addressed in this chapter). Now it is finally time to discuss the self itself and to clearly outline what exactly I mean by self and ego within the neuronal, psychological, and psychodynamic contexts.

Chapter 9

The Self and its Brain

Concept of self in psychoanalysis and neuroscience

The concept of self is crucial in psychoanalysis. Although Freud focused more on the ego than on the self, Kohut as one of his successors shifted the self into the centre of psychoanalysis (Kohut, 1971, 1977) by founding self psychology (for an overview, see Milch, 2001). The self has recently also become a major focus in neuroscience. Several authors have been searching for the neural correlates of the self and assume that the self may play a central role in the neural architecture of the brain (Panksepp, 1998; Kelley et al., 2002; Damasio, 2003b; Northoff and Bermpohl, 2004; Northoff and Panksepp, 2008; Panksepp and Northoff, 2008). Although coming from opposite directions, the usage of the concept of self in both neuroscience and psychoanalysis suggests a possible overlap between the two disciplines with regard to the self.

Milrod (2002, pp. 21–22) has pointed out several similarities between neuroscientific and psychodynamic concepts of self, including self–object differentiation, linkage to consciousness, representation of the self, and the bodily based nature of the self. However, he also points out a major difference between neuroscientific and psychodynamic concepts of self. While neuroscience is concerned with a brain-based universal and objective self, psychoanalysis focuses more on a psyche-based individual and subjective self (see Figure 9.1): "Neuroscientists strive to explain fundamental phenomena such as perception, consciousness, emotion, memory, etc., including the subtleties of their integration, and in this way build up an understanding of the basic functioning of the organism. In recent years they have included a study of the self as it integrates with consciousness, emotion, and awareness of the object. They try to discover the general truth, and often use phylogenetic concepts in arriving at an understanding of what a particular concept (e.g. the self) is, what it interrelates with, and what it influences, and is influenced by. In short, they concern themselves with the universal and objective. Psychoanalysis, which has historically focused on the individual and has been more interested in ontology, has as its goal the understanding of protracted intrapsychic, interpersonal, and subjective functioning of the individual. It was in order to better understand that functioning that psychoanalysts had to deal with the self and its representation. In dealing with the self, the psychoanalyst is more likely to focus on the contents of the self and its representation, the state of stability or fragility it may possess, and under what circumstances these characteristics shift back and forth in the course of treatment. In other words, they focus on those elements that make each individual different from one another" (Milrod, 2002, pp. 22–23) (see also Figure 9.1).

How can we bridge this primary gap between brain-based concepts of the self in neuroscience (see Figure 9.2a) and psyche-based concepts of the self in psychoanalysis (see Figure 9.2b)? One option is to modify the psyche-based concept of self in psychoanalysis and adapt it to the brain-based concept of self. However, this means giving up some essential characteristics of the psychodynamic concept of the self, such as its aforementioned focus on "intrapsychic, interpersonal, and subjective functioning of the individual." Alternatively, one may modify the brain-based concept of the self, which is possible only by abandoning the characterization of the self as "universal and objective." This in turn will allow us to account for the subjective and individual

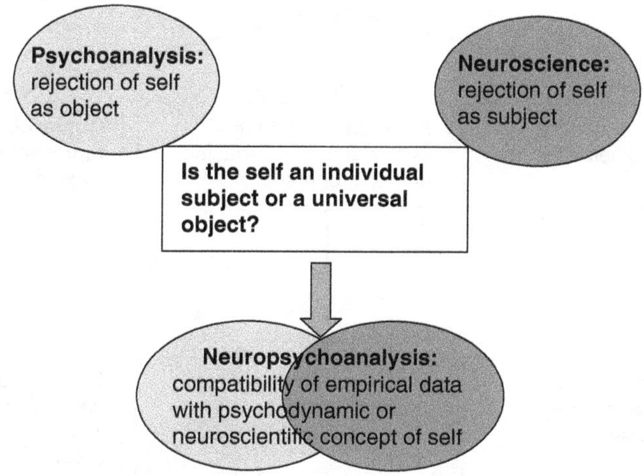

Fig. 9.1 Is the self an individual subject or a universal object?

character of the self as presupposed in psychoanalysis. Thus we seem to have no choice but to abandon either the objective brain-based aspects of the self, that are the focus of neuroscientific accounts, or its subjective component, as highlighted in psychoanalysis.

Current neuropsychoanalysis most often prefers the first option (i.e. adapting psychodynamic concepts to the concepts in neuroscience, especially those in cognitive neuroscience). In contrast, I here want to pursue the second strategy, namely adapting neuroscientific concepts of self and brain to the psyche-based psychodynamic concept of the self. This means that, methodologically, I shall not be following the path of reductionism, which means that psychodynamic concepts are reduced to (and ultimately eliminated in favor of) neuroscientific descriptions. Instead, I test for what may be called plausibility and compatibility between psychodynamic concepts and neuroscientific observations (Northoff, 2004a,b; Northoff et al., 2007).

The aim of this chapter is to directly compare (rather than to reduce) psychodynamic and neuroscientific concepts of the self. The comparison will focus on three main characteristics that are

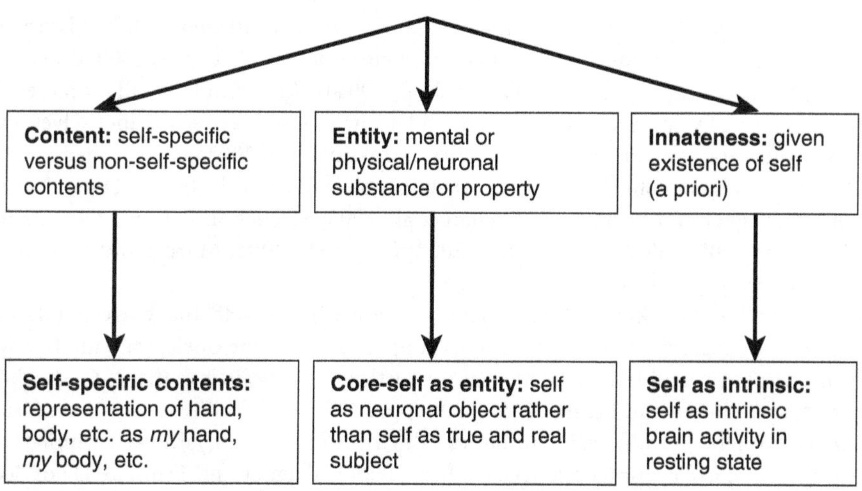

Fig. 9.2a Neuroscience: brain-based concept of the self as special object.

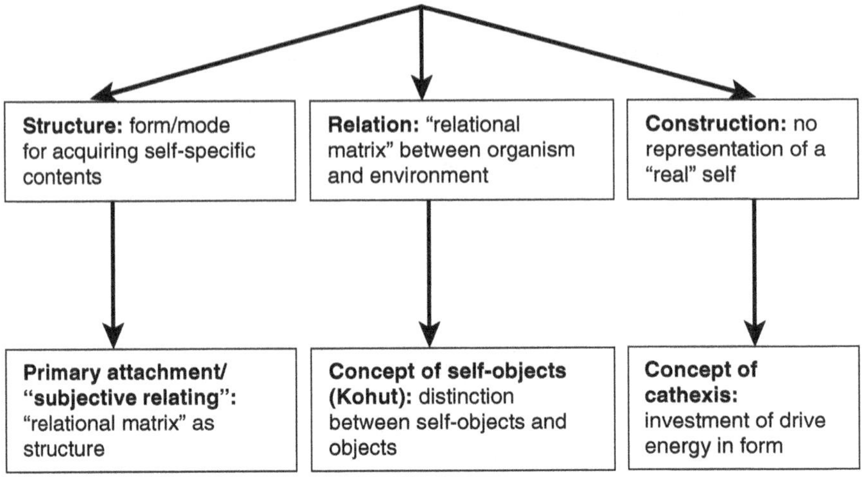

Fig. 9.2b Psychoanalysis: psyche-based concept of the self as individual subject.

controversially presupposed in neuroscientific and psychodynamic concepts of the self, namely content versus structure, innateness versus construction, and entity versus relation. This will be followed by a discussion of the implications for a neuropsychodynamic account of the self, including related psychodynamic concepts such as self-objects, the ego and the id. Thus my strategy here is slightly different to that in the preceding chapters. Rather than discussing neuroscientific and psychodynamic considerations sequentially, as in previous chapters, I here directly confront and compare them with each other in my neuropsychodynamic hypotheses. I focus on specific controversial features of the self, namely content versus structure, innateness versus construction, and entity versus relation in my three neuropsychodynamic hypotheses of the self.

Subcortical–cortical midline structures and the self as structure rather than content

Neuroscientific account: self as content

The question of the self has been one of the most salient problems throughout the history of philosophy, and more recently also in psychology and neuroscience (Gallagher, 2000; Gallagher and Frith, 2003; Metzinger and Gallese, 2003; Northoff, 2004). For example, William James distinguished between a physical self, a mental self, and a spiritual self. These distinct selves are now related to distinct brain regions (Panksepp, 1998; Damasio, 1999, 2003a,b; Gallagher, 2000; Stuss et al., 2001; Churchland, 2002; Kelley et al., 2002; Lambie and Marcel, 2002; LeDoux, 2002; Turk et al., 2002; Gallagher and Frith, 2003; Keenan et al., 2003; Kircher and David, 2003; Panksepp, 2003; Turk et al., 2003; Vogeley and Fink, 2003; Dalgleish, 2004; Marcel and Lambie, 2004; Northoff and Bermpohl, 2004).

Damasio (1999) and Panksepp (1998, p. 2003) suggest a "proto-self" that corresponds more or less to James' physical self. The "proto-self" is supposed to outline one's body in affective and sensory–motor terms, and is associated with subcortical regions such as the PAG, the colliculi, and the tectum (Gillihan and Farah, 2005; Panksepp, 2007). Such bodily self-related sensorimotor contents strongly resemble William James's description of the physical self. Similarly, what has been described as "minimal self" (Gallagher, 2000; Gallagher and Frith, 2003) or "core or mental self" with mental contents (Damasio, 1999) might correspond more or less to James' concept of mental self. The "core or mental self" builds upon the "proto-self" in mental terms and is associated with regions like the thalamus and the ventromedial prefrontal cortex (see, for instance, Damasio, 1999, 2003a). Finally, the "autobiographical self" of Damasio (1999) and the "narrative self" of Gallagher (Gallagher, 2000; Gallagher and Frith, 2003) strongly rely on linking past, present, and future events, resembling James' concept of a spiritual self. The "autobiographical or extended self" that allows one to reflect upon one's "proto-self" and "core or mental self" is associated with cortical regions like the hippocampus and the cingulate cortex.

In summary, the self is here defined on the basis of different contents. The proto-self presupposes bodily contents, namely those of one's own body. The mental self is determined by specific mental contents, namely one's own mental states as distinct from those of other people. Finally, the autobiographical self presupposes autobiographical contents and distinguishes them from heterobiographical contents. William James extended the concept of the self of Descartes by assuming contents other than purely mental contents (i.e. physical–bodily and autobiographical–spiritual contents); this amounts to an amplification of contents with the basic, i.e. content-based definition of the self remaining. These different contents provide the very basis for current neuroscience to "neuronalize" the self, and the different contents are then associated with different brain regions. Thus the neuroscientific concept of self remains essentially content based, with the different contents being mirrored in different brain regions.

Psychodynamic account: self as structure

Freud himself did not assume a self separate from the ego, the id, and the superego. However, as pointed out by Milrod (2002b, p. 8), he used the concept of ego in an ambivalent way. Milrod argues that Freud's concept of ego referred to the individual, to his character, to the executive part of the psychic apparatus, or to the psychic representation of the individual (see also Meissner 2003a,b, 2007). By referring to the individual, the ego points out the subjective experiential or phenomenological aspect. When describing the character, the ego refers to objective psychological personality traits and is thus determined psychologically rather than phenomenologically. When being associated with the executive part of the psychic apparatus, the ego refers to a seemingly objective structure of the mind, namely its functional aspects. Finally, the ego as psychic representation points to a higher-order cognitive–reflective function that is necessary to represent something. These different characterizations of the ego, be they phenomenological, psychological, functional, or representational, open a vast area of possible ambiguities. These ambiguities in the concept of the ego make it in turn possible, if not necessary, to introduce the concept of the self (see Figure 9.2b).

Although others (e.g. Winnicott, Klein, Gede, Hartman, and Jacobson) have also used the concept of self in the gestalt of, for example, self-schemata or self-organization (Richards, 1982; Milch, 2001; Milrod, 2002), Kohut (1977) is considered to have been the first to give the concept of self a central role in psychoanalysis. He founded what has been called self psychology. What does Kohut mean by self? He points out that in order to access another person's intra-subjective experiences, we can rely on empathy that enables us to understand the other's subjective experiences. This shifts the subjective–experiential and phenomenological aspects from the ego to the self.

The self then becomes the placeholder for the subjective–experiential and phenomenological aspects within the concept of the ego. The concept of self in this (narrow) sense describes the specifically subjective and concrete self-specific contents within the overarching objective and abstract structure of the ego. Later, however, Kohut shifts the concept of self even further towards the centre. He releases the self from the overarching structure of the ego and considers it as a fourth psychic structure alongside the id, the ego, and the superego (see Milch, 2001 for an account of the development of the concept of self after Kohut). The concept of self is then no longer restricted to the subjective–experiential and phenomenological aspects of a person. Instead, the self provides the very ground and basis of the person as such (i.e. their mind).

What is the self and how can we determine it more specifically? First and foremost, it should be pointed out that despite all discussions and contradictory positions within psychoanalysis, the self should not be understood and conceptualized as specific content in an ontological sense. Rather than inferring from psychological appearance of the self to its ontological characteristics, one may assume a clear distinction between psychological and ontological characterization of the self. Psychologically, we can apprehend the self only via empathy and introspection, which shifts the focus to subjective–experiential and phenomenological aspects. This designates the self as necessarily subjective, as distinct from any kind of objective content, be it self-specific or otherwise.

However, what is the self if not some specific content (i.e. self-specific content)? The self may be characterized as structure. What does the term "structure" mean? It does not refer to specific contents such as self-specific contents. Instead, the concept of structure refers to the organization and formatting of contents by means of which contents can be constituted as such and linked and adjusted to each other in a regular, constant, and predictable way (see also Milrod, 2002b, p. 7, in particular with regard to the emphasis on constancy, predictability, and regularity). Such organization and structure make it possible to constitute contents as subjective–experiential and phenomenological, rather than as objective–observational and psychological/neuronal.

Comparison between neuroscientific and psychodynamic accounts: self as content or structure?

Recent imaging studies in humans have shown that various cortical regions, but predominantly the so-called cortical midline structures (CMS), are involved in what is called self-related processing (SRP) (Northoff and Bermpohl, 2004; Northoff et al., 2006; Uddin et al., 2007), and these are integrated with subcortical processes to yield an integrated subcortical–cortical midline system (SCMS). The lowest regions of this distributed SCMS network include the periaqueductal gray (PAG), the superior colliculi, and the adjacent mesencephalic locomotor region (MLR), as well as preoptic areas, the hypothalamus, and the dorsomedial thalamus (DMT) (Holstege et al., 1996; Panksepp, 1998). The cortical regions include the ventromedial and dorsomedial prefrontal cortex (VMPFC and DMPFC), the pregenual and supragenual anterior cingulate cortex (PACC, SACC), the posterior cingulate cortex (PCC), and the medial parietal cortex (MPC) (Northoff et al., 2006). As such, the concept of the SCMS describes those subcortical and cortical regions that were designated as paracore–midline regions as distinct from lateral regions (see Chapter 6).

The assumption of an integrated SCMS is consistent with various lines of research that have shown that core self-related functioning involves both cortical and subcortical regions (Panksepp, 1998; Phan et al., 2004; Northoff et al., 2008). Many imaging studies in humans have focused on cortical regions, in part because of the involvement of strong cognitive components, such as evaluative judgments, in the respective tasks (Northoff et al., 2006). This is well reflected in the experimental paradigms of most imaging studies of the self. They compare the evaluation of self-related stimuli with the evaluation of non-self-related stimuli. However, this raises the question of whether the neural activity in the SCMS is associated with the cognitive functions implicated in the evaluation of stimuli as self-related, or in the self-relatedness of the stimuli themselves.

In order to eliminate cognitive functions that are implicated in the evaluation of self-related stimuli, we applied pure perception without any cognitive evaluation task and correlated the observed signal changes with the subjective ratings obtained in a post-scanning session (Heinzel et al., 2006; Northoff et al., 2008; Schneider et al., 2008). In addition to the aforementioned cortical midline regions, this also yielded subcortical midline regions such as the hypothalamus, the DMT, the PAG, and the superior colliculi. Taken together, these results provide evidence in support of the SCMS being crucially involved in self-relatedness, rather than some cognitive function such as evaluation that is necessary to bring the self-relatedness of stimuli into awareness, resulting in self-consciousness.

Are these results more compatible with the concept of the self as content or structure? In the case of a content-based concept of self, one would expect the involvement of specific brain regions which may be associated with self-specific contents as distinct from others that process non-self-specific contents. One would consequently assume processing of different contents and domains and the awareness (or representation) of them in different regions. However, in the case of a structure-based concept of self, one would expect that self-relatedness involves a whole network of brain regions (i.e. a functional system). Such a functional system may enable the organization and structuring of the various stimuli along higher and lower degrees of self-relatedness. Such organization and structuring would concern all kinds of stimuli, thus remaining independent of their modality and domain.

The SCMS must be regarded as a functional unit. As such it might predetermine and predispose the brain's neural activity to process incoming stimuli in a certain way (see also Northoff et al., 2006). This is especially true given that the high resting-state activity of the SCMS may influence the neural activity in other regions, such as sensory regions, and their predisposition to be activated by incoming sensory stimuli. Accordingly, the SCMS is not so much an "executive system" as an "enabling system" (Northoff et al., 2009).

What do I mean by this? An "enabling system" provides the neural ground for specific kinds of neural processing by, for instance, setting the appropriate level of resting-state activity that may be necessary to process specific goal orientations. As such, the "enabling system" organizes and structures the ground on which the stimuli and their actual contents can be processed. This distinguishes an "enabling system" from "executive systems" that carry out and realize the processing of the stimuli themselves and their various contents including, for example, an actual goal orientation.

The SCMS may indeed be an "enabling system" rather than an "executive system." It may then predispose and predetermine the neural activity of other brain regions, including the stimulus processing in the various modalities and domains. This may allow the SCMS as an "enabling system" to provide the structure and organization by means of which the brain may be able to process self-specific contents and to distinguish them from non-self-specific contents. The basic organization and structuring of stimuli may then be associated with the SCMS as an "enabling system," while the actual distinction between self-related and non-self-related contents may be accounted for by the "executive systems" such as the lateral cortical regions.

I therefore postulate that the SCMS as corresponding to the median–core and paramedian–paracore cortical and paracortical regions (as distinct from lateral ones) is what I call a "neural predisposition of the self" (i.e. an enabling and predisposing condition). This amounts to what I call the hypothesis of "neural predisposition of the self" (NPS). The NPS distinguishes between enabling and executing neural conditions, and postulates that both correspond to distinct aspects of the self, namely the structure and organization of the self and the contents of the self. To characterize the self exclusively by contents would thus mean to ignore the SCMS as an "enabling system" and/or confuse it with "executive systems." Thus the distinction between enabling and executive systems as mirrored in the distinction between SCMS and lateral cortical regions is not compatible with the characterization of the self by contents. Instead, the characterization of the SCMS as an "enabling system" involves determination of the self as structure rather than content.

If the SCMS is indeed an "enabling system" rather than an "executive system," neural activity within it should not be associated with particular contents, such as specific sensory modalities (e.g. visual, auditory, gustatory) or domains (e.g. sensory–motor, spatial, social, emotional). Stimulus-induced neural activity in the SCMS should thus remain independent of both modalities and domains, signaling what is called domain independence (Northoff et al., 2006; Walter et al., 2008). Such domain independence would mirror the fact that our self spans different sensory modalities (e.g. olfactory, gustatory) and different domains (e.g. sensory–motor, spatial, social, emotional).

The domain independence of the self has indeed been demonstrated in a recent meta-analysis by our research group (Northoff et al., 2006). We investigated and compared all imaging studies on self-relatedness conducted in different sensory modalities (e.g. auditory, visual, olfactory) and different domains (e.g. verbal/vocal, emotional, social, facial, spatial, memory) (Northoff et al., 2006). Common to the different studies of self-relatedness in different modalities and domains was the recruitment of neural activity in the SCMS. This suggests that neural activity in the SCMS with regard to self-relatedness remains domain independent. What is crucial in determining the degree of neural activity in the SCMS is thus not so much the origin of the stimuli (i.e. their modality or domain), but rather its degree of self-relatedness independent of its modality and domain.

In summary, the characterization of the SCMS in terms of both being an "enabling system" and having domain independence argues against the determination of the self by specific contents. Instead, both features of the SCMS are more compatible with the concept of the self as structure or organization. The NPS postulates that the SCMS (in conjunction with functional principles such as difference-based coding and neuronal contextualization) allows for the kind of neural structure and organization that enables and predisposes to the subsequent constitution of a self. For that, however, the self needs first and foremost to be distinguished from the brain itself, thus involving what I have described as brain–self differentiation (see Chapter 6). I therefore postulate that the SCMS may enable and predispose to possible brain–self differentiation, which in turn provides the ground upon which an actual self can be constructed. This leads us to the question of whether the self is innate or constructed. This will be dealt with in my second neuropsychodynamic hypothesis.

High resting-state activity and the self as constructed rather than innate

Neuroscientific account: self as innate

The concept of self in current neuroscience may also be characterized by what may be called innateness, namely the intrinsic and inborn character. Our body and our mental states are supposed to be inborn and intrinsic, and must as such be distinguished from the body and mental states of other people. Thus our own body and our own mental states are the basis of our self. Since our own body and our own mental states are supposed to be innate, the self itself is regarded as innate (i.e. intrinsic and inborn).

If the self is innate, the neural activity that supposedly underlies the self must be intrinsic and inborn, too. Neuronal activity that remains independent of any kind of stimulus-induced activity must be considered intrinsic. This is the case in neuronal activity in the resting state that in certain regions is indeed high. High resting-state activity has been associated with the so-called default-mode network (DMN) (Reichle et al., 2001; Buckner et al., 2008). The DMN has been observed both in humans (Raichle et al., 2001; Fransson, 2005; Damoiseaux et al., 2006; Fox and Raichle, 2007) and in chimpanzees and macaques (Rilling et al., 2007; Vincent et al., 2007).

The DMN largely consists of medial cortical regions such as the ventromedial and dorsomedial prefrontal cortex, the anterior and posterior cingulated, the superior temporal gyrus. and the hippocampus (see Buckner et al., 2008 for an overview). Since the initial discovery of the DMN (Raichle et al., 2001), regions outside it have also been reported to show high resting-state activity, including various subcortical and cortical medial regions (Buckner et al., 2008 ; Northoff et al., 2008). Thus there seems to be a strong neuroanatomical overlap between the DMN and the SCMS, with the latter also showing high resting-state activity.

Is such high resting-state activity in the SCMS related to the self? High resting-state activity is most probably related to the observation of predominant deactivation, so-called negative BOLD responses, in these regions in functional magnetic resonance imaging (fMRI) (Raichle et al., 2001; Fox and Raichle, 2007; Northoff et al., 2007). Thus high resting-state activity in this so-called DMN is supposed to be associated with introspection of or attention to one's self-specific contents which are consecutively represented as internal and thus mental contents. The self is then characterized by higher-order cognitive function (e.g. by attention) and meta-representation of specific contents as self-related in awareness (Wicker et al., 2003). While the self itself is considered innate and related to intrinsic resting-state activity in the DMN/SCMS, representation of the self is associated with different degrees of stimulus-induced deactivation in precisely the same regions.

Psychodynamic account: self as constructed

Current attachment theory assumes the concept of primary intersubjectivity, which presupposes that human infants are born with the capacity to identify subjective mental states in themselves and other people, and to be aware of them (Fonagy et al., 2007, pp. 291–293). Mentalization of one's own and others' states as mental states is consequently considered to be as innate as attachment. The concept of self must then be considered innate, as it can be traced back to attachment and mentalization.

However, the intrinsic and inborn nature of our mental states has been questioned by Fonagy et al. (2007). Taking a developmental perspective, they argue that our self and its mental states are not innate but rather constructed in relation to its caregivers and environment. Providing an abundance of empirical evidence in this direction, Fonagy et al. (2007) reject any assumption of innateness and opt rather for what they call a social constructivist approach to the self (see also Kirshner, 1991, who comes to the same conclusion, although from a philosophical perspective, citing Hegel and Lacan as support).

The self and its mental states are not innate, and they are not based upon some innate mental, psychic, or neuronal contents. Instead the self, including its subjective and objective contents, is socially constructed in relation to its environment. This implies that the self can no longer be characterized by innateness but by construction. Moreover, what we as external observers call content may then be characterized rather as relation (i.e. the relation between organism and environment). Different contents presuppose different kinds of relation between organism and environment. The self's determination by specific contents (i.e. mental, psychic, or neuronal) must be replaced by the concept of relation between organism and environment and thus by a "relational matrix." Taken together, this means that the self is characterized by construction and relation rather than by innateness and content.[1]

[1] Characterizing the self by construction within the relation to the environment implies a wider notion of psychic reality that includes both inner and outer realities. Relying on his developmental approach, Fonagy et al. (2007) expressed this nicely: "In many ways following in the footsteps of relational theorists . . . we now recognize that the intrapsychic and interpersonal domains of psychoanalysis come together in the

The shift from innateness and content to construction and relation raises the question of how the self can be constructed in relation to the environment. What is the energy that drives the relation to the environment with the consecutive construction of the self first and foremost possible? Freud introduced the concept of cathexis (from a Greek word meaning "occupied" or "filled") to refer to the investment of drive energy (Chapters 4 and 5 for further characterization of drive energy) in a person, object, or idea. Milrod (2002b, p. 9), for instance, assumes that drive energy is invested in the representation of subject and object (i.e. object representation and self representation). Drive energy though may also be withdrawn, implying what is called decathexis (Milrod, 2002b, footnote 5 on p. 9), or shifted to other objects, involving recathexis (see Chapters 4 and 5 for an extensive discussion of the concept of cathexis).

Freud and his successors characterized drive energy as primarily libidinal (and aggressive) and thus sexually based. However, this view was challenged by Kohut, who introduced a new form of drive energy, namely narcissistic energy, which he considered to be separate and different from libidinal drive energy. Narcissistic drive energy is supposed to be associated with the self itself in that it is considered to constitute and maintain the self itself as distinct from its representation (i.e. self-representation).[2] This marks an important distinction from libidinal drive energy, which is invested in the self-representation rather than the self.

However, one may need to associate both narcissistic and libidinal energy with specific contents in which the energy is invested. Thus what underlies both contents, be they narcissistic or libidinal, is cathexis as the investment of energy. The investment of energy independent of the kind of contents in which the energy is invested is by itself of interest for the neuroscientist, as it must be located right at the border between the brain's neural structure and the psyche's structure. More specifically, due to cathexis as investment of energy in whatever content, psychological structure and organization is first and foremost constituted and constructed as distinct from that of the brain. In other words, cathexis may be a necessary empirical condition of possible brain–self differentiation.

---

intersubjective, in which reality is defined as a relational matrix that incorporates both the internal and the external world. If we look outside ourselves, we do not simply see an external world; what attracts our attention are other minds, even though these are external to us. What concerns us, in both the internal and the external world (for the most part) is subjectivity. Thus, it is a mistake to contrast subjectivity (the internal domain of psychoanalysis) with externals and, by implication, objectivity (the domain of other disciplines). The critical developmental dimension is not individual and subjective, but shared. The external comes to be something inherently "other" (as opposed to self), but this is a developmental achievement, not accessible to all of us all of the time. In this sense, the external world is a concept that could never be restricted to physical reality. We attempt to show that developmentally the external is inherently subjective, and the self necessarily differentiates itself from this larger subjectivity of "othernesses." So, in agreement with Freud, and somewhat different from common parlance, we assume both external and internal to be part of psychic reality" (Fonagy et al., 2007, p. 919).

2 See also Meissner (2003a,b, 2007), who analogously distinguishes the self-as-agent from the self-representation. However, we should be careful, as what he calls self-as-agent does not correspond to what I here mean by narcissism, the main difference being that the self-as-agent is characterized in cognitive terms, as distinct from narcissism, which I describe more in affective and intero-exteroceptive terms (see Chapter 7). The same applies to Meissner's concept of the self-as-object, which he characterizes in cognitive (i.e. linguistic) terms (Meissner, 2003a,b, 2007).

Comparison between neuroscientific and psychodynamic accounts: self as innate or constructed?

The concept of an innate self implies prediction of an innate or inborn self-specific neural activity that remains (more or less) independent of bodily, cognitive–mental, or environmental stimuli and thus of non-self-specific contents. Such neural activity that underlies an innate self must be distinguished from the neural activity that may be crucial in representing such an innate self in the gestalt of self-representation. One may consequently assume a distinction between the innate self and the representation of that innate self, with the former being intrinsic and the latter remaining extrinsic.

In contrast, the construction-based concept of self presupposes some neural activity that is necessary to construct the self in relation to the environment. Due to its essential relational nature, this neural activity should be modulated by stimuli both from one's body and from the environment. Since in this case construction replaces representation, no essential difference is assumed between self and self-representation or, to be more precise, between self and self-construction. The self is nothing but construction, and this construction does not need to be represented as such in order to be accessible to the subject or person itself. There is consequently no longer any distinction between an intrinsic self and an extrinsic representation of that self.

Is the high resting-state activity in the SCMS related to the self as innate and its modulation to the representation of the self? The high resting-state activity and the predominant modulation of negative BOLD responses (NBR) are a characteristic of the SCMS. The proponent of the innate-based concept of self may argue that this proves his point, namely that this neural activity is innate, as otherwise it could not occur in the resting state. Since the neural activity must be regarded as innate, one cannot do other than assume an innate self, given the association of the SCMS with self-relatedness. Any stimulus-induced modulation in resting-state activity of the SCMS is then associated with the representation of the self (i.e. self-representation), rather than with the self itself.

However, this neglects a crucial point. If the high resting-state activity in SCMS is indeed innate, it should not be possible to modulate it by actual exteroceptive stimuli and their degree of self-relatedness. The only possible modulation of the SCMS activity level should then be associated with self-representation that in turn presupposes cognitive functions such as the evaluation of self-relatedness. However, a recent study by our research group demonstrated that the mere perception of different degrees of self-relatedness modulates the activity level of the SCMS in the resting state. Schneider et al. (2008) investigated whether the degree of resting-state activity in the SCMS is modulated by the level of self-relatedness by the prior perception of an emotional stimulus. Interestingly, we observed that resting-state activity in various cortical midline regions (the VMPFC, DMPFC, and PCC) was significantly higher when the previously perceived stimulus showed a higher degree of self-relatedness than when it showed low self-relatedness. Most importantly, this change in subsequent resting-state activity level could be traced back specifically to self-relatedness rather than some other stimulus feature, such as emotional intensity and valence.

These results demonstrate that the resting state's neural activity level is not fixed and static but rather changeable and dynamic. If the degree of high resting-state activity in the SCMS can be changed in accordance with the degree of self-relatedness of the previous stimulus, high resting-state activity in the SCMS can no longer be regarded as innate, but rather as constructional. I consequently assume that the resting-state level of activity in the SCMS constitutes and constructs the actual relation between organism and environment. This relation is constituted in orientation on the closeness of the latter and its stimuli to the organism, thus signifying what I call self-relatedness. Therefore high resting-state activity in the SCMS is not as isolated from the environment (and thus purely intrinsic) as is presupposed in the accounts of the self as innate, but rather it mirrors what phenomenological descriptions refer to as "being in the world" (see, for instance, Legrand, 2007, p. 590).

Based on these studies (see also Northoff et al., 2008; Schneider et al., 2008) and the research of others (D'Argembeau et al., 2005; McKiernan et al., 2006), I assume that such high resting-state activity in the subcortical–cortical midline network may constitute continuous self-related processing. Such continuously ongoing processing of self-relatedness even in the absence of new streams of exteroceptive stimuli might allow us to maintain a continuous and temporally extended "sense of relatedness" and thus a "core self." While high resting-state activity may be regarded as the "physiological baseline," continuous self-related processing (SRP) even in the absence of exteroceptive stimuli may then be described as the "psychological baseline" (Northoff and Bermpohl, 2004; Northoff et al., 2006).

High resting-state activity in the SCMS may enable and predispose to the constitution of a relationship between organism and environment via self-related processing. Self-related processing in turn makes it possible to constitute different kinds of contents, whether they are categorized as subjective–experiential and phenomenological or as objective–observational and psychological/neuronal. Rather than executing these different contents, high resting-state activity in the SCMS may then enable the constitution of these contents by providing relation between organism and environment with different degrees of relatedness (i.e. self-relatedness). Thus the high resting-state activity level may be a crucial feature that makes it possible for the SCMS to take on the functional role as an "enabling system" as distinct from an "executive system."

Self-related processing via the SCMS as an "enabling system" may be crucial in constituting and constructing the self as such, independent of any specific content, while the actual contents of the self may then be processed in other brain regions that allow for the different "executive systems." The conceptual distinction between self and self-representation is then replaced by that between self and its different contents which may mirror the neural distinction between "enabling and executive systems." However, if the distinction between self and self-representation is eliminated in favor of construction of self and contents (i.e. "enabling and executive systems"), the concept of an innate self must be abandoned in favor of the self as construction.

What is innate is no longer the high resting-state activity itself in the SCMS, but the process of constructing. Due to the high resting-state activity and its modulation and subsequent adaptation to the environmental stimuli and their degree of self-relatedness, the SCMS may prove crucial in constructing a self. The process of construction may be innate, while high resting-state activity and the self may be considered the results of this process. If the process itself changes, the result becomes a different one. This clearly indicates that it would be wrong to characterize the results, namely the high resting-state activity and the self, as innate by themselves. One may consequently claim that an innate-based concept of the self confuses process and results when it considers the results as given (i.e. as innate), thereby neglecting the process (i.e. self-related processing) that leads to these results.

I have so far discussed what is innate and what is constructed within the psychological context of the psychic apparatus. What, though, is innate within the neuronal context of the brain? I postulate that the kinds of neural code and associated neuronal mechanisms which the brain employs to structure and organize its neural activity are innate. More specifically, I postulate that difference-based coding and neuronal contextualization (see Part II for details) are innate rather than constructed. Both difference-based coding and neuronal contextualization enable and predispose to the subsequent construction of a certain kind of neuronal organization and structure that, for instance, allows for rest–stimulus and stimulus–rest interaction as well as for trilateral interaction and intero-exteroceptive linkage as previously described. Such neural organization and structure enables and predisposes in turn to the construction of specific contents (i.e. mental contents) (see Chapter 5), which may then characterize the self as content based. My neuropsychodynamic hypothesis about the constructed rather than innate nature of the self thus operates exactly at the border between brain and self, where both differentiate from each other as described in the gestalt of brain–self differentiation. Therefore my hypothesis must be considered a hybrid hypothesis that must be "located" right at the junction between neuronal and psychological/psychodynamic contexts.

Self–other continuum in neural activity and the self as relation rather than entity

Neuroscientific account: self as entity

What concept of self is presupposed in these neuroscientific descriptions? Conceptually, the current neuroscientific descriptions may be traced back to Descartes (see also Bennett and Hacker, 2003). Descartes assumed that there was a special entity or substance underlying the self and, due to his famous insight "I think, therefore I am," he characterized this entity or substance as non-physical and thus as mental. Current neuroscientific approaches to the self are still very much in the ontological tradition of Descartes in assuming that a mental entity or substance underlies the self.

However, there are some important differences between Descartes and current neuroscientific accounts. First, neuroscientific accounts no longer presuppose that a mental entity underlies the self. Instead they assume the self to be neuronal. However, this changes only the characterization of the entity, which is no longer mental but neuronal, while the characterization of the self in terms of an underlying entity, be it mental or neuronal, remains (see also Bennett and Hacker, 2003; Northoff et al., 2009). Secondly, neuroscientific accounts might want to put forward that they no longer assume the self to be a single self. Instead, as previously described, there might be multiple selves with different contents, as suggested by William James with his distinction between physical, mental, and spiritual selves.

However, the mere multiplication of the self does not eliminate the assumption of an underlying entity, but rather it multiplies the number of entities (see also Bennett and Hacker, 2003; Northoff et al., 2009). Thus the current neuroscientific accounts still remain within the Cartesian framework of an entity- or substance-based concept of self, although in a neuronal and multiplied gestalt.

Psychodynamic account: self as relation

One of the main purposes of introducing the concept of the self was to characterize subjective experience and thus subjectivity in a truly individual and personal way. As such the self must be distinguished from the ego which, with its superordinate and overarching function and structure, is rather objective, general, and impersonal (Richards, 1982; Milch, 2001, p. 56; Milrod, 2002, pp. 7–8). Defining the self as the structure of subjective experience implies reference to the individual or person. The self may then be regarded as a substructure of the ego. As such, the self bridges the gap between the individual subjective person on the one hand and its universal objective psychological apparatus, namely the ego, on the other (Milrod, 2002, p. 9). The self may then be understood as essentially subjective, so that one may also want to speak of a "subjective self" (Meissner 2003a,b, 2007).

However, when extending the concept of the self to the person as a whole as being the basis of their mind, the difference between ego and self becomes at least questionable. This is because the person and their mind include both subjective and objective aspects. The self is then the bridge between subjective and objective aspects of the person and their mind, although this makes distinction between the self and the ego rather difficult. This may have led Kohut in his later work to regard the self as a fourth superordinate structure alongside the ego, the id, and the superego.

These different definitions of the concept of self reveal the same ambiguities as the concept of the ego, because the duality between subjective and objective realms is reinforced rather than being bridged. The concept of self must thus be considered a "conceptual tranquilizer" rather than a true bridge: "As a result of these considerations, I have argued against the use of self as a superordinate concept in psychoanalytic theory and have focused on the shortcomings of three self psychologies that use the self in this way. I believe that Klein, Gedo, and Kohut all offer the self as a kind of conceptual tranquilizer for the philosophical, theoretical, and clinical dualities that are inherent in psychoanalytic work" (Richards, 1982, p. 956).

How can we build a true bridge between subjective and objective realms in the concept of self without confounding it with the ego? Kohut's concept of self-objects may represent such a bridge. He considered self-objects to be crucial in constituting the self, with the self-objects being to the self what the air is to breathing (Kohut, 1971, 1977). Self-objects are a function that can be induced by subjective experience of particular objects or by inner bodily states and emotional stability (Kohut, 1984). The self-objects describe an aspect of object relation that is experienced subjectively. As such, self-objects designate the subjective intra-psychic experience of the objects (i.e. self-objects), which must be distinguished from the objective inter-psychic or interpersonal relationship between self and object. As well as this subjective–experiential aspect, the second main characteristic of self-objects is that they are experienced as stabilizing for the self. As such they strengthen the coherence of the experience of one's self (e.g. by inducing an existential feeling of the self and its narcissism).

However, the concept of self-objects has a certain ambiguity. On the one hand, self-objects describe an intra-psychic subjective experience of an object. On the other hand, self-objects refer to objects that by definition are objective (Milch, 2001, pp. 66–67). Subsequent authors have emphasized either the subjective or the objective aspect (for an overview, see Milch, 2001, pp. 65–84). For instance, Wolf regards the concept of self-object as a purely subjective aspect of a self-maintaining or self-preserving function (Wolf, 1980, 1988; Wolf et al., 2000). In contrast, Stolorow et al. (1993, 1997) and Bacal (1987) determine the process by means of which the self becomes linked to objects as truly relational process. Here the relation to the object must be specific in order for it to induce the subjective experience of a self-maintaining function. Self-objects are then considered to be relational rather than subjective.

The relational approach to self-objects implies that self psychology and object relation theory (Klein, Winnicott, Bowlby, Sutties, Rycroft) can no longer be regarded as contradictory but rather as complementary (see also Bacal, 1987; Milch 2001, pp. 274–282). The relational determination of self-objects traces both self and objects back to a common underlying feature, namely the relation, that as such bridges the gap between subjective and objective realms. This implies co-occurrence between self and object in different gestalts. Milrod (2002b, pp. 11–12), for instance, states that "awareness of something outside the self" is inextricably linked to simultaneous "awareness of the self," resulting in co-occurrence of self-constancy and object constancy.

Comparison between neuroscientific and psychodynamic accounts: self as entity or relation?

How must the neural activity in the SCMS be related to self-relatedness in the case of the self as entity and relation? If the concept of the self is presupposed as entity, one would expect a clear-cut distinction between self and other in terms of both brain regions and neuronal modulation.

Different brain regions should mediate self and other, and the neural activity in the regions that mediate the self should make a categorical distinction between self and other in an all-or-nothing manner. Either these regions are activated by the self, or they remain silent when the self is not involved.

This contrasts with the concept of the self as relation. There one would expect overlap between self and other in the very same regions, because it is the relation rather than either the self or other that determines the respective regions. Moreover, self and other should be modulated by the neuronal activity in these regions in a continuous and therefore parametric way, thus showing continuous transitions between self and other.

What do the imaging findings tell us? Are they more compatible with the self as entity or as relation? Several imaging studies show involvement and recruitment of the SCMS during both self and other (Gillihan and Farah, 2005; Northoff et al., 2006; Legrand and Ruby, 2009). Therefore the SCMS cannot be considered a specific region for the self that is exclusively activated and recruited during self but not during non-self (i.e. other). Instead, neural activity in the SCMS seems to be induced by both self and other.

The regional overlap between self and other in the SCMS raises the question of how the SCMS can then differentiate between self and other. A categorical distinction, i.e. all-or-nothing, between self and non-self is excluded as that implies exclusive recruitment of the SCMS during self, rather than regional overlap. Is the neural distinction between self and other via the SCMS therefore an illusion that does not correspond to the brain function? Some recent studies (Phan et al., 2004 ; Moran et al., 2006; Northoff et al., 2008; Schneider et al., 2008) have investigated different degrees of self-relatedness (high, low, or medium) and correlated this with the neural activity in various SCMS. All of the studies observed that the degree of self-relatedness correlated and corresponded to the degree of neural activity in various SCMS. This demonstrates that self-relatedness is mediated in the SCMS in a continuous and parametric way rather than in a categorical all-or-nothing manner. What is mediated in the SCMS is thus not so much the self as distinct from the other as the relation between self and other, which may be weighted either towards the self (indicating high self-relatedness) or towards the other (indicating low self-relatedness).

Both of the empirical conditions presupposed by the concept of the self as relation are thus supported by empirical findings. There is regional overlap between self and other in the SCMS which modulates both in a parametric rather than categorical all-or-nothing manner. This provides strong support for the concept of the self as relation rather than entity. In the case of the self as entity, one would have expected regional dissociation rather than regional overlap, and categorical rather than parametric modulation.

How can we account for such overlap between self and other within the neural context of the brain? The hypothesis of NPS argues that this may be related to the specific neural code that the brain applies to its own neural processing of intero- and exteroceptive stimuli. Rather than processing each stimulus by itself independent of the others, in a stimulus-based way, the brain may in fact code each stimulus in relation to others (i.e. in terms of the difference between them), which amounts to what I earlier called difference-based coding. This means that each incoming stimulus is related to others, including extero- and interoceptive and neural stimuli, resulting in rest–stimulus interaction, stimulus–rest interaction, intero-exteroceptive linkage and trilateral interaction (see Part II). By coding the different stimuli with regard to their difference, a relationship between them is established. Each stimulus is thus coded in relation to the others. This in turn makes it impossible to clearly isolate and characterize each stimulus by itself independent of the other stimuli.

What are the implications of such difference-based coding for the interaction between different regions and networks? They are that each region and network constitutes its neural activity only in relation to the respective others, which I described as "neuronal contextualization" (see Chapter 6). Due to such neuronal contextualization, the brain's neural activity itself can only be considered within its own neuronal context. What are the implications of such neuronal contextualization for the psychological context of the distinction between self- and non-self-related stimuli and thus between self and other? Corresponding to the impossibility of neuronal isolation, self-related stimuli cannot be isolated from non-self-related ones either within the psychological context. However, this means that there must be a continuum between self- and non-self-related stimuli and thus between self and other, and this is exactly what can be observed in current imaging studies.

I postulate that the overlap between self and other as observed in imaging studies may correspond within the neuronal context to the application of difference-based coding and neuronal contextualization to the brain's own stimulus processing. In other words, neuronal contextualization within the context of the brain may correspond to the observed psychological contextualization of the self by the other, and vice versa, within the psychological context. As the self is supposed to be intrinsically relational to the other within the psychological context, the brain itself (i.e. its neural structure and organization) is intrinsically relational by itself.

What are the implications of the intrinsically relational nature of the brain's neural structure and organization for the psychodynamic context? As demonstrated in Chapter 6, the constitution of self and objects can be characterized by what I called "co-constitution and co-occurrence." I postulate that the "co-constitution and co-occurrence" of self and objects, the self–object differentiation within the psychodynamic context, corresponds to difference-based coding and neuronal contextualization within the neuronal context of the brain. As our brain's neural structure and organization are supposed to be intrinsically relational, I postulate that self and object are intrinsically relational although distinct. This is mirrored by the aforementioned psychodynamic concepts of "self–object differentiation" and "co-constitution and co-occurrence of self and objects" (see Chapter 6).

Neuropsychodynamic concepts

Self-related processing as a basic process underlying the self

I postulate that difference-based coding is an enabling and predisposing condition of possible brain–self differentiation with the subsequent constitution of a self. This pertains to the neuronal context of the brain. However, it remains unclear which process within the psychological context corresponds to difference-based coding within the neural context of the brain. What I call self-related processing (SRP) within the psychological context is supposed to correspond to difference-based coding within the neuronal context. SRP describes the coordination of various basic emotional processes and bodily interoceptive stimuli (e.g. emotional, motivational, homeostatic, and bodily need states) with exteroceptive stimuli (e.g. sensory stimuli) in relation to the organism's goal-directed activities (Ikemoto and Panksepp, 1999; Kelley et al., 2002; Northoff et al., 2006; Alcaro et al., 2007).

A recent imaging study has shown that an apparently purely interoceptively based process such as body perception can be traced back to the relationship between intero- and exterceptively induced activity rather than to interoceptive activity itself (Wiebking et al., 2010). Although this is a preliminary finding, it supports the assumption of self-related processing as a relational process that by relating intero- and exteroceptive stimuli constitutes mental states (i.e. subjective–experiential and phenomenological contents) (see Northoff et al., 2009 for further support for and more extensive discussion of this view).

SRP is consequently not a higher-order cognitive function by itself, but rather a very basic one that enables and conditions rather than actually executes cognitive functions. As such SRP constitutes the intrinsic linkage and the common coding of sensory and motor stimuli with regard to their importance and meaning for the respective organism within its actual environment (i.e. its relatedness), while SRP in this sense cannot be equated with any kind of contents like self-specific as distinct from non-self-specific contents. Instead, SRP may rather be conceptually defined as the process that first and foremost distinguishes between different degrees of self-relatedness, including high and low self-relatedness (the latter being what we, as external observers, call "other").

The relation between organism and world and thus between neural and intero- and exteroceptive stimuli is strongly affectively colored. This may be because it allows external events to be linked to the internal organism's basic needs (Panksepp, 1998; Northoff and Panksepp, 2008; Panksepp and Northoff, 2009). Neuoroimaging studies in humans have demonstrated that precisely those words and pictures that are highly related to the respective individual's self are considered to be more emotional than those that show rather low degrees of self-relatedness (Northoff and Bermpohl, 2004; Northoff et al., 2006; Northoff et al., 2008). The affective "coloring" of the world may yield phenomenal experience of survival-relevant objects and events as "mine" or as "belonging to me," reflecting what has been called "belongingness" or "mineness" (Lambie and Marcel, 2002).

What does this characterization of self-related processing as an affective, non-cognitive, basic, and underlying process imply for the concept of the self? The self can no longer be characterized by specific contents, but rather by a specific process (i.e. self-related processing) that allows or enables the constitution of different kinds of contents. This implies that the self can also no longer be considered a mere object at all, including even specific objects, as Legrand (2007, p. 589) described it nicely with the term "self-as-object." This is for instance the case in the neuroscientific concepts of the self as bodily self, mental self, and autobiographical self, where the self is defined by objective psychological functions and contents.

However, the rejection of the "self-as-object" does not mean the assumption of the "self-as-subject" (see also Meissner 2003a,b, 2007 for a more or less analogous distinction), for that characterizes the self as subject (rather than object) from first-person perspective as distinct from third-person perspective. This is for instance the case in phenomenological accounts (see, for instance, Legrand, 2007; Legrand and Ruby, 2009) that define the self as a specific subjective perspective with subjective–experiential and phenomenological contents. This means that the self is still characterized by contents, although not objective–observational ones, as in neuroscientific concepts of the self, but rather subjective–experiential and phenomenological ones.

SRP, as presupposed here, cannot be associated with either the "self-as-object" or the "self-as-subject." Instead, it makes this distinction first and foremost possible by making it possible to distinguish between subject and object and thus between both concepts of the self. SRP must consequently be regarded as more basic and fundamental than either subjective (i.e. phenomenological) or objective (i.e. neuroscientific) concepts of the self. As we shall see in the following account, characterization of SRP as non-cognitive, affective, basic, and fundamental to both subjectivity and objectivity is central to developing neuropsychodynamic concepts of self, self-objects, and objects. Neither SRP nor the implied sense of self can be equated with any kind of content, such as self-specific contents as distinct from non-self-specific contents, or subjective–experiential contents as distinct from objective–observational contents. Instead, SRP may be conceptually determined as a process that first and foremost makes the distinction between different kinds of contents with different degrees of self-relatedness possible.

Self-related processing and self-objects

Psychodynamically, self-related processing in the sense already described may be associated with a very basic sense of attachment (see Figure 9.3). Attachment can then no longer be regarded as

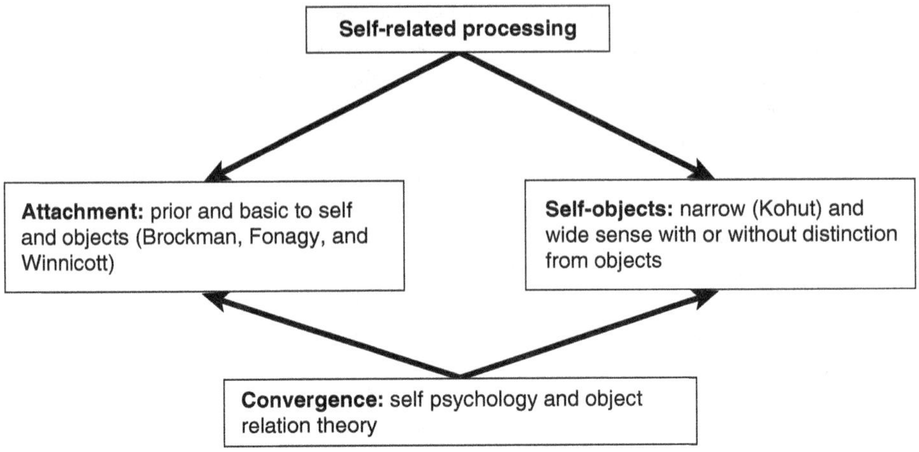

Fig. 9.3 Self as structure: self-related processing and self-objects.

secondary and as building upon some self or object, but must rather be considered primary, as it first and foremost makes any kind of self and its distinction from objects possible. Interestingly, Brockman (2002) appears to opt for such a basic and primary sense of attachment. Relying on Darwin's evolutionary theory (rather than on Freud and his emphasis on the narcissistic state being primary), Brockman characterizes attachment as primary and describes it as an instinctive priming of the newborn to attach to objects in the environment. The first outward object to which a newborn attaches is of course the mother, and this is then followed by attachment to other objects.

I consider self-related processing or relational processing to correspond on the functional level to what Brockman describes psychodynamically as a basic and primary sense of attachment. What Brockman describes as attachment, namely the primary outward orientation to attach to objects, comes very close to what I described as relational processing (i.e. self-related processing). Similar to his concept of attachment, I assume self-related processing to be necessary to differentiate between self and object. Since both attachment and self-related processing enable the constitution and differentiation between self and object, they must be regarded as preceding both self and object. This is directly expressed by Brockman with regard to attachment: "Attachment begins before any sense of self and before any sense of object to attach to" (Brockman, 2002, p. 90).

Another commonality between self-related processing and attachment consists of their close linkage to emotions. The constitution and differentiation between self and objects is made possible by attaching emotions to them and thus affectively coloring them. However, one should not consider emotions to be the source and origin of self and objects, since that would be to confuse the necessary condition (i.e. self-related processing or attachment) with what it conditions (i.e. self and objects, including their affective coloring). Thus, analogous to self and objects, self-related processing and attachment must be regarded as preceding emotions: "Attachment begins before emotion, and indeed belongs to a different biological system than emotion" (Brockman, 2002, p. 90).

Another psychodynamic concept that comes close to what I call self-related processing is the concept of "subjective relating" as described by Winnicott (1965, 1971, 1975). "Subjective relating" refers to the ability to relate to (and transitorily control) objects according to the infant's needs. Such "subjective relating" constitutes the connections between the infant and its environment, which in turn enables the infant to acknowledge and recognize the objects' separate existence. Thus, in a similar way to self-related processing, "subjective relating" precedes any possible differentiation between self and object.

What are the implications of the concept of self-related processing for the psychodynamic concept of self-objects? Self-objects are supposed to be distinguished from mere objects by their relation to the self and their intra-psychic subjective experience. However, if any sense of object can be traced back to construction and constitution on the basis of self-related processing, primary attachment, or "subjective relating," the distinction between self-objects and objects becomes fragile. Based on self-related processing and its "relational matrix," any object must be considered a self-object in a very basic sense, since any object is then somehow related to the organism and its developing self. This implies that the distinction between self-objects and objects, as purported by self psychology, can no longer be considered primary and categorical but rather as gradual and continuous. Self-objects and objects reflect different degrees of relatedness between organism and objects in the environment, rather than being primarily different categories.

If the relationship between the organism and the related objects is particularly close, the self-object may be strongly constitutive and extremely stabilizing for the developing self (i.e. the newborn). Consider the example of the mother as first self-object. If the mother is loving and caring, the relationship to the mother will be very close and tight, colored with positive emotions, and highly stabilizing for the developing self. If, in contrast, the mother is rather non-loving and neglecting, the mother as self-object may not be as strongly constitutive and stabilizing for the developing self, and rather negatively charged emotionally. This example demonstrates that the constitution of self-objects very much depends on the relationship between the organism and the environment that is constituted as such by self-related processing, attachment, or "subjective relating."

One notable instance of such "subjective relating" concerns one's own body. Since the interoceptive stimuli from one's own body provide a continuous steady input, one's own body may be constituted as the first self-object (see Chapter 7 for further details). And since the constitution of the self is very much based on self-objects, self and body are closely intertwined with each other without, however, being identical. This close relationship between self and body as self-object may correspond on the neuroscientifically to the concept of a bodily based "proto-self" (see Panksepp, 1998), and phenomenologically to what is called the "lived body" and the "embodied self" as distinct from the "objective body" (Legrand, 2007; see also Meissner 2003a,b, 2007 for an investigation of the relationship between self and body within the psychodynamic context).

What does the notion of "subjective relating" imply for the concept of the self-object? The concept of self-objects needs to be extended to include both a narrow sense (only close relation and high degree of self-relatedness) and a wider sense (both close and loose relation, with lower and higher degrees of self-relatedness). The wider sense of self-objects is very compatible with Brockman's concept of primary attachment, and also with the concept of self-related processing as relational processing that has been suggested here.

This implies that the conceptual (and psychodynamic) boundaries between self-objects and objects are blurred and undermined by tracing both back to a common underlying condition (i.e. self-related processing, attachment, or "subjective relating"). More generally, self psychology and object theory will be closely linked together and regarded as distinct aspects of the same underlying process, rather than as two separate disciplines. This contrasts with the scenario when one presupposes the narrow meaning of self-objects. In this case, self-objects (in a narrow sense) must primarily be distinguished from objects, which implies that self psychology and object theory are separate disciplines with no major links between them.

If self-related processing is regarded as the necessary (i.e. enabling and predisposing) condition of self-objects in a wider sense, self-objects must be characterized as being relationally based.

No object is processed in isolation from its relation to the organism, which implies that we remain in principle unable to distinguish the object as a distinct and separate element independent of its relation to the organism. Thus a relationally based characterization of self-objects excludes an elementally based account. What is processed is thus not the object as a distinct and separate element, but its relation to the organism. Rather than the object as element, the relation of that object to the organism is processed which, as previously demonstrated, is highly consistent with our brain's neuronal processing. This implies that, psychodynamically, self-objects are inherently (or intrinsically) relationally based rather than elementally based, as is well reflected in Brockman's concept of primary attachment and Winnicott's concept of "subjective relating."

Relational concepts of self and ego

How does such a relationally and construction-based concept of self stand in relation to what Freud described as ego? Let us start with the concept of self (see Figure 9.4). I want to distinguish between a narrow and wide concept of self. The narrow concept of self concerns those self-objects that show a high degree of self-relatedness and are stabilizing, indicating continuity, predictability, and stability. The narrow concept of self thus concerns specific contents (i.e. self-objects in a narrow sense), which makes it a content-based concept of self.

Such a narrow concept of the self predominates in the current neuroscientific account, where a bodily self is often distinguished from a mental self and a social self. This is for instance most apparent in Feinberg (2010), who assumes a multilayered concept of self when he distinguishes between a bodily self, a relational self, and a narrative self. The bodily self is the self that is based upon our body, the relational self is the self that emerges in interaction with others, and the narrative self is the self that develops and narrates its own history over time. This differs from the concept of self that is presupposed here. When I speak of the self as relational, I do not restrict the concept of relation to a particular relation, namely the social relation to others, as Feinberg does. Instead, I presuppose the concept of relational in a wider sense than Feinberg, including any kind of relation, such as one's relation to one's own body (mirroring what Feinberg calls the bodily self) and one's relation to oneself (as in Feinberg's narrative self). The wider and more inclusive concept of relation means that my concept of the relational self is no longer based upon contents at all, be they bodily, social, or narrative. This allows me to pursue a form- and

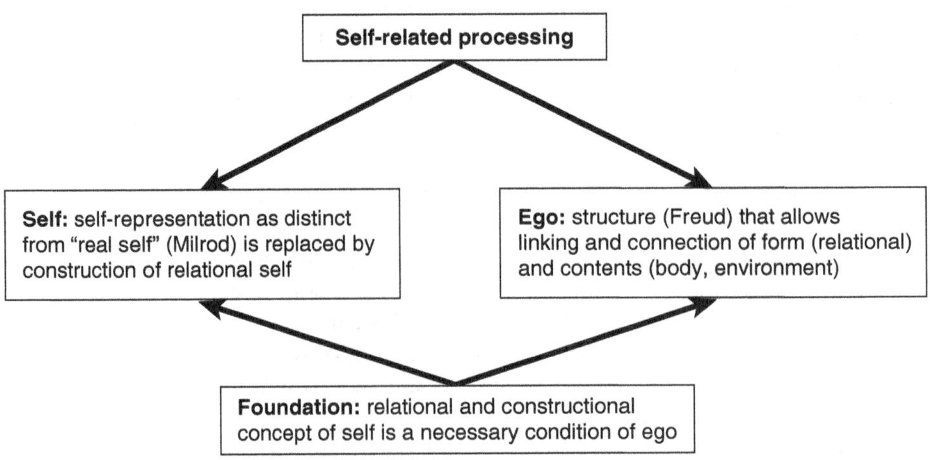

Fig. 9.4 Self as relation: self as relational form and ego as structure.

organization-based concept of self that determines the self by a specific form or organization rather than by specific contents.

Such a wider concept of relation is presupposed in a wider concept of self that concerns any object which is related to the organism (i.e. self-objects in a wide sense). This implies that, in contrast to the narrow concept of self, the wide concept of self explicitly includes the relation between organism and environment. In contrast to the narrow concept, the wide concept of the self is therefore no longer based on contents but on relation. This implies a form- or mode-based concept of self rather than a content-based one. The self can consequently be characterized by a specific organization or structure rather than by specific contents (i.e. self-specific contents).

Where do the narrow and wide concepts of self stand in relation to the concept of the ego? The ego has been described as an overarching structure. Although there has been discussion about the different and ambiguous meanings of the ego in the psychodynamic literature, the exact meaning of the concept of "structure" itself remains unclear. "Structure" is often supposed to describe organization or organizational principle as opposed to mere content, but what makes such organization possible and how are contents constituted? Since the linkage between intero- and exteroceptive stimuli is crucial in self-related processing, the ego as structure may be associated with linking the various bodily and environmental stimuli (i.e. intero- and exteroceptive stimuli) to a specific form or mode, namely the relational organization.

The concept of the ego as structure may be defined in two ways. First, the ego may be defined by both intero- and exteroceptive stimuli whose linkage allows the constitution of specific contents or self-objects in a wider sense. Secondly, the definition of the ego needs to include the organization that makes such linkage first and foremost possible. To enable such intero-exteroceptive linkage, the organization must be relational, as otherwise contents and self-objects cannot be constituted on the basis of intero- and exteroceptive stimuli. To neglect either aspect in the definition of the ego means making the ego either blind (as when neglecting the second aspect, namely the organization) or empty (as when neglecting the first aspect, namely intero- and exteroceptive stimuli). Thus the ego may be considered the structure that enables the integration and convergence between these two aspects, namely intero- and exteroceptive stimuli and relational organization.

Is this structure-based ego still overarching as originally intended by Freud? The ego as structure based is no longer an overarching structure when compared with the wider concept of self, because both describe the same process (i.e. basic attachment, self-related processing or "subjective relating") that allows the constitution and differentiation of self and objects. However, the structure-based ego may be regarded as overarching when compared with the narrow concept of the self as content based. The specific contents of the self (self-objects in a narrow sense) can be constituted only when there is prior linkage between form and content by the structure of the ego.

Construction-based concepts of cathexis and the id

Brockman (2002, p. 90) considers primary attachment as the "basic drive." This basic drive makes attachment possible and provides the energy that is necessary to establish relation of the organism to the environment and its possible objects. Energy is necessary to make any kind of primary attachment and self-related processing possible. Without energy, possible construction of the "relational matrix" remains impossible. Psychodynamically, this energy is characterized as libidinal and/or narcissistic, and its investment has been described by the term cathexis. However, neither libidinal nor narcissistic cathexis can account for the energy that must be invested in order to enable self-related processing, basic attachment, or "subjective relating." Although both libidinal and narcissistic concepts of cathexis presuppose that energy is cathected into sexual life or the

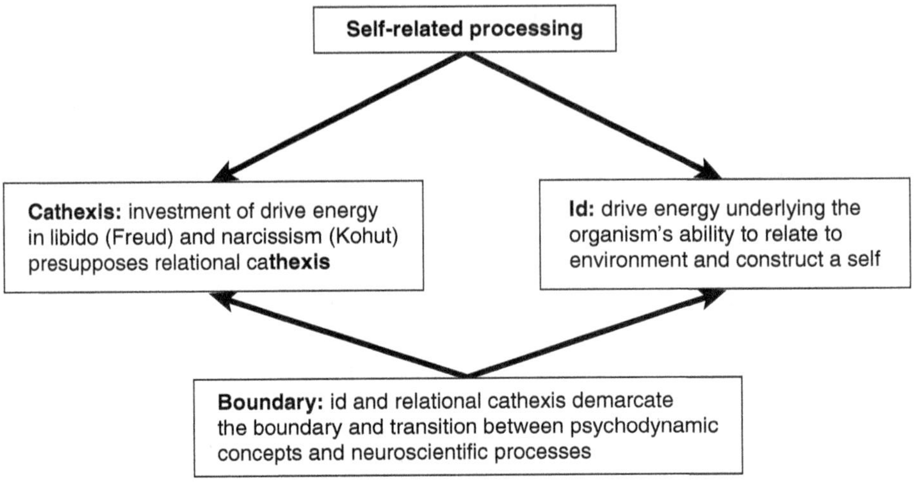

Fig. 9.5 Self as constructed and id as drive energy of relation/construction.

narcissistic self, energy in the relational approach suggested here is invested primarily in the organism's relation to the environment. Thus energy is presupposed in both cases, but its outlet and investment (i.e. its cathexis) differ.

Psychodynamically, cathexis is often assumed to concern specific contents, be they libidinal (according to Freud) or narcissistic (according to Kohut). However, in the relational approach suggested here, energy is no longer invested in specific contents, be they libidinal or narcissistic, but in a specific form or mode. More specifically, energy is invested in constructing the relational mode, namely the relationship between organism and environment. This in turn makes it possible to acquire any kinds of contents, including libidinal or narcissistic contents. I call this concept of cathexis "relational cathexis" (see Figure 9.5).[3]

Based on the aforementioned findings, I postulate that the high resting-state activity of the subcortical–cortical midline structures may provide the energy that is necessary for establishing organism–environment relation (i.e. the "relational matrix"). If this is true, the subcortical–cortical midline structure's high resting-state activity should be crucial in enabling self-related processing. As demonstrated by the aforementioned findings, this is supported by empirical evidence, albeit preliminary. What is described neuroscientifically as high resting-state activity may then be conceptualized neuropsychodynamically as relational cathexis. Relational cathexis must be considered more basic and primary than libidinal and narcissistic cathexis, as the former enables the constitution of the latter.

[3] However, one may argue that the former includes the latter, so that relational cathexis is a specific subset of libidinal or narcissistic cathexis, rather than being a separate and distinct form of cathexis. It is certainly true that both libidinal and narcissistic contents presuppose relation between organism and environment, so that both may be considered to be closely connected to relational cathexis. However, close connection is not the same as inclusion. Something can only be included in something else if both concern some kind of content. However, this is not the case, as the concept of relation is not a specific content like a relational content that can possibly be included in, for instance, non-relational or elemental content. Thus the argument for possible inclusion of relational cathexis within libidinal or narcissistic cathexis is not valid, and if advanced can only lead to confusion of form/mode and content.

Where does the concept of "relational cathexis" as advanced here stand in relation to the concept of the id? The id has been regarded in psychoanalysis as the source of drive energy (i.e. cathexis), which was originally described by Freud as libidinal, and later, by Kohut and his followers, as narcissistic. The relational–constructional concept of cathexis extends the concept of the id by going beyond the boundaries of exclusive definition of the id by specific contents, be they libidinal or narcissistic. This opens the concept of the id to a specific form or mode (i.e. the relational organization).

However, at the same time, the concept of the id is limited by the demarcation of its boundaries. The concept of the id within the relational framework also demarcates the boundary of the brain and thus the border between neuroscience and psychoanalysis. Beyond this point, the characterization of the id by relational cathexis and thus by a psychodynamic (or psychological) concept becomes impossible. This is because no psychodynamic or psychological content can now account for the investment of energy in the organism–environment relation. If there is no longer any content (including libidinal and narcissistic content), but only mere energy and drive (without content), nothing can now be described in either psychodynamic or psychological terms.

Psychodynamic (and psychological) concepts and descriptions thus reach their boundaries at the points of relational cathexis and the id, beyond which they become impossible. Beyond this point, both relational cathexis and the id can be described only in terms of neuroscientific concepts. This is because the methodological boundary between psychodynamic/psychological concepts and neuroscientific concepts can be traced back to the boundary between person and brain, where both converge and are simultaneously differentiated from each other.

Thus both the id and relational cathexis cannot be described in psychodynamic detail, but must be further specified in terms of neuroscientific concepts such as high resting-state activity. The concepts of the id and relational cathexis make a truly neuropsychodynamic account necessary, as otherwise, in the case of a purely psychodynamic description, they would remain unspecified. The concepts of cathexis and the id must thus be regarded as being located right at the boundary between neuronal and psychological/psychodynamic contexts. Metaphorically speaking, they can be regarded as border stations between the neuronal and psychodynamic territories. As such they require description and characterization from both sides of the border stations and thus from both territories, otherwise they remain open and unspecified. And it is such double description and characterization that I here and in Chapter 8 aimed to develop in my neuropsychodynamic hypotheses, which basically describe the transition between the two territories and can therefore themselves be regarded as border stations.

The need for double description (i.e. in neuroscientific and psychodynamic terms) makes it clear that psychoanalysis needs neuropsychoanalysis if it wants to describe and specify its own concepts in full detail. Neuropsychoanalysis is thus confronted with the task of revealing those processes that allow for convergence and transformation between psychodynamic and neuroscientific concepts, and thus laying bare the boundary between the two disciplines. Let me return to the example of the self. I assume self-related processing, primary attachment, or "subjective relating" to be characterized by processes that are located and operate at precisely this boundary. As such they are supposed to enable the convergence and transformation of psychodynamic and neuroscientific concepts. Self-related processing, primary attachment, or "subjective relating" must be considered to be truly neuropsychodynamic concepts rather than being either psychodynamic or neuroscientific. As such they must be considered necessary conditions of the possible differentiation between brain and self (i.e. brain–self differentiation), and ultimately the differentiation between the neural apparatus of the brain and the psychic apparatus of the psyche, and thus between brain and mind.

Concept of the brain

What does the relational–constructional concept of the self imply for the concept of the brain? The brain may be considered to provide the necessary conditions of what I described as the form or mode, rather than sufficiently constituting specific contents by itself. The form or mode of the self was characterized as relational rather than elemental. This relational form or mode of the self is supposed to correspond psychodynamically to primary attachment, and to correspond functionally–neuronally to self-related processing as mediated by the SCMS. If self-related processing is supposed to be the necessary functional–neuronal condition of the distinction between subjectivity and objectivity, the SCMS underlying self-related processing must be considered to be the necessary condition of the distinction between subjectivity and objectivity, and thus between self and object in a psychodynamic context. In short, the SCMS and SRP may be considered necessary (or enabling) empirical conditions of possible brain–object and brain–self differentiation, and thus of self–object differentiation.

This implies that the concept of self cannot be characterized either by subjective contents, as in the psychoanalytic and phenomenological traditions, or by objective contents, as in current neuroscience. Instead, the concept of self must designate those processes that first and foremost enable the differentiation between subjective and objective contents of the self. Rather than determining the concept of self by contents, it may be better determined by a specific organization, namely the relational one as distinct from an elemental one.

Such a relational concept of the self can be linked to the brain only if one presupposes a corresponding concept of the brain. An organization-based concept of the brain focuses on structure, relation, and construction rather contents, entity, and innateness as in a content- or substance-based concept of the brain. Rather than investigating the neuronal correlates of self-specific contents, one may focus on revealing those principles of the brain's neuronal organization that make possible the construction and differentiation of self-specific and non-self-specific contents such as different self-objects (i.e. in a narrow or wider sense). This is exactly what I stated in my hypothesis of the self which postulates that the processes correspond to difference-based coding and neuronal contextualization within the neuronal context of the brain's neural structure and organization.

Such an organization-based concept of the brain may also help to further specify the concept of the brain as functioning, as discussed in Part I of this book. The concept of the brain as functioning was distinct from that of the brain as observed and the brain as experienced. Following our account of the self, we may now determine the brain as functioning in more detail. The brain as functioning may well correspond to what I have here described as the organization-based concept of the brain. Thus the brain as functioning may empirically be characterized by structure, relation, and construction rather than by contents, entity, and innateness. But what enables and predisposes us to develop a content- or substance-based concept of the brain as characterized by contents, entity, and innateness? I assume that this is due to the bias in our observation of the brain that focuses on contents, entities, and innateness. Thus the brain as observed may be characterized by a content- or substance-based concept of the brain.

If one now does not distinguish between the brain as functioning and the brain as observed, one is prone to project the properties and features of the latter onto the former and to characterize the brain itself, as it is independent of our observation, in a content- or substance-based way. However, our account of the concept of the self and, even more important, the empirical data on the relationship between brain and self, render such identification and subsequent reduction of the brain as functioning to the brain as observed implausible. More specifically, as was pointed out in Part I, reducing the brain as functioning to the brain as observed is conceptually implausible, while the example of the self that has been discussed here shows such reduction to be empirically implausible as well. Thus we have no choice but to follow the threads of conceptual and empirical plausibility and to target the brain as functioning rather than the brain as observed.

What are the implications of the presupposition of the brain as functioning for the empirical account and investigation of the self? The self and its distinct characteristics can now no longer be associated with neuronal activity in specific brain regions, specific genes or proteins, etc., but rather they are associated with specific principles according to which our brain's neuronal activity is organized. Thus the focus shifts from the neuronal activity in specific brain regions to its underlying and guiding principles of neuronal organization that constitute neuronal activity across the various brain regions.

Such a focus on the principles whereby our brain structures and organizes its neuronal activity will also lay bare the boundary at which neuroscientific and psychodynamic concepts converge and are differentiated from each other. The crucial relevance of this border territory and its border traffic at the junction between neuronal and psychodynamic contexts will become even more apparent when considering neuropsychiatric disorders such as depression. They will show that traffic jams, detours or short-cuts in the traffic in this border territory at the junction between neuronal and psychodynamic contexts can lead to disastrous consequences. This will be the focus of Part IV of this book.

Part IV

Disordered Equipment

Freud regarded psychological disturbances as the starting point for his assumptions about the psychodynamic mechanisms in healthy subjects. I here follow the same path by now considering depression and psychosis as litmus tests for the neuropsychodynamic hypotheses in healthy subjects previously postulated. Let me be a little more specific. I considered the neuronal mechanisms of rest–stimulus and stimulus–rest interaction, difference-based coding, intero-exteroceptive linkage and trilateral interaction to enable and predispose to the constitution of psychodynamic features like internalization/externalization, self-objects/narcissism, unconscious/dreams, and self/ego. If my neuropsychodynamic hypotheses about the relationship between the aforementioned neuronal mechanisms (i.e. the neural equipment) and these psychodynamic features in healthy subjects hold, changes in the latter as observed in depression and psychosis should be associated with alterations in the former. In other words, the neural and mental equipment already discussed should be disordered in depression and psychosis, leading to what I call "disordered equipment."

One may even go one step further and postulate that alterations in the neuronal mechanisms should enable and predispose to the kind of psychodynamic changes that characterize depression and psychosis. If so, my neuropychodynamic hypotheses in healthy subjects gain some indirect support from cases with disordered equipment. Most generally, I postulate that depression can be regarded as a disorder of brain–self differentiation, whereas psychosis concerns abnormal changes in brain–object differentiation. This rather general hypothesis will be discussed at the end of each chapter, as well as in the Appendix that follows Chapter 12.

Before discussing depression, I should like to make a brief methodological comment. My focus here is not so much on the neuronal and psychodynamic mechanisms that induce depressive and psychotic symptoms themselves. These sufficient (but non-necessary conditions) may be called "neural and psychodynamic correlates" (see also the Introduction). Instead, I shall focus on the necessary and non-sufficient conditions of possible depression and psychosis and thus "neural predisposition" rather than "neural correlates." My account here aims to reveal what kind of neuronal and psychodynamic mechanisms must be set in motion and thus occur prior to the outbreak of depressive and psychotic symptoms in order to enable and predispose to the latter. This may be described as a "mechanisms-based approach." I intend to show that the depressive and psychotic symptoms are the consequence or effect of the recruitment of normal functioning neuronal and psychodynamic mechanisms within an abnormal context.

Finally, I here consider both depression and psychosis to be "system disorders" (for the case of depression, see Mayberg, 2009; Northoff et al., 2011). System disorders affect the whole brain (including all of its levels), rather than causing a deficit in a single region or a specific gene. Within the psychodynamic context such a system disorder may correspond to the recruitment and involvement of all of the psychodynamic features discussed (i.e. internalization/externalization, narcissism/self-objects, unconscious/dreams, and self/ego). Being characterized as system disorders in this sense may distinguish depression and psychosis from other disorders such as stroke which show more isolated and regionally limited deficits in both neuronal and psychodynamic mechanisms.

Chapter 10

Depression and the Brain

Reactivation of early object loss

There have been many psychodynamic accounts of depression by various authors (for excellent overviews, see Böker et al., 2000; Böker and Northoff, 2005; Gabbard, 2005a,b). Rather than describing the details of the different psychodynamic theories, I want to focus on some key features that ultimately can be traced back to Freud and his description of depression in his classic work *Mourning and Melancholia* (Freud, 1917). I here focus on three psychodynamic key features of depression, namely reactivation of early object loss, loss of actual object relations, and increased introjection coupled with negative affect (see Figure 10.1). While focusing on these three features, I am fully aware that this list is by no means exhaustive and thus neglects many other issues, such as the complexity of psychodynamic constellations (Bleichmar, 1996, 2010) and the psychotherapeutic modulation of depression (Taylor, 2008).

Let me start with the first psychodynamic key feature, namely reactivation of early object loss. Freud (1917) pointed out that early object losses in childhood lead to vulnerability to depression in adulthood. The central role of early object losses in depressed individuals has been emphasized since, and has surfaced in various life events in early childhood, such as maternal or paternal separation, parental divorce, and physical or sexual abuse (for overviews, see Bleichmar, 1996, p. 947; Böker et al., 2000; Gabbard, 2005a, pp. 213–215). What is common to these different life events? The early object loss almost always threatens some core aspect of the identity and self-worth of the child, thus disrupting its objects relations or, to put it in the language of self psychology, the self-objects and their preserving and maintaining function for the self. This means that the early object loss must be subjectively perceived and experienced as such and thus concern those self-objects that are relevant to the respective self. Otherwise, in the case of object losses that do not concern self-objects relevant to the respective self, the early object loss may not be threatening to the identity and self-worth of that particular self.

How can this early object loss as disruption of the child's self-objects affect the adulthood of that person? It can often be observed that the onset of depressive symptoms is triggered by life events that objectively may remain unrelated to the life event(s) in early childhood. However, subjectively they may be perceived and experienced by the affected individual as similar and analogous to the event(s) in early childhood (Böker et al., 2000). Imagine the following scenario. The boss does not enjoy his coffee in the morning because he has had a bad night. His devoted secretary who makes the coffee for him every morning may, though, perceive and experience his dislike of the coffee as personal rejection, which may trigger the outbreak of a depressive episode. Although her boss's comments about the coffee were not meant as a personal rejection, she takes them personally because they remind her of her father's rejection of her in her early childhood.

How is it possible that an object loss in early childhood can affect subjective perception and experience of seemingly unrelated objects in adulthood? Bleichmar (1996) points out that the desire and longing to recapture the lost object must accompany the early object by itself: "For depression to appear, the loss of the object must be accompanied by a persistence of the desire for

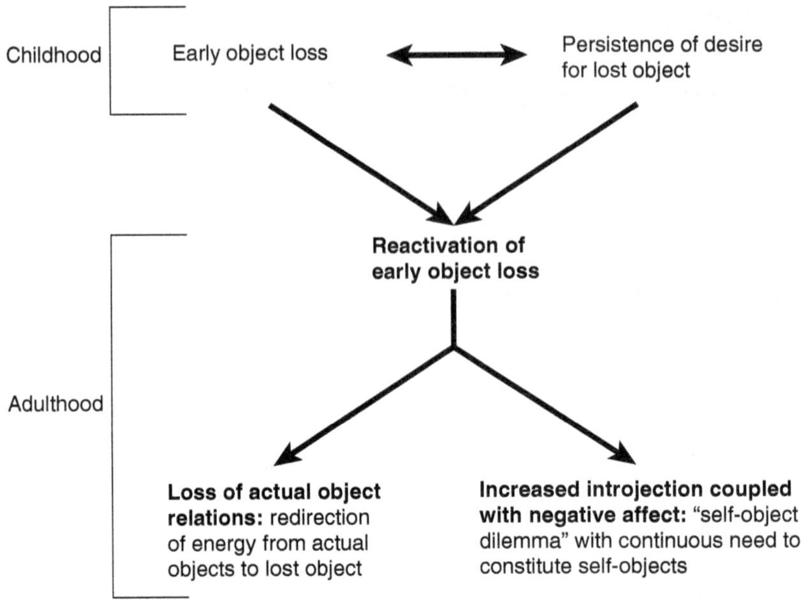

Fig. 10.1 Psychodynamic key features in depression.

that object and by a representation of its attainability" (Bleichmar, 1996, p. 935). Interestingly, Freud makes the same point when he speaks of an "unsatisfiable cathexis of longing" in his *Inhibitions, Symptoms and Anxiety* (Freud, 1926, p. 172). The persistence of the desire for the early lost object may thus mediate the transfer of the early lost object into adulthood. If, for instance, there is early object loss but no persistence of the desire for the lost object from childhood, reactivation of the early object loss in later adulthood is rather unlikely. To put it in the more formal terms of the philosopher, early object loss may be considered a necessary but non-sufficient condition of the later reactivation of early childhood loss in adulthood, with the consecutive outbreak of depressive symptoms, while persistence of the desire for that object is a sufficient but non-necessary condition by itself.

The reactivation of early childhood loss in adulthood shifts the lost object back into the current focus of the adult subject. This shift may trigger a regressive process with strong fixation on the mental representation of the loved object. Such fixation on the mental representation of the lost object consumes high levels of energy, which Freud (1917) compared with an open wound that draws energy from all sides. Let us further develop the analogy with the wound. The reactivation of the early object loss is like a wound that starts bleeding again after a long period when it had been almost forgotten and was thought to have healed completely.

However, the early healing process was incomplete, which became clear once a life event occurred that was subjectively analogous or similar to the one that first caused the wound, much earlier on. And in the same way that the early outbreak of bleeding was compensated for by devoting high energy resources to heal the wound and stop the bleeding, the outbreak of the bleeding in later adulthood is associated with recruitment of all the resources currently available to immediately stop the wound bleeding. The recruitment and full devotion of the resources to stop and heal the reactivated bleeding of the wound as residual of the person's past draws heavily on the energy that the person usually invests in their present self- and object representations.

Loss of actual object relations, increased introjection coupled with negative affect, and the "self-object dilemma"

The focus on the early lost object in adulthood consumes energy, and this necessitates the withdrawal of energy from the constitution of the current objects and self-objects in adulthood. Rather than internalizing and introjecting current objects so as to stabilize (i.e. maintain and preserve) the self by specific self-objects from the present and thus from adulthood, the past object from early childhood (i.e. the early lost object) now becomes internalized and introjected within the present context in adulthood. This involves two key psychodynamic features, namely loss of actual object relations and increased introjection coupled with negative affect, which will be described in more detail below (see also Figure 10.1).

If all energy is consumed by the increased introjection of and identification with the mental representation of the lost object, there is no energy left to be invested in other objects, such as the constitution of objects (and possible self-objects) related to the present in adulthood. The focus on the lost object (or the mental representation of it) from the past leads to the neglect of all other present objects as possible self-objects in the actual environment in adulthood. This means that present objects cannot be subjectively perceived and experienced independently in their own terms, but only in terms of (or, as Freud would probably say, in the shadow of) the lost object from the past.

The loss of actual object relations as the withdrawal of energy from objects in the present was nicely described by Freud himself in a famous passage from *Melancholia and Mourning*: "But the libido was not displaced onto another object, it was withdrawn into the ego. There, however, it was not employed in an unspecific way, but served to establish an identification of the ego with the abandoned object. Thus the shadow of the object fell upon the ego, and the latter could henceforth be judged by a special agency, as though it were an object, the forsaken object. In this way, an object-loss was transformed into an ego-loss and the conflict between the ego and the loved person into a cleavage between the critical activity of the ego and the ego as altered by identification" (Freud, 1917, p. 249).

Why, though, is the shift to the early lost object in the present context of adulthood with its increased internalization and introjection coupled with negative affect, which is the third key psychodynamic feature? By reactivating the lost early object from the past in the present context, the affects assigned to the early object are also transferred from the past to the present. More specifically, as in early childhood, the reactivation of the early object loss is painful, and is consequently assigned negative affects. There is therefore negative affective assignment in the present context to the reactivated lost early object from the past. Since the reactivated early lost object is not only reactivated but also serves, as self-object, a self-stabilizing function, the self itself is assigned a negative affect, and as a result the self is perceived in terms of guilt, sadness, rage, or anger (Gabbard, 2005a, pp. 216–221).

The assignment of negative affect to both objects (i.e. self-objects) and self distinguishes melancholia from mere mourning, as was well expressed by Freud himself: "It is well worth notice that mourning involves grave departures from the normal attitude to life. It never occurs to us to regard it as a pathological condition and refer it to medical treatment. We rely on it being overcome after a certain lapse of time, and we look upon any interference with it as useless or even harmful. The distinguishing mental features of melancholia are a profoundly painful sense of dejection, a cessation of interest in the outside world, loss of capacity to love, inhibition of all activity . . . a lowering of the self-regarding feelings to a degree that finds utterance in self-reproaches and self-revilings, and culminates in a delusional expectation of punishment" (Freud, 1917, p. 249).

However, the impact of the early lost object goes far beyond the assignment of negative affect in that it influences and strongly shapes the self itself. Let me return to my distinction between possible and actual self-objects in Chapter 7 on narcissism and self-objects. The number of possible self-objects in the present of the depressed person is limited and, in the most extreme scenario, restricted to the early lost object as the only possible self-object. The depressed patient then has no choice but to constitute and stabilize his own self by integrating (i.e. internalizing and introjecting) the early lost object from the past as present self-object. However, this means that he attempts to stabilize (i.e. preserve and maintain) his own present self on the basis of a past loss to his self.

This leads to what I call the "self-object dilemma" for the depressed self. The concept of this dilemma is based on the assumption that the depressed self, like any self, aims to stabilize its own self by constituting self-objects which are usually taken from its present context. However, as already discussed, access to objects from the self's present context is blocked and redirected to the early lost object. The only way for the depressed self to constitute self-objects in order to stabilize (i.e. preserve and maintain) its present self is to recruit and thus to internalize and introject the early lost object, but this is not stabilizing, and in fact is destabilizing. However, as the depressed person has no way of constituting self-objects other than by recruiting the early lost object, he has no choice but to continue doing so and thus to constitute destabilizing rather than stabilizing self-objects.

The "self-object dilemma" is caused by the continuous need to constitute self-objects in order to stabilize the self, while at the same time no longer having access to any kind of stabilizing self-objects. The only way for the depressed self to stabilize its own self in the present is by constituting rather destabilizing self-objects on the basis of the early lost object from the past. This leads to increased introjection and internalization of the lost early object as actual self-object which, due to its destabilizing nature, further reinforces the need to constitute self-objects, with the subsequent need for further introjection and internalization, and so on. Accordingly, the "self-object dilemma" consists of the persistent need for self-stabilizing self-objects even if only destabilizing self-objects are available. The depressed person is consequently confronted with two undesirable alternatives. Either they "choose" to no longer stabilize the self by self-objects, which results in the destabilization of the self, or they opt to stabilize the self, which however can only be achieved in a destabilizing way (i.e. by destabilizing self-objects). In short, the depressed subject is confronted with a dilemma.

How can the depressed self escape when there is such a lack of stabilizing self-objects? It does exactly the same as it would do in the healthy state, getting even closer to self-objects and identifying the self with them. In the most extreme case, the depressed patient does indeed identify his own self with the present self-object as constituted on the basis of the past lost early object, so that he subjectively perceives and experiences his own self solely and exclusively in terms of the lost object. This is the moment when introjection becomes associated with identification (Böker and Northoff, 2005), and more specifically when the present self identifies itself with the past lost object. I call such identification of the present self with the past lost object "self-object confusion." The concept of "self-object confusion" describes how the present self takes and identifies its own self with the past lost object. However, such self-object confusion means that the depressed self identifies its own self with a lost object and thereby aggravates even further the negative affect that is already associated with the lost object, by now transferring it to the self itself (via what I earlier described as "self–self-object integration"; see Chapter 7).

The identification of the present self with the lost object from the past is as painful in adulthood as it was in childhood. The only way to avoid this is to split the self into two parts, namely the part of the self that is identified with the lost object, and the part of the self that experiences the self as the lost object. This may be the basis of the often described split within the self or ego itself.

The original ego is split into the ideal ego and the ego as identified with the object loss, with both standing in opposition to and thus in conflict with each other (Böker and Northoff, 2005): "Melancholia shows us the ego divided, fallen apart into two pieces, one which rages against the second. This second piece is the one which has been altered by introjection and which contains the lost object. But the piece that behaves so cruelly is not unknown to us either. It comprises the conscience, a critical agency within the ego, which even in normal times takes up a critical attitude towards the ego, though never so relentlessly and so unjustifiably" (Freud, 1921, p. 120).

In summary, I have here singled out three key psychodynamic features in depression, namely reactivation of early object loss, loss of actual object relations, and increased introjection coupled with negative affect. However, it is important to be aware that these three key features describe only some aspects (although crucial ones, I assume) of depression while neglecting others. It should therefore be mentioned that I have here omitted other key psychodynamic features such as the exact account of the ideal ego, subtypes of depression (e.g. anaclitic and introjective depression) (Blatt, 1998), different psychodynamic constellations related to different symptoms (Bleichmar, 1996, 2010), the psychodynamics underlying suicide (Gabbard, 2005a, p. 221ff), and psychotherapeutic intervention (Bleichmar, 1996, 2010).

Finally, I should also mention that I have here omitted the psychodynamic relationship of melancholia/depression to manic and psychotic features (see, for instance, Böker et al., 2000; Böker and Northoff, 2005). This means that when I use the term "depression" I refer to melancholia or unipolar depression while excluding other forms of depression, such as bipolar depression, schizoaffective disorder, etc. However, a full psychodynamic account of all the facets of depression, including the development of corresponding neuropsychodynamic hypotheses, is beyond the scope of this book.

Increased self-focus and decreased environment focus

I have so far described three key psychodynamic features of depression, namely reactivation of early object loss, increased introjection coupled with negative affect, and loss of actual object relations. Although these psychodynamic key features describe the kind of psychodynamic mechanisms that enable and predispose to depression, they do not focus so much on the subjective experience and perception of the depressive symptoms themselves. Instead this is dealt with in phenomenology which, loosely (and rather broadly) defined (for details, see Northoff, 2004a,b), describes the subjective experience of the depressive symptoms from the first-person perspective. Such a phenomenological approach must be distinguished from the psychopathological approach that targets the depressive symptoms in the rather objective terms of observation from third-person perspective. Since the three psychodynamic key features described clearly point to an altered self in depression, the following phenomenological and psychopathological account focuses on the self. For this purpose, I want to start with a quote from a recent paper that nicely describes the alterations of the self in depression:

> She sat by the window, looking inward rather than looking out. Her thoughts were consumed with her sadness. She viewed her life as a broken one, and yet she could not place her finger on the exact moment it fell apart. "How did I get to feel this way?" she repeatedly asked herself. By asking, she hoped to transcend her depressed state; through understanding, she hoped to repair it. Instead, her questions led her deeper and deeper inside herself—further away from the path that would lead to her recovery.
>
> (cited by Treynor et al., 2003, p. 247)

This description of a depressed patient shows three crucial characteristics of the self which will be conceptualized as increased self-focus, association of the self with negative emotions, and increased cognitive processing of the self (see Figure 10.2 and Northoff, 2007).

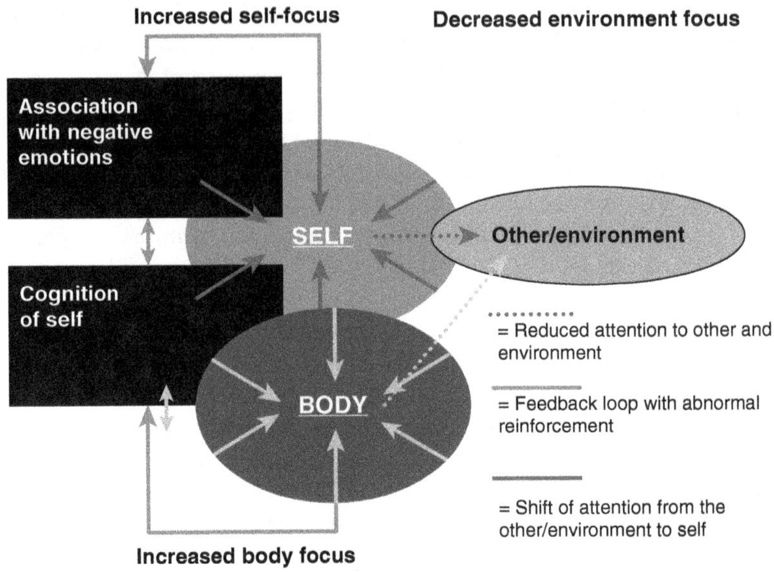

Fig. 10.2 Phenomenological account of the self in depression (modified version of the figure in Northoff 2007).

Let me start with the increased self-focus. Like our patient, almost all depressed patients look inward rather than outward, and focus very much on themselves while no longer being able to shift their focus on to others (see the arrows pointing towards the self in Figure 10.1). Social–psychological theory describes self-focused attention as a focus on internal perceptual events (i.e. information from those sensory perceptions that react to changes in bodily activity) (Ingram, 1990). The self-focus may also involve enhanced awareness of one's present or past physical behavior (i.e. heightened cognizance of what one is doing or what one is like). In addition to such increased self-focus, the depressed patient's focus is often also on their own body. Depressed patients show heightened awareness of their own body which phenomenologically results in the subjective perception of diffuse bodily symptoms (Wiebking et al., 2010). The increased self-focus may therefore be accompanied by what I call "increased body-focus."

The increased self-focus and body-focus imply that the depressed person's attention is no longer focused on their relation to the environment and environmental events, as in healthy individuals, but rather on itself as the prime focus, with the environment shifting into the background. Thus the increased self-focus is associated with what one may call "decreased environment-focus." The concept of the decreased environment-focus describes how the depressed patient's subjective experience and perception are no longer directed towards the environment and its respective persons and events. Instead, the patient's subjective perception and experience are directed towards their own body and their own cognition, resulting in what I here describe as "increased self-focus." This means that in depression the balance between the environment-focus and the self-focus is unilaterally shifted towards the latter, at the expense of the former. Therefore the decreased environment-focus involves an increased self-focus, and vice versa (see also Figure 10.2).

This is also supported by recent empirical data. Empirical research clearly indicates that there is heightened self-focused attention in depression. A variety of studies that have assessed self-focused attention with diverse measures and methodologies all converge on the finding of an increased and perhaps prolonged level of self-focused attention in depression (Ingram, 1990).

What remains unclear, though, is whether this increased self-focus is purely explicit and thus conscious, or whether it is already present at an implicit and thus unconscious level.

Another characteristic is the attribution of negative emotions to the self (see the arrows pointing from negative emotion to self in Figure 10.2), which is the association of the self with negative emotions. The self is associated with abnormal sadness, guilt, mistakes, inabilities, death, illness, etc. which may ultimately result in paranoid delusions. A recent study investigating symptom clusters in the Beck Depression Inventory (BDI) observed three BDI factors, including the self-blame factor (Grunebaum et al., 2005). Interestingly, depressed patients with a history of previous suicide attempts showed significantly higher scores on the BDI self-blame factor than those without a history of suicide attempts. Moreover, the self-blame factor correlated significantly with the total number of suicide attempts and with known risk factors for suicidal behavior (for a review, see Northoff, 2007). Such self-blame possibly results from the association of the self with predominantly negative emotions in depression, while at the same time these patients remain apparently unable to attribute any kind of positive emotions to their self.

Finally, there is also increased cognitive processing of the self. The patient typically suffers from increased cognitive processing. She thinks about herself and her mood and tries desperately to discover the reasons for her depression, but as a result only sinks deeper and deeper into the depressed mood (see the arrows pointing from negative cognitions to the self in Figure 10.1). This cognitive processing of the self is described as rumination, and is often considered to be a method of coping with negative mood that involves increased self-focused attention and self-reflection (Ingram, 1990).

On the basis of a rumination scale, Treynor et al. (2003) suggest a two-factor model of rumination. They call the first factor "reflection," which describes a purposeful turning inward to engage in cognitive problem solving in order to alleviate one's depressive problems. The second factor is best described as "brooding," which is passive comparison of one's current situation with some unachieved standard oriented towards others and thus on what can psychodynamically be termed the ideal self (Böker et al., 2000). Corresponding to the first factor, namely reflection, Rimes and Watkins (2005) suggest an increase in what they call "analytical self-focus" in depression which they define as thinking analytically "about" oneself and one's symptoms. Such increased analytical self-focus is related to the increased ratings of the self as worthless and incompetent and to ratings of depressed mood.

Those researchers distinguish between the increased "analytical self-focus" in depression and what they call "experiential self-focus." The latter is supposed to describe the focus on the direct experience of one's thoughts, feelings, and sensations in the present moment. In contrast to the increased analytical self-focus, the experiential self-focus tends to be rather low and decreased in depression, as these patients no longer experience themselves and their self as such. There is thus an imbalance between analytical and experiential self-focus in depression. The question of whether this imbalance corresponds to the imbalance between an "increased self-focus" and "decreased environment-focus" must remain open here, and requires further conceptual and empirical investigation.

In summary, I have here described three characteristics of the self in depression from a phenomenological and psychopathological perspective, namely increased self-focus, association of the self with negative emotions, and increased cognitive processing of the self. How are these phenomenological characteristics related to the psychodynamic key features? The increased self-focus may be assumed to correspond to the psychodynamic feature of increased introjection of the lost object. The psychodynamically postulated couplings of both introject and self with negative affect may be mirrored in the phenomenological description of the association of the self with negative emotions. Finally, the psychodynamic feature of the loss of actual object relations may

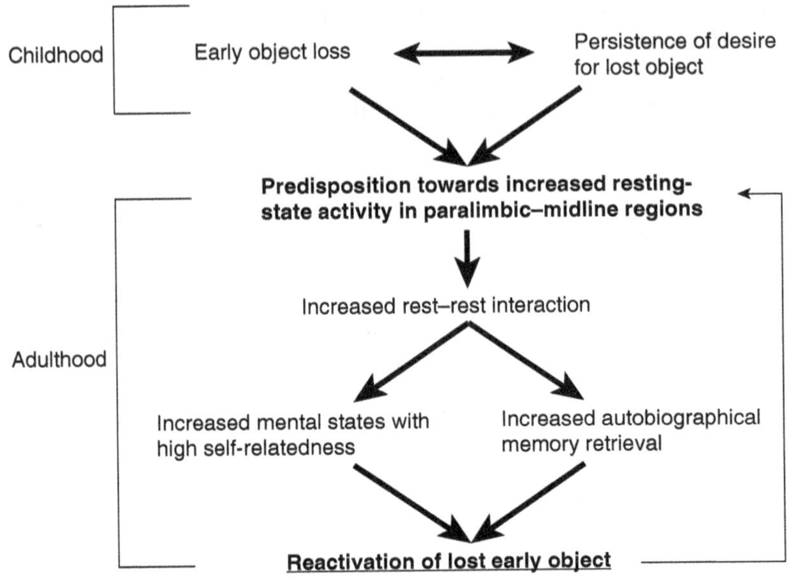

Fig. 10.3 Increased resting-state activity and reactivation of early object loss.

be assumed to correspond to the phenomenological feature of increased cognitive processing of the self. The cognitive processes no longer concern actual objects but only the lost objects as identified with the self.[1]

Elevated resting-state activity and the reactivation of early object loss

Neuropsychodynamic hypothesis

My first neuropsychodynamic hypothesis assumes that the psychodynamic feature of reactivation of early object loss may be related to an abnormal change in the resting-state activity in predominantly midline regions. There is solid empirical evidence for abnormally elevated resting-state activity, in anterior cortical and subcortical regions in particular, in depression. I postulate that such increased resting-state activity may enable and predispose to the reactivation of early object loss (see Figure 10.3).

The task of demonstrating support for this neuropsychodynamic hypothesis is twofold. I need to demonstrate how the persistence of the desire for the lost object that is supposed to bridge the gap from early childhood to adulthood may be related to the abnormal elevation of the resting-state activity in adulthood. However, this can only be achieved by revealing the psychological processes associated with the resting-state activity in general and its abnormal elevation in depression in particular. First, however, I need to demonstrate the empirical support for the abnormally elevated resting-state activity, including its anatomical details.

1 I am fully aware, though, that these correspondences between phenomenological and psychodynamic features are at best speculative at this point, and that they need to be elaborated in much more detail. However, that is beyond the scope of this book.

Empirical findings

Several reviews have been published about the structural and functional changes in major depressive disorder (MDD) (Mayberg, 2002, 2003, 2009; Phillips et al., 2003; Drevets et al., 2008a,b; Savitz and Drevets, 2009; Alcaro et al., 2010; Price and Drevets, 2010). Therefore I shall here briefly summarize the main findings and conclusions from the various reviews, and then relate them to functional anatomy as delineated in the normal healthy brain.

Alcaro et al. (2010) conducted a meta-analysis of all imaging studies of human MDD that had focused on resting-state activity. This yielded hyperactive regions in the PACC, the VMPFC, thalamic regions such as the dorsomedial thalamus and the pulvinar, pallidum/putamen, and midbrain regions like the VTA, substantia nigra (SN), the tectum and the PAG. In contrast, resting-state activity was hypoactive and thus reduced in the dorsolateral prefrontal cortex (DLPFC), the posterior cingulate cortex (PCC) and the adjacent precuneus/cuneus (Alcaro et al., 2010). These results are in close agreement with other meta-analyses (Fitzgerald et al., 2006, 2007; Savitz and Drevets, 2009; Price and Drevets, 2010). Also, Price and Drevets (2010) and Savitz and Drevets (2009) emphasized the role of the hippocampus, the parahippocampus and the amygdala where resting-state hyperactivity was also evident in MDD. Interestingly, the very same regions and the PACC also show structural abnormalities with reduced gray matter volume in imaging studies, and reduced cell counts for markers of cellular function in post-mortem studies (Savitz and Drevets, 2009; Price and Drevets, 2010).

Involvement of these regions in MDD is further corroborated by the investigation of resting-state activity in animal models of MDD. Reviewing the regions that showed resting-state hyperactivity in the different animal models yielded the anterior cingulate cortex, the central and basolateral nuclei of the amygdala, the bed nucleus of the stria terminalis, the dorsal raphe, the habenula, the hippocampus, the hypothalamus, the nucleus accumbens, the PAG, the DMT, the nucleus of the solitary tract, and the piriform and prelimbic cortex (Alcaro et al., 2010). In contrast, evidence of hypoactive resting-state activity in animal models remains sparse, with no clear findings (Alcaro et al., 2010).

These findings indicate that there is abnormally high resting-state activity in extended subcortical and cortical medial regions of the brain. This has allowed authors such as Phillips (Phillips et al., 2003), Mayberg (2002, 2003, 2009) and Drevets (Savitz and Drevets, 2009; Price and Drevets, 2010) to assume dysfunction in the limbic system in depression or more specifically in the "limbic–cortico–striato–pallido–thalamic circuit," with the interaction between medial prefrontal and limbic regions being crucial (Price and Drevets, 2010). This though may need to be extended to include subcortical primary-process emotional regions at the mesencephalic level as suggested by the animal data. However, one human study has found concurrent involvement of both cortical and subcortical regions in resting-state hyperactivity (Grimm et al., 2009). Relying on pure perceptual rather than cognitive tasks, these investigators (indirectly through analysis of stimulus-induced activity, i.e. deactivation, in resting-state regions) demonstrated concurrent resting-state hyperactivity within medial cortical structures (i.e. paralimbic and midline regions), as well as at subcortical levels (e.g. PAG, thalamus, tectum).

How do these findings fit within the anatomical characterization of the healthy brain that I already delineated? Briefly, I there delineated three anatomical territories in the brain that extend from subcortical to cortical regions in a medio-to-lateral direction (see Chapter 6 for full details). There are core regions in the subcortex directly adjacent to the aquaduct which extend to the cortex where they have been called paralimbic regions, as including the PACC, the SACC, and the PCC. Directly adjacent to these core regions, there are median and lateral paracore regions in the subcortex that are supposed to resurface in the cortex as the midline regions (i.e. cortical midline structures, or CMS, including the VMPFC, DMPFC, and MPC). Finally, there are regions lateral to the median and lateral paracore in the subcortex that correspond to lateral cortical regions such as the DLPFC.

The paralimbic areas and the cortical midline structures, e.g. the inner and middle ring (see chapters 4 and 5 for details) generally show hyperactivity during the resting state in people with MDD. This is convergent with data from animal models of depression that show exactly these regions, the limbic forebrain regions like the amygdala, the hypothalamus, and the hippocampus as well as the mesencephalic core–paracore areas like the raphe nuclei, the locus coeruleus, the habenula, and the nucleus of the solitary tract to be hyperactive in the resting state. Why do these regions rather than others as for instance more lateral regions show resting state hyperactivity in MDD? I assume that the anatomical, i.e. subcortical-cortical distribution of resting state hyperactivity in MDD may simply arise from the fact that the mesencephalic core–paracore system is extended anatomically and functionally into the forebrain (i.e. the limbic system) and associated cortical regions (the paralimbic areas and the CMS) (see Figure 10.3).

Another observation fits well with this anatomical model in the healthy brain. The outer ring covers the lateral regions at the cortical level such as the DLPFC and the sensory and motor regions. In the DLPFC in particular and in part also in the motor cortex (Alcaro et al., 2010), resting-state hypoactivity has been consistently observed, especially in people with MDD. Another region that shows reduction of activity is the visual cortex, where the concentration of GABA in particular has been found to be reduced (Sanacora et al., 1999; Maciag et al., 2010). Mirroring the distinction that Mayberg (2002, 2003, 2009) made between ventral and dorsal systems, these findings suggest that there are resting-state activity imbalances between the inner/middle and outer cortical rings and thus between the paralimbic/midline and lateral cortical regions.

Taking these findings together, resting-state activity in people with MDD may be characterized by a subcortical–cortical imbalance between the inner/middle and lateral rings. More specifically, the subcortical core–paracore regions appear to be hyperactive in the resting state, which is extended over the forebrain regions to the cortical level of the paralimbic areas and CMS. In contrast, the subcortical and especially the cortical regions of the lateral ring, such as the lateral prefrontal cortex and the sensory–motor cortex, appear to show hypoactivity in the resting state. From a purely anatomical perspective, one may consider this imbalance to gradually spread between the inner/middle and outer rings in the radial–concentric organization across subcortical and cortical regions (see Figure 10.3).

Neuropsychodynamic account

What are the implications of such increased resting activity in the paralimbic and midline regions for the psychological and psychodynamic context of depression? Let me start with the psychological implications. The resting-state activity in the paralimbic and midline regions in healthy subjects is psychologically often associated with retrieval of autobiographical memories (Northoff and Bermpohl, 2004; Buckner et al., 2008). Autobiographical memory retrieval has been shown to implicate many midline regions, such as the PACC, VMPFC, DMPFC, PCC, and precuneus, which makes it likely that the same psychological process may also be associated with the resting-state activity in the same regions (Buckner et al., 2008). In the same way that autobiographical memory retrieval may be recruited in association with events or people in the actual environment, precisely the same process may be instantiated in the resting state. Psychologically, this may be manifested as, for instance, mind-wandering, daydreaming, or dreams (see Chapter 8), in which retrieval of the lost object will predominate.

Although more empirical data are needed to support the linkage between resting-state activity and autobiographical memory retrieval, this provides one possible explanation for what could happen psychologically in the resting state in depression. The abnormal increase in resting-state activity in depression may psychologically enable and predispose to the increased retrieval of autobiographical memories associated with past events, such as those of the lost objects. Since autobiographical memory retrieval reactivates the past memories within the present context, their abnormal increase in the context of the elevated resting-state activity may enable and predispose to the reactivation of the lost early object from the past within adulthood in the present. However, empirical data to support such a hypothesis are currently lacking, and although plenty of evidence suggests that there is increased resting-state activity in depression, its effect on psychological functions such as autobiographical memory retrieval has yet to be investigated.

One may question why autobiographical memory retrieval concerns predominantly the memories of the lost object rather than autobiographical memories of other objects that were not lost. Why is there such a predisposition to apparently autobiographically retrieve the lost object rather than other non-lost objects? Psychodynamically, this was related to the persistence of the desire for the lost object. Why is there such persistence of the desire to recapture the lost object? The lost object concerns traumatic events that affect the identity and worth of the self. Taken within the psychological context this means that the traumatic event and thus the lost object must show an extremely high degree of self-relatedness (see Chapter 9 for the exact definition of this), which however was lost. One may now assume that this high degree of self-relatedness may trigger and mediate autobiographical memory retrieval and, more specifically, that it may bias and direct the latter towards the lost object rather than other objects (be they lost or non-lost) with lower degrees of self-relatedness.

What does the assumption of the relationship between autobiographical memory retrieval and self-relatedness entail empirically? First, one would assume that self-relatedness recruits many of the regions that are implicated in autobiographical memory retrieval, and would thus involve the midline and paralimbic regions. This is indeed the case, as was discussed in detail earlier on (see Northoff et al., 2006 and Chapter 9 of this book). Secondly, on the basis of such regional–anatomical overlap, one would assume that there is direct interaction between autobiographical memory retrieval and self-relatedness, with enhancement of the former by higher degrees of the latter. Such interaction effects may be manifested psychologically as increased retrieval rates, as has indeed been demonstrated by the so-called self-reference effect. Neuronally, interaction may be manifested as enhancement of midline activity, as has indeed been shown in a recent imaging study (Sajonz et al., 2010).

However, the empirical data have one shortcoming. They all concern autobiographical memory retrieval and self-relatedness in relation to exteroceptive stimuli, thus presupposing stimulus-induced activity. However, such stimulus-induced activity must be distinguished from the resting state and its intrinsic activity as targeted here on the basis of the increased resting-state activity in depression. There is no direct evidence yet for autobiographical memory retrieval and self-relatedness, including their interaction, occurring in the resting state. Although some indirect evidence for an effect of autobiographical memory retrieval and self-relatedness on the resting-state activity level has been obtained (d'Argembeau et al., 2005; Schneider et al., 2008; see Chapter 9 for details), direct evidence is still lacking.

So far I have discussed how the increased resting-state activity in the paralimbic–midline regions may enable and predispose to the reactivation of the lost object via the psychological functions of autobiographical memory retrieval and self-relatedness. However, this leaves open the question of what are the exact neuronal mechanisms. Since we are here dealing only with the resting state, there can be no rest–stimulus and stimulus–rest interaction yet. The only kinds of interactions that are possible are rest–rest interactions. How is the increased resting-state activity in the paralimbic–midline system related to rest–rest interaction? Is rest–rest interaction abnormal in depression, and if so, how is it related to the reactivation of early object loss?

I demonstrated in Part II of this book that neuronal activity is coded in terms of differences, amounting to difference-based coding. Moreover, such difference-based coding was supposed to enable and predispose to neuronal–mental transformation, with larger neuronal differences being associated with a higher probability of them being transformed into mental states (see Chapter 5). Increased resting-state activity may now lead to increased neuronal differences and thus to an increase in rest–rest interaction. Such increased rest–rest interaction may occur either within one particular region or between different regions, as for instance between the various paralimbic and midline regions. However, this has yet to be investigated.

How does such increased intra- and transregional rest–rest interaction in the paralimbic–midline system influence the respectively associated mental states? The reader may recall that difference-based coding was supposed to enable and predispose to the constitution of mental states (i.e. neuronal–mental transformation) (see Chapter 5). As previously stated, there is solid empirical evidence for paralimbic–midline activity being associated with mental states that concern highly self-related stimuli. If, due to the increased resting-state activity and increased rest–rest interaction, the neuronal differences that are coded in the resting state are supposed to be larger than usual, one would expect an increased likelihood of mental states, and in particular of mental states involving highly self-related stimuli. One would consequently postulate not only increased rest–rest interaction in the paralimbic–midline regions but also increased mental states relating to events and people of high self-relatedness. This is highly consistent with the phenomenological feature of rumination already described. Most importantly, these mental states relating to events and people of high self-relatedness may bias and direct autobiographical memory retrieval towards retrieving events with high self-relatedness, such as the lost early object.

What does this imply for the psychodynamic context? Increased intra- and transregional rest–rest interaction within the paralimbic–midline system may enable and predispose to the reactivation of early object loss. More specifically, increased rest–rest interaction may induce mental states with a high degree of self-relatedness which may bias and direct autobiographical memory retrieval towards retrieving events and people with extremely high degrees of self-relatedness. This in turn makes it more likely, indeed probable, that the events and people associated with the trauma and thus the early object loss from the past will be autobiographically retrieved and set in the present context of the adult person. The increased likelihood of autobiographical memory retrieval of highly self-related events may be assumed to correspond, at least in part, to what psychodynamically has been described as the "persistence of the desire" to recapture the lost object.

Based on these considerations, I offer the following psychological and neurological hypotheses (see also Figure 10.3). I postulate that the reactivation of early object loss is closely related to increased self-relatedness of mental states and thus to ruminations. Moreover, I assume that the lost early object and its reactivation in adulthood may bias autobiographical memory retrieval towards highly self-related events. Finally, I postulate that the reactivation of early object loss may be accompanied by even further increases in paralimbic–midline resting-state activity, due to the extremely high degree of self-relatedness of the lost early object (given the well-known effects of self-relatedness on increasing neural activity in the CMS). However, although all three hypotheses seem to be plausible, given the empirical data demonstrated, they have yet to be tested directly and experimentally.

Finally, one may want to raise the question of the origin or cause of the increased resting-state activity in the paralimbic–midline regions in depression. There is much debate these days about whether depression is a neurodevelopmental or neurodegenerative disorder, with the final verdict remaining open (for recent overviews, see Drevets et al., 2008a,b). Neuropsychodynamically, one may postulate that the traumatic event itself and thus the early object loss may be crucial in enabling and predisposing to the abnormal elevation of the paralimbic–midline resting-state activity in depression. Individuals at risk for depression also appear to show a tendency towards increased paralimbic–midline resting-state activity (for recent reviews, see Mayberg, 2009; Price and Drevets, 2010), which would at least indirectly support our assumption.

Reduced rest–exteroceptive stimulus interaction and abnormal affective assignment of actual objects

Neuropsychodynamic hypothesis

The second key psychodynamic feature concerns the loss of actual object relations, with an abnormal shift and redirection towards the early lost object. Object relations in depression are no longer characterized by the actual objects of the present environment, but by the lost early object from the past environment. The question is now how the abnormal elevated resting-state activity I discussed influences the neural processing of exteroceptive stimuli such that it enables and predisposes to the loss of actual object relations (see Figure 10.4).

Based on the considerations in the healthy brain (see Part II of this book), I postulate that an abnormal decrease in the resting-state activity in the lateral system strongly influences and shapes stimulus-induced activity. More specifically, I postulate that the impact of the stimulus and thus the neural activity that it induces (i.e. stimulus-induced activity) is reduced in depression. Thus what I referred to as rest–stimulus interaction may be reduced in depression on the basis of abnormally reduced resting-state activity in the lateral system.

What does such reduced rest–stimulus interaction imply for the psychodynamic context? I postulate that reduced rest–stimulus interaction concerns mainly exteroceptive stimuli, and

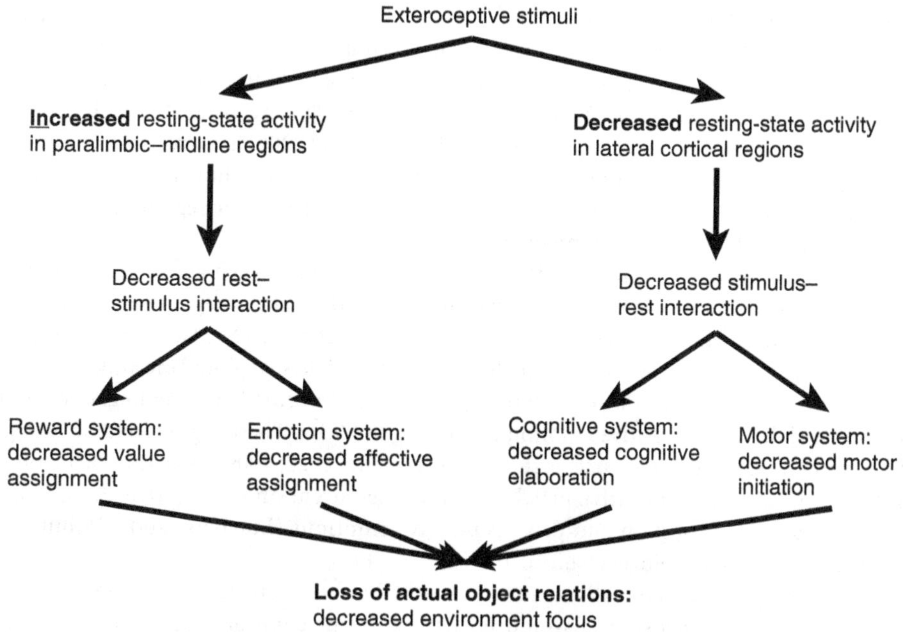

Fig. 10.4 Reduced neural processing of exteroceptive stimuli and loss of actual object relations.

may thus be specified as reduced as rest–extero interaction. Such rest–extero interaction enables and predisposes to the constitution of objects (see Chapter 6). Reduced rest–extero interaction as assumed in depression may subsequently lead to reduced constitution of objects within the actual context, which may well correspond to the loss of actual object relations. I therefore postulate that reduced rest–extero interaction enables and predisposes to the loss of actual object relations.

In the following account, I intend to develop the second neuropsychodynamic hypothesis in two steps. First, I shall focus on how negative affect is assigned to objects, and then I shall discuss how the constitution of actual objects is almost lost in depression.

Empirical findings

How does the abnormal resting-state activity influence the neural processing of external stimuli (i.e. stimuli from outside the brain)? External stimuli are processed in both lateral cortical and paralimbic–midline regions, depending on the respective functional system (e.g. affective, reward, sensory, cognitive, etc.). The external stimuli that encounter the brain are thus not only confronted with resting-state hypoactivity in the lateral cortical system but also, at the same time, are exposed to abnormally elevated resting-state activity in the paralimbic–midline system. This may result in a net effect that consists of reduced external interactions with cognitions, especially in the paralimbic–midline systems, for which there is now consistent empirical support.

Functional brain activation studies, especially with emotional stimuli, show predominantly hypoactivity in the PACC, the VMPFC, and the medial orbital prefrontal cortex (MOFC) (Elliott et al., 1998, 2002; Mayberg et al., 1999, 2000; Canli et al., 2004, 2005; Rose et al., 2006). More recent fMRI studies have demonstrated predominantly reduced activation in these regions in subjects with depression during emotional tasks. Keedwell et al. (2005) demonstrated abnormal signal increases in the PACC, the MOFC, and the VMPFC during happy stimuli in depressed patients and lower signal increases in these regions during sad stimuli. Other researchers have shown lower MOFC and subgenual anterior cingulate cortex (SUACC) decreased signal changes when subjects with MDD were exposed to emotional stimuli (Elliott et al., 2002; Liotti et al., 2002; Kumari et al., 2003; Lawrence et al., 2004), whereas higher signal increases were observed in the PACC and VMPFC (Elliott et al., 2002; Fu et al., 2004), which were associated with good treatment responses (Davidson et al., 2003) (see also Fitzgerald et al., 2007 for a recent meta-analysis). Taken together, the findings obtained during emotional stimulation show reduced stimulus-induced activity, especially in the anterior cortical midline and paralimbic regions, thus suggesting that there is reduced rest–stimulus interaction.

Other regions that show abnormalities include reward regions such as the ventral striatum/nucleus accumbens and the right/left amygdala during positive and/or negative emotional stimulation in subjects with MDD (Kumari et al., 2003; Canli et al., 2004 ; Lawrence et al., 2004 ; Surguladze et al., 2005). This suggests that changes in both of these regions (the amygdala and the ventral striatum/nucleus accumbens) participate in the abnormally strong negative emotion processing and decreased positive emotion processing in MDD. These regions and other regions of the reward-seeking circuitry, such as the VMPFC (Heller et al., 2009), may thus be involved in what has been called the "negative bias," which focuses attention on negative emotions, with diminished capacity to sustain and process positive emotions (Mayberg, 2003; Phillips et al., 2003; Heller et al., 2009; Heinzel et al., 2010).

One of our recent studies directly focused on rest–stimulus interaction in depression (Grimm et al., 2009). We found that stimulus-induced activity during emotional stimulation was indeed influenced, with a decreased negative BOLD response (i.e. deactivation) in those regions with high resting-state mid-cortical activity (e.g. the PACC, VMPFC, and PCC) (Grimm et al., 2009; Sheline et al., 2009). Interestingly, reduced deactivation in these cortical and subcortical regions predicted the severity of depression and degree of hopelessness, thus establishing a direct relationship between reduced rest–stimulus interactions and the severity of depression (Grimm et al., 2009).

In addition to an altered affective state, rest–stimulus interactions have also been shown to be reduced for rewarding stimuli. The reward-seeking system includes subcortical regions such as the VTA and the ventral striatum (including the nucleus accumbens), and cortical regions such as the VMPFC, whose neural activity valuates external stimuli by assigning positive valence to them. Moreover, various studies have now shown reduced stimulus-induced activity during reward in depressed patients (Steele et al., 2004; Kumar et al., 2008 ; Hasler et al., 2009; Heller et al., 2009 ; Pizzagalli et al., 2009; Eshel and Roiser, 2010). Such decreased stimulus-induced activity is, as before, indicative of reduced rest–stimulus interactions in cortical midline regions.[2]

Neuropsychodynamic account

What does such decreased rest–exteroceptive stimulus interaction entail in the psychological and psychopathological context? On the basis of the findings described, one may assume that decreased affect is assigned to exteroceptive stimuli as mirrored in decreased stimulus-induced activity during emotional stimulation. This may correspond symptomatically to the lack of feeling and emotional involvement in environmental events and people (i.e. actual objects) that is often reported by depressed patients. Furthermore, decreased value or reward may be assigned to exteroceptive stimuli due to decreased stimulus-induced activity in the reward system, which is often assumed to be closely related to anhedonia (the inability to experience pleasure). Finally, fewer cognitive stimuli may be generated in the lateral cognitive system, which may lead to the well-known cognitive deficits in depressed patients (Mayberg, 2009).

What does reduced rest–stimulus interaction entail for the psychodynamic context? I showed that rest–stimulus interaction is closely associated with externalization and projection (see Chapter 6). By means of the resting state influencing stimulus-induced activity, the self can ingrain or impose itself on the respective stimulus, and thus project or externalize itself on its self-objects. If the resting-state activity is too high and the stimulus-induced activity is too low, any subjective perception and experience of exteroceptive stimuli may be more strongly influenced by the self and its imposition, paving the way for projection and externalization. Such abnormally strong imposition of the increased resting state upon subsequent stimulus-induced activity may enable and predispose to the imposition of the past object loss (as being closely related to the self) on the subject's perception and experience of present objectively unrelated people and events (i.e. objects).

---

[2] Interestingly, however, there appears to be hyperactivity in the very same regions during punishing stimuli (for a review, see Eshel and Roiser, 2010). This is consistent with the symptomatic patterns in depressed patients, which show increased guilt and feelings of punishment, while neuronally it suggests increased rather than decreased rest–stimulus interaction for affectively negative inputs. One would consequently assume that the increased resting-state activity in the reward regions biases and directs stimulus-induced activity toward a negative affect and thus punishing direction. This may account for the negativistic information-processing bias and concurrent anhedonia (the inability to experience pleasure) in depression. This needs to be analyzed in more detail in future studies.

Reduced rest–stimulus interaction, reduced goal-oriented cognitions, and the loss of actual object relations

Neuropsychodynamic hypothesis

I now want to discuss the second part of the hypothesis of the neuronal mechanisms underlying the loss of actual object relations. I postulate that reduced rest–stimulus interaction also applies to the cognitive domain, where it leads to the reduced initiation and generation of goal-oriented cognitions, with the subsequent loss of actual object relations.

Empirical findings

In addition to investigating emotional stimuli, imaging studies of external stimuli related to cognitive tasks such as judgment, executive functions, or working memory have been conducted. These have yielded reduced activity particularly in the DLPFC in depression, which is quite a robust finding (Fitzgerald et al., 2006; Grimm et al., 2008). Reduced stimulus-induced responses in more purely cognitive brain regions such as the DLPFC suggest that diminished resting-state activity in these regions may not promote "psychologically healthy" internal generation of cognitive stimuli for the anticipation of many positive events, as commonly occurs in euthymic people. If so, it would be predicted that there will be reduced rest–stimulus interaction in the lateral cortical regions such as the DLPFC as well.

Based on these observations, Phillips et al. (2003) and Mayberg (2003, 2009) suggest a model of altered reciprocal functional relationship between the VMPFC and DLPFC in MDD. This model of ventral–dorsal dissociation is predominantly based on findings in the resting state (and only partially on functional activation studies) showing hyperactivity in the ventral prefrontal cortex (VMPFC and PACC) and hypoactivity in the dorsal prefrontal cortical regions (left DLPFC) in patients with acute MDD (Elliott et al., 1998, 2002; Keedwell et al., 2005; Kumari et al., 2003; Fu et al., 2004; Davidson et al., 2003; Canli et al., 2004; Lawrence et al., 2004; Surguladze et al., 2005), yielding a ventral–dorsal dissociation with abnormal reciprocal modulation, but this has yet to be demonstrated in MDD.

Such ventral–dorsal dissociation is quite compatible with the more general model of medial–lateral reciprocal modulation, indicating that stimulus-induced deactivation in medial regions is accompanied by increased activation in the lateral frontal regions, and vice versa (Goel and Dolan, 2003; Northoff et al., 2004). Medial deactivation and lateral activation in the prefrontal cortex are thus assumed to reciprocally modulate each other during affective and cognitive tasks. This reciprocal modulation has also been shown to be diminished in depression, with decreased medial deactivation occurring together with decreased activation in the lateral cortex (Grimm et al., 2008, 2009; Bermpohl et al., 2009).

The deactivation in the medial regions may be abnormally reduced because of elevated resting-state activity and reduced rest–stimulus interaction. Due to reduced rest–stimulus interaction, the incoming external stimuli may not even get so far into hierarchical processing as to induce deactivations in the medial cortical regions. And if there is concurrently decreased exteroceptive input from other regions, deactivation in medial regions remains diminished from the very beginning, leaving individuals to repeatedly ponder the same issues that are troubling them. With regard to diminished reciprocal modulation, this means that the system as a whole can only do what it has always done, namely modulate and adapt the activity in lateral regions in an obsessional repetitive way.

Neuropsychodynamic account

What does the reduced rest–stimulus interaction entail for the psychodynamic context of the loss of actual object relations? Reduced stimulus-induced activity means first and foremost that the external stimulus is not processed as much as it used to be, and that therefore, metaphorically speaking, it does not leave a strong trace in the brain. More specifically, it cannot even induce stimulus-induced activity in the lateral regions of the brain which are in charge of generating goal-oriented cognitions, working memory, and other cognitive functions. However, if these cognitive functions cannot even be instantiated, relation to the actual objects cannot be built up. The loss of goal-oriented cognition may thus entail loss of actual object relations.

Reduced rest–stimulus interaction in the DLPFC may enable and predispose to a reduction in what Carhart-Harris et al. (2008, p. 3) call "object cathexis," which they associate neuropsychologically with goal-oriented cognitions. Due to abnormal or reduced medial–lateral reciprocal modulation, neural activity in the DLPFC and thus goal-oriented cognition are reduced in depression, which psychodynamically may lead to reduced object cathexis (Carhart-Harris et al., 2008, pp. 3, 9–11).

Is medial–lateral reciprocal modulation disturbed and therefore the cause of the loss of actual object relations with reduced goal-oriented cognitions as Carhart-Harris et al. (2008) suggest? It may look as if this is indeed the case. One may associate reduced object cathexis as reduced goal-oriented cognition with decreased deactivation in the anterior medial regions as triggered by the incoming exteroceptive stimulus (and its associated object). Due to their reciprocal relationship, decreased medial prefrontal deactivation (i.e. negative BOLD response) may induce less neural activation (i.e. positive BOLD response) in lateral prefrontal regions, which psychologically may correspond to decreased constitution of goal-oriented cognitions.

This may make it appear as if reciprocal modulation is disturbed by itself, which leads to reduced object cathexis as is assumed by Carhart-Harris et al. (2008). However, I would rather assume the opposite. Reciprocal modulation is preserved well in depression. That is demonstrated by the fact that the apparently reduced medial deactivation is accompanied by reduced activation in lateral regions. Rather than assuming that there is a lesion or disturbance in reciprocal modulation itself, we have to look for the reasons for the reduced medial deactivation which in turn triggers reciprocal modulation to be abnormal.

I postulate that deactivation in the medial regions may be abnormally reduced because of elevated resting-state activity and reduced rest–stimulus interaction. Due to reduced rest–stimulus interaction, the incoming exteroceptive stimuli may not even manage to induce deactivation in the medial cortical regions. And if there is decreased exteroceptive input from other regions, deactivation in the medial regions is decreased from the very beginning. With regard to reciprocal modulation, this means that it can only do what it always does, namely modulate and adapt the activity in the lateral regions accordingly. This means that the apparent abnormality in reciprocal modulation may be nothing but an adaptive process, to account for and adapt the lateral prefrontal cortex to reduced deactivation in the medial regions. Thus what appears to be disturbed (i.e. reduced) is nothing but the manifestation of the normal adaptive function of reciprocal modulation in a different context (i.e. reduced exteroceptive input to the anterior midline regions with reduced deactivation).

Rather than being abnormal itself, the apparently abnormal reciprocal modulation is the consequence or effect of a normal functioning mechanism that aims to adapt the brain to a changing context (i.e. reduced rest–stimulus interaction). The problem in depression is thus not a lesion or abnormality in reciprocal modulation but, on the contrary, the preservation of its normal and adaptive function. Unlike Carhart-Harris et al. (2008), I therefore postulate that it is specifically the normal functioning of reciprocal modulation that enables and predisposes to the loss of actual object relations which may hold at least for the case of cognitive objects (e.g. goal-oriented cognitions).

Imbalance between intero- and exteroceptive processing and increased introjection coupled with negative affect

Neuropsychodynamic hypothesis

I demonstrated that increased resting-state activity in the core–paralimbic and paracore–midline systems could possibly enable and predispose to reactivation of the lost object. This was followed by a consideration of the neuronal consequences of increased paralimbic–midline and decreased lateral cortical resting-state activity for stimulus-induced activity as related to exteroceptive stimuli. Based on recent imaging findings, I postulated that rest–stimulus interaction is reduced in the different systems, including affective, reward, and cognitive systems. This was supposed to allow a decreased neural processing of exteroceptive stimuli and subsequently decreased affective and value assignment and reduced cognitive elaboration. Neuropsychodynamically I postulated that such reduced rest–stimulus interaction (i.e. rest–extero interaction) could enable and predispose to what is described as loss of object relations within the psychoanalytic context.

How does such reduced rest–extero interaction influence subsequent neural processing? The neuronal mechanisms of intero-exteroceptive linkage and trilateral interaction are still at work and, metaphorically speaking, they need to be provided with input, in particular with specific stimuli that can be processed. This raises the question of how the loss of exteroceptive stimuli in rest–stimulus interaction can be compensated for. I postulate that the loss of exteroceptive stimuli is compensated for by increased recruitment of those stimuli that are still available, notably interoceptive stimuli and those exteroceptive stimuli that are stored in the person's own autobiographical memory. One may consequently postulate a neuronal shift from reduced rest–present extero interaction to increased rest–intero and rest–past extero interaction.

What does such a neuronal shift from reduced rest–present extero interaction to increased rest–intero and rest–past extero interaction entail for the psychodynamic context? Instead of present exteroceptive stimuli, now past exteroceptive and present interoceptive stimuli are not only the target of rest–stimulus interaction, but also leave their trace on the resting state itself via stimulus–rest interaction. Analogous to rest–stimulus interaction, there is a shift from present exteroceptive to past exteroceptive and present interoceptive stimuli in stimulus–rest interaction, with the latter thus structuring and organizing the brain's resting-state activity.

However, increased stimulus–rest interaction of past exteroceptive and present interoceptive stimuli means that the very same stimuli are also those candidates that enable and predispose to subsequent introjection and internalization. What is introjected and internalized is thus no longer the objects related to present exteroceptive stimuli, but rather the objects associated with present interoceptive stimuli (i.e. the person's own body as object), and past exteroceptive stimuli related to the lost object. I consequently postulate that the neuronal shift in rest–stimulus and stimulus–rest interaction from present exteroceptive stimuli to present interoceptive and past exteroceptive stimuli enables and predisposes to what psychodynamically is described as increased introjection coupled with negative affect (see Figure 10.5).

If this hypothesis holds, one would expect abnormal neuronal processing of interoceptive stimuli in depression in both the interoceptive network (i.e. the core–paralimbic regions) and the

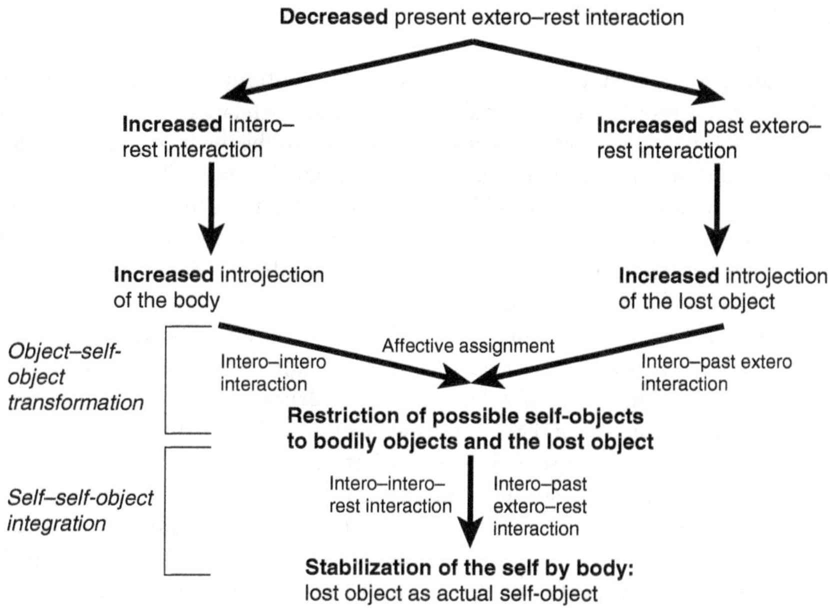

Fig. 10.5 Altered stimulus–rest interactions, introjection, and the self-object dilemma.

regions with high resting-state activity (i.e. the paracore–midline systems). The former may be related to abnormal rest–intero interaction while the latter may enable and predispose to intero–rest interaction. This will be the first part of the third neuropsychodynamic hypothesis.

Empirical findings

Patients with MDD often suffer from generalized bodily symptoms such as heart pounding, increased breathing (with yawning), and multiple diffuse bodily aches. This seems to go along with abnormally increased awareness of their own bodily processes (i.e. body perception), including sensitivity to stress and autonomic–vegetative changes, as demonstrated in a recent study (Wiebking et al., 2010). The same study also investigated the neuronal activity during exteroceptive and interoceptive awareness (tone and heartbeat counting) in relation to the brain's resting-state activity. Interoceptive stimuli by themselves (e.g. the heartbeat) induced a "normal" degree of brain signal changes (i.e. activation) in the bilateral anterior insula[3] in depressed patients when considered relative to the preceding resting-state activity levels. This suggests that there is no abnormality in interoceptive stimulus processing.

However, in contrast to stimulus-induced activity during interoceptive stimuli, we observed abnormally reduced activity during exteroceptive stimuli. More specifically, we observed that exteroceptive stimuli induced decreased stimulus-induced activity in the insula in depressed patients as compared with healthy subjects. This led us to further question whether such reduced activity is related to the exteroceptive stimulus itself or to differences in the resting-state activity levels. The latter was indeed the case, as we observed increased resting-state activity in the insula

[3] The abnormalities in depression are not confined to the insula, but also occur in typical exteroceptive regions such as the visual cortex (Desseilles et al., 2009; Golomb et al., 2009; Keedwell et al., 2010), which further supports our assumption of abnormalities in exteroceptive stimulus processing in depression.

itself, which is consistent with the resting-state hyperactivity in the core–paralimbic system to which the insula belongs.

To test for independent changes in exteroceptively related stimulus-induced activity, we then calculated the exteroceptively related stimulus-induced activity relative to the preceding resting-state activity level. Interestingly, when calculated in relation to the preceding resting state activity level, the observed difference in exteroceptively-related signal changes disappeared; this means that the difference between depressed and healthy subjects is not due to the exteroceptive stimuli but rather to the preceding resting state activity level. Thus, when considering this resting-state activity level, there was no longer any difference between healthy and depressed subjects in signal changes during exteroceptive processing.

In contrast to the exteroceptive stimuli, we did not observe any difference between healthy and depressed subjects during the interoceptive stimuli in both analysis ('absolute' as independent of the resting state and 'relative' in dependence of the preceding resting state). This difference between intero- and exteroceptive stimuli with regard to relative and absolute signal changes suggests that there is differential interaction of both kinds of stimuli with resting-state activity. Either rest–stimulus interaction is reduced during exteroceptive stimuli or rest–stimulus interaction is increased during interoceptive stimuli, but this cannot be differentiated on the basis of our findings. However, it is clear that there is an imbalance of activity between intero- and exteroceptive stimulus processing, including their respective interaction with the resting-state activity level. Because of the paucity of research in this area, additional imaging studies are needed to investigate changes in interoceptive processing in depression.

The study by Wiebking et al. (2010) also investigated psychological measures of body perception, employing the Body Perception Questionnaire (BPQ). They found that the BPQ scores were significantly increased in depressed patients, indicating increased bodily awareness. Most interestingly, in contrast to healthy subjects, the increased BPQ scores no longer correlated with the signal changes during the resting state and the exteroceptive condition. This suggests that depressed patients no longer properly modulate their degree of neuronal activity so as to down-modulate the perception and awareness of their own body and shift attention from the body to the environment. This may help to explain the many somatic complaints that characterize MDD. Although tentative, such a lack of correlations with abnormally increased neuronal activity has also been reported for other measures of abnormal psychological states, such as excessive negative affect, self-relatedness problems, and negativistic temporal projections to future possibilities in depression (Grimm et al., 2009; Wiebking et al., 2010).

In a subsequent study, Wiebking et al. (2011) also investigated what happens in the DMN, the paracore–midline system, during intero- and exteroceptive stimuli so as to explicitly address the question of how these stimuli interact with the high level of resting-state activity that characterizes this network. Again they observed no difference in signal changes (i.e. deactivation) induced by interoceptive stimuli in the DMN, while they observed decreased deactivation related to exteroceptive stimuli. Thus in both the insula and the DMN there appears to be decreased neuronal reactivity to exteroceptive stimuli by comparison with interoceptive ones.

Taken together, these findings are indicative of an imbalance in the neural processing between intero- and exteroceptive stimuli, with the latter but not the former inducing decreased neural activity. This may lead to relatively increased neural processing of interoceptive processing and rest–intero interaction compared with the apparently absolutely reduced exteroceptive processing and rest–extero interaction. As already noted, this abnormal shift toward interoceptive processing may psychopathologically promote increased bodily awareness and subsequent concerns about undesirable bodily symptoms. Phenomenologically one may here speak of increased body-focus. Meanwhile, the decreased exteroceptive processing may be accompanied by reduced awareness of and concern with environmental changes, especially positive events. This may correspond phenomenologically to what I described as decreased environment-focus.

Neuropsychodynamic account

We could present some empirical evidence, albeit tentative, in favor of reduced rest–extero and extero–rest interaction as well as relatively increased rest–intero and intero–rest interaction in depression. Furthermore, the finding suggests that the reduced deactivation in the paralimbic–midline system is apparently closely related to abnormally high degrees of self-relatedness. And we present some data which support the view that the elevated resting-state activity and reduced extero–rest interaction in the anterior paralimbic–midline system appears to be related to the restriction of subjective time to the past and present, while being blocked for the future.

What does this entail for the psychodynamic context of increased introjection coupled with negative affect? Let me consider the aforementioned imbalance between intero- and exteroceptive stimulus processing with the apparent relative increase in intero–rest interaction. The reader may recall that stimulus–rest interaction was supposed to enable and predispose to internalization and more specifically to introjection of objects (see Chapter 6). The abnormal shift of stimulus–rest interaction from exteroceptive to interoceptive stimuli enables and predisposes to the introjection of predominantly the person's own body as related to the interoceptive stimuli. In contrast, objects from the environment as based on exteroceptive stimuli have a much lower likelihood of being introjected because of reduced extero–rest interaction.

Based on these considerations, I postulate that the imbalance between extero–rest and intero–rest interaction with the abnormal shift to the latter enables and predisposes to the increased introjection of predominantly the person's own body as bodily object at the expense of other objects stemming from the environment (i.e. environment objects). As I mentioned, this shift from environment to bodily objects in introjection is highly consistent with the increased focus of depressed patients on their own body and their various rather non-specific bodily symptoms (for further details with regard to the body, see Böker and Northoff, 2010). In order to shed more light on the high importance of the body and the various bodily symptoms in depression, one would need to investigate the relationship between interoceptive processing and self-relatedness in more detail. However, such investigations have not yet been reported in either healthy or depressed subjects.

Let me dwell a little longer on the phenomenological and psychodynamic implications of the imbalance between intero–rest and extero–rest interaction. If the intero–rest interaction is increased at the expense of the extero–rest interaction (be it relatively or absolutely), one would also expect a shift in the subjective experience and thus in the contents of the mental states. The mental states' contents may then no longer concern events and people in the environment, and thus environment objects, but rather the person's own body as the main object. This may be phenomenologically well reflected in what I described as the decreased environment-focus that accompanies an increased self- and body-focus. Therefore I postulate that what phenomenologically is described as the imbalance between decreased environment-focus and increased self- and body-focus may be enabled and predisposed to by the neuronal imbalance between extero–rest and intero–rest interaction.

Increased paralimbic–midline activity and the "self-object dilemma"

Neuropsychodynamic hypothesis

One crucial characteristic of the third psychodynamic feature of depression is the "self-object dilemma," which refers to the continuous need for self-stabilizing self-objects even if the available self-objects are in fact destabilizing rather than stabilizing. I here postulate that the self-object dilemma is closely related to the altered neural processing of the self as well as the shift in subjective time in depression.

More specifically, I expect depressed patients to show abnormally high degrees of self-relatedness both behaviorally–psychologically and neuronally as related to their increased introjection. One would also assume that the increased resting-state activity in the paracore–midline system in particular is related to the shift in subjective time from the present to the past as manifested in the shift from present to past extero–rest interaction. As we shall see, there is some evidence to support both assumptions although, due to the limited number of studies, it must be regarded as preliminary at this stage.

Empirical findings

A key feature of MDD is hopelessness. Hopelessness is closely related to the sense of time, and more specifically to the ability to extend one's hopes and expectations into the future (or to use Endel Tulving's terms, autonoetic consciousness). If one is no longer able to extend and project oneself into the future, one is no longer hopeful, focusing on positive life options and possibilities. This is an accurate description of MDD. Almost all of the items in the Beck Hopelessness Scale (BHS) monitor one's ability to anticipate positive future events. Recent research highlights the finding of elevated scores on the BHS in depressed individuals (Grimm et al., 2009; Wiebking et al., 2011). Moreover, these studies found that elevated resting-state activity in the PACC and the VMPFC correlated with the BHS scores. The higher the resting-state activity in the PACC and the VMPFC, the higher were the scores on the BHS, and this relationship was specific for hopelessness rather than other symptoms (see also Heinzel et al., 2009; Yao et al., 2009).

These findings highlight the psychopathologically specific nature of the relationship of hopelessness with resting-state activity in the VMPFC. This is highly consistent with observations in healthy subjects where the VMPFC and the PACC rather than more posterior midline regions were found to be associated with the extension (i.e. slowing) of time in subjective perception, such as during anticipation or prospection (Addis et al., 2007; Schacter and Addis, 2007a,b). The abnormally elevated resting-state activity in the VMPFC appears to impair anticipation and thus one's experiences of extending hopes into the future. The abnormal resting-state activity level also appears to block the ability of MDD patients to project hope into the future, thereby promoting hopelessness and ultimately helplessness. Although it is speculative at this point, this view leads to various hypotheses.

Let us consider this issue in a little more detail. The aforementioned studies evaluated brain signal changes induced by external stimuli in resting-state regions such as the PACC and the VMPFC. Wiebking et al. (2011) were even able to demonstrate that signal changes in the same regions that were induced by interoceptive stimuli did not correlate with the BHS scores. This suggests that external stimuli may be specifically related to the extension of the neural (and thus experiential) time window. Neurally, this means that the external stimulus may no longer influence the resting-state activity level in the PACC and VMPFC in MDD, leading to what we call reduced stimulus–rest interaction (Northoff et al., 2010). Therefore self-related neural processing

(Northoff et al., 2006) may be increasingly constrained to shorter time windows where internally generated stimuli begin to prevail because of the reduced impact of external events. Depressed patients may thereby be locked into narrow psychological time windows (which is commonly observed in MDD), further promoting feelings of hopelessness, as they can no longer envision positive future possibilities, but only the sustained limbo of ongoing negative affect and feelings of punishment. The narrowing of the subjective time window may thus lead to an increased focus on the self, coupled with an extremely negative affect.

The symptom of increased self-focus concerns the heightened awareness of their self-image in depressed patients, even more so as they are no longer absorbed by positive interpersonal interaction, and ongoing positive environmental objects and events (Northoff, 2007). The increased self-focus accompanies the abnormal experiencing of predominantly negative emotions and attribution of negative characteristics (e.g. worthlessness) to one's self, which fuel seemingly endless negativistic ruminations (Northoff, 2007 for details), with little modulation by external events. Convergent evidence from various imaging studies (Lemogne et al., 2009, 2010; Grimm et al., 2010) has consistently demonstrated reduced signal changes during presentation of either self-related emotional words or emotional pictures, with abnormal activity changes in the anterior paralimbic–midline regions such as the PACC, the VMPFC, and the DMPFC.

These anterior paralimbic–midline regions have long been associated with self-related information processing that describes how feelings of personal relevance or meaning are projected onto various external stimuli and internal states (Northoff et al., 2006; Enzi et al., 2009). One would consequently expect that elevated resting-state activity in these brain regions would lead to increased self-related processing and thus to abnormally increased self-relatedness in MDD patients. This is exactly what one study reported when it showed significantly increased scores for self-relatedness with regard to especially negative emotional pictures (Grimm et al., 2009).

Since self-related mental focus is already high during the resting state, one would expect it to be comparatively low with regard to external stimuli, with even reduced rest–stimulus interaction, perhaps surprisingly, even in the case of highly self-related external stimuli. In contrast, internally generated stimuli as related to the abnormally elevated resting-state activity may become salient and be assigned an abnormally high degree of personal relevance, as is routinely evident in the obsessive ruminations of MDD patients. Accordingly, depression is characterized by an imbalance of relations between internally and externally generated stimuli and states of mind, with the former being hijacked by negative affect. We postulate that symptomatically this imbalance in rest–stimulus interactions between internally and externally generated stimuli promotes increased self-focus and a reduction in the influence of positive environmental events.

Neuropsychodynamic account

What enables and predisposes to such increased introjection of the lost object? To answer this question we need to consider the concept of extero–rest interaction. Reduced extero–rest interaction concerns present exteroceptive stimuli. However, this does not exclude the possibility that the reduced present extero–rest interaction is substituted and compensated by increased recruitment of past exteroceptive stimuli to interact with the resting state (i.e. past extero–rest interaction). I therefore postulate that there is also a neuronal imbalance between present and past extero–rest interaction, with the latter being abnormally increased.

What does such abnormally increased past extero–rest interaction entail for the psychodynamic context? It means that those objects which are related to past exteroceptive stimuli are preferentially recruited to be internalized and introjected. And since there is reactivation of the lost object, such introjection predominantly concerns the lost object. Psychologically, this means that one may assume that the past exteroceptive stimuli related to the lost object are attributed an abnormally high degree of self-relatedness, although this has yet to be tested empirically. This attribution of abnormally high degrees of self-relatedness may be associated with reduced deactivation in paralimbic–midline regions which, due to stimulus–rest interaction, may further enhance the resting-state activity level in these regions. This in turn may reduce present extero–rest interaction and thus aggravate the imbalance between present extero–rest and intero–rest interaction. One may consequently postulate that there is a vicious cycle at the neuronal level which may enable and predispose to the kind of circular mechanisms described in the phenomenological and psychodynamic context.

How does this relate to the "self-object dilemma" that is postulated? The reader will recall that the "self-object dilemma" is caused by the fact that is the availability of stabilizing possible self-objects is reduced in depression, while there is nevertheless an ongoing and continuous need to integrate possible self-objects as actual ones into the self in order to stabilize the latter. How is the "self-object dilemma" related to the neuronal mechanisms described here? The transformation of objects to possible self-objects (i.e. object–self-object transformation) was supposed to be enabled and predisposed to by intero-exteroceptive linkage and subsequent affective assignment (see Chapter 7). Due to the imbalance between present extero–rest and intero–rest interaction, the only objects available in depression that can enter object–self-object transformation are the person's own body and the lost object. Thus only these undergo intero-exteroceptive linkage which, due to the absence of a current exteroceptive linkage, is transformed into intero–intero or intero–past-extero linkage. However, this means that the negative affect from the lost object is also transferred and imposed upon the present affective assignment and thus the person's own body and the self.

The restriction of possible self-objects means that only the person's own body and the lost object can be integrated into the self (self–self-object integration) (see Chapter 7 for details) and thus converted into actual self-objects. I assumed (see Chapter 7) that self–self-object integration is enabled and predisposed to by the trilateral interaction between rest and intero- and exteroceptive stimuli (i.e. rest–intero-extero interaction), yielding what I called "hybrid neural activity." This is now manifest in either rest–intero-intero or rest–intero-past extero interaction, which entails the predominance of interoceptive or past exteroceptive stimuli in the hybrid neural activity, although that enables and predisposes to the integration of either the person's own body or the lost object as actual self-object into the self. However, this means that the self is forced to stabilize itself on the basis of the lost object and/or the own body which, as described, is in fact destabilizing rather than stabilizing. Thus what is described as the "self-object dilemma" in the psychodynamic context may be enabled and predisposed to by the imbalance between extero–rest and intero–rest interaction and its subsequent effects on intero-exteroceptive linkage and trilateral interaction (see also Figure 10.5).

Neuroexistential account

At a more general level, the "self-object dilemma" demonstrates how normally adaptive mechanisms such as introjection, self–self-object integration and object–self-object transformation can become disruptive in an abnormal context. And, most importantly, it is the recruitment of these otherwise normal well-functioning mechanisms that induces the symptoms of depression. What does this mean for the neuronal context? It means that the increased resting-state activity is by itself not the cause or neural correlate of the depressive symptoms. Instead it only enables and predisposes to the recruitment of the neuronal mechanisms of rest–stimulus and stimulus–rest interaction which, due to their preserved and thus normal function, properly adapt to the changed neuronal context. More specifically, reduced extero–rest interaction is compensated for by increased intero–rest and past extero–rest interaction.

Why does the brain recruit its neuronal mechanisms in the changed context? Why does the brain not simply shut off and stop everything, including the recruitment of the subsequent neuronal mechanisms? The reader will recall that introjection and internalization of objects were assumed to enable and predispose to the constitution of the self and its differentiation from the brain (i.e. brain–self differentiation) (see Chapter 6). Stopping the neuronal mechanisms that enable and predispose to introjection would thus stop the constitution of the self and its differentiation from the brain. However, the depressed patient cannot afford to do this, as that would be to succumb to the subjective experience and perception of the death of their own self. In other words, the depressed patient and their brain aim to maintain the self and thus brain–self differentiation by every means possible in order to preserve the subjective existence of the self as distinct from their brain as a mere organic mass (see Appendix for further details).

Acknowledgments

At this point I particularly wish to thank my long-standing collaborator and friend Heinz Böker for introducing me to psychoanalysis in general and the psychodynamic processes in depression in particular. This gave me a deeper understanding and guidance in my exploration and investigation of the brain.

Chapter 11

Psychosis I[1]: Psychodynamics and Phenomenology[2]

Lack of energy investment in objects ("decathexis of objects")

In his writings about psychosis, Freud considered the notion of cathexis to be crucial in understanding the psychodynamic processes underlying psychosis[3] (Freud, 1924). Let me briefly recall what is meant by the concept of cathexis, which will then serve as the basis for understanding its changes in psychosis. Cathexis describes the investment of energy in something in general and in another person, event, or idea in particular (see Chapter 4). This entails three characteristic features of cathexis, namely non-specificity with regard to contents, constitution and construction of objects by means of establishing a special relationship between self and content, and multiple psychological manifestations.

The concept of cathexis remains non-specific with regards to the contents, as the energy can possibly be invested in any kind of content, be it other people and their self, the person's own self, ideas, bodies, cognitions, events, etc. While remaining non-specific with regard to the content, the concept of cathexis must be regarded as specific for the constitution and construction of objects. The characteristic of objects, including self-objects,[4] is that they are (or represent) specific contents for a specific person, meaning that they are specifically related to the self of the person in question (i.e. they are, as I said, self-related). Cathexis as the investment of energy is supposed to enable and predispose to self-related processing by means of which stimuli, i.e. person, idea, cognition, body, events, etc. that do not bear any special relation to the person's self are transformed into personally specific objects that as such bears a special relation to the respective person's self (e.g. "self-relatedness") (see Chapter 9).

Cathexis as the investment of energy may be crucial in the constitution and construction of personally specific and self-related objects from personally non-specific and non-self-related

[1] I here and in the following account (in both this and the next chapter) deliberately use the term "psychosis" to cover the broad range of symptoms, rather than focusing on a specific etiology or cause of these symptoms as indicated by the term "schizophrenia," which refers to a specific disorder with a specific (although as yet unknown) cause as distinct from other causes of psychosis, such as infections or drugs.

[2] I here devote an entire chapter to the psychodynamics (and phenomenology) of psychosis, which I deem to be necessary in view of the complexity of psychosis. Discussion of the neuropsychodynamic hypotheses will follow in Chapter 12.

[3] I shall use the terms "psychosis" and "schizophrenia" interchangeably, since I here focus more on the mechanisms underlying the symptoms shared between them, while my focus is less on the etiology, which differs between psychosis and schizophrenia.

[4] In this first section I do not distinguish between the terms "object" and "self-object," thus listing them together. However, they will be further distinguished in my account of psychosis in the third section here, which is explicitly devoted to self-objects as distinct from mere objects, thereby referring to my earlier distinctions between objects and self-objects as discussed in Chapter 7 on narcissism and self-objects.

people, ideas, or events. Rather than accounting for specific contents, cathexis thus enables and predisposes to the construction and constitution of a special relationship between self and content. This, as we have already seen in Chapter 6 on defense mechanisms, changes not only the content, which is transformed from a mere object into a self-object, but also the person itself, who is transformed from an objectively non-specific person or ego into a subjectively specific self. Taken together, this leads to a double characterization of cathexis. Cathexis enables and predisposes to the transformation of mere stimuli into objects (stimulus–object transformation) (see Chapters 4 and 5), and the subsequent transformation of objects into self-objects (see Chapter 7).

The third characteristic feature of cathexis pertains to the psychological manifestations of the investment of energy in whatever content. Depending on the kind of content in which energy is invested, one may distinguish between different psychological manifestations of cathexis. Let me give some examples (see McGlashan 2009, p. 477, for some hints in this direction). If the energy is invested strongly in the relationship with other people, one may speak of attachment. If the energy is invested in the person's own arousal and ambition, one may speak of motivation. If the energy is strongly invested in the person's own sexual drive, one may speak of libidinal energy. If the energy is invested in the person's own self and ego, one may speak of narcissistic energy. If the energy is invested in the perception of external stimuli, one may speak of salience and attention, and so on. This demonstrates that psychologically (i.e. within the psychological context) the concept of cathexis may take on different gestalts which though, energetically and thus within a psychodynamic–energetic context, may be traced back to one and the same function, namely the investment of energy.

I have now prepared the ground for the characterization of psychosis by the redirection of object cathexis. Let us go back to Freud and how he considered psychosis in a psychodynamic context. Freud (1924a, b) assumed that the quantity of energy invested in people, ideas, events, etc. from the external world is significantly reduced in psychosis. The cathexis is withdrawn from objects of external origin, so that he spoke of an "object decathexis." Since the energy is withdrawn or decathected from the people, ideas, or events in his external environment, they cannot be constituted and constructed as objects (or as self-objects, as the self psychologist would add). The lack of investment of energy means that no special relationship between the person's self and contents as a relationship of self-relatedness can be constituted and constructed, which then results in the loss and absence of objects (including self-objects) of external origin.

Due to their lacking transformation into objects, the people, ideas, or events that the psychotic person perceives and experiences no longer bear any special and self-related relationship to him. If, however, the people, events, or ideas that he perceives and experiences are no longer related to his self, he remains indifferent and no longer shows any interest. This is nicely expressed in Freud's description of the case of Schreber: "The patient has withdrawn from the people in his environment and from the external world generally the libidinal cathexis which he has hitherto directed onto them. This everything has become indifferent and irrelevant to him" (Freud, 1911, p. 74).

The withdrawal of object cathexis, or "decathexis of objects" as Freud calls it, concerns not only the real people, ideas, or events in the external environment but also the mental representation of any kind of objects previously constructed and constituted (Freud himself seemed to remain ambivalent about this, as has been pointed out by Gabbard, 2005a, pp. 183–184). This means that the psychotic person becomes alienated not only from the external, social world but also from his own mental world and its mental representations of objects. Applying the quote by Freud means that the person's own mental representation of objects and self-objects becomes "indifferent and irrelevant." One may therefore want to speak of what I call "indifference to any kind of object/self-object[5] be it of social or mental origin."

Volatile and unstable inner and outer ego boundaries

The "indifference to any kind of object/self-object be it of social or mental origin" leads to the second step of psychosis as described by Freud. The first step consisted of the withdrawal of any kind of object cathexis (including self-object cathexis), be it mental or social, and is now followed by a second step, namely the denial of any reality whatsoever (which distinguishes psychosis from neurosis). If objects (and self-objects), be they mental or social, are no longer constituted and constructed, one may therefore assume that there are no objects (and self-objects) at all and if so, one may also doubt that there is a reality at all including objects (and self-objects). Thus the loss of social and mental objects (and self-objects) accompanies the denial of reality as associated with the lost objects (and self-objects).

It is important to be clear that it is not only the social reality and its objects that are denied, but also the mental reality, and worse still the existence of objects and self-objects that the person experiences and perceives. The psychotic patient experiences and perceives his own mental representations of objects and self-objects as non-specific and non-self-related and therefore as indifferent and irrelevant to him. In other words, using the language of self psychology, the psychotic patient is no longer able to transform objects into self-objects (object–self-object transformation), and thus perceives and experiences objects, if at all, as remaining unrelated (i.e. non-self-related) to him.

Both of these aspects, namely social and mental decathexis as one might call them, were developed in subsequent psychodynamic accounts of psychosis. While Sullivan (1962) focused on the social aspect when emphasizing the abnormal nature of interpersonal relationships in psychotic patients, Federn (1952) shifted the focus more to the mental aspect and more specifically to the ego or self who has those mental states. Rather than assuming mere decathexis of objects, as Freud did, Federn assumed that energy was no longer invested in the boundaries of the ego, thereby defining the ego by subjective experience, and thus coming close to what I earlier called self (or sense of self).

Federn distinguished between outer and inner ego boundaries. The outer ego boundary refers to the sense organ of the ego (or self), and if it is weak, as in the case of decreased or lacking energy investment, one remains unable to differentiate between what belongs inside and what belongs outside, and thus between ego (self) and environment (world). The inner ego boundary refers to the differentiation between conscious and unconscious states, and relates to the difference between what can be subjectively (and thus consciously) experienced by the self and what cannot.

Dümpelmann (2004) specifies the concept of inner and outer boundaries in more detail. He considers outer boundaries to be physical in that they presuppose the outer physical space that allows to both relate and differentiate one person from (an) other person(s); the boundary demarcates here the distinction between outer physical space of the other person and the inner mental space of the initial person. While he assumes the inner boundaries distinguish between self and objects within the inner mental space of the person; this inner purely mental boundary is

[5] As already indicated, I do not distinguish here between objects and self-objects with regard to psychosis. However, I shall do so in Chapter 12.

supposed to correspond to what psychologically I described as the distinction between self- and non-self-related stimuli (see Chapter 9).

Changes in inner and outer boundaries as described by Federn and later by Dümpelmann may be closely related to what I described as "indifference to any kind of object be it of social or mental origin." Disturbances in the outer boundaries, as due to decathexis as lacking energy investment, may lead to the inability to construct and constitute objects of external and thus social origin, since that presupposes some distinction from the internal and thus non-social ego or self. Changes in the inner boundaries may lead to the inability to distinguish the mental representations of objects, including self-objects, from the (mental representation) of the person's own self (the ego), which may make the psychotic patient indifferent to the objects of mental origin, including their own self-objects.

In summary, I have here described how psychotic patients may suffer from a lack of energy investment in objects, including self-objects, of both social and mental origin. The apparent deficit in constructing and constituting objects and self-objects of either mental or social origin may be traced back to disruptions in the inner and outer boundaries of the ego (i.e. self). Phenomenologically, the lack of energy investment may be manifested by indifference to and irrelevance of any kind of object, including self-objects, be it of social or mental origin, which I described as the "indifference to any kind of object/self-object be it of social or mental origin."

Attunement and "crisis of common sense"

The reader may be surprised to see a phenomenological account at this stage. The phenomenological account refers to the way that the patient subjectively perceives and experiences psychosis and its various symptoms. This distinguishes it from the psychodynamic account, which targets the psychological mechanisms underlying subjective experience and perception, rather than subjective experience and perception themselves. However, if the psychodynamic account aims to be empirically plausible, it must be compatible not only with the neuronal data but also with the phenomenological account and its description of subjective experience and perception.

As with the case of depression, I shall complement the psychodynamic account of the underlying psychological mechanisms with a phenomenological account of the surfacing subjective experience and perception. While I am aware that such an assumption of correspondence between psychodynamic and phenomenological mechanisms requires much more extensive conceptual elaboration and theoretical justification, this would be beyond the scope of this book. All I can do here is therefore to point out intuitive analogies and converging similarities between phenomenological and psychodynamic accounts.

As pointed out, the phenomenological account focuses on the subjective experience of the objective stimulus from the perspective of the person who is experiencing it. For instance, it describes how we subjectively experience incoming sensory stimuli as exteroceptive auditory stimuli or interoceptive stimuli from our own body in a subjective way, and thus how we can relate them to our own self. The linkage of sensory stimuli to our own self in subjective experience thus concerns what phenomenological accounts (Blankenburg, 1969; Bin Kimura, 1997; Parnas et al., 1998, 2003; Sass, 2000; Sass and Parnas, 2001; Parnas, 2003; Sass, 2003) describe as "attunement." The concept of "attunement" describes, most broadly, the relation of the self to the world (i.e. how the self adjusts and adapts to the various objects, events, and other people in its environment). Most importantly, phenomenologists point out that "attunement" in this sense already operates at a pre-reflective, implicit, or preconceptual level, which Parnas (2003) and Sass (2003) describe as pre-reflective or preconceptual attunement. In the following account I shall use the concept of attunement to describe such pre-reflective, pre-conceptual, and implicit adaptive processes.

The phenomenological description points out the close relationship of the subjective experience of the self with the other and the world. In the same way that the other is ingrained in subjective experience of the self, the self is infused in subjective experience of the world and its objects. This means that the self is what the phenomenologists call "present" in our subjective experience of the world, implying that there is not yet a clear-cut distinction between subject and object at this early stage. "Presence" describes how there is an implicit and pre-reflective self-awareness in the subjective experience of the objects in the world, so that pre-reflective self-awareness and pre-reflective world-awareness appear to go hand in hand (Merleau-Ponty, 1965; Zahavi, 2005). What I pre-reflectively perceive and subjectively experience in the world gives me not only the presence of the world and its objects, but also the presence of myself as self. Thus there is what is called a "first-person givenness" of my self in my subjective experience of the world.

Parnas describes nicely how this presence is altered in schizophrenia (Parnas et al., 2001; Parnas, 2003). The subjective experience of the world and its objects is no longer accompanied by a pre-reflective self-awareness. The person's own self, the self that experiences the world, is no longer included in that very subjective experience: "The prominent feature of altered presence in the pre-onset stages of schizophrenia is disturbed ipseity, a disturbance in which the sense of self no longer saturates the experience. For instance, the sense of myness of experience may become subtly affected: one of our patients reported that this feeling of his experience as his own experience only 'appeared a split-second delayed'" (Parnas, 2003, p. 225).

These patients can no longer refer to themselves in their subjective experience of the world. It is as if the subjective experience of the world is no longer their own subjective experience. It may belong to and be experienced by someone, but it is not their own self who makes and has those experiences. Due to the absence (as opposed to presence) of the schizophrenic patient's own self in their subjective experience of the world, they become detached, alienated, and estranged from their own subjective experience. Due to such detachment of their self from their own experience, their subjective experience from their own first-person perspective is no longer experienced as subjective and thus as lived. The experiencing self is consequently no longer affected by its own experiences, which Sass (2003) describes as disorder of self-affectivity. The person's own self is no longer experienced as their own self, and most importantly is no longer experienced as the vital centre and source of their own experiences, actions, perceptions, thoughts, etc. This reflects what Sass (2003) calls the diminished self-affection or auto-affection, mirroring the fact that the self is no longer affected by its own subjective experiences.

However, if the self is no longer affected by its own subjective experience, it stands apart from the objects in the world that are experienced. A gulf, or a phenomenological distance as Parnas calls it (Parnas, 2003, p. 225), opens up between the world and the self, with the objects of the world no longer making intuitive sense and thus being meaningful to the person who is experiencing them. The person's own self thus becomes almost objective and mechanical in their subjective experience and perception of the world. It seems to be at this point that schizophrenic patients lose their common sense, which Stanghellini and Ballerini (2007, 2008) refer to as a "crisis of common sense" or "loss of pre-reflective operative common sense that disrupts self-experience in the context of relatedness." This may also be the point at which schizophrenic patients begin to question the meaning and nature of the very objects and events of the world that they experience, as well as the nature of their own self, which they can apparently no longer access in their subjective experience. They thus become hyper-reflective, as Sass (1996, 2003) would say.

In summary, psychosis may be characterized phenomenologically by disrupted attunement which makes it impossible for the psychotic patient to relate to his environment in a subjective–experiential way. This converges nicely with what psychodynamically is described as the loss of both object relations and ego boundaries. The loss of object relations means that the psychotic patient no longer has access to objects, including self-objects, in inner mental representations, which may be related to blurred (inner) ego boundaries. As such the loss of object (and self-object) relations and ego boundaries in subjective perception and experience may correspond (intuitively) to what phenomenologically has been described as disturbed pre-conceptual attunement, altered presence (ipseity), diminished self-affection, and "crisis of pre-reflective operative common sense."

Self-objects and affects

What are the implications of the loss of object relations (including self-object relations) and ego boundaries for the constitution of the self? Let us go back briefly and consider how the self is constituted and constructed. As was pointed out in Chapter 7 on narcissism and the self, one essential feature is the constitution and construction of self-objects (see Chapters 7 and 9). Self-objects are neither pure objects nor pure self. Instead they describe those objects that are central for the self in constituting and constructing as well as maintaining the existence of the self as distinct from other selves and the person's own brain and body. One may thus say metaphorically that self-objects provide the bridge between self and objects, and I therefore characterized them as relational. Being the bridge between self and objects means that self-objects also delineate the boundary between self and objects, and more specifically what I described as the inner or mental boundary between self and objects.

How can we distinguish self-objects from mere objects? One essential feature of self-objects is their association with affect and emotions, or "affective assignment" as I called it. Self-objects are thus objects that can be characterized by strong emotional involvement. In addition, they presuppose linkage between extero- and interoceptive stimuli (see Chapter 7) which means that they presuppose linkage of the object with the respective person's body. Consideration of affective assignment and intero-exteroceptive linkage as crucial features of self-objects raises the question of their alterations in psychosis. Let me make a brief methodological comment. So far I have not distinguished between objects and self-objects in my psychodynamic account of psychosis. However, as indicated in the last two paragraphs (and described in detail in Chapters 7 and 9), self-objects may need to be distinguished from mere objects. If so, this distinction may be relevant in psychosis, too, as will be discussed below.

However, before dwelling further on self-objects, let me briefly return to the even more basic relationship between self and objects in psychosis. As previously described, the mutually dependent co-occurrence and co-constitution between self and objects is no longer balanced in acute psychosis. It is shifted either towards the self at the expense of objects/self-objects or towards the objects at the expense of the self, resulting in what I shall, like Mentozs (1991), describe as an "existential dilemma" (see next section for details). The existential dilemma describes how for the psychotic patient there is only the alternative between constituting and constructing objects, including self-objects or his own self: Either he invests energy in the constitution of the self or, alternatively, he directs his energy investment to the constitution of objects. In contrast to the healthy subject, co-constitution of both self and objects at the same time remains impossible for the psychotic person. Even worse, the constitution of objects is only possible at the expense and ultimately the denial of the own self meaning which also holds in the reverse way with the constitution of the own self being possible by denying and negating objects. There are no longer any bridges or boundaries between self and object that could eventually balance the constitution and construction of self and objects. However, this prevents the psychotic patient from taking an intermediate stance and leads him instead to oscillate between the two extremes, namely the mutually exclusive construction and constitution either of the self or of objects/self-objects.

What does this entail for self-objects? It means that the constitution of self-objects as distinct from both mere objects and the self itself remains impossible. Objects are no longer assigned a specific affect or an emotion that distinguishes them from others and transforms mere objects into self-objects. This lack of affective assignment accompanies a lack of intero-exteroceptive linkage, which means that objects are no longer properly associated and linked to the body of the respective person. However, lack of both affective assignment and intero-exteroceptive linkage means the absence of self-objects.

Let me make a brief conceptual comment concerning the term "self-object." The term "impossible" in the context of the constitution of self-objects does not here refer to the non-existence of self-objects as such. Rather it indicates that self-objects as truly stabilizing (i.e. maintaining and preserving) the self can no longer be constituted. In other words, the psychotic patient is unable to constitute "good self-objects," and this is compensated for by constituting self-objects that are no longer self-stabilizing and which are thus "bad self-objects" as one might call them. This is so because, as Kohut has pointed out, self-objects are to the self what breathing is to the body (see Chapter 7). Thus when I describe self-objects as being impossible (in this and the next chapter), I refer to "bad self-objects" as distinct from "good self-objects," rather than meaning the absence of any self-objects at all (i.e. their non-existence) in psychosis.[6]

The disturbance in the constitution of self-objects and their crucial role in mediating psychotic affects and symptoms is nicely described in a recent paper by Rosenbaum and Harder (2007, p. 18): "One self-object function is to recognize, distinguish, and respond accordingly to distinct affect states. If this cannot be accomplished at all, the individual's sense of self-boundary suffers. Another self-object function is to facilitate the tolerance, acceptance, comprehension and intelligibility of intense and/or contradictory affect states. An inability to accomplish this function may lead to psychic functions in which disavowal may be employed. Clinically, a primitive state of projective identification or delusion may ensue. A third self-object function helps the individual to attend to his/her own stimulus barrier insofar as intense affect can be used as a signal of a changing internal state, rather than as an indicator of impending traumatic fragmentation. If this is not accomplished, then the emergence of intense affect states may be experienced as shame and evoke self-hatred and further psychotic defense utilization" (Rosenbaum and Harder, 2007, p. 18).

The close linkage of self-objects with affect poses a problem for the psychotic patient. If he is no longer able to constitute and construct self-objects, he remains unable to assign affects to objects and therefore remains unable to transform objects into possible self-objects (object–self-object transformation) (see Chapter 7). What can he do then with the affective states if they can no longer be assigned to self-objects and thereby appropriately channeled? He can assign them to either his self or the objects that he constitutes and constructs. Assignment of affects to his own self means that his self may become identified with and overwhelmed by the affects, which in turn may induce existential anxiety and fear of self-loss.

Dümpelmann (2010) describes this as "the lack of affect tolerance and the experience of the overwhelming power of affects," meaning that the self remains unable to distance itself from its own affects and cannot do other than identify its own self with them. To use more cognitive terminology, one may want to say that the psychotic person (or its self) may remain unable to properly regulate and control the affects and therefore becomes over-flooded and overexcited.[7]

---

6 I owe this remark to M. Dümpelmann, who made me aware of the difference between the non-existence of self-objects and "bad self-objects."

7 Relying on Stern and his description of infants, Garfield (1995, 2001) speaks of so-called "vitality affects" that describe non-specific, basic, body-oriented affects which are assigned to the experience of motion and kinematics of the person' own body. These vitality affects must be distinguished from categorical affects

In contrast, assignment of affects to the objects means that the affects have nothing to do with the person's own self, are distanced and unrelated to it, and thus cannot be experienced and perceived as such leading to what Dümpelmann (2009) has described as "disturbance in the perception of affects."

Self-objects and the fragmentation of the self

What does the failure to constitute and construct self-objects in psychosis mean for the self and its relationship to objects? If there are no longer any self-objects, the constitution and construction of the self, including its distinction from mere objects, become fragile and incoherent. The absence of self-objects means that the self no longer has any bridge or boundary to mere objects. The self is either totally isolated from the objects, which results in autism, or it is totally immersed in the objects, which results in fusion (see Mentzos, 1991). This leads the psychotic patient to oscillate between the two extremes of the constitution and construction of either the self or the object, which he experiences and perceives to be mutually exclusive, thus posing him with a dilemma.

The absence of self-objects as a bridge between self and objects in psychotic patients is associated with their inability to establish some boundary between self and objects. Due to the absence of such boundaries, the constitution and construction of the self is identical with that of objects, and vice versa. This means that the constitution of objects concerns not only the objects themselves but also the person's own self and its mere existence as subjectively perceived and experienced. In other words, the (subjectively perceived and experienced) existence of the person's own self is at stake when constituting objects, which makes the dilemma existential.

How does the failure to constitute and construct self-objects affect the gestalt of the self? The lack of self-objects means that the self can no longer be properly constituted and constructed in a coherent and unified way. The self consequently becomes fragmented, incoherent, and disintegrated, which in turn will lead to symptoms such as depersonalization, derealization, delusions of alien control, and other typical ego disturbances. Most importantly, as pointed out by Benedetti (1983), the perception and experience of the person's own self as incoherent, fragmented, and disintegrated becomes associated with predominantly negative affects. He speaks of "inner death landscapes" and a "negative existence" of the psychotic self (in subjective perception and experience).

How can we account for such "negative existence" of the self within our framework of affective assignment and the existential dilemma? Negative affects predominate in psychosis because, due to the existential nature of the dilemma, there is a continuous threat of the self to not exist (in subjective perception and experience), leading to what one may describe as "existential anxiety." And since the constitution of the (subjective) existence of the self is identical with the constitution of objects, the existential anxiety is affectively assigned not only to the objects but also, to complete the vicious circle, to the self itself. This makes any assignment of positive affects and thus positive affective experience of the person's own self impossible. Once the constitution of objects

that describe more specific and less body-oriented affect states. Following Garfield, vitality affects are crucial in allowing for affective attunement and constitution of self-objects, and thus for integrating affect and self (Garfield, 1995, pp. 119–123; Dümpelmann, 2009). By using the concept of vitality affects, Garfield presupposes the need to consider what I earlier called intero-exteroceptive linkage when describing self-objects. Garfield assumes that psychotic patients are no longer able to develop categorical affects as specific affects related to specific external events and thus to exteroceptive stimuli. Instead they are dominated by vitality affects that are bodily related, with the focus on the interoceptive aspect of the intero-exteroceptive linkage.

is perceived and experienced as being completely impossible, resulting in the total loss of object relations that was already described, even such negative affect can no longer be assigned to the self. However, this means that the self is perceived and experienced as existing (if at all) in a negative mode (i.e. as "negative existence," as Benedetti calls it), which in turn results in what he describes as "inner death landscapes" (Benedetti, 1983).

How can the self react to the perception and experience of itself as fragmented, disintegrated, and incoherent? The answer is very simple. It can try to get rid of the fragments and parts of itself that are associated with negative affects. This it does by projecting those parts that pertain to either the self or the body to the environment. More specifically, it can project its own affects, cognitions, and eventually bodily parts onto objects within its external environment. Such projection[8] allows the psychotic patient to attribute his own predominantly negative affects, cognitions, and bodily parts to objects outside his own self (e.g. people or events in the environment), by means of which he tries to distance his self from his own negative affects and their existential anxieties that he subjectively perceives and experiences. However, the negative affects are not lost by such projection, but are only shifted from the self to the object and more specifically to external people and events. This means that the psychotic patient cannot avoid encountering them again in the gestalt of delusions of persecution, delusions of alien control, hallucinations, and bodily or brain manipulation (Benedetti, 1983; Dümpelmann, 2009).[9]

By projecting its own affective, cognitive, and bodily states onto objects in the external environment, the psychotic self can avoid not only re-encountering its own affective states but also paying a high price itself. The psychotic self itself becomes empty and thus devoid of any content when projecting its own affective, cognitive, and bodily states outside itself. Metaphorically speaking, projection means that the self itself becomes empty. How does such emptiness of the self manifest itself in subjective perception and experience of the self? If the self empties itself, there is no longer anything that the psychotic patient can associate with his self when perceiving and experiencing himself. His existence is not only negative but also empty or non-existent in his subjective perception and experience of himself. This may be the moment when the psychotic patient feels urged to take on the existence or identity of another person, resulting in identity disturbance.

However, the new identity of the self may still be subjectively perceived and experienced as insufficient, as the new identity is still not yet completely identical in a one-to-one way with the old one. There may therefore be a split within the self between its "old" and "new" identities and thus between "normal" and "abnormal" parts, as was nicely described by Freud himself in his later work: "Even in a state so far removed from the reality of the external world as one of hallucinatory confusion, one learns from patients after their recovery that at the time, in some corner of their mind (as they put it), there was a normal person hidden who, like a detached spectator, watched the hubbub of illness go past him. . . . In all psychosis—even in the most extreme cases—two mental attitudes are to be found: . . . the normal one, which takes account of reality, and another one, which under the influence of the instincts, detaches the ego from reality" (Freud, 1940, pp. 201–202).

---

8 One should be careful, though. Due to the reasons discussed below (e.g. the breakdown of defense mechanisms as such), this is, strictly speaking, no projection in the proper sense of the term (e.g. as defense mechanisms). One may rather describe it as "concretization, magic or condensation," to use the words of M. Dümpelmann.

9 Due to their strong affective assignment, the projected parts may be prone to be introjected again and thus be related to the self, where they are then again projected to objects in the external environment, and so on. Thus there may be what Volkan called an "introjective–projective cycle" or relatedness that describes all inner and outer flow through the outer ego boundary (Volkan, 1976, 1981).

Subjective and objective self and body

The central role of disturbances of the self in psychosis was pointed out by early psychiatrists such as Kraepelin and Bleuler. More specifically, Kraepelin (1913, p. 668) characterized schizophrenia as "the peculiar destruction of the inner coherence of the personality" with a "disunity of consciousness" ("orchestra without conductor"). Bleuler (1911, p. 58) also pointed out that schizophrenia is a "disorder of the personality by splitting, dissociation" where the "I is never completely intact." Joseph Berze, a contemporary of Bleuler and Kraepelin, even referred to schizophrenia as "basic alteration of self-consciousness" (Berze, 1914). Jaspers (1962b, p. 581) also noticed "incoherence, dissociation, fragmenting of consciousness, intrapsychic ataxia, weakness of apperception, insufficiency of psychic activity and disturbance of association, etc." as being the basic unifying "central factor" in schizophrenia.

These early descriptions of a disrupted self are complemented by current phenomenological accounts that focus predominantly on the subjective experience and perception of the self, thus pertaining to what is also called the "subjective self" (Rosenbaum and Harder, 2007, pp. 16–18). One focus is how psychotic patients experience and perceive their own body. Describing a specific case, Parnas (2003, pp. 226–228) points out the tendency of schizophrenic patients no longer to experience their own body as subjective (i.e. what the phenomenologists such as Merleau-Ponty (1965) call "lived body"), but rather as objective and merely physical body. The schizophrenic patient feels detached and alienated from his own body, and he can no longer make sense of the body and his perception of it. This may result in the early stages in the subjective experience of bodily incoherence, with parts of the body being experienced as disconnected, alienated, and ultimately no longer belonging to the person or self, amounting to what Parnas (2003, pp. 227–228) calls a "disturbance in the subjectivity of the body."

Stanghellini (2009) describes the subjective experience of the body as mere object in terms of the concept of disembodiment. The schizophrenic patient no longer has access to his own body as lived body. Since the body as lived body mediates any meaningful experience of the self, the objects in the world and the others in the world, the schizophrenic patient no longer has any meaningful and significant experience of either. Disembodiment in schizophrenia thus leads to the disruption of any meaningful encounter with one's own self, other selves, and objects in the world. Stanghellini consequently regards the lack of subjective access to the body as lived body as crucial for the schizophrenic changes in self-awareness, object-awareness, and intersubjectivity.

The central role of the body in schizophrenia has also been postulated by the Swiss psychopathologist Scharfetter (1996, 2003). He considers the body, and the subjective experience of the body, to be crucial in constituting what he calls the "empirical ego," which he regards as the center of self-awareness or self-experience. Schizophrenic patients suffer from what he calls "lack of ego-vitality," namely decreased awareness and experience of being alive. This mainly refers to basic bodily functions such as a reduction in or absence of perception of breathing, heartbeat, etc. Another concept, namely ego activity, describes the subjective experience of the self as self-directing activity that governs motor actions and cognitive processes such as thoughts in an active and unifying way. Schizophrenic patients often feel alienated not only from their own body, which can be described as a lack of ego-vitality, but also from their actions, emotions, and thoughts, thus mirroring what Scharfetter describes as a lack of ego-activity.

Scharfetter also introduces other categories to describe self-awareness that go beyond the body. One of these is ego-consistency or ego-coherence, which refers to the experience of a united, coherent, and consistent being. Schizophrenic patients often experience themselves as dissolving and dissoluting, thus no longer being able to maintain a sense of a unified self in their experience and perception of themselves. A further category is ego-demarcation, namely the ability to distinguish oneself from other objects and people and to demarcate boundaries between one's own ego/self and other selves. Scharfetter strongly bases it upon the perception and experience of one's body, with for instance schizophrenic patients experiencing parts of their body as being outside their self. Finally, Scharfetter also introduces the category of ego-identity, which describes the sense of remaining one and the same person across time, including its physical and mental changes.

These aspects of the self may be subsumed under the concept of the "objective self" as used by Rosenbaum and Harder (2007, p. 18). They characterize the objective self by functions such as self-agency, self-coherence, self-affectivity, and self-history, all of which may be disturbed in psychosis. Interestingly, they assume that the disturbances in the various functions of the objective self and the subjective self (and the self-objects) can be traced back to "difficulties distinguishing boundaries of the inside/outside-structure and self/other-structure concerning perceptions, feelings, thoughts and actions" (Rosenbaum and Harder, 2007, p. 22).

Taken together, both psychopathological and phenomenological accounts emphasize the central role of disturbances in the self and its body in psychosis. This clearly converges with the psychodynamic account of lacking self-objects, such as the body as one's own body, which may lead to the kind of subjective experiences and perceptions that are described in the phenomenological accounts. Moreover, the phenomenological description of a disrupted subjective and objective self may correspond closely to the psychodynamic description of the fragmentation of the self and its negative existence. More specifically, what is described phenomenologically as "altered subjective self" may correspond to what is conceptualized psychodynamically as "negative existence," while the psychodynamic counterpart of the "objective self" within the phenomenological/psychopathological context may be found in the fragmented, disintegrated, and incoherent self or ego.

Finally, there appears to be almost one-to-one correspondence between phenomenological and psychodynamic accounts with regard to inner and outer ego boundaries. Phenomenological accounts speak of (the subjective experience and perception of) attunement between the inner mental world of the self and the outer physical world (i.e. the "inside/outside-structure") (Rosenbaum and Harder, 2007, p. 22), which seems to correspond closely to what is psychodynamically conceptualized as the outer ego boundary, namely the boundary between the mental self and the physical world. Meanwhile, what is described psychodynamically as the inner ego boundary, namely the boundary between self and object within one's mental representations, may on the phenomenological side (more or less) correspond to the boundary within consciousness between self and other (i.e. the "self/other-structure") (Rosenbaum and Harder, 2007, p. 22).

"Existential dilemma" and self–object differentiation

In his early writings, Freud (1911) coupled the loss of (mental and social) object relatedness with the regression to an autoerotic stage of earlier development, a hitherto suppressed and hidden homoerotic homosexual stage. He illustrates this by the case of Schreber, where he assumes the reactualization of earlier homosexual attraction which is now no longer directed towards his father but the treating physician (e.g. Flechsig). Later, however, after developing his structural model, Freud (1924a, b) revised his view. While his early view may be considered to presuppose a deficit model, he now shifts the view of psychosis as conflict into the centre when he regards psychosis as conflict between the ego and the external world. The conflict between the ego and the external world distinguishes psychosis from neurosis as characterized by a conflict between the id and the ego. Freud's assumption of a conflict between the ego and the external world raises the question of how the constitution and construction of the self are affected by the lack of investment in the constitution and construction of objects/self-objects[10] of either social or mental origin.

The withdrawal of energy from the constitution and construction of objects/self-objects raises the question of what happens to that energy. I described earlier (see Chapter 6) how the constitution and construction of objects and self-objects are associated with the construction and constitution of the self. I henceforth spoke of "co-constitution and co-construction of self and objects" that allows for a balance between internalization/introjection and externalization/projection and thus between "giving out" and "taking in." Mentzos (1991) assumes that such a balance between internalization and externalization (see Chapter 6) involves a fundamental bipolarity between self- and object-related tendencies, or between subject and object constitution.

The constitution and construction of the self remove resources and energy for the constitution and construction of objects/self-objects, and vice versa. On the other hand, both self and objects/self-objects are mutually or bilaterally dependent on each other in their respective constitution and construction. The constitution and construction of the self are dependent on the constitution of objects/self-objects in the same way as the constitution of the latter is dependent upon the former. This results in what I earlier called "co-constitution and co-construction of self and objects/self-objects" (see Chapter 6).

Due to the fact that self and objects are constituted and constructed in a mutually interdependent way, both may therefore be regarded as twin siblings. Changes in this balance with extreme shifts away from one of the twin siblings may then lead to grave consequences for the other one, and vice versa. If, for instance, the energy is no longer invested in the constitution and construction of objects and self-objects, its investment must be shifted towards its sibling or counterpart, namely the self. The lack of investment of energy in the constitution and construction of objects of either mental or social origin may affect the constitution and construction of the self in two ways. First, the energy investment will be redirected from the object/self-object to the self. The balance is thus shifted unilaterally from one of the twin siblings (the objects/self-objects) towards the other one (the self). This leads to the denial of any kind of object/self-object and ultimately of reality altogether. If any kind of objects/self-objects and reality are denied, there is no longer any need to invest energy in them, which symptomatically may result in total withdrawal from reality and consequently in autism.

Secondly, since the constitution and construction of the self is dependent on that of objects/self-objects (and vice versa), the lack of energy investment in the latter also affects the former. This means that despite the abnormally high energy investment in the constitution and construction of the self, the self will nevertheless remain unstable and fragile (see next section for details). To stabilize his self, the psychotic patient has no choice but to direct his energy investment into the constitution and construction of objects which, however, as indicated is identical with the constitution and construction of his own self. The patient may therefore have no choice but to undergo what Mentzos (1991) describes as "fusion" of his self with the objects, which symptomatically results in the perception and experience of persecuting objects.

Following Mentzos (1991, 1992), the rather complex relationship between self and objects/self-objects confronts the psychotic patient with what Mentzos calls an "existential dilemma." The psychotic patient can either opt to direct his energy completely into the constitution and construction of his self, although this means that he can no longer invest any energy in the constitution and construction of objects/self-objects. However, this means that even if he invests energy in his self it will nevertheless remain unstable and fragile due to the lack of objects/self-objects.

---

10 I again use both terms, "objects" and "self-objects," in parallel, while being aware of their distinction.

In other words, the psychotic patient has the option of investing energy in his own self, but that means that he will have neither a stable self nor an existing external world. Thus the energy investment in the self is possible only at the expense of the existence of an external world (in his mental states and representations and thus as subjectively perceived and experienced).[11]

Alternatively, the psychotic patient can direct his energy completely into the constitution and construction of objects/self-objects, although this means that he can no longer invest energy in the constitution and construction of his self. However, this means that the objects/self-objects will nevertheless remain unstable and fragile due to the lack of a self. In other words, the psychotic patient has the option of investing his energy in the objects/self-objects, but that means that he will have neither stable objects/self-objects nor a self. Thus the energy investment in objects/self-objects is possible only at the expense of the existence of his own self (in his mental states and representations and therefore as subjectively perceived and experienced).

Taken together, this makes it clear that whether the psychotic patient chooses to invest energy in his own self or in objects/self-objects, this is only possible at the expense of the respective other, thus resulting in a dilemma (the need to choose between two undesirable options). By choosing one of these options the psychotic patient aims to repair the deficits in the other option, and vice versa. However, he is bound to fail because for such repair to be possible he would need to choose the other option, which would lead him back to the same dilemma. Due to the mutual dependence in the constitution and construction of self and object/self-objects, the patient has no choice but to stabilize at least one of them (i.e. self or object/self-objects) by investing energy in the other, and vice versa.

The dilemma in psychosis is not about the choice between two undesirable contents or objects, as at the neurotic level, but rather about the choice between the (mental) constitution and construction of the self and that of (mental) objects and reality. The existential dilemma is thus not about a conflict between two different objects and their relation to the self, as in the case of neurosis, but more basically (and hence existentially) about the self and its existence as distinct from objects, and vice versa. Psychosis is thus about conflict with regard to contents but about existence of the self as distinct from mere objects/self-objects. Mentzos (1991) characterizes this existential dilemma as a conflict in a wider sense, an existential conflict that concerns the lack of differentiation between self and objects. He distinguishes such conflict in a wider sense from a more narrow meaning of the concept of conflict, pertaining to drives and motivation as in neurosis.[12] Clinically, the distinction between wider and narrow meanings of the concept of conflict is thus supposed to correspond to the distinction between psychosis and neurosis.

---

11 What Mentzos describes as an existential dilemma within the existential context, Burnham (1969) puts into the psychological context where it resurfaces as the "need–fear dilemma." The need–fear dilemma describes the need of the psychotic patient for object relationships which he lacks. At the same time, though, he fears nothing more than those object relationships, because they threaten the existence of his own self, so he is caught in a dilemma with regard to object relationships. As Dümpelmann (2004, p. 149) says, "The more object relationships are needed, the less object relationships can be tolerated. And this is why psychotic patients construct and constitute themselves a different, foreign and private world." In other words, it is so that they have some substitute object relationships but not real objects.

12 Phenomenologically and thus in subjective experience and perception, the existential dilemma may be traced back to disruption of a basic relation between self and other. If the subjective experience of the other is ingrained and infused by the self and its subjective experience of the world, the other becomes automatically related and linked to the self, resulting in what Parnas (2003, p. 233) calls "basic relation." Such "basic relation" has phenomenologically been described by the concept of "fungierende or operative intentionality" (Merleau-Ponty, 1965), which refers to the pre-reflective non-objective tacit relation to

"Existential dilemma" and compensatory mechanisms

The existential dilemma in psychosis points out the crucial relevance of compensatory mechanisms. While any balanced constitution of self and objects/self-objects fails, the redirection of energy investment towards the constitution and construction of an auxiliary or delusional self or objects/self-objects may be considered a compensatory attempt by the psychotic patient to overcome his deficit in the constitution and construction of real objects. And such compensatory but ultimately flawed attempts may well account for the symptoms observed in psychosis, as has already been observed by Freud himself in his case of Schreber: "The delusional formation, which we take to be the pathological product, is in reality an attempt at recovery, a process of reconstruction. Such a reconstruction after the catastrophe is successful to a greater or lesser extent, but never wholly so: in Schreber's words, there has been a 'profound internal change' in the world. But the human subject has recaptured a relation, and often a very intense one, to the people and things in the world, even though the relation is a hostile one now, where formerly it was hopefully affectionate" (Freud, 1911, p. 74).

Subsequent authors writing on schizophrenia, including Bleuler, Benedetti, Scharfetter, and Mentzos, followed this line of thought and also regarded the psychotic symptoms as compensatory attempts to restructure and reorganize the constitution and construction of self and objects/self-objects, including self–object differentiation (for a good overview, see Hartwich, 2006, p. 172). In a psychodynamic context, these compensatory mechanisms cannot be regarded as defense mechanisms in the proper sense of the term, meaning that they are mechanisms to defend "something," be it a self or objects/self-objects. The problem is that the "something" (the objects/self-objects or the self) is almost completely broken down and disrupted, so that there is no longer anything at all that can be defended in psychosis.

The recruitment of defense mechanisms presupposes the existence of a certain psychological structure and organization, such as the differentiation between self and objects/self-objects (see also Dümpelmann, 2004; Hartwich, 2006, pp. 173–175). However, precisely these structural and organizational differentiations collapse in the acute psychosis, and the compensatory attempts are aimed at re-establishing the psychological structure, including the differentiation between self and objects/self-objects. Most importantly, the lack of structural differentiation between self and object means that the mechanisms that can be recruited to restore (and defend, in a wider sense) that very differentiation are also not clearly structured and organized by themselves. In other words, the defense mechanisms themselves, whose development from immaturity to maturity is strongly dependent on the constitution of self and objects/self-objects and thus their structural differentiation, are broken down in psychosis (or at least the more mature ones are). The psychotic patient thus has no choice but to revert to the very early defense mechanisms of internalization and externalization, such as introjection and projection.

However, the situation may get worse. Even the distinction between internalization and externalization may break down, which means that the two may strongly overlap or even be identical to each other. This means that when recruiting compensatory mechanisms to restore self–object differentiation and their co-constitution and co-construction, the psychotic patient can only rely

the world. Operative intentionality allows for the co-occurrence of both object and self in subjective experience, which makes our automatic pre-reflective attunement to the world first and foremost possible and ultimately what is described as "common sense." And, as we saw, this common sense is disturbed in psychosis, which has also been described in terms of the concept of autism, which describes the psychotic patient's lack of vital (i.e. subjective experiential and perceptual) contact with the world (Parnas, 2003).

on non-specific mechanisms without clearly distinguishing between, for instance, introjection and projection. Due to a lack of psychological structure and organization, the rather amorphous compensatory mechanisms in psychosis do not yet reach the level of defense mechanisms in a proper or narrow sense, and may thus be conceptualized as either "paraconstruction" or defense mechanisms in a wider sense (Hartwich, 2006, pp. 173–176).

In summary, the bipolarity and mutual dependence of the constitution and construction of objects/self-objects and self involves what can be described as an "existential dilemma" in psychosis. The psychotic patient is confronted with the choice of either restoring his self at the expense of objects/self-objects, or of restoring the latter at the expense of the former. However, since the constitution and construction of self and objects/self-objects are mutually dependent on each other, neither redirection of energy investment will work (i.e. establish and stabilize the mental representation of either the self or the objects/self-objects). Since this dilemma is not about the relationship of the self to different objects/self-objects and their respective contents, as in neurosis, but about the very (mental) existence of the self as distinct from mere objects/self-objects, the dilemma is existential. The symptoms in psychosis may thus be regarded as compensatory attempts to resolve this existential dilemma, for which the psychotic patient cannot even recruit proper defense mechanisms because of the breakdown of his own psychological structure and organization.

Volatile self–object boundaries and early traumatic experiences

I characterized the acute psychosis as lack of energy investment in object/self-object relations, which was supposed to be related to unstable ego boundaries that could be described phenomenologically by the concept of attunement. A more detailed examination demonstrated that psychotic patients are confronted with an existential dilemma in that constitution and construction of their self is possible only at the expense of that of objects/self-objects, and vice versa. This puts them in a situation where they have to choose between either the existence of their own self without any objects/self-objects, or having relations to objects/self-objects without any self.

Due to the mutual dependence of the constitution and construction of the self on the constitution of objects/self-objects, and vice versa, the dilemma becomes existential. The psychotic patient is not only forced to make a choice between two alternatives, self or objects/self-objects, but by making such a choice he is also confronted with the threat of the non-existence of his own self if he opts for the constitution of objects/self-objects. Since there is no escape from this existential dilemma, any compensatory attempt must necessarily lead to some perceptual, experiential. and behavioral adaptations outside the common and usual realm. While psychodynamically the existential dilemma may be traced back to the lack of relationship between self and object/self-objects, it may correspond to what phenomenologically has been described as autism, namely the absence of any relatedness to the world, including not only other people but also the person affected and thus their own self.

Further considerations clearly showed that psychotic patients suffer from a deficit in constituting and constructing truly self-stabilizing self-objects (i.e. "good self-objects" as distinct from "bad self-objects") as both bridge and boundary between self and objects. Due to the lack of self-object constitution, the psychotic patient is confronted with freely floating predominantly negative affects, especially existential anxieties, which he is now forced to assign to either self or objects. This leads to what Benedetti described as the "negative existence" of the psychotic self which, due to the lack of emotional feelings towards objects (including self-objects), becomes incoherent, disintegrated, and fragmented. I complemented the psychodynamic account here by phenomenological and psychopathological accounts that described the disturbance of the self in psychosis in its various symptomatic facets.

This brief summary raises the question of the origin of the lack of relationship between self and objects which lies at the core of the existential dilemma and the subsequent fragmentation of the self. The concept of ego boundaries provides one clue in this direction. The psychotic patient is no longer able to experience self-objects as stabilizing for the own self and remains therefore unable to distinguish them from both self and mere objects in his subjective experience, the psychotic patient's inner mental ego boundaries between (his mental representations of both) self and objects become volatile and increasingly fragmented. This in turn means that his outer physical ego boundaries between his mental world (including both his self and objects/self-objects) and the outer physical world (of mere objects and events) becomes fragmented as well (in his subjective perception and experience). One may consequently diagnose a disturbance of inner and outer ego boundaries in psychosis, which may be specified as unstable or volatile self–object boundaries. What does that imply for the self-object boundaries? The concept of object is here to be understood in a wide sense that includes both objects (in a narrow sense as distinguished from self-objects) and self-objects; this means that the self-object boundaries include both inner and outer self-object boundaries as described above. In the following account I shall use the term "self–object boundaries" in this double sense, including both objects (in a narrow sense) and self-objects as well as inner and outer boundaries, unless otherwise specified.

What is the origin of such volatile self–object boundaries? Mentzos (1991) points out that the psychotic patient's early relationships (e.g. in infancy and childhood) do show unstable representations of both self and objects/self-objects. Their object (and self-object) relations remain volatile, which puts the people at risk for later psychosis in an ambivalent and uncertain situation. On the one hand they feel urged to search for object (and self-object) relations in order to stabilize the constitution and construction of their own self. On the other hand they are afraid of any kind of object (and self-object) relation, because it may threaten their own self and its constitution as subjectively perceived and experienced. Mentzos (1991) therefore argues that what he described as the "existential dilemma" in acute psychosis may already be predisposed to by unstable self and object/self-object representations in the early infancy and childhood of these patients. The self–object boundary in both of its manifestations (inner and outer) may already be constituted as volatile and unstable in early childhood in those subjects who later become psychotic. In other words, the early constitution and construction of volatile and unstable self–object boundaries may predispose to and thus be what one could call a "psychodynamic risk factor" for the subsequent manifestation of psychosis in early adulthood.

Dümpelmann (2004, 2009) goes one step further and argues that the early traumatic experiences that led to the constitution and construction of such volatile self–object boundaries are often re-experienced and reactualized in the actual psychotic contents (although in a distorted form). For instance, Freud's famous case of Schreber suffered from many hypochondrial ideas (e.g. that his brain was softened, that his skin was cut off his skull, and that his body was being handled, manipulated, and changed for a holy purpose). He also had the feeling that he suffered from the plague, and ultimately he experienced and perceived himself as dead. In addition, Schreber showed religious delusions of a God and his radiations who would tear out the soul from himself and all other people. Dümpelmann (2009, 2010) points out that these psychotic contents are not pure coincidence because they represent a re-actualization of his early childhood traumata where his god-alike (subjectively) experienced and perceived father, who was a doctor, used to punish and correct him by special orthopedic manipulative (and rather torturing, as we would say today) devices he, i.e. the father, had developed in order to fix his son's head (for other case histories where early trauma is reactualized in the later psychotic content, see Dümpelmann, 2002, 2009).

Most importantly, it is not the application of the orthopedic manipulative tools that is relevant here, but the fact that they were applied by Schreber's father, a very close person, which is crucial in constituting and constructing stable self–object boundaries (Dümpelmann, 2009). If his own father as a highly relevant object (i.e. self-object) for Schreber's constitution of the self violates the boundaries of his son's self, this transgression of the outer self–object boundaries will be internalized by Schreber himself. The violation of the outer self–object boundary leads to the constitution and construction of a rather fragile and thus volatile inner self–object boundary. Such a volatile inner self–object boundary may then predispose the person concerned to react to subjectively analogous events in adulthood in the same way as they reacted in early childhood when the outer self–object boundary was violated. The contents in the existential dilemma in adulthood may thus resemble (symbolically) the early violation of the outer self–object boundary. Schreber, for instance, reactualizes his father's tormenting manipulations of his skull in the gestalt of a God who tears out his and others' souls, and as a result subjectively perceives and experiences his brain to be softening (Dümpelmann, 2009).

Volatile self–object boundaries and brain–object differentiation

Is there any predisposition to develop such volatile self–object boundaries in early childhood? There has been much biological and psychodynamic research on disturbing events in early infancy of psychotic patients. Various factors, such as viral infection, pregnancy and delivery complications, genetic abnormalities, and altered neuromotor and neurocognitive development have been discussed on the biological side (for an overview, see Mäki et al., 2005), while social factors such as an abnormal relationship to an over-suppressive and dominant mother, separation of parents with the absence of a triangulating father, sexual and non-sexual abuse have been raised on the psychodynamic side (Dümpelmann, 2002, 2004, 2009, 2010) (for a general overview, see Mäki et al., 2005).

Is psychosis ultimately a biological or a social disorder? Is it due to vulnerability as biologically caused or to abnormal social interaction? Dümpelmann (2004) points out that presenting such an alternative is misguided by describing the following case. There may, for instance, have been an increased perceptual sensibility of the infant that makes it experience the mother's touching as intrusive, painful, and threatening. This leads to the infant's withdrawal from the mother's touching, which in turn increases the mother's efforts to provide care for and touch her infant, resulting in even more withdrawal by the infant, and thereby inducing a vicious cycle. If it is ongoing and continuous, the infant will perceive and experience the mother's touch and care as a violation of the outer self–object boundary. This in turn may affect the constitution of its inner self–object boundary by becoming fragile and volatile.

If, for instance, a person other than the infant's own mother or father touches the infant in the same way (as the mother does), the infant may not perceive and experience this as painful and intrusive. This may be because the other person, being a less relevant object (and not an important self-object like the mother) may not induce the same anticipations or expectations in the infant. However, this may change if that other person takes over (socially) the role of the mother, and the real mother disappears. Thus, as in the case of Schreber, it is the relevance (i.e. self-relatedness) of the other person and their possible role as self-object (rather than mere object) for the infant that makes the difference to the latter's subjective perception and experience. More specifically, it is the association of the event with a specific and personally relevant person as self-object, rather than the event itself or the mere object, that transforms the event and its violation of the outer self–object boundary into a traumatic experience which then cannot do other than affect the constitution of the inner self–object boundary.

What is social and what is biological in this case? What is vulnerability and what is interaction? This example makes it clear that the biological and social aspects remain inseparable at this level. One therefore needs to assume an interaction model between biological and social factors to account for the constitution and construction of volatile self–object boundaries. This is, for instance, well reflected in the often cited Finnish epidemiological adoptee studies (Tienari, 1991; Wahlberg et al., 1997; Mäki et al., 2005). Those researchers investigated people with a genetic risk of developing psychosis/schizophrenia (i.e. increased incidence of psychosis/schizophrenia in the family) who were later adopted by families other than their biological ones. Most importantly, they also investigated the adoptive families themselves, visited their homes, and measured family dysfunction. Low levels of family dysfunction protected the adoptees from later psychosis, even though they were genetically vulnerable. However, genetically vulnerable adoptees who ended up in a highly dysfunctional adoptive family tended to develop psychosis subsequently. Yet adoptees with low genetic vulnerability did not tend to develop psychosis at all, even if they were raised in a highly dysfunctional adoptive family. These findings thus support the notion that there is an interactive effect of genetic risk and rearing environment in psychosis.

What are the implications of this for psychosis in general and for schizophrenia as a specific disorder? It means that schizophrenia as a disorder lies at the border between the biological and the social. It is neither purely biological nor purely social, but must apparently be "located" at exactly the point where the social world is constituted and constructed as distinct from the biological world. One may thus "locate" it at the point where the biological and social worlds differentiate from each other. I call this "biological–social differentiation."

What does this location at the biological–social crossroads mean for the psychodynamic context? Psychosis may be regarded as a truly psychosomatic disorder that can be traced back to an abnormal interaction between psychological and somatic factors which lead to volatile self–object boundaries (see, for instance, Volkan 1976, 1981). More specifically, psychosis (or schizophrenia) may be considered a psychosomatic disorder of the brain, as Mentzos (1991) has stated, where psychological and neuronal (i.e. somatic) processes converge and are simultaneously differentiated from each other. What does the description of psychosis as psychosomatic disorder of the brain mean for its psychodynamic characterization? First, taking a psychoanalytic perspective, it means that psychoanalysis may need to look toward neuroscience to better understand the mechanisms and processes underlying the constitution and construction of volatile self–object boundaries. This in turn means that the investigation of psychosis requires convergence and integration between psychoanalysis and neuroscience, and thus neuropsychoanalysis.

Secondly, now taking a neuroscientific perspective, it means that we need to better understand those neuronal mechanisms that lead to the differentiation between brain and object/self-objects (i.e. brain–object differentiation), as well as to the differentiation between brain and self (i.e. brain–self differentiation). Only if we gain some insight into these mechanisms can we better understand what goes wrong, and why, in the constitution of self and objects (i.e. self–object differentiation) in psychosis. Thus we have to focus on neuronal mechanisms such as difference-based coding and associated rest–stimulus and stimulus–rest interaction, intero-exteroceptive linkage, and trilateral interaction, as discussed in Parts II and III of this book.

What exactly do I mean by brain–object differentiation? The concept of brain–object differentiation describes those conditions that first and foremost make the constitution and construction of objects as distinct from the person's own brain (and body) possible. Thus neuroscience may reveal those neuronal mechanisms and processes in the brain itself that are necessary (but not sufficient by themselves) to enable and predispose to the differentiation between brain (body) as merely physical and objects as mental. I demonstrated that difference-based coding and subsequent rest–stimulus interaction enable and predispose to such brain–object differentiation.

Taken within the psychodynamic context, brain–object differentiation is assumed to correspond to externalization and more specifically to projection within the psychodynamic context (see Chapter 6). As such brain–object differentiation pertains to the boundary between the mental world, including both self and objects/self-objects (in a psychodynamic and thus mental sense),

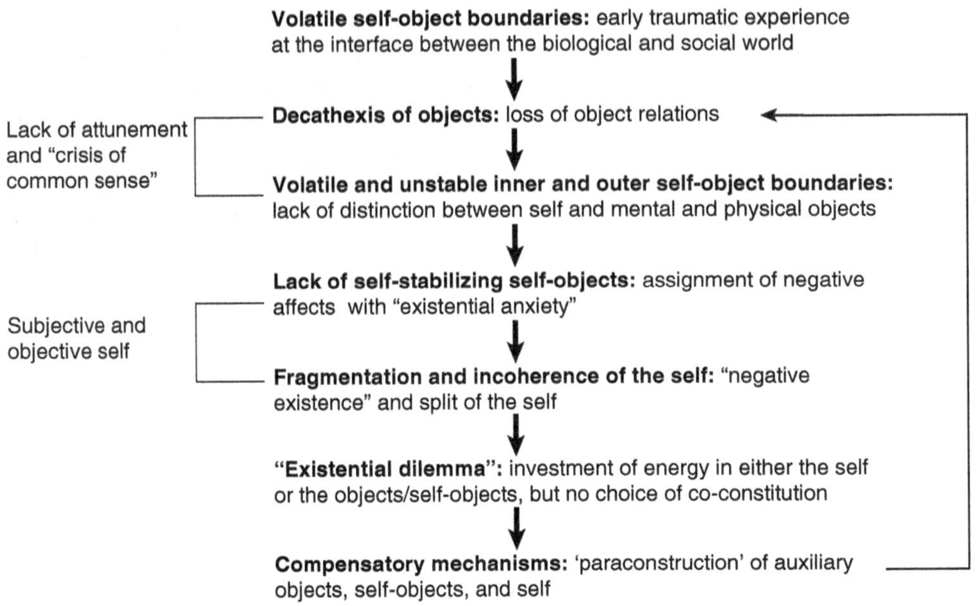

Fig. 11.1 Psychodynamic features in psychosis.

and the physical world, including objects, events, etc., and thus objects in a physical sense. In other words, brain–object differentiation may enable and predispose to the constitution and construction of the outer self–object boundary. This in turn, (see Chapter 6), also influences the constitution of the inner self–object boundary as based on the differentiation of the self from both objects (i.e. self–object differentiation) and the person's own brain (i.e. self–brain differentiation).

How does this relate to psychosis? As we have seen, the outer self–object boundary may be violated in psychosis, which in turn leads to a volatile inner self–object boundary. And we have seen that the disturbance of the outer self–object boundary may be related to biological–social interaction (or biological–social differentiation as I called it), which may be traced back to changes in brain–object differentiation. Taken together, this means that volatile inner and outer self–object boundaries in psychosis may be traced back to changes in brain–object differentiation and its impact on brain–self differentiation. Abnormal brain–object and brain–self differentiation may then provide access to a better understanding of the processes underlying the abnormal constitution of self and objects in psychosis (i.e. self–object differentiation as described in this chapter) (for an overview, see Figure 11.1). Therefore we have no choice but to turn our attention to possible neuroscientific and neuropsychodynamic mechanisms if we want to better understand the abnormal self–object differentiation in psychosis, including its alteration of the different psychic boundaries (i.e. inner and outer self–object boundaries) (see also Figure 11.1).

Acknowledgments

I wish to thank Michael Dümpelmann for his wonderful introduction to and teaching about the psychodynamics of psychosis/schizophrenia, which was essential not only for writing this chapter but also for a collaborative brain imaging study on the neural effects of psychodynamic psychotherapy in schizophrenia. He also read previous drafts of this and the next chapter on psychosis, and I am grateful to him for making some important comments.

Chapter 12

Psychosis[1] II: Neuropsychodynamic Hypotheses[2]

At this stage, I suggest a specific neuronal and neuropsychodynamic hypothesis in psychosis. I assume that difference-based coding itself is altered in psychosis. Difference-based coding is still considered to function in psychosis, although in an abnormal way. Based on the detailed characterization of difference-based coding in the healthy brain (see Parts II and III), I now assume specific changes in the respective neuronal mechanisms which I postulate can be traced back to changes in difference-based coding. I therefore speak of an "altered difference-based coding" hypothesis in psychosis (ADC). Analogous to the case of depression, the ADC presupposes psychosis as a system disorder that as such is supposed to affect various if not all regions and neural networks in the brain and all related psychological systems underlying the different affective and cognitive functions. At this time I shall focus on the neuronal mechanisms that enable and predispose the system to be dysfunctional, the "neural predisposition" as I called it earlier. The search for "neural predisposition" as necessary neural conditions that enable and predispose to psychosis as a system disorder must be distinguished from the sufficient conditions and thus the "neural correlates" of the various psychotic symptoms. Thus my focus here is more on what must happen prior to the outbreak of psychotic symptoms, rather than on what occurs in the brain during the occurrence of psychotic symptoms.

Loss of object relations and altered neural processing in the sensory cortex

Neuropsychodynamic hypothesis

The first psychodynamic key feature in psychosis is the loss of object relations which Freud described as the withdrawal of energy from objects, or "decathexis of objects" (see Part I). Thus the concept of object refers to objects in both the social world (i.e. the outside physical world, including other people) and the mental world (i.e. the inner mental world of the subject as characterized by their mental representations). As described in Part I, cathexis of social and mental

1 As in the previous chapter, since my focus is primarily on the symptoms rather than a specific etiology, I here prefer to speak of psychosis rather than schizophrenia.

2 There is an enormous abundance of different types of data available on psychosis. Needless to say, I shall not be able to refer to all of the data. I will therefore focus on the main issues relevant here and will draw upon the respective data. Moreover, I shall not be able to discuss all of the controversial findings in detail, and will therefore rely on reviews and summaries of the main findings in the field. I will fall short especially with regard to the details of the vast number of studies on cognitive functions, predominantly due to the fact that I here advocate a sensory-based bottom-up approach to psychosis rather than the still predominant cognitive-based top-down approach. Some of these theories will be discussed in the footnotes, which obviously cannot do justice to all of the details in these accounts.

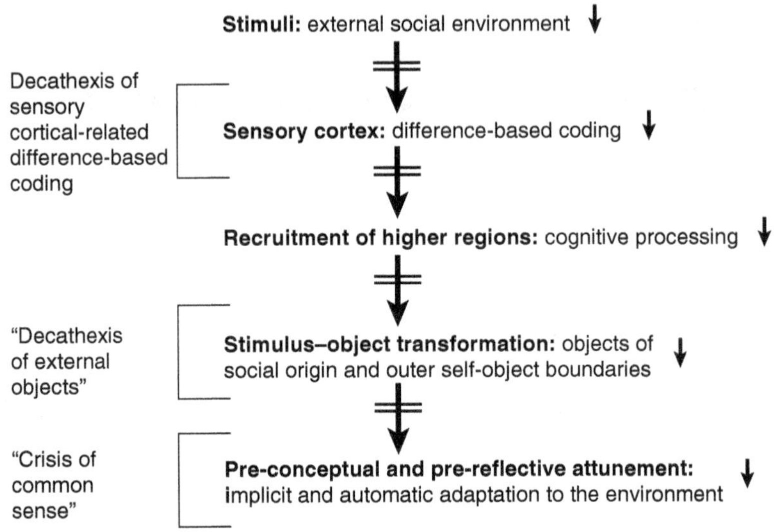

Fig. 12.1 Altered sensory processing and loss of object relations.

objects is associated with the constitution of outer and inner ego boundaries or, as I call them, "self–object boundaries."

Based on these distinctions, my first neuropsychodynamic hypothesis about the loss of object relations will include two aspects. The first aspect of the hypothesis (Ia) concerns the loss of object relations with regard to social objects, which I assume to be related to altered neural processing in the sensory cortex (see Figure 12.1). The second aspect of the hypothesis (Ib) focuses on the loss of object relations with regard to mental objects that accompanies disruption of the inner self–object boundaries which I postulate to correspond to reduced stimulus–rest interaction in the sensory cortex. I shall here focus on the first hypothesis (Ia), while the second one (Ib) will be discussed in the next section.

How do social objects affect the brain? They come in the gestalt of what the neuroscientist calls stimuli that are associated with a specific event, thing, person, etc. as subsumed here under the concept of the social object. If there is a loss of social object relations, one would postulate that the (exteroceptive) stimuli associated with the social object are no longer properly processed in the brain, as for instance in the sensory cortex (e.g. the auditory or visual cortex). If this is so, one would expect alterations in the sensory cortex in psychosis. Interestingly, there are indeed many empirical findings of altered neural processing of sensory information in psychosis (for excellent overviews, see Javitt, 2009). Due to the prominence of auditory hallucinations, the focus has been mainly on the auditory cortex, where early electrophysiological potentials have been investigated.

Empirical findings

One such early electrophysiological potential related to sensory processing in the auditory cortex is the mismatch negativity (MMN). The MMN is an electrophysiological potential that can be measured when an oddball or deviant auditory stimulus occurs embedded in a stream of familiar or standard auditory stimuli. The MMN can be measured in both EEG and MEG as a negative waveform that results from subtracting the event-related response to the standard event from the response to the deviant event. Elicited by sudden changes in auditory stimuli, the MMN occurs about 100–250 ms after the onset of the deviant stimulus, and is strongest over the frontal and temporal regions. Although the MMN is primarily an auditory potential (for a review, see Naatanen et al., 2007), it is matter of debate whether potentials analogous to the MMN also occur in other sensory modalities (e.g. in visual and somatosensory modalities) (for a review, see Garrido *et al.*, 2009).

In psychosis, numerous studies have demonstrated changes and deficits in the MMN (for reviews, see Garrido *et al.*, 2009; Javitt, 2009). More specifically, these studies have demonstrated a reduction in the amplitude of the MMN in psychotic patients for both frequency and duration deviants. Moreover, certain amplitudes of the MMN even correlate with disease severity and cognitive dysfunction in these patients, further highlighting the apparently crucial relevance of the MMN as a marker of altered auditory processing in psychosis.

In addition to the MMN, other markers of early sensory processing in the auditory cortex have been observed to be altered in psychosis. These include reductions in the amplitude of early electrophysiological potentials such as P50 and N100, which are elicited by simple repetitive stimuli. This distinguishes them from the MMN, which is induced by a deviant stimulus after a series of repetitive stimuli. Both potentials (P50 and N100) are assumed to be generated in the auditory cortex, including both primary and secondary auditory regions (see Javitt, 2009 and Turetsky *et al.*, 2007 for recent overviews, as well as Tregellas et al., 2009 for a recent fMRI study). Taken together, these findings point to deficits in the early stages of auditory sensory processing, when the stimulus starts to be evaluated.

What do these deficits in early electrophysiological potentials mean in the psychological and functional context? Let us go back to the MMN. Psychologically, the MMN has been associated with implicit and thus automatic processing, as it occurs independent of and thus prior to attention (Garrido *et al.*, 2009). For instance, the MMN is induced when subjects do not pay any attention at all to the stimuli, be they standard or deviant, or when they perform a task that is completely unrelated to the stimuli. Thus pre-attentive cognitive processes that allow for the detection of the deviant stimulus have been assumed to underlie the MMN. The independence of the MMN from attention is further highlighted by its occurrence in sleep and even in comatose patients in vegetative states who have no consciousness (Qin *et al.*, 2010). This strongly indicates that the MMN not only occurs prior to attention but also remains completely independent of consciousness. Thus psychologically the MMN appears to mirror the early stages of sensory processing prior to attention and consciousness, and it is these early implicit and automatic processes that seem to be disrupted in psychosis.

Functionally, the MMN results from the mismatch between the deviant (i.e. present) and standard (i.e. previous) auditory stimuli, hence the name "mismatch negativity." This implies, most importantly, that the MMN must be considered a potential that results from a difference between different stimuli (i.e. deviant and repetitive stimuli), rather than from a single stimulus alone. And it is this difference that induces an electrophysiological potential, namely the MMN. The MMN may consequently be considered a nice example of the neural coding of stimuli in terms of differences between different stimuli rather than in terms of the isolated stimulus itself (see Chapter 4). Hence the fact that the amplitude of the MMN is reduced in psychosis tells us that these patients are apparently no longer able to generate proper neural differences between different stimuli involving altered difference-based coding.

Let us now be more specific. How exactly is the difference between deviant and repetitive stimuli generated in the MMN? Neural coding of the difference between the deviant and repetitive stimuli presupposes first inhibition of any response to the repetitive stimulus and secondly identification of the deviant stimulus as deviant (Turetsky et al., 2007). While the inhibition of repetitive stimuli is mirrored by early potentials such as the P50 and N100 (and the prepulse inhibition), detection of a deviant stimulus is related to the MMN and even later potentials such as the P300. The reductions in the various electrophysiological potentials previously described lend evidence to the view that psychosis can be characterized by failures in both inhibition of repetitive stimuli and detection of deviant stimuli in early automatic sensory processing (Turetsky et al., 2007; Javitt, 2009). The deficits in both inhibition and detection may lead to reduced neuronal differences in early processing in the sensory cortex in psychosis. This leads me to assume that the automatic generation of early neural differences between different stimuli in terms of difference-based coding may be reduced and thus abnormal in psychosis.

These changes in the difference-based coding of early automatic sensory processing are not limited to the auditory cortex, but are also observed in other sensory systems such as the visual cortex. The amplitudes in early visual electrophysiological potentials such as the steady-state visual-evoked and auditory-evoked potentials (ssVEP, ssAEP), the N100 and the P100 have been found to be reduced in psychotic patients (Javitt, 2009). At the same time, low- and high-frequency visual stimuli induced significantly lower neural activity in primary and secondary visual cortex in psychotic patients when compared with healthy subjects (Martinez et al., 2008). Taken together, these findings indicate deficits in early automatic sensory processing of visual stimuli that, as I assume analogously to the auditory cortex, may eventually be traced back to the inability to generate proper neural differences in the gestalt of difference-based coding. Since the other sensory systems (olfactory, somatosensory, and gustatory) also show physiological and phenomenological abnormalities (Javitt, 2009), one may assume that there is a general alteration in early automatic processing of sensory stimuli and thus of difference-based coding in the sensory cortex of patients with psychosis.

Neuropsychodynamic account

What do these alterations in the difference-based coding of early automatic sensory processing entail for the constitution and construction of objects?[3] How are they related to what I described as the loss of object relations? Freud assumed that energy is withdrawn from the constitution and construction of objects, which he described as the "decathexis of objects." How can we relate such withdrawal of energy to the neuronal level?

As we have seen, psychotic patients are apparently no longer able to generate proper neural differences during early automatic sensory processing. More specifically, they seem to remain unable to properly suppress or inhibit neural activity changes to repetitive stimuli and to detect deviant stimuli, which taken together results in an inability to generate proper neural differences between different stimuli. However, if the neural differences are reduced, subsequent stimulus–object transformation as based on difference-based coding and subsequent rest–stimulus interaction are impeded, which means that the incoming exteroceptive stimuli are no longer transformed into objects. Thus reduced neural differences in difference-based coding involve decreased stimulus–object transformation (see Chapters 5 and 6 for details of stimulus–object transformation).

How is the neural processing of reduced neural differences and the subsequently decreased stimulus–object transformation related to the energy (i.e. cathexis)? Both inhibition of repetitive stimuli and detection of deviant stimuli require energy, as is well reflected in the induction of

[3] I here explicitly focus on objects rather than self-objects. I postulate that the constitution of what are called objects within the psychodynamic context is closely related to early sensory processing, while their transformation into possible self-objects may be more related to affective assignment and intero-exteroceptive linkage in those regions where the exteroceptive stimuli are processed subsequently to the sensory cortex (see also Chapter 1).

corresponding electrophysiological potentials (i.e. P50, N100, and MMN). The reduction of amplitudes in these early potentials indicates that the psychotic patients no longer invest as much energy in the early automatic processing of sensory stimuli. In other words, the energy investment is withdrawn from the early automatic sensory processing stages in the sensory cortex, so that one may speak of "decathexis of sensory processing."

However, the question now arises as to whether the assumed "decathexis of sensory processing" corresponds in the neural context to what Freud described as the "decathexis of objects" in the psychodynamic context. Let me go into some detail here. The "decathexis of sensory processing" refers to the neural processing of different stimuli (i.e. repetitive and deviant) and their relation to each other in terms of difference-based coding. One may consequently specify our concept as "decathexis of sensory cortical-related difference-based coding." The decreased investment of neural energy in the coding of early neural differences in the sensory cortex may lead to decreased stimulus–object transformation, with the subsequent loss of object relations as has been so well described by Freud. Based on these considerations, I postulate that the "decathexis of sensory cortical-related difference-based coding" makes the constitution of objects impossible, which in turn leads to the aforementioned loss of object relations as "decathexis of objects."

If this holds true, one would expect, at the empirical level, that the deficits in early sensory processing of stimuli affect later more associative (and cognitive) stages where the different stimuli are fused and synthesized (and thus transformed, as I call it) into an object. Javitt (2009) does indeed show that the deficits in early sensory processing have major implications for later processing, or "upward consequences" as he calls them. The early auditory processing deficits affect subsequent phonetic processing with auditory comprehension, emotional/prosodic and social processing of auditory processing with interpretation and recognition of tones, and ultimately auditory attention (Javitt, 2009), while early visual deficits may affect subsequent motion detection, perceptual openness/closure, and face recognition/reading. These functions that are related to the "upward consequences" are disturbed in psychosis and may therefore be related and traced back to the deficits in early sensory processing rather than to higher-order cognitive deficits in, for instance, executive functions, verbal learning, or working memory (Javitt, 2009).

What does this mean psychodynamically? Our hypothesis that "decathexis of sensory cortical-related difference-based coding" is related to the "decathexis of objects" may be supported by the data and thus empirically plausible. However, caution is needed. First, I do not argue that "decathexis of sensory cortical-related difference-based coding" is the neural correlate and thus the sufficient condition of the "decathexis of objects." Instead, I only argue that the former, "decathexis of sensory cortical-related difference-based coding" is a necessary, enabling and predispositional condition for the latter to be possible and thus a necessary but not sufficient condition. Secondly, this is assumed to hold only for what I called "social objects" (i.e. those objects that can be traced back directly to the external environment and thus the physical world), while it is not assumed to hold for those objects that are mentally represented, namely mental objects. Thus the current hypothesis is supposed to hold for social objects and therefore for what I described as the outer self–object boundary (i.e. outer ego boundary), while it does not apply to mental objects and the inner self–object boundary.

How is our hypothesis related to the previously described phenomenological features? I characterized the early sensory processing by difference-based coding, meaning that the relation between different stimuli rather than each stimulus individually is coded. This orientation on the relation between stimuli involves on a larger scale an adaptive or better attuning function. The neural activity to the one stimulus is attuned to the one related to the previous stimulus, and so on. Difference-based coding may consequently correspond well to what phenomenologically is described as "attunement." And the alleged alteration in difference-based coding, with smaller differences in psychosis, may then well correspond to the phenomenological diagnosis of a disturbance in attunement.

Let us be more specific. Phenomenological description characterized the attunement as pre-reflective and pre-conceptual. By "pre-reflective" it is meant that it is not yet accessible in reflective consciousness (but, if at all, only in phenomenal consciousness), and the term "pre-conceptual" means that it is prior to concepts and the verbal level. Both features, pre-reflective and pre-conceptual, are highly compatible with the implicit and automatic nature of difference-based coding in early sensory processing that occurs prior to any consciousness and verbalization. As such, the deficits in early sensory processing may also correspond to the phenomenological description of the "crisis of common sense," in that psychotic patients do seem to remain unable to account for sensory-based common-sense assumptions.

Taken together, I postulate that what is psychodynamically described as loss of object relations (i.e. the "decathexis of objects") may be related in the neural context to the loss of energy investment in early sensory cortical processing (i.e. the "decathexis of sensory cortical-related difference-based coding"). Phenomenologically, the "decathexis of sensory cortical-related difference-based coding" may correspond to the loss of pre-reflective and pre-conceptual attunement and the "crisis of common sense."

Loss of inner self–object boundaries and abnormal rest–rest interaction in the sensory cortex

Neuropsychodynamic hypothesis

I have so far developed a neuropsychodynamic hypothesis about the social objects and the outer self–object boundary, while I have ignored mental objects and the inner self–object boundary. These are the focus of this section. I postulate that the loss of mental objects and inner self–object boundaries corresponds to reduced rest–stimulus interaction in the sensory cortex. Since one of the most prominent and best researched symptoms in psychosis is auditory verbal hallucinations, I shall focus on these in order to illustrate the here postulated neuropsychodynamic hypothesis (see Figure 12.2). In auditory verbal hallucinations (AVH), psychotic patients often hear other people's voices speaking. AVH can therefore be defined as the perceptual experience of voices experienced as true perception in the absence of any external stimulus.

Psychodynamically, one may characterize AVH as representation of an object in one's own mental states (i.e. a mental object) that most notably is not directly related to the external environment and thus to what I called social object. AVH thus reflect a change in the inner self–object boundary. Usually the inner self–object boundary is drawn between the self on the one hand, which is presumed to have no direct connection at all to the external physical environment, and the mental representations of objects on the other, which in some way or other can be traced back or related to the external physical environment. This seems to be no longer the case in AVH. Here the inner mental boundary between self and objects appears to be disrupted, as otherwise the psychotic patient could not "locate" a purely inner mental object as exclusively related to his self and its mental states, the AVH, within the outer physical environment.[4] Thus the

[4] One may want to argue that AVH still show a connection to the external physical world in that they represent the voices of other people. However, this concerns only the content of the voices, not their origin, and it is the latter (and not the former) that is crucial in defining the object as object in the case of the inner mental self–object boundary.

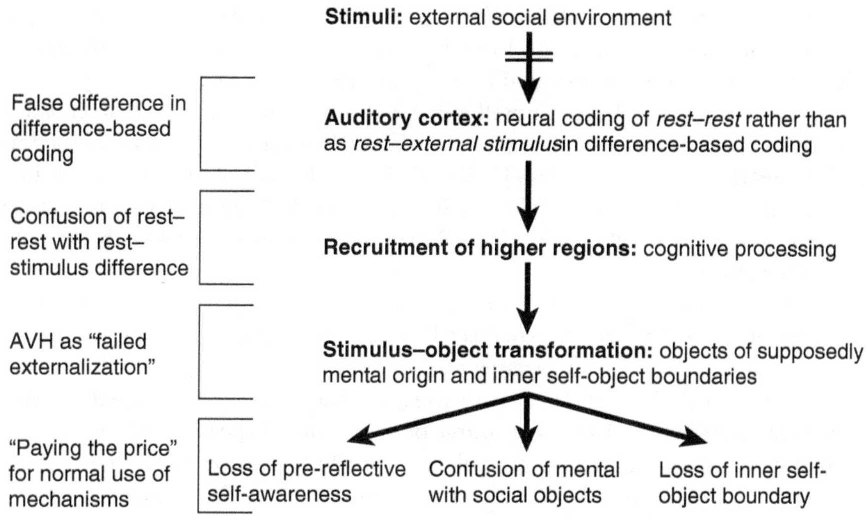

Fig. 12.2 Abnormal sensory rest–rest interaction and loss of inner self–object boundaries in auditory verbal hallucinations (AVH).

case of AVH can be regarded as an illustrative example of the loss of inner self–object boundary. Having clarified the psychodynamic ground, I now turn my attention to the recent empirical findings in AVH.

Empirical findings

Recent brain imaging studies have explored the structural and functional neural underpinnings of AVH (for an excellent review, see Allen et al., 2008). Structural studies observed altered (most often reduced, rather than increased) gray matter reduction in the superior temporal gyrus (STG), the planum temporale and Heschl's gyrus (and other regions such as the lateral prefrontal cortex and the thalamus) in psychotic patients with AVH when compared with those without AVH (Shin et al., 2005; Lin et al., 2006; Allen et al., 2008). Since these regions include both the primary and secondary auditory cortex, these findings are consistent with lesion studies that showed alterations in sensory regions in hallucinations (Braun et al., 2003).

In addition to the structural studies, there have been many functional studies which tested for neural activity in psychotic patients with AVH during some task, such as inner speech (i.e. imagining speaking particular sentences) or verbal imagery (i.e. imagining sentences spoken in another person's voice). Patients with AVH showed reduced activity in various regions involved in the inner monitoring of speech, such as the middle temporal gyrus (MTG), the thalamus, the parahippocampal gyrus, the left or right inferior frontal cortex (including Broca's regions on the left), the parietal cortex (including Wernicke's area) and the supplementary motor area (SMA) during verbal imagery tasks in which spoken sentences of other people are imagined (or word generation tasks) (McGuire et al., 1996a; Stephane et al., 2001; Shergill et al., 2003, 2004; Hoffman et al., 2007; Kumari et al., 2008; Sommer et al., 2008; Raij et al., 2009; Wible et al., 2009). These regions are involved in language processing, and more specifically in monitoring the generation of verbal utterances (i.e. verbal self-monitoring). Since studies in healthy subjects have also implicated these regions in verbal self-monitoring (McGuire et al., 1996b; Fu et al., 2006), a reduction in activity in these regions suggests that there is a deficit in the monitoring of inner speech (i.e. verbal self-monitoring) in AVH.

These results have been taken as support for the theory of impaired self-monitoring in psychosis (Frith and Done, 1988; Frith et al., 1992; for a recent account in neurobiological terms, see Stephan et al., 2009). Due to impaired inner monitoring of his own cognitive processes, the psychotic patient is not aware that he himself initiated the thoughts and the verbal sentences, and assumes therefore that someone else from the external world makes his thoughts and actions and monitors his intentions. In the case of AVH, the deficit in verbal self-monitoring of the internally generated speech cannot be recognized as such (i.e. as internally generated), and is therefore misattributed to some external person (Frith, 1992, p. 73; Blakemore and Frith, 2003; Fu and McGuire, 2003, p. 426).[5]

However, the theory of impaired inner speech monitoring (i.e. verbal self-monitoring) has aroused some criticism, as it neither explains why one experiences a particular external source during AVH nor accounts for the involvement of other regions, including the auditory cortex, in AVH (Jeannerod, 1999; Pacherie et al., 2006; Allen et al., 2007a,b, 2008). Using a speech-monitoring task in which subjects had to decide the source of pre-recorded speech, Allen et al. (2007a,b) observed that patients with AVH were more likely to attribute their own speech externally than patients without AVH. Such misidentification of self-generated speech in patients with AVH was accompanied by altered activity in the STG and the anterior cingulate cortex (ACC) (Allen et al., 2007a,b, 2008; Mechelli et al., 2007).

Based on their own and other researchers' results, Allen et al. (2008) put forward a more sophisticated theory of AVH, thereby integrating the assumption of impaired self-monitoring into a wider context. The frequently reported observation of spontaneous hyperactivity of the secondary auditory cortex in particular during AVH (in the absence of any external stimulation and thus during the resting state) may prime the subject to perceive complex auditory objects such as words or sentences in a more intense way. Since the secondary auditory cortex may be involved in object perception as distinct from perception of single aspects such as tones as processed in the primary auditory cortex, resting-state hyperactivity in this region indicates what Allen et al. (2008) call "over-perceptualization." There may thus be abnormal or increased bottom-up modulation from the auditory cortex to the other cortical regions, which allows the subject to experience and perceive their own internal auditory activity in a more vivid way.

Abnormally strong bottom-up modulation from the auditory cortex may be accompanied by reduced top-down modulation from the cortical regions involved in speech and language processing (Allen et al., 2008). These include Broca's area and Wernicke's area, underlying the generation and perception of speech, as well as regions such as the ACC, the SMA, and the DLPFC, which are involved in monitoring speech. Thus Allen et al. (2008) regard the altered balance between reduced top-down modulation and increased bottom-up modulation as crucial in generating AVH (they also include regions implicated in emotion processing in their model; this will be discussed in the next section, on interoceptive processing) (see also Figures 12.3a and 12.3b). However, the origins of abnormal hyperactivity in the auditory cortex remain unclear.

Early functional studies observed an association of hallucinations/delusions as reality distortions with increased resting-state activity (e.g. metabolism or perfusion) in the superior or middle temporal lobe that includes the auditory cortex (Liddle, 1992; Gur et al., 1995). Comparing the

[5] Frith argues that such a deficit in self-monitoring can explain a particular group of psychotic symptoms, namely those that are characterized by confusion between self and others. These symptoms include auditory hallucinations as well as delusions of control and the so-called passivity phenomena, in which the patient's own actions, emotions, and thoughts are experienced as estranged from and not belonging to himself, as if they are made by someone else (see also Schneider, 1959 who subsumed them under the concept of first-rank symptoms).

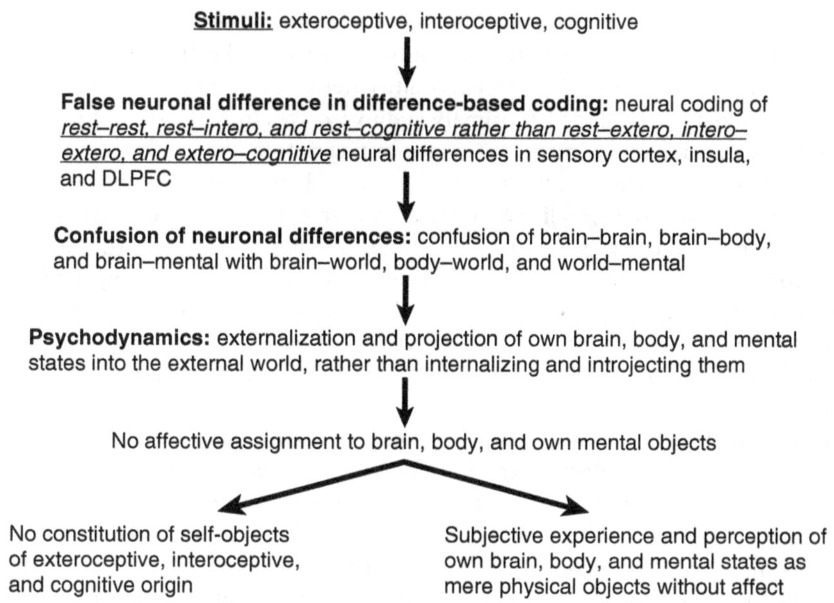

Fig. 12.3a False neuronal differences in difference-based coding and the loss of self-objects.

resting-state activity during AVH with that during the absence of AVH within the same patients, Dierks et al. (1999) observed hallucination-related activity in Broca's area, the temporal gyrus and the primary auditory cortex (see also Lennox et al., 1999). Taken together, these findings lend support to the assumption that the auditory cortex may show increased resting-state activity in the acute psychotic state when the patient is suffering from auditory hallucinations.

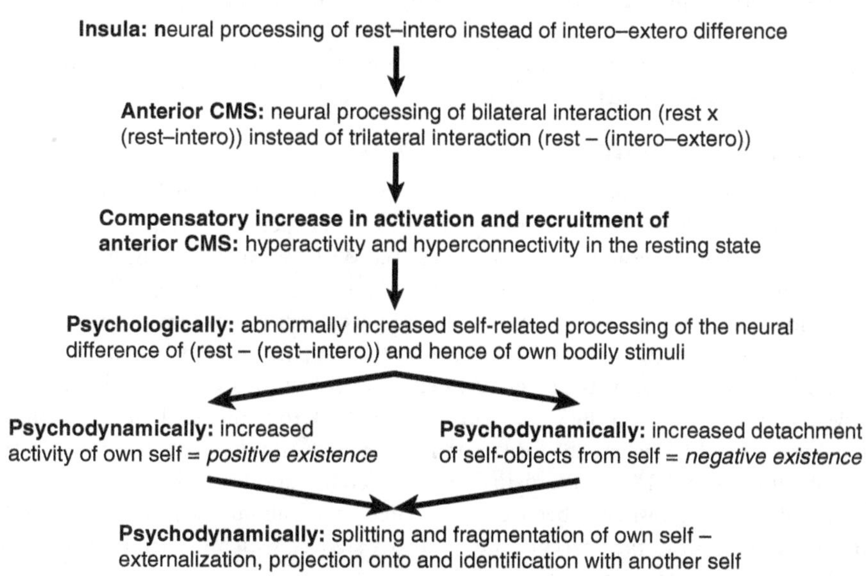

Fig. 12.3b Fragmentation of the self and bilateral neural interaction in anterior cortical midline structures.

The assumption of abnormal endogenous activity in the auditory cortex during AVH (for an investigation of resting-state activity in auditory cortex in healthy subjects, see Hunter et al., 2006) is further supported by a recent multicenter study. Ford et al. (2009) reported less activation to external tones in the left primary auditory cortex in AVH, suggesting that the primary auditory cortex is already "turned on" in the resting state by showing increased endogenous activity. This in turn leads to what the authors describe as being "tuned in," namely the orientation of perception towards internally generated activity rather than towards externally generated activity (i.e. stimulus-induced activity). More support for abnormal resting-state activity comes from electrophysiological studies in psychotic patients. These demonstrate impaired generation of oscillatory activity in the gamma frequency (i.e. at 40 Hz),in the resting state (and during stimulus-induced activity) in the auditory and visual cortex when measuring so-called steady-state evoked potentials (SSEPs) that reflect intrinsic oscillatory activity (for a review, see Uhlhaas and Singer, 2010).

What does the finding of a seemingly increased resting-state activity in the auditory cortex imply for rest–stimulus interaction? The study by Ford et al. (2009) suggests that there is decreased rest–stimulus interaction in the auditory cortex in psychosis. This means that the external stimulus elicits less (or no) activity in the auditory cortex when compared with its resting-state activity level. Such reduced rest–(external) stimulus interaction no longer leads to the generation of a neural difference between resting-state activity level and stimulus-induced activity. This may also account physiologically for the aforementioned reductions in the amplitudes of early electrophysiological potentials during stimulus-induced activity.

The assumption of reduced rest–stimulus interaction is empirically supported by the observation that psychotic patients show reduced amplitudes and reduced phase synchronization in the beta- and gamma-frequency bands during stimulus-induced activity in the auditory and visual cortex (for a review, see Uhlhaas and Singer, 2010). Since phase synchronization in particular is supposed to account for the integration of different neural activities (i.e. neural differences) within a local cortical network (Uhlhaas and Singer, 2010), impaired phase synchronization could indicate that the stimulus itself is not properly processed in relation (i.e. in terms of differences) to other stimuli (e.g. neural stimuli from the brain's intrinsic activity in the auditory cortex involving reduced rest–stimulus interaction).

Neuroscientific account

What does the reduced presence or even complete absence of rest–external stimulus interaction in the auditory cortex imply for subjective perception and experience? Why does the psychotic patient nevertheless perceive and experience voices (i.e. AVH) even though the external auditory stimuli may no longer be processed properly? What is the source of the stimuli that the patient perceives and experiences as voices in AVH? As previously indicated, many cognitively oriented theories assume that the stimuli in AVH come from the patient's own cognitions that are misattributed and then false-positively perceived as external voices. More specifically, the theory of predictive coding assumes that the voices can be traced back to a mismatch between anticipated and actual stimuli in AVH where the former, the anticipated stimulus is confused with a real and thus actual stimulus. AVH (and also delusions; see Fletcher and Frith, 2009) in psychosis may then be traced back to a mismatch between anticipated beliefs about a sensory input and the real auditory–sensory input in the perception of auditory signals, which in turn leads to abnormal hyperactivity in the auditory cortex as already described.

How does my hypothesis of altered difference-based coding in psychosis stand in relation to that of predictive coding? Let us change perspective from that of the observer and his cognitive account in terms of predictive coding[6] to that of the brain itself. What must it be like for the brain itself to show higher resting-state activity in the auditory cortex?[7] The brain itself does not distinguish between resting-state activity and stimulus-induced activity as we as observers do. Instead, all that the brain processes are changes in neuronal activity, which are usually associated with changes in its relation to the external physical environment and thus with what we as observers call stimuli. The brain may therefore register that there are changes in the neural activity of the auditory cortex which then, as it usually does, associates with external stimuli from the physical environment. From the perspective of the brain, metaphorically speaking, there is thus some rest–stimulus interaction, although not with external stimuli but with internal stimuli as we as observers call them. Therefore, from the perspective of the brain, there is rest–internal stimulus interaction which we from the observer's perspective describe neurally as an increase in resting-state activity and thus as rest–rest interaction and, phenomenologically, as AVH.[8]

---

6 More specifically a disturbance in what is called corollary discharge or predictive coding is assumed in psychosis. A self-generated movement is associated with signals to the sensory cortex from motor regions informing it about the intended movement. This in turn allows the sensory regions to predict the possibly incoming sensory signals as the sensory consequences of that movement. This signal has been called "corollary discharge" or "efference copy" (Blakemore et al., 1998; Stephan et al., 2009). which allows sensory and motor processing to be coordinated in a fine-tuned way. In psychosis, many authors assume that there is disconnection between sensory and motor processing and thus abnormal corollary discharge, with a resulting mismatch between the predicted or anticipated auditory consequences of self-generated speech and the actual auditory experience (Feinberg and Guazzelli, 1999; Ford et al., 2002, 2007, 2008; Bennett, 2008).

The assumption of abnormal corollary discharge may be considered a specific case of the more general assumption of altered predictive coding (Frith et al., 1992; Friston, 2005; Stephan et al., 2009). Predictive coding provides the general framework for minimizing predictive errors when comparing the predicted or anticipated signal with the actual incoming stimulus. Imagine the case of an abnormal balance between anticipated predictions and actual sensory input with, for instance, too much emphasis on the anticipated input, such as the expectation that a particular person will speak. The actual incoming sensory input is then perceived and experienced only in the light of the anticipated prediction (i.e. the predicted source and thus the speaker), while leaving no room for the actual source or speaker, which is then no longer perceived as such (Friston, 2005; Stephan et al. 2009; see also Fletcher and Frith, 2009 for a detailed application of the theory of predictive coding to hallucinations and delusions in psychosis).

7 I am well aware that I here confuse the brain with a person by attributing abilities to the brain (e.g. having a perspective, what it must be like for the brain, etc.), thus committing what Bennett and Hacker (2003) called "mereological fallacy" (see Part I of this book for details). However, this is meant only metaphorically, not literally, in order to illustrate my point about shifting perspectives.

8 The reader may wonder how my account stands in relation to the theory of predictive coding that has recently also been applied to psychosis and more specifically to positive symptoms such as delusions and hallucinations (see footnote 5 as well as Fletcher and Frith, 2009, for a psychodynamic criticism). As I discussed (see Chapter 4), the theory of predictive coding focuses on stimulus–stimulus interaction, and specifically on the interaction between cognitive and sensory stimuli, and thereby considers at best only the activation baseline while ignoring rest–rest, rest–stimulus and stimulus–rest interaction as well as what I called the neural and arousal baseline. As such it focuses only on anticipation and prediction and thus on the executing (or sufficient) conditions of predictive coding, while it ignores the enabling or predisposing conditions that first and foremost generate anticipations and predictions (i.e. difference-based coding and rest–stimulus and stimulus–rest interaction). My concept of difference-based coding can thus be seen to provide a larger framework and can therefore easily accommodate predictive coding, which may then be regarded as a special instance of the former with regard to stimulus–stimulus interaction. My account of AVH in particular and of psychosis in general focuses on these enabling and predisposing conditions

What does the replacement of rest–external stimulus interaction by rest–internal stimulus interaction imply for difference-based coding? The brain still generates a neural difference. However, rather than coding the neural difference between the resting-state activity and the external stimulus-induced activity, the brain now codes the neural difference between the previous and the current resting-state activity level (i.e. before and after the elevation of its resting-state activity). Thus the neural difference is no longer between resting-state activity and stimulus-induced activity but between two different levels of resting-state activity.

Since the brain has already generated a neural difference based on the two different resting-state activity levels, it no longer reacts to external stimulus-induced activity, hence the reduction in the amplitudes of early electrophysiological potentials. This means that difference-based coding is functioning, but only that it is redirected from internal–external differences within the auditory cortex itself to internal–internal differences. The neural shift from internal–external to internal–internal neuronal differences in difference-based coding implies that what we call "rest–external stimulus interaction" is replaced by what we may better call "rest–rest interaction." One may consequently postulate an abnormal increase in rest–rest interaction and decreased rest–stimulus interaction in the auditory cortex of patients with psychosis.

Is there any empirical evidence to support such increased rest–rest interaction in psychosis? If there is indeed increased rest–rest interaction in psychosis, one would expect increased amplitudes and phase synchronization in the beta and gamma bands that account for the integration of different neural activities within a local cortical network, such as the auditory or visual cortex. Interestingly, this is exactly what several studies have found, when they reported that positive symptoms and in particular auditory and visual hallucinations correlate with enhanced amplitudes and phase synchronization of beta- and gamma-band resting-state activity in psychotic patients (for a review, see Uhlhaas and Singer, 2010).

What is described electrophysiologically as enhanced amplitude and phase synchronization may then correspond at the functional level to the neural coding of increased neuronal differences in the intrinsic oscillations or fluctuations within the resting-state activity itself. Instead of integrating resting-state activity and external stimulus-induced activity mirroring rest–stimulus interaction, phase synchronization in the auditory cortex (and other sensory cortices) now integrates the neural difference between two different resting-state activity levels, leading to what I described as increased rest–rest interaction.

However, this leaves open the question of how such abnormal rest–rest interaction transforms into the above-postulated rest–internal (i.e. mental) stimulus interaction and thus the subjective perception and experience of auditory stimuli and hence voices? The answer is very simple. Since the neural differences in the auditory cortex are usually associated with internal–external differences rather than with internal–internal differences, the brain takes the latter to be an instance of the former. In other words, it confuses its current neuronal difference (the internal–internal difference) with the usual and prior difference (the internal–external difference), which makes the respective subject prone to assume that an auditory stimulus is "located" within the external environment, resulting in what we (as observers) call AVH. In other words, metaphorically speaking the brain confuses mere rest–rest interaction with rest–external stimulus interaction and therefore induces all of the processes that it usually sets in motion when processing external

(i.e. difference-based coding as well as rest–rest, rest–stimulus, and stimulus–rest interaction), which may then generate the kind of stimulus–stimulus interactions that have been so well described in the theory of altered predictive coding in psychosis (Fletcher and Frith, 2009). Therefore I consider the two accounts to be complementary rather than contradictory.

(i.e. physical and exteroceptive) stimuli. Rest–rest interaction is consequently treated as if it were rest–stimulus interaction, and may therefore trigger the reactivation or read-out of stored experiences as internal or mental representations of voices that have been encountered in the past.

Let me be more specific and postulate how such read-out of stored experiences of voices from the past may be generated. Since inner speech in healthy subjects is not associated with neural activity in the primary auditory cortex (Aleman et al., 2005), abnormally high resting-state activity in this region suggests that it may be perceived and experienced as if there is a real external sound. This in turn may prompt the respective subject to infer an external perceptual source, thereby turning the perception of their own resting-state activity (i.e. rest–rest interaction) into what we call hallucinations (McGuire et al., 1996a; Dierks et al., 1999; Frith, 1999). Activity changes 6–9 seconds before the onset of AVH were observed in the left inferior frontal cortex, the ACC, and the right middle temporal gyrus, while during the AVH itself activity was present in the bilateral temporal gyrus (and the insula) (Shergill et al., 2000a,b, 2004; van de Ven et al., 2005; Hoffman et al., 2007, 2008). There is also ample evidence for altered (i.e. reduced) functional/effective connectivity from prefrontal regions like the DLPFC, the ACC and the left inferior frontal cortex to regions in the temporal cortex, including the auditory cortex, during verbal self-monitoring (Mechelli et al., 2007) or AVH itself (Hoffman et al., 2007; Raij et al., 2009). This is further corroborated by the observation of abnormal structural fronto-temporal connectivity in Diffusion Tensor Imaging (DTI) in psychotic patients with AVH (Hubl et al., 2004; Shergill et al., 2007).

These data enable me to refine my hypothesis of altered difference-based coding in psychosis (ADC) by assuming a two-stage process. First, I postulate abnormally elevated resting-state activity in the auditory cortex with subsequent abnormal rest–rest interaction in this region. While elevated resting-state activity in this region is supported by the data, abnormally increased rest–rest interaction has yet to be demonstrated. The assumption of abnormally increased rest–rest interaction entails reduced rest–external stimulus interaction during, for instance, verbal stimuli. This has indeed been postulated (see, for example, Allen et al., 2008), although it has not yet been experimentally demonstrated.

Secondly, I postulate that the brain may confuse abnormal rest–rest interaction with rest–external stimulus interaction, thereby inferring externally located voices with subsequent AVH as the second stage, which in turn leads to the voices heard in AVH. This inference of external voices may be neurally related to subsequent stages of auditory processing (i.e. "upward consequences," as I called them earlier), which empirically may be supported by the abnormalities in the higher cortical regions observed during AVH. In terms of neural timing, one would thus expect abnormal rest–rest interaction to precede the involvement of higher cortical regions, which in turn may occur prior to the onset of AVH, which is at least in part in accordance with the findings already described.

Finally, one may want to raise the question of the origin of the abnormally increased rest–rest interaction in the auditory cortex and, more specifically, why there is such abnormally elevated resting-state activity in the auditory cortex, as for instance during (or before) AVH. While this issue is crucial to understanding the causal chains, I leave it open here, and raise it later in the context of the other sections.

Neuropsychodynamic account

What does my ACD hypothesis about AVH imply for the neuropsychodynamic context? What psychodynamically is described in AVH as the loss of the inner boundary of the self with objects that have their ultimate origin in the external physical world may neuronally correspond to abnormal rest–rest interaction and its subsequent confusion with rest–external stimulus interaction in the auditory cortex. Why, though, does the patient hear external voices in AVH?

Psychodynamically this may be because they associate a neural difference in the auditory cortex coded in rest–rest interaction with previously heard auditory objects encountered in their past. Based on the inference and recruitment of their previous mental auditory object representations of verbal utterances from the past, the patient anticipates and hears verbal voices and thus what we as observers describe as AVH.[9] And in the same way that an external stimulus is anticipated in the case of rest–stimulus interaction, the neural difference as coded in rest–rest interaction is also anticipated as a possible external stimulus and thus as what Arieti described as "listening attitude" (Hoffman, 2010).

Let me recount in neuropsychodynamic terms what may happen in AVH. The patient applies the same kind of neural mechanisms that he always relies on, namely difference-based coding of changes in neural activity in the auditory cortex and subsequent processing in higher regions to infer the kind of object related to the stimulus that induced the neural activity changes in the auditory cortex. Both mechanisms, difference-based coding and subsequent processing in higher regions, normally allow for the location of the object associated with the respective stimuli in the external physical environment, and thus for what psychodynamically is called externalization. The psychotic patient does both difference-based coding in the auditory cortex and subsequent processing in higher regions, but fails nevertheless. This is because the auditory cortical neural difference that he codes in terms of difference-based coding can no longer be traced back to the difference between rest and external stimulus, but to that between two different resting-state activity levels.

However, even though the neural difference stems from a different origin, this does not affect subsequent neural processing in the auditory cortex itself, which henceforth codes neural differences nevertheless and thereby induces difference-based coding and subsequent processing in higher regions. Therefore the same processes as in healthy subjects are set in motion but, due to the different origin of the auditory cortical neural difference, this does not lead to the same result. Rather than leading to real externalization of the respectively resulting objects as in healthy

---

9 The theory of impaired inner monitoring has been criticized for remaining unable to explain that AVH are usually associated with a particular external author (see Pacherie et al., 2006; Allen et al., 2007a,b, 2008). While it can explain that AVH are not associated with the affected person, as traced back to impaired inner self-monitoring, it cannot explain why AVH are associated with another person in general and a specific person in particular. This is so because the former does not imply the latter. Thus the attribution of the source to another particular person remains unexplained in impaired inner monitoring.

This is also the point at which the theory of predictive coding falls short. If the hypothesis of predictive coding claims that the anticipated prediction, the possible source of the anticipated auditory signal, is the determining factor, it must assume that it is supposed to be another person, and a specific person at that. The question though is why the psychotic patient assumes another particular person rather than their own person to be the anticipated source. If the imprint of the anticipation is too strong, one would rather assume that the person in question assumes him- or herself to be the source of the anticipated incoming auditory signal. This however is not so. AVH are rarely, if ever, related to the person'sown voice, and thus the person speaking to him- or herself, as if from their own first-person perspective to their own first-person perspective as a form of a dialogous monologue. Instead, AVH are usually attributed to another particular person and occur consecutively from second- or third-person perspective (Fu and McGuire, 2003, p. 427). This however remains unexplained in the hypothesis of altered predictive coding, which would at least be equally compatible with experience of AVH of the person him- or herself from first-person perspective. The neuroscientific and neuropsychodynamic hypotheses suggested here can thus go beyond both accounts, namely inner monitoring and predictive coding, by being able to account for association of voices with other persons and their appearance from second- or third-person perspective (rather than first-person perspective).

subjects, the psychotic patient's attempt to externalize the object fails because he cannot associate it with specific external stimuli that may have caused the neural difference in auditory cortex. One may consequently speak psychodynamically of what I call a "failed attempt at externalization." The concept of a "failed attempt at externalization" describes the attempt to externalize and thus locate a specific object in the environment.

As in the case of an external stimulus and subsequent rest–stimulus interaction, there is an attempt to externalize, i.e. project, the stimulus, i.e. object to the environment, but this fails because the object's origin is internal, i.e. the own mind, rather than the environment. I therefore postulate that AVH may result from the normal use of the seemingly well-functioning mechanisms of both difference-based coding and subsequent neural processing in higher regions in an abnormal neuronal context (i.e. rest–rest rather than rest–extero difference in the auditory cortex). Thus it is not that energy is not invested, but rather that it is invested in a false-positive or wrong neural difference in the auditory cortex. What Freud called the "decathexis of objects" may consequently be reformulated as a "failed cathexis of absent objects" or even better as "cathexis of false (neuronal) difference."

The psychotic patient pays a high price for his use of otherwise "normal" neuronal mechanisms, since his confusion of rest–rest interaction with rest–stimulus interaction entails the "failed cathexis of absent objects" as "cathexis of false difference." He loses (the representations of) his mental objects as he no longer subjectively perceives and experiences them as objects of mental origin but as objects of social origin. He thereby loses his inner purely mental self–object boundaries which he now subjectively perceives and experiences as outer rather than inner self–object boundaries across the mental–physical divide.

This leads to his ultimate denial of external physical reality, since he false-positively identifies his mental reality with the external physical reality, where they resurface in what we call AVH (because he constituted the former by the same neuronal mechanisms, i.e. the normal mechanisms, as the latter in a "normal" context). Such confusion between inner and outer self–object boundaries may correspond within the phenomenological context to the loss of mineness, presence, and self-affectivity by events and persons in the environment, thereby undermining and disrupting what the phenomenological account describes as pre-reflective self-awareness.

Lack of self-objects and confusion of neural differences in interoceptive, sensory, and cognitive regions

Neuropsychodynamic hypothesis

Besides the loss of object relations, another psychodynamic key feature in psychosis is the loss of self-objects, i.e. those self-objects that are stabilizing for the self and that are thus "good self-objects" as distinct from destabilizing self-objects or "bad self-objects"[10] (see also Chapter 11 for more details). Psychotic patients are no longer able to properly constitute and construct truly self-stabilizing self-objects, i.e. "good self-objects" (see Chapter 11), including both their own body as first self-object as well as non-bodily based self-objects. The constitution and construction of self-objects is supposed to be based upon affective assignment and intero-exteroceptive linkage, both of which are assumed to be mediated by neural activity in the interoceptive network

[10] I thank Michael Duempelmann for making me aware of the difference between "bad self-objects" and "non-existent self-objects" (see Chapter 11 for more detailed discussion).

with the insula as key region (see Chapter 7 about narcissism). This sets the stage for the second neuropsychodynamic hypothesis.

The general neuropsychodynamic hypothesis is that the lack of self–object constitution and construction in psychosis is related to altered neural coding of the relation between neural, interoceptive, and exteroceptive stimuli within the interoceptive network, including the insula as key region. More specifically I postulate that, analogous to the case of the auditory cortex, difference-based coding in the insula relies on coding a false-positive neuronal difference that confuses the neural difference between intero- and exteroceptive stimulus-induced activity with that between resting-state activity and interoceptive stimuli (see also Figure 12.3a). However, before considering possible abnormalities in difference-based coding, we need to gain some background information about the insula.

Empirical findings

Recent imaging studies using fMRI investigated neural activity during interoceptive stimulus processing such as evocation of blood pressure changes during isometric and mental tasks, heartbeat changes and perception, anticipatory skin conductance during gambling, and heart rate modulation during presentation of emotional faces (Craig, 2002, 2003; Critchley et al., 2005; Pollatos et al., 2007a,b,c). These studies observed neural activity changes in the right insula, the anterior cingulate cortex extending from supragenual to dorsal regions (SACC/DACC), and the amygdala. This led to the assumption that specifically the right insula and the SACC/DACC integrally represent autonomic and visceral responses that are transferred from the spinal cord through the midbrain, the hypothalamus and the thalamocortical pathway to the right insular cortex (Craig, 2003, 2003; Critchley et al., 2005).

Based on these results, the insula, especially the right anterior one, is assumed to be involved in re-presenting the autonomic and visceral state of the body and thus its interoceptive processing, which is then transferred to higher cortical centres like the prefrontal cortex (Craig, 2002, 2003, 2009). This is well in accordance with the connectivity pattern of the insula that receives much afference from lower vegetative centers and sends much efference to the prefrontal cortex and other cortical regions (Craig, 2003, 2003, 2009). This allows the insula to give rise to a "mental image of one's physical state" which, according to Craig (2009), provides the basis for subjective awareness of one's self as "material (or bodily) me."

In addition to interoceptive sensory afferences, the insula also receives many exteroceptive afferences, e.g. from the auditory, visual, olfactory, and gustatory cortices (Craig, 2002, 2009). One would therefore expect exteroceptive stimuli to also modulate neural activity within the insula, which has indeed been shown in recent studies (Critchley et al., 2005). More specifically, a recent study of ours (Wiebking et al., 2011) demonstrated that both intero- and exteroceptive stimuli, e.g. heart beat and tone awareness, induce neural activity changes in the insula. Most importantly, it was the neural difference between intero- and exteroceptive stimulus activity changes rather than interoceptively induced activity by itself that correlated with measures of stress-related behavior. This provides some empirical support for the hypothesis that the behaviorally relevant neural activity in the insula presupposes the coding of the neural difference between intero- and exteroceptive stimuli, the "intero–extero difference" as I shall call it in the following.

How does this relate to psychosis and its apparent changes in difference-based coding? We have already seen in the case of the auditory cortex that difference-based coding relies on the false difference in psychosis. Rather than coding the difference between exteroceptive stimuli and the resting-state activity, i.e. the rest–extero difference, difference-based coding was here based on coding the neural difference between two different resting-state activity levels, i.e. the rest–rest difference. How can we apply this to the insula? Here the case is a little more complicated because we are also confronted with interoceptive stimuli from the person's own body that are continuously coming in and interacting with the brain's neural stimuli from its resting-state activity and the exteroceptive stimuli. Thus, in addition to the rest–extero and possible rest–rest differences, we are also confronted with rest–intero and intero–extero differences.

What does this imply for the case of psychosis? Let me first consider the literature, which indeed shows some abnormalities in the insula in psychosis. Several investigations also report structural and functional changes in the insula as well as in other regions. Interestingly, investigations during AVH reveal abnormalities not only in the auditory cortex but also in the insula. Many studies report mostly increased activity in the insula, either before or during AVH (Dierks et al., 1999; Shergill et al., 2000a,b, 2004; Sommer et al., 2008). This suggests that auditory stimulus processing and stimulus-induced activity in the auditory cortex may also affect neural activity in more interoceptive regions like the insula.

Neuroscientific account

Based on these findings of abnormal insula activity in psychosis, I postulate the following. Analogous to the auditory cortex, difference-based coding in the insula may be based on a false difference. Rather than coding the intero–extero difference and taking it as the behaviorally relevant difference, behaviorally relevant neural activity changes in the insula in psychosis may rely on the rest–intero difference. Instead of the intero–extero difference, the rest–intero difference is now taken as the behaviorally relevant neural activity change. Difference-based coding is thus still at work and still functions normally but, as in the case of the auditory cortex, it is applied to the false difference, i.e. rest–intero instead of intero–extero.

What are the implications of this for subjective perception and experience of the body? This means that the intero–extero difference and its associated perception and experience of the body in relation to the environment are now replaced by the rest–intero difference, resulting in perception and experience of the brain in relation to the body. However, since due to autoepistemic limitation we cannot perceive and experience our own brain as brain (see the theoretical part), the now coded rest–intero difference is perceived and experienced as if it is the usual intero–extero difference. Interoceptive stimuli and hence our own body are then perceived and experienced as if they were exteroceptive stimuli and thus mere physical objects from the external environment, while the resting-state stimuli are associated with the body.

How does this relate to phenomenological descriptions? The reader may recall that the phenomenological account pointed out the loss of mineness of the body, with the psychotic patient no longer experiencing their own body as lived body but rather as merely objective body. The loss of subjective experience of the patient's own body as lived body corresponds nicely to what psychodynamnically is described as the body as self-object. I postulate that the reliance of the neural activity in the insula on a false neural difference, i.e. rest–intero rather than intero–extero difference, and the subsequent confusion of the former difference with the latter in subjective perception and experience may account for the loss of the lived body and the body as self-object. If the interoceptive stimuli are no longer associated with the body but with external physical objects, one of course no longer subjectively experiences one's own body as lived but as rather objective. Therefore I assume that the reliance on a false neural difference in the neural coding of neural activity in the insula may account for the phenomenological feature of the loss of the lived body.

Neuropsychodynamic account

How does this relate to the loss of the body as self-object as a psychodynamic key feature? Psychodynamically, this means that the interoceptive stimuli are externalized rather than internalized, and are thus, to be more specific, projected to the outer physical world rather than being introjected and linked to the inner mental self. Therefore interoceptive stimuli and hence the body have no chance of being constituted as self-stabilizing self-object and thus as what I called "good self-object." Instead, all efforts and attempts are directed towards their externalization, thus resembling what in the case of the auditory cortex I described as a "failed attempt at externalization." Unlike the case of the auditory cortex, the "failed attempt at externalization" no longer concerns absent exteroceptive stimuli but rather the continuously present interoceptive stimuli from the person's own body in the neural processing of the insula.

The insula's neural coding of the false neural difference, i.e. rest–intero instead of intero–extero, may enable and predispose to the confusion of the person's own brain and body in subjective perception and experience with a merely external, i.e. physical, object. This can for instance be nicely seen in the case of Schreber, who perceived and experienced his own brain and body as mere physical objects that as such were exposed to physical and tormenting influences of the environment. Thus the same mechanism, namely normally functioning difference-based coding within an abnormal neuronal context, is assumed to be at work in both the auditory cortex and the insula. The only difference may consist in the content that is externalized. Difference-based coding of rest–rest interaction in the auditory cortex leads to false-positive externalization of mental objects, while in the insula it entails false-positive externalization of one's own body.

I have so far focused on the changes in intero-exteroceptive linkage as one characteristic feature in the constitution of self-objects, while neglecting the other feature of self-objects, namely affective assignment (see Chapter 7 on narcissism). The insula is a region that is assumed to be crucial in processing interoceptive stimuli from one's own body and linking them to emotions in general and emotional feelings in particular (Craig, 2002, 2003, 2009; Critchley et al., 2004, 2005; Paulus and Stein, 2006; Pollatos et al., 2007a,b,c). Affective assignment may still be functioning normally in psychosis but, due to the change in the neurally coded difference, it becomes redirected.

Instead of the intero-extero difference being subject to affective assignment, it is now the rest–intero difference to which the affects and emotions are assigned. Since the rest–intero difference is associated with externalization and subsequent loss of the body as self-stabilizing and thus "good self-object," predominantly negative affects like fear and existential anxiety are assigned. This may result in the subsequent perception and experience of the body as not only physical and thus in a purely objective way, but also as associated with threat, fear, and anxiety. Hence, as in the case of difference-based coding, the functional mechanisms of affective assignment may still function normally but are apparently redirected, and it is this redirection that may enable and predispose to the patient's abnormal perception and experience of their own body.

However, besides the body there are also other self-objects that may originate from either cognitive or sensory stimuli and hence from either mental or social objects. As demonstrated in the case of the auditory cortex, the neural difference between the resting state and exteroceptive stimuli, the rest–extero difference, may be replaced by the rest–rest difference enabling and predisposing to auditory hallucinations. Since due to increased rest–rest interaction exteroceptive stimuli cannot induce sufficient neural activity changes and consecutively proper neuronal differences, i.e. rest–extero, in the neural processing of the auditory cortex, exteroceptive stimuli have no chance of being further processed in subsequent higher regions. Put into the psychodynamic context this implies that they have no chance of being assigned the respective affects and hence ultimately being constituted as self-objects. The lack of affective assignment to stimuli may result in the often observed affective apathy, emotional indifference, blunted affect, and total indifference to the environment that psychopathologically is often subsumed under the concept of negative symptoms. This means that the loss of intero-exteroceptive linkage may be related neuronally to reduced investment of the brain's resting-state activity in the subsequent neural processing of exteroceptive stimuli, which psychodynamically may correspond to the loss of energy investment in affective assignment. One may consecutively speak not only of a "decathexis of objects" but also of what I describe as "decathexis of affect."

How about self-objects being constituted on the basis of cognitive stimuli and hence from what one may call "mental objects"? For this, difference-based coding would be necessary and would be employed to higher cognitive regions like the dorsolateral prefrontal cortex (DLPFC), which is found to be severely disturbed in psychosis (Lewis, 2009; Gonzalez-Burgos, 2008). However, difference-based coding in these higher cognitive regions may also be disturbed, so here again the false neuronal difference is coded. For instance, the neural coding of the extero–cognitive difference may be replaced here by the rest–cognitive difference so that one's own cognitive stimulus is false positively taken as an exteroceptive stimulus.

One's own cognitive stimulus and its respectively associated mental object are subsequently externalized and projected to the outer world and thus perceived as physical object rather than as mental object. Needless to say this makes the constitution of one's own mental object as self-object impossible, since for that the reverse process, internalization with introjection, would be necessary. Hence it is again the application of a normally functioning mechanism, difference-based coding, to the false difference, i.e. rest–cognitive rather than extero–cognitive, that makes the constitution of self-objects of mental origin impossible.

In summary, I postulate that the constitution and construction of self-objects of exteroceptive, interoceptive, or cognitive origin may be disrupted in psychosis because of the application of normally functioning difference-based coding to the false neuronal difference. Difference-based coding is applied to the neural differences of rest–rest, rest–intero, or rest–cognitive rather than those of rest–extero, intero–extero, or extero–cognitive differences. By confusing the former with the latter, the subsequent constitution of the different kinds of proper self-objects is disrupted from the very beginning.

This reverses the psychodynamic processes from internalization and introjection to externalization and projection, so that instead of "mental self-objects," mere "physical objects" are constituted as self-objects in subjective perception and experience. And since they are constituted as mere "physical objects," neither any affect nor any personal relevance is assigned to them, resulting in what I called "decathexis of affect." Most importantly, such self-objects as merely affect-free physical objects cannot be stabilizing, i.e. maintaining and preserving, for the self and must therefore be regarded as what I earlier (in Chapter 11) called "bad self-objects," which unlike "good self-objects" destabilize rather than stabilize the self.

Fragmentation of the self and bilateral neural interaction in anterior cortical midline regions

Neuropsychodynamic hypothesis

I have so far described the changes in transforming objects into self-objects in psychosis and assumed it to be enabled and predisposed to by the confusion of neural differences. This first step, i.e. object–self-object transformation, is followed by a second step of self-object constitution with the integration of the self-object into the self, i.e. self–object-self integration (see Chapter 7). In order to understand how the fragmentation of the self in psychosis is generated, we need to understand what happens or better still what does not happen in self–object-self integration.

In Chapter 7 I postulated that neuropsychodynamically self–object-self integration may be associated with the coordination of neural activity between the insula and the anterior cortical midline structures (aCMS) that include regions like the anterior cingulate cortex (ACC) and the ventro- and dorsomedial prefrontal cortex (VMPFC, DMPFC). Psychologically, these regions are associated with linking emotions and self-relatedness so that the affective assignment of self-objects may be crucial for their integration into the self. Based on these prior assumptions, I here postulate the fragmentation of the self in psychosis to be related psychodynamically to abnormal self–object-self integration which neuronally is assumed to correspond to altered interaction between insula and aCMS. Furthermore I postulate that the aCMS may show abnormal hyperactivity which psychodynamically may correspond to the compensatory mechanisms to repair the fragmented self with subsequent projection, identification and split (see Part I for details) (see also Figure 12.3b).

Empirical findings

Let me start with the empirical findings. Various studies investigated recently the DMN in psychosis. Some studies showed abnormal connectivity between resting-state regions, i.e. the DMN, during the resting state. Recent imaging studies in psychosis reported indeed abnormal resting-state activity and connectivity in the aCMS. One study (Whitfield-Gabrieli et al., 2009) demonstrated that the aCMS (and posterior CMS like the PCC/Precuneus) show decreased task-induced deactivation (TID) during a working memory task in both psychotic patients and their relatives when compared to healthy subjects. This is indicative of decreased task-related suppression and possibly increased resting-state activity.

Furthermore, the very same psychotic subjects also showed increased connectivity of the aCMS with other posterior regions of the CMS, the PCC. Both hyperconnectivity and decreased TID correlated negatively with each other meaning that the more decreased task-related suppression, the more increased the connectivity. Finally, both decreased TID and increased connectivity in aCMS correlated with psychopathology, i.e. predominantly positive symptoms as measured with the PANS scale.

Decreased TID in aCMS were also observed in an earlier study that also investigated working memory (Pomarol-Clotet et al., 2008). Similar to the study described, they let subjects perform a working memory task and observed abnormally decreased TID in aCMS in psychotic patients when compared to healthy subjects. And similar to the other study, they also observed abnormal task-related activation in the right dorsolateral prefrontal cortex in psychotic patients. Another study (Mannell et al., 2010) also observed abnormal TID in aCMS as well as abnormal connectivity from aCMS and posterior CMS to the insula in psychotic patients thus lending further support to the other findings (see also Williamson, 2007; Calhoun et al., 2008; Jafri et al., 2008; Park et al., 2009). In addition to TID and connectivity, another abnormal measure of resting-state activity is the temporal features, more specifically fluctuations or oscillations in certain temporal frequencies. For instance, Hoptman et al. (2010) demonstrated that low frequency fluctuations in the resting state were increased in the aCMS (and the parahippocampal gyrus) in psychotic patients while they were decreased in other regions including the insula. Abnormally increased low frequency oscillations in the aCMS (and posterior CMS regions and the auditory network) and their correlation with positive symptom severity were also observed in another study on psychotic patients (Rotarska-Jagiela et al,. 2010).

Taken together, even though not being fully consistent at this point (for reviews see Broyd et al., 2009; Greicius et al., 2009), the empirical findings demonstrate clear resting-state abnormalities in the aCMS in psychosis which often also concern the posterior CMS and the insula. More specifically, the findings suggest that the resting-state activity in the aCMS is abnormally increased as indicated by the observations of decreased TID, increased connectivity, and increased low-frequency fluctuations in psychotic patients.

How is the apparently abnormally increased resting-state activity in the aCMS related to biochemical mechanisms? As described previously (see Chapter 5), task-induced deactivation (TID) as indirect measure of resting-state activity are closely related to GABA as the main inhibitory transmitter (Northoff et al. 2007b; Muthukumaraswamy et al., 2009). As such GABA seems to be central in mediating rest–stimulus interaction and hence ultimately the neural coding of neural differences, i.e. difference-based coding. The empirical findings of decreased TID and altered difference-based coding in psychosis lets one suggest abnormalities, e.g. hypofunction of GABA in aCMS. Interestingly, this is exactly what has been observed.

The group around F.Benes conducted several post-mortem studies where they observed indeed altered gaba-ergic interneurons in typical DMN regions like the anterior cingulate cortex (ACC) and the hippocampus (Benes, 2007; Benes et al., 2008; Woo et al., 2008; Berretta et al., 2009). Such alteration of gaba-ergic interneurons holds also true for other regions like the sensory cortex (see Yoon et al., 2010 who observed reduced GABA in visual cortex in psychosis that also correlated with a behavioral measure of visual inhibition) and the DLPFC (Lewis et al., 2005). More specifically, a specific subset of gaba-ergic interneurons has been found to be altered in psychosis, particularly those that contain and express mRNA for NMDA receptors (i.e. the NR2A subumit) (Lisman et al., 2008; Woo et al., 2008; Berretta et al., 2009). GABA-ergic interneurons in upper cortical layers (like layer 2, 3, and 4 as observed by Benes) modulate and control long-range glutamatergic cortico-cortical connections in lower layers (i.e. layers 3–5). The deficits in GABA may thus affect cortico-cortical connectivity via glutamatergic mediation and NMDA-receptors which have also been observed to be hypofunctional in psychosis (Lisman et al., 2008; Corlett et al., 2009).

One may speculatively assume that the reported correlation between reduced TID and increased connectivity in aCMS may be mediated by the interplay between GABA and NMDA-receptors. More specifically, based on the findings in healthy subjects, reduction in GABA may lead to decreased TID and increased resting-state activity as it is indeed observed in psychosis (see also Lisman et al., 2008). At the same time, reduced GABA in upper cortical layers may release NMDA-receptors in lower cortical layers from their GABA-ergic mediated neural inhibition; this may increase neural excitation in cortico–cortical output layers and consequently increase the degree of cortico–cortical connectivity (see also Corlett et al., 2009 who assume similar dependency of cortico–cortical connectivity on NMDA-receptors and frame it within the context of generating predictions as postulated in predictive coding). As such decreased TID as possibly related to reduced GABA co-occur with abnormally increased cortico–cortical connectivity, i.e. hyperconnectivity, which then can be traced back to abnormal interaction between GABA and glutamate via NMDA-receptors. However, direct evidence for such relationships is still lacking so that this hypothesis must be considered preliminary at this point.

Neuroscientific account

What do these findings entail for the neural processing and subsequent constitution of the self? Let me first start with briefly characterizing the aCMS. As I described (see Chapter 7), the insula and the aCMS are closely connected with each other. As such the aCMS receive interoceptive input in addition to the neural input, i.e. the own resting-state activity, and the exteroceptive input coming from the various sensory regions that are directly connected via afferences and efferences with the aCMS. This characterization of the aCMS as "convergence region" (Shore, 2003; Heinzel et al., 2009) provides the anatomo-structural ground for what I described (see Chapter 6 on defense mechanisms) as "trilateral interaction" between neural and intero- and exteroceptive stimulus processing.

How can such trilateral interaction in aCMS be related to the here supposed neural coding of the false neuronal difference? I postulated that neural activity in the insula codes no longer the neural difference between intero- and exteroceptive stimuli, the intero-extero difference, but rather the one between resting-state activity and interoceptive stimuli, the rest–intero difference which is taken false positively as the former, i.e. the intero-extero difference. Since the insula is closely connected to the aCMS, the false neuronal difference, i.e. rest–intero difference, is transmitted to the aCMS where it is false positively taken as intero-extero difference.[11]

Since the rest–intero difference is taken as intero-extero difference in aCMS, the incoming neural signals are treated accordingly meaning that the processes underlying trilateral interaction are set in motion entailing activation and recruitment of the aCMS. However, what is supposed to be an input signal of intero-extero difference upon which trilateral interaction is employed turns out to be an input signal of rest–intero rather than intero-extero difference. Hence, what seemed to be a trilateral interaction between neural and intero- and exteroceptive stimuli, i.e. rest–(intero-extero), turns out to be merely a bilateral interaction between neural and interoceptive stimuli (since the difference to the exteroceptive stimuli was not coded in the insula), i.e. rest–(rest-intero). The neural coding of the false positive neuronal difference is thus transferred from the insula to the aCMS where it lets the originally trilateral interaction degenerate into a bilateral one.

What does this mean within the psychological context of the self? I characterized the aCMS psychologically by self-related processing (see Chapter 9 and Qin and Northoff, 2010; Qin et al., 2011 for recent papers). The recruitment and activation of the aCMS during the bilateral interaction of neural and interoceptive stimuli, excluding exteroceptive stimuli means that self-relatedness is assigned false positively to the false neuronal difference (and hence to the false stimuli). This entails that the difference rest–intero is assigned an abnormally high degree of self-relatedness while exteroceptive stimuli are not related at all to the self. This though means that psychologically

---

[11] Whether this confusion of difference is related to the described deficit in GABA-ergic mediated neural inhibition can only be speculated at this point. The deficit in GABA may lead to decreased neural inhibition and, in conjunction with faulty glutamatergic mediation via NMDA-receptors to neuronal disinhibition (Lismann et al., 2008). Decreased neural inhibition and subsequent neuronal disinhibition in aCMS entail that the incoming signals from for instance from the insula, the rest-intero difference, can no longer be properly inhibited by the aCMS; this in turn may pave the way for confusing it with the normal input, the intero-extero difference, which, as in the normal case, is not inhibited but rather coded in relation, e.g. difference, to the actual resting-state level in the aCMS. Hence, I postulate that the well established deficit in GABA-ergic mediated neural inhibition in the aCMS may enable and predispose to the confusion of the intero-extero with the rest-intero difference which the aCMS receive as false input from the insula. However, one should be aware that this hypothesis is speculative at this point. Moreover, besides GABA and Glutamate, other transmitter systems like dopamine, serotonin, and acetylchline may also play a major role (see for instance Lisman et al., 2008; Corlett et al., 2009).

Some evidence for a role of Glutamate in mediating neural processing of altered difference-based coding comes from the Mismatch negativity (MMN) as described in the first section here. Studies with NMDA-receptor antagonists like Ketamine reported significant reductions of the MMN amplitudes. The apparent NMDA-ergic dependence and thus the glutamtergic modulation of the MMN is also in accordance with the fact that NMDA-receptor dysfunction is implicated in psychosis. This may explain why both NMDA-receptor antagonists like Ketamine and psychosis lead to more or less similar reductions in the amplitudes of the MMN. In contrast to NMDA-receptors and Glutamate, there is no solid, robust and consistent evidence for the involvement of other transmitter systems like GABA, Dopamine, Serotonin and Acetylcholine in constituting and modulating the MMN (Garrido et al., 2009). Based on these findings one may speculatively assume the glutamatergic modulation of the neural coding of the false difference, i.e. rest-rest rather than rest-extero, in the difference-based coding of the auditory cortex.

the own self is detached almost completely from exteroceptive stimuli and hence from any contents, i.e. objects, in the environment.

Neuropsychodynamic account

Psychodynamically, the detachment of the self from exteroceptive stimuli and their associated contents and objects entails that the integration of the possible self-object as actual self-object into the self becomes impossible. The neural coding of the false positive neuronal difference, rest–intero, and its subsequent processing in the supposedly trilateral interaction that turns out to be a mere bilateral interaction, i.e. rest–(rest-intero), makes any integration of self-objects (if there were some) into the self impossible. Hence, one may postulate neuropsychodynamically that the self–object-self integration may be blocked by the degeneration of the trilateral interaction into a bilateral one in aCMS. The self remains consequently detached, devoid or empty of self-objects.

The lack of self–object-self integration however leads to further efforts to constitute and integrate self-objects into the self (because "good self-objects" are so essential for the self to stabilize it). Neuronally, this may be reflected in increased recruitment and activation of the aCMS as it is well supported by the previously described indicators of increased resting-state activity in these regions in psychosis, e.g. reduced TID, increased connectivity, and low-frequency fluctuations. Psychologically, this means that there may be abnormally increased (Qin et al., 2010) self-related processing which, put into the psychodynamic context, no longer enables and predisposes the desired effect of self–object-object integration.

Instead, it contributes to further detach and empty the self from self-objects since if the rest–intero difference is assigned higher and higher degrees of self-relatedness, the psychological difference to the exteroceptive stimuli (e.g. the rest–extero and intero-extero) becomes greater and greater. One may therefore assume that larger differences between intero- and exteroceptive stimuli with regard to the assigned degree of self-relatedness are accompanied by lower likelihoods of the resulting self-objects to be stabilizing for the self thus being "good self-objects." In other words, the more intero- and exteroceptive stimuli divert from each other with regard to self-relatedness, the more likely the resulting self-object may destabilize the self thus being a "bad self-object."

Based on the assignment of abnormally high degrees of self-relatedness to the own body, one would assume that the patients' experience and perceive their own body as intimate self-object and thus as highly self-related. This however is not the case; one can often observe rather the opposite with the psychotic patients experiencing this difference and thus their own body (and thoughts) as non-self-related. How can that be? I already described that the psychotic patient confuses and takes the rest–intero difference to be an intero-extero difference, similar so in the case of the aCMS. Here he false positively takes the bilateral interaction of neural and interoceptive stimuli, i.e. rest–(rest–intero), for a trilateral interaction between neural and intero- and exteroceptive stimuli, i.e. rest–(intero-extero). He thus does what he always does (i.e. in the normal healthy state), he externalizes and projects what he takes for a self-object to the external environment. In this case he however does not project a self-object with high self-relatedness to the external environment but his own self and his own body and "locates" them as mere physical objects (rather than as self and self-object) in the environment. He consequently perceives and experiences himself and his body (and his thoughts) as another person corresponding well to what is described as first-rank symptoms of Schneider, the passivity phenomena, or ego-disturbances.[12]

[12] Relying on Freud, Carhart-Harris and Friston (2010) characterize psychosis by increased primary process thinking and reduced secondary process thinking (see also Rosenbaum et al., 2005). Based on their approach, they relate the predominance of primary process thinking in psychosis with a failure of

However, his perceptions and experiences of his own self are split. On the one hand, he projects and externalizes his own self to the external environment where he reencounters his projected and externalized self in his perceptions and experiences of others. All his efforts to integrate self-objects into his self for which he increases neural activity in his aCMS have thus the opposite effects, they alienate his own self from himself by detaching and emptying his own self of its self-objects. On the other hand, there is still the increased self-related processing in his own brain which (due to autoepistemic limitation; see part one) he cannot project and externalize to the external environment. Taken together, this results in a split within one's own self (or ego if one wants to say so). He cannot other than to subjectively perceive and experience a split between the increasing presence of the activity of self-related processing on the one hand and the increasing absence of (self-stabilizing) self-objects on the other. What I described as psychological difference or gulf between self- and non-self-related stimuli and hence between intero- and exteroceptive stimuli may consequently surface in subjective perception and experience as (phenomenological) difference or split within the own self.

Following Benedetti's framework, the increasing absence of self-objects may correspond to what he described as negative existence. Since the self-objects are for the existence of the self what breathing is for an organism (see Chapter 7 on narcissism), the detachment of the self from (truly self-stabilizing and thus "good") self-objects detaches the self from its own existence resulting in what Benedetti describes as "negative existence." And this detachment and loss of (self-stabilizing) self-objects induces anxiety and fear, the externalized and projected self-objects (and the other) are consecutively assigned a predominantly negative affect. The concept of "negative existence" may then be understood in a double sense as both existential, concerning the absence of (self-stabilizing) self-objects, and affective referring to the assignment of negative affects to the projected and externalized self-objects.

However, even though negative there is still some existence, and that may be due to the increased presence of self-related processing which analogously may be described as "positive existence." There is still activity of the self, or as one may say continuous if not increasing attempts to construct the (subjective perception and experience of the) existence of the own self via self–object-self integration. This though is doomed to fail. Instead of constructing the own self via self–self-object integration, another person's self is created since what is falsely taken as self-objects is false

the default-mode network to inhibit and exert control of regions outside the default-mode network and hence to suppress primary process thinking in favor of secondary process thinking. This stands opposite to the hypothesis suggested here. Rather than assuming loss of control, I postulate that it is the control that the aCMS exert is the problem, the increased neural processing of the rest-intero difference and the bilateral interaction which is false positively taken for the intero-extero difference and a trilateral interaction. The more this processing, e.g. self-related processing takes place, the more the self is supposed to become detached from its self-objects and the more the split within the own self increases. Hence, rather than assuming decreased control I assume increased control of the aCMS. This corresponds well to the observation of apparently increased resting-state activity in the aCMS in psychosis. If Carhart-Harris and Friston are right in that the DMN exerts less inhibitory control over the systems outside the DMN, one would expect decreased resting-state activity in the aCMS which though is contradictory to the empirical findings that indicate the opposite, increased resting-state activity. This makes their assumption empirically rather implausible. What, admittedly, remains unclear in my account though is how it relates to the distinction between primary and secondary process thinking and the apparent increase of the former in psychosis. I touch upon this issue in my chapter above (see Chapter 4) with an extensive elaboration of it especially in the case of psychosis being though beyond the scope of this book.

positively externalized and projected to the external environment. This results in the subsequent perception and experience of the projected and externalized self-objects as belonging to another person's self. Hence, the own self-objects that are supposed to support and maintain the existence of the own self are now perceived and experienced as if supporting and maintaining the existence of another person's self.

However, due to the high degree of self-relatedness, the other person's self is experienced and perceived as closely related to the own self. Due to the attribution of such high degree of self-relatedness to the projected and externalized self-object, the psychotic patient's self cannot do other than to identify himself and thus his own self with the other person's self thereby enabling and predisposing what psychopathological is described as identification or more generally as ego-disturbance. And since the detachment of the self-object is associated with negative affects, e.g. existential fear and anxieties, the other person is subjectively perceived and experienced as threatening and intrusive. This in turn may enable and predispose to the occurrence of predominantly negative affects in delusions of alien control, first-rank symptoms, ego disturbances or passivity phenomena.[13]

"Existential dilemma" and abnormal cortico-cortical neural coupling

Neuropsychodynamic hypothesis

I so far demonstrated how the psychodynamic features of loss of object relations, loss of inner self–object boundary, the loss of self-objects and the fragmentation of the self may be enabled and predisposed by the "normal" employment of well-functioning neuronal mechanisms in an abnormal neuronal context. How does this apply to what I called "existential dilemma"? As previously described, the psychotic patient encounters what psychodynamically was conceptualized an "existential dilemma," the choice between preservation of the (subjective) existence of the own self at the expense of objects versus the constitution of objects at the expense of the (subjective) existence of the own self (see Chapter 11 for details). Since however the preservation of the self requires the constitution of objects and ultimately of self-objects (and vice versa), the choice is not only an existential one that concerns the existence of the own self but also a dilemma since on the basis of that constellation the existence of the own self is at stake whatever choice the patient makes.

Mentzos characterized the constitution of self and object by a fundamental bipolarity and mutual dependency (see Chapter 6). The question is now how these psychodynamic features can be related to specific neuronal mechanisms and in particular to those we already described so far, i.e. neural coding of false positive neuronal difference leading to systematic alterations in difference-based coding and rest–rest/stimulus interactions. Neuronally, I highlighted the anterior CMS in the last section associating them with abnormally increased neural activity and self-related processing and subsequent fragmentation of the self in psychosis. In contrast to the aCMS

---

[13] One may also discuss how the three-stage model of psychosis by the former German psychopathologist Klaus Conrad relates to our distinct stages of neuronal processing as postulated here (see Mishara, 2010 for a nice description). Klaus Conrad described three stages, Trema, Apophany, and Anastrophe which are consecutive stages accompanied by the development of specific symptoms, e.g. delusional mood, delusions as revelations, and delusions of references. While the first stage, Trema and delusional mood, may be related to the loss of object relations, the last stage, Apophany and delusions of reference, may be due to the alterations in aCMS and the loss of self-objects including the previously described compensatory processes.

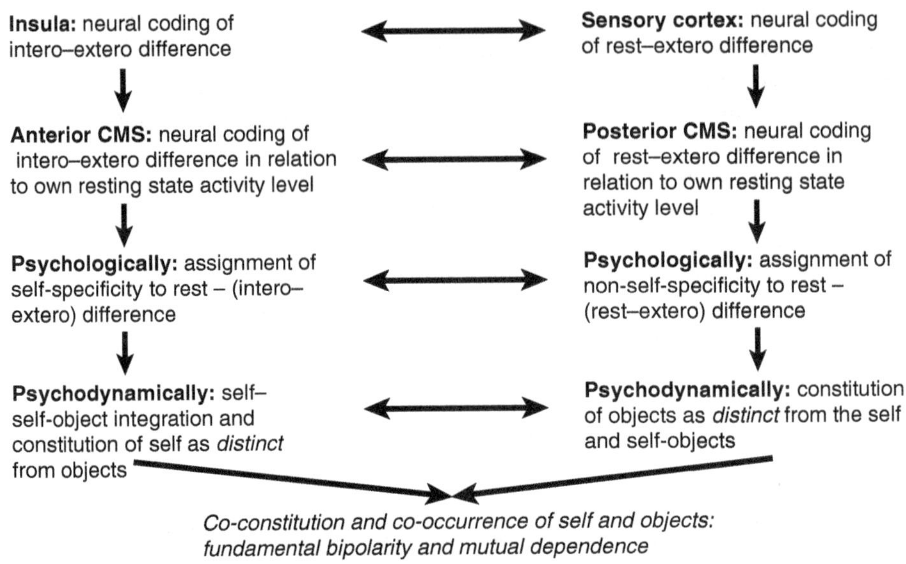

Fig. 12.4a Existential dilemma and abnormal cortico-cortical synchronization in healthy subjects.

that have been specifically associated with the neural processing of self-specific stimuli, the posterior regions of the CMS like the posterior cingulate cortex (PCC) and the precuneus do seem to be recruited and activated specifically during non-self-related but familiar stimuli (Qin and Northoff, 2011; Qin et al., 2011). While being functionally and anatomically closely connected to each other, anterior and posterior CMS, i.e. aCMS and pCMS, seem to dissociate from each other when it comes to neurally process stimuli with different degrees of self-relatedness, i.e. self-specific and non-self-specific (see Figure 12.4a).

Based on these observations in healthy subjects, I postulate the following in psychosis. I postulate that there may be an abnormally tight relationship between anterior and posterior CMS in psychosis that couple and link the neural processing of self- and non-self-specific stimuli together to such a degree that the one relies on and ultimately becomes identical with the respective other and vice versa. More specifically, I postulate that neural processing in the anterior and posterior CMS remains unable to dissociate from each other when processing self- and non-self-specific stimuli so that the latter are taken for the former or vice versa. This in turn may enable and predispose to what I described as "existential dilemma" where self-constitution is considered to be identical with object-constitution and vice versa (see Figure 12.4b). In short, I postulate that the "existential dilemma" within the psychodynamic context may correspond within the neural context to an abnormal relationship between anterior and posterior CMS.

Empirical findings

I already described some of the fMRI resting-state studies in psychosis in the previous section thereby putting a specific focus on the aCMS. Most of these studies demonstrated increased functional connectivity between anterior and posterior midline regions, i.e. of the aCMS with the pCMS, in the resting state in psychosis (Whitfield-Gabrieli et al., 2009; Mannell et al., 2010; Rotarska-Jagiela et al., 2010). The presumed increases in resting-state functional connectivity between aCMS and pCMS have however not been as consistently observed as the previously described findings in aCMS (see also Williamson, 2007; Calhoun et al., 2008; Jafri et al., 2008; Kim et al., 2009; Park et al., 2009 for findings in this direction though not always being fully consistent).

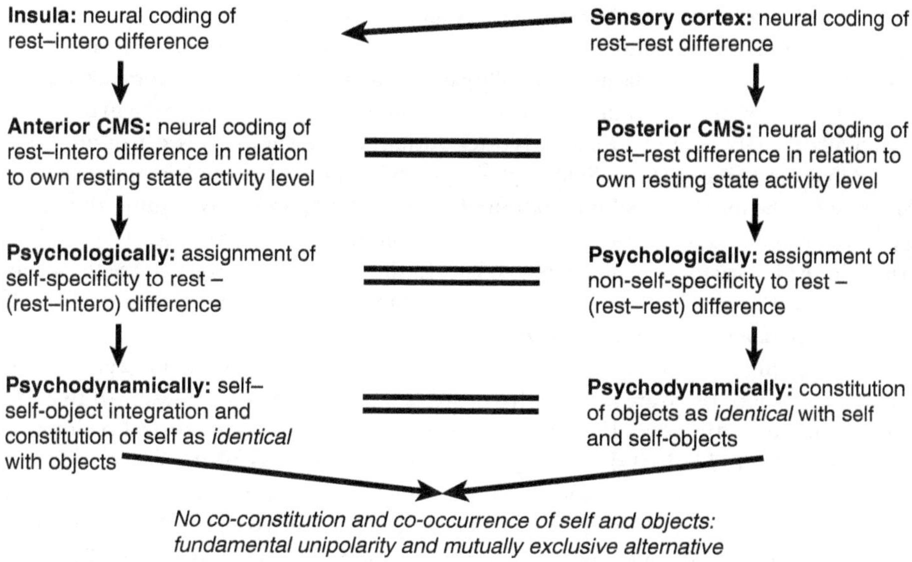

Fig. 12.4b Absent existential dilemma and normal cortico-cortical synchronization in healthy subjects.

Besides decreases in aCMS-pCMS connectivity, decreases in connectivity from aCMS to temporal cortex including the auditory cortex, i.e. fronto-temporal dysconnectivity, and decreased connectivity from pCMS to the temporal cortex and especially the hippocampus have been observed (for a recent review, see Stephan et al., 2009). This has led to what is called the "dysconnectivity hypothesis" that assumes reduced or altered long-range connections between different cortical regions to be crucial in the pathogenesis of psychosis (Stephan et al., 2009).[14]

---

[14] How does my hypothesis of altered difference-based coding relate to the dysconnectivity hypothesis? The shift of the focus from stimulus-induced activity and stimulus-stimulus interaction to rest-rest and rest-stimulus interaction entails that I do not take one particular region or a specific network into account but the interaction between different networks. I here for instance paradigmatically focused on the interaction between the DMN and the sensory corticae, i.e. insula and auditory cortex. One may consecutively assume that the "basic disturbance of the self" may be traced back to altered connectivity and functional integration between different regions. This puts the here suggested hypothesis in a certain closeness to the "dysconnectivity hypothesis" (Stephan et al., 2009).

One may want to compare the here advanced hypothesis can be regarded as instance of the "dysconnectivity hypothesis" in psychosis (see Stephan et al., 2009 for a recent review and Ellison-Wright and Bullmore, 2009 for a meta-analysis of structural connectivity data). In a nutshell the "dysconnectivity hypothesis" postulates abnormal functional interaction and integration between different regions in the brain. Formulated in such general way, my hypothesis can also be regarded as a version of the "dysconnectivity hypothesis" since it also presupposes abnormal functional interaction and integration between different brain regions.

There is one principal difference though. As formulated the "dysconnectivity hypothesis" focuses on the functional integration of stimulus-induced activity across different brain regions and consecutively on interaction between different stimuli, i.e. stimulus-stimulus interaction. My hypothesis, in contrast, shifts the focus from stimulus-induced activity to resting-state activity and more specifically to rest-rest interaction and its interaction with the former, i.e. rest-rest interaction and rest-stimulus interaction.

This focus on rest-rest interaction and rest-stimulus interaction is necessary, as my hypothesis claims, to understand the subsequent dysconnectivity during stimulus-induced activity and the resulting

How does such dysconnectivity impact physiological mechanisms of integrating neuronal activity within one and across different regions? In the sections about the sensory cortex, I already discussed findings of altered beta and especially gamma frequency bands in circumscribed regions like the auditory or visual cortex in psychosis. Analogous findings were also made in higher cortical regions like the DLPFC and the parietal cortex during both rest and task (e.g. working memory tasks) (Uhlhaas and Singer, 2010). Since high frequency bands like beta and gamma account predominantly for the local cortical integration of neural activity, one may assume alterations in local cortical integration of neural activity in these regions in psychosis. And based on the findings described in the last section, i.e. increased activity in the aCMS, I would postulate the same, i.e. abnormal cortical integration as reflected in beta and gamma bands, to hold for the anterior CMS which though remains to be demonstrated.

In addition to high frequency fluctuations that account for local neuronal integration within one particular region, there are also low frequency oscillations which are assumed to establish neuronal synchronization over longer distances, i.e. between different cortical regions. Some resting-state studies did indeed show an increase in low frequency oscillations and concomitant reductions in high frequency oscillations in the resting state (Uhlhaas and Singer, 2010). This is in accordance with two fMRI studies (Hoptman et al., 2010; Rotarska-Jagiela et al., 2010) showing increased low-frequency fluctuations (<0.06 Hz) in psychotic patients across especially the aCMS and the pCMS. However, the evidence for low frequency abnormalities and hence of abnormal long-range cortico-cortical neuronal synchronization between especially the aCMS and the pCMS in psychosis is rather sparse at this point. Moreover, it remains unclear how the apparent abnormalities in high-frequency oscillations and local cortical integration are related to the low-frequency oscillations and cortico-cortical synchronization.

Interestingly, cortical synchronization via high-frequency oscillations like beta and gamma bands has been shown to be related to GABA-ergic neurotransmission (Uhlhaas and Singer, 2010). More specifically, GABA-ergic interneurons act apparently as pacemaker by producing rhythmic fast inhibitory postsynaptic potentials (IPSP) in pyramidal neurons in the cortical layers; this may be sufficient to synchronize the firing of a large population of pyramidal cells that determine the dominant frequency in a larger network. There is indeed some evidence from both humans and animal models that the altered high frequency oscillations in for instance the gamma range may be related to the well-established deficits in GABA in psychosis (Uhlhaas and Singer, 2010). In addition to GABA, NMDA-receptor hypofunction and thus changes in glutamatergic

---

symptoms as postulated by the "dysconnectivity hypothesis." Consider the previously described distinction between enabling and executing conditions. One may argue that altered difference-based coding may be regarded an enabling but not an executing condition which as such may be related to the risk or resilience factors rather than the psychotic symptoms themselves. While the dysconnectivity as postulated in the dysconnectivity hypothesis must be considered an executing rather than an enabling condition which as such may be closely related to the psychotic symptoms.

Since they seem to focus on distinct neural conditions of psychosis, enabling and executing conditions and the here advanced of hypothesis of altered difference-based coding and the dysconnectivity hypothesis may be regarded as complementary and inclusive rather than competitive and exclusive. More specifically, the hypothesis of altered difference-based coding aims to bring to light those changes in the brain that must be there in presymptomatic stages in order for the dysconnectivity during the acute symptomatic outbreak to be possible. The complementary character of the dysconnectivity hypothesis and my hypothesis of altered difference-based coding is even more likely given that both aim to explain changes in the self though in different stages, i.e. symptomatic and presymptomatic. Hence one may link both hypotheses in the future in more detail.

neurotransmission may also play a role here which is rather plausible given the close connections between gaba- and glutamatergic systems in the cortex.

Taken together, the empirical findings provide some rather preliminary evidence of increased cortico-cortical coupling especially between anterior and posterior CMS while the coupling of both aCMS and pCMS with the temporal cortex seems to be decreased. And one may speculatively assume that these alterations in cortico-cortical coupling may be related to the well established gaba-ergic and NMDA-receptor mediated glutamatergic hypofunction with subsequent alteration in the balance between neuronal inhibition and neuronal disinhibition. However, the empirical findings must be considered preliminary entailing that the neuropsychodynamic hypothesis generated must be regarded at best tentatively.

Neuropsychodynamic account

What does the apparently increased coupling between aCMS and pCMS entail for the processing of neural differences and thus for difference-based coding? Let us briefly recount the situation in the healthy brain. Due to its strong connections and input from the insula, the aCMS may process predominantly the intero-extero differences in relation to their own resting-state activity level, i.e. rest–(intero-extero). The pCMS, in contrast, normally account for and receive input from regions strongly involved in autobiographical/episodic memory retrieval like the precuneus and the hippocampus; this means that functionally they receive predominant cognitive stimuli in the gestalt of rest–(rest-cognitive/extero) differences from the past within the present context, i.e. their present resting-state activity level. Relying again on difference-based coding and the cortico-cortical connectivity between aCMS and pCMS, both differences, rest–(intero-extero) and rest–(rest-cognitive/extero), may be compared with each other yielding novel neuronal differences; this is a continuously ongoing process of mutual matching and comparison that ultimately may lead to the distinction between self-specific and non-self-specific stimuli in the healthy subject.

How is the situation in psychosis? First and foremost, the aCMS and pCMS do not receive the proper neural differences anymore; instead of rest–(intero-extero) the aCMS receive rest–(rest-intero) while the pCMS encounter rest–(rest-rest) rather than rest–(rest-cognitive). What happens now? Due to the apparently abnormally tight cortico-cortical coupling and synchronization between aCMS and pCMS, rest–(rest-intero) and rest–(rest-rest) differences are not processed as different differences anymore; they can consequently no longer be related in different ways to the resting-state activity in the CMS thereby yielding novel neural differences (on the basis of difference-based coding). Instead, aCMS and pCMS are now functioning as one functional unity that as such is able to neurally process only one neuronal difference, either the rest–intero difference or the one of rest–rest.

Putting it figuratively, neural activity in the abnormally tightly coupled aCMS and pCMS is confronted with the choice between the neural processing of one of the neuronal differences, either the rest–intero difference as coming in through the aCMS or the rest–rest difference as entering through the pCMS. This is an All-or-Nothing decision with no alternative option as for instance comparing and matching both differences with each other as to yield novel neuronal differences and so on as in the healthy brain. In psychosis there seems to be All-or-Nothing decision: Either the rest–intero difference is processed or the rest–rest difference is processed. This means that psychologically the choice is between the neural processing promoting self-specificity or the one enabling and predisposing non-self-specific specificity. While simultaneous and thus parallel processing between both with their subsequent comparison and matching, as in the healthy brain, is supposed to remain impossible in the psychotic patients.

Taken this within the psychodynamic context, this means that the choice is either between the constitutions of the self, as based on the neural processing of self-specificity in aCMS, or the constitution of objects as based on the neural processing of non-self-specificity in pCMS. Either the self or the object can be constituted with parallel constitution of both, i.e. co-constitution and co-occurrence between self and objects (see Chapter 6) remaining principally impossible. And most importantly, the choice for the self-constitution is possible only at the expense of processing the object-constitution since only one of the neuronal differences, rest–intero or rest–rest, can be processed in the abnormally tightly coupled aCMS and pCMS. What is psychodynamically described as "existential dilemma" may then correspond and be traced back within the neuronal context to the abnormal tightly coupling between aCMS and pCMS; the latter makes parallel processing between different neural differences and thus between self- and non-self-specific stimuli impossible which instead is converted into an All-or-Nothing dilemma within the neuronal context. In short, the "existential dilemma" in the psychodynamic context may correspond to an "All-or-Nothing dilemma" within the neural context.

This entails that the concepts of fundamental bipolarity and mutual dependency between self- and object-constitution may also need rephrasing in the context of psychosis in order for them to be empirically plausible. Self- and object-constitution can no longer be characterized by fundamental bipolarity in psychosis; their relationship may instead be better described by "fundamental unipolarity" accounting for the fact that either self- or object-constitution rather than parallel constitution is possible. Moreover, the concept of mutual dependency needs to be reformulated as "mutual exclusivity" with either self-constitution being possible at the expense of object-constitution or vice versa. Taken together, one may re-describe the "existential dilemma" in the neuropsychodynamic context as "All-or-nothing dilemma" that can be characterized by "fundamental unipolarity" and "mutual exclusivity" between self- and object-constitution.

Volatile self–object boundaries and unstable neural differences in difference-based coding

Neuropsychodynamic hypothesis

So far, I described that the constitution of both self and objects is significantly altered in psychosis. I associated the disruption in the constitution of self and objects with the neural coding of false differences, rest–rest, rest–intero, and rest–cognitive instead of rest–extero, intero–extero, and extero–cognitive, across different neural regions and networks, e.g. sensory cortex, interoceptive network, and cortical midline regions. I postulated that "normal" neural coding, i.e. difference-based coding, of false positive neuronal differences reverses the constitution and construction of both self and objects. Instead of internalizing and introjecting (stimuli and) objects on the basis of neural coding of the "right" neural differences, difference-based coding of false positive neuronal differences reverses this process by supporting externalization and projection. This makes the constitution and construction of both self and objects impossible so that the psychotic patient employs the "right" neural mechanisms to false positive neuronal differences. As indicated at the beginning of this chapter I call this hypothesis of psychosis "Altered Difference-based Coding" (ADC).

What I left open are the conditions that enable and predispose to difference-based coding to be altered in psychosis. For this I turn back to psychoanalysis. The psychodynamic account assumes that there is a certain predisposition for such impairment in the constitution and construction of self and objects in the gestalt of volatile self–object boundaries (see Chapter 11 for details). The hypothesis is that the self–object boundaries are volatile because of some trauma in early childhood with the trauma being apparently situated right at the boundaries between biological and social domains. One may consequently assume not only an early developmental but also a biopsychosocial origin of volatile self–object boundaries.

How can such volatile self–object boundaries be related to the here supposed mechanisms of neural coding of false positive neuronal differences in difference-based coding? Based on the previously described neuropsychodynamic mechanisms supposedly underlying the constitution and construction of self and objects (see Chapter 6 on defense mechanisms), I postulate that volatile self–object boundaries may correspond in the neural context to unstable neuronal differences in difference-based coding.

More specifically, I postulate that the neural distinction between rest–extero and rest–rest in sensory cortex, between intero-extero and rest–intero in the interoceptive network, and between extero–cognitive and rest–cognitive in cortical regions may be not as clear-cut and distinctive in the to-be psychotic patients as in healthy subjects. And I postulate that the seemingly unstable nature of these neural differences in differences-based coding may be related to early developmental biopsychosocial traumata; the biopsychosocial traumata may account for the blurring of the neural differences between resting-state activity (levels) and stimulus-induced activity (levels) which enables and predisposes subsequent confusion of different neuronal differences. Taken together, this leads me to the following neuropsychodynamic hypothesis: I postulate that the volatile self–object boundary may be enabled and predisposed by the neural coding of unstable neural differences in difference-based coding which may be traced back to early biopsychosocial trauma (see also Figure 12.5).

Empirical findings

Let me now venture briefly into the empirical findings. There is much discussion about a neurodevelopmental hypothesis in psychosis (Lewis and Levitt, 2002). The neurodevelopmental hypothesis argues that there may be some biological (or social) alteration in early infancy (or even prenatally) due to some genetic changes, viral infection, obstetric or gestational complications, or otherwise which may predispose one to develop psychosis. However, psychosis does not

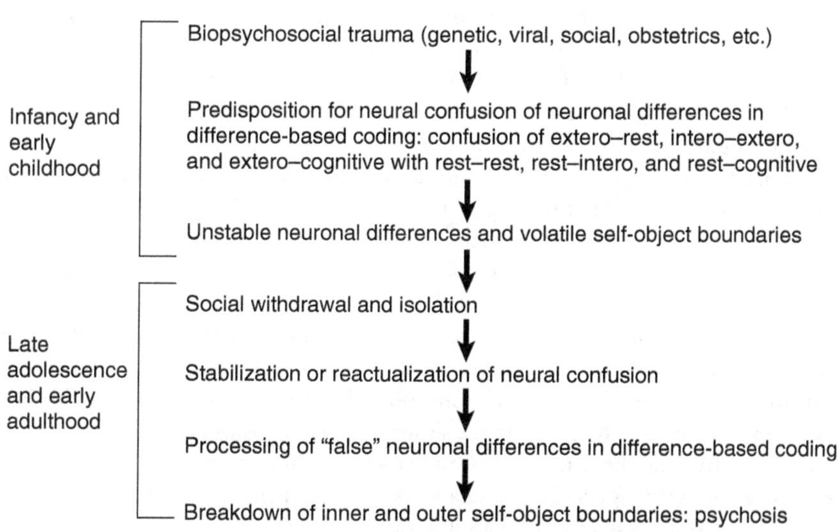

Fig. 12.5 Volatile self–object boundaries and unstable neuronal differences in difference-based coding.

become manifest before late adolescence or early adulthood, raising the question for some later developmental process in the human brain around that time.

One empirical suggestion in this regard is that the outbreak of psychosis may be related to the overpruning of synaptic contacts in late adolescence and the subsequent reorganization of the brain's wiring and its neural networks (see Lewis and Levitt, 2002 for an overview). Core regions and circuits implicated in psychosis such as the DLPFC, the hippocampus, the sensory cortex and cortical midline structures have been shown to be sensitive to developmental changes in the period of late adolescence or early adulthood (Lipska and Weinberger, 2000; Lewis and Levitt, 2002; Fransson and Marrelec, 2008; Lewis and Gonzales-Burgos, 2008; Lewis, 2009). This lets one suggest that the developmental changes in these regions may be disturbed in psychotic patients which may prone or predispose them to the outbreak of the respective symptoms.

A recent study used EEG to study neural synchrony in healthy subjects ranging from 6 to 21 years (Uhlhaas et al., 2009a,b). They observed that in early adulthood theta, beta, and gamma oscillations and their long-range synchronization increase to an enormous extent. This increase in synchronization in early adulthood is though preceded by a significant reduction of beta and gamma oscillations during late adolescence that follows continuous increases in synchronization from childhood to late adolescence.

This suggests that the maturation of cortical synchronization and neural networks in early adulthood goes through a period of transient destabilization in late adolescence before being organized in the most stable and mature way. Late adulthood may thus be considered a critical period for constituting stable and more precise cortico-cortical synchronization. Some preliminary evidence suggests that the processes of transient destabilization and mature stabilization in late adolescence and early adulthood may be related to changes in cortico-cortical myelination and GABA which both undergo changes in exactly these time periods and may thereby significantly impact neural wiring and synchronization in early adulthood (Di Cristo, 2007; Uhlhaas et al., 2009a,b).

Neuroscientific account

What do these findings entail for our hypothesis of unstable neural differences in difference-based coding? If the cortico-cortical synchronization is not as well developed, neural activities stemming from different stimuli, i.e. neural, intero, and extero and cognitive, may not be as well coordinated and integrated. Instead of synchronizing neural differences within the resting state itself across different subregions in for instance the auditory cortex, different resting-state activity levels in distinct auditory cortical subregions may be treated as distinct activities stemming from distinct stimuli, i.e. exteroceptive. This destabilization or desynchronization may then predispose the neural coding of the different resting-state activity levels in the different auditory cortical subfields as rest–extero difference rather than as mere fluctuation or oscillation within the resting state itself and thus as mere rest–rest interaction to which abnormal neuronal synchronization is employed.

The neuronal desynchronization may predispose the confusion of rest–rest differences with rest–extero differences which may also apply to other regions in the brain. And as I postulate the same confusion may be at work in other regions too. While this hypothesis is plausible it requires further experimental support by for instance showing a predisposition towards (i) desynchronization of neural activity, (ii) neural confusion between resting-state activity and exteroceptive stimulus-induced activity especially in late adolescence, and (iii) increased confusion of rest–(intero-extero) with rest–(rest-intero) in adulthood.

How does the predisposition towards desynchronization and neural confusion relate to the biopsychosocial trauma in early infancy? Hoffman (2007) recently postulated what he calls the "social deafferentiation hypothesis" (SAD). The SAD postulates that high levels of social withdrawal and isolation in vulnerable individuals prompt the predisposition to generate psychotic symptoms. Following Hoffman, the SAD relies on several observations of the crucial nature of social isolation and withdrawal in psychosis. First, often the onset of auditory hallucinations, delusions and other psychotic symptoms is preceded by reduced interpersonal interactions and social isolation. Thereby social isolation should not be understood in an absolute sense but rather relative to the person's standard or usual degree of social contact and hence its own prior baseline of social involvement.

Second, delusions and hallucinations produce socially and emotionally meaningful contents of intra- or interpersonal contents that can be considered as substitutes of the real world. They may therefore, regarded as compensatory attempts to escape social isolation and withdrawal by producing some kind of relation to a world though an imaginary one as substitute for the real world (see also McGlashan, 2009, p. 479). Third, sensory deafferentiation of for instance the visual cortex produces neuronal reorganization and complex hallucinations; analogously social withdrawal may also prompt neuronal reorganization with the subsequent generation of psychotic symptoms. Fourth, social withdrawal and isolation in critical developmental periods as in early infancy or late adolescence may significantly impact the changes in the processes of neuronal synchronization and wiring going on anyway in these critical time periods.

Neuropsychodynamic account

How can we relate the SAD hypothesis to the psychodynamic assumption of the relation of the volatile self-object boundaries to early biopsychosocial trauma? Social isolation and withdrawal predisposes one to relocate inner and outer self-object boundaries. Increased social withdrawal and isolation (relative to one's baseline of social involvement) predisposes one to shift from the outer boundary between self and the objects of social origin towards the inner boundary between self and the objects of mental origin. This however means that inner and outer self–object boundary may more and more overlap and at worst become indistinguishable.

Neuronally, one may postulate that the shift of the outer self-object towards the inner one may go along with increased likelihood to confuse the rest–extero differences with rest–rest differences in sensory cortex. The SAD hypothesis may thus be complemented in neural regard by the assumption of social withdrawal and isolation triggering the predisposition to neural confusion between different neuronal differences in sensory cortical different-based coding.

How is this related to early biopsychosocial trauma? Early biopsychosocial trauma as for instance a genetic predisposition to abnormally intense perception or an intrusive mother or father (as in the case of Schreber) may destabilize the neural constitution and construction of neural differences. If for instance the perception is abnormally intense, the exteroceptive stimuli may be blocked (see part one for details) and its neural coding may be replaced by the neural coding of internally generated neural differences with rest–rest replacing rest–extero. The same may hold in for instance a physically or mentally over intrusive mother or father whose exteroceptive input may either be blocked or abnormally processed as either an interoceptive or cognitive stimulus rather than exteroceptive. This predisposes the neural confusion of the normally processed neuronal difference of intero-extero with the one of rest–intero in for instance the insula and its subsequent reverberations on neural coding of the rest–(rest-intero) difference in the aCMS and self-related processing.

Although they're general, unspecific, and not sharp, these examples at least indicate how the presumed unstable neuronal differences in difference-based coding may be related to biopsychosocial trauma in both early and late development and consecutively to the constitution and construction of volatile self–object boundaries. This means that my hypothesis of altered difference-based coding (ADC) in psychosis is a truly biopsychosocial hypothesis rather than being either a purely biological or social hypothesis. By being both biological and social at the same time, my ADC hypothesis is truly biopsychosocial that focuses on abnormal differentiation between biological and social dimensions in psychosis. In short ADC is about what one may want to call "Biological–Social Differentiation" as enabled and predisposed by the brain which makes it well compatible with Mentzos' concept of "psychosomatosis of the brain."

I would though like to specify the concept of "psychosomatosis of the brain" in both neuronal and psychodynamic regard. Neuronally, one may want to speak of a "differentiosis of the brain" since the process of differentiating and thus of constituting the "right" neuronal differences in difference-based coding seems to be affected in psychosis. It is not that neural differences cannot be generated anymore; rather the process of their constitution as differences is affected or better even infected by a virus that forces the brain to code and spread false positive neuronal differences with the subsequent confusion between rest–extero and rest–extero. The "differentiosis of the brain" may thus be compared with an infection by a virus that changes the direction of otherwise healthy and well functioning physiological mechanisms like difference-based coding.

Rather than disturbing and destroying difference-based coding, it is the way that difference-based coding is used and applied when it is redirected from the "right" to the "false" neural differences, i.e. from rest–extero to rest–rest. In other words, the virus affects only what is coded as neuronal difference in difference-based coding and thus its contents while leaving the coding itself (as distinct from its contents) basically intact. And as we could see especially in the case of the sensory cortex, this redirection is "located" right at the interface between brain and world and hence between biological and social dimensions thus being truly and literally biopsychosocial. In other words, the "differentiosis of the brain" affects the differentiation between biological and social worlds and thus the "Biological–Social Differentiation."

The "differentiosis of the brain" may well correspond within the psychodynamic context to what may be called "differentiosis of the psyche." In the same way the constitution of neuronal differences between different kinds of neural activities, i.e. rest, intero and extero, is affected by metaphorically speaking of a "neural virus," the latter also enables and predisposes abnormal constitution of self–object boundaries as being crucial for the co-constitution and co-occurrence of self and object and subsequent self–object differentiation in the psychodynamic context. Since the boundaries between self and object are volatile, their difference and consecutively self and objects themselves can no longer be properly co-constituted and co-constructed. Analogously to the brain, the construction of self and objects including their boundaries is not by itself faulty, it is only the way these mechanisms are used and applied. In the same way the "neural virus" redirects difference-based coding to the false differences, its psychodynamic twin, the "psychodynamic virus," redirects the constitution of outer self–object boundaries towards the inner self–object boundaries. This though makes the outer and consecutively inner self–object boundaries rather volatile which in turn strongly affects the constitution and construction of self and objects and subsequently self–object differentiation.[15]

[15] One may want to raise the question for the difference between psychosis and dreams since the latter can also be characterized by violations of the self–object boundary (see chapter two in part three). In the same way I characterized dreams by the neural confusion between rest-extero and rest-rest, I considered the same confusion to be crucial in psychosis. Is psychosis thus nothing but a specific of form of dreams? While many authors including Freud himself (see Llewellyn, 2009 for an overview) have argued in this direction, the resemblance between dreams and psychosis applies predominantly to positive symptoms, e.g. delusions and hallucinations, while it does not pertain to negative symptoms, ego-disturbances and religious contents (see Llewellyn, 2009 for a nice review). The main neural differences, from my point of

Neuroexistential account

One may finally want to raise the question why the neuronal and psychodynamic mechanisms of difference-based coding and self- and object constitution are still used and applied in a normal way. If there are false neuronal differences and volatile self–object boundaries, one could simply diagnose them as such and not even start to activate and recruit the mechanisms of difference-based coding and self- and object constitution. In such case, I postulate that their disastrous consequences, i.e. the outbreak of psychotic symptoms, would not occur. Hence it would be much for easier for both brain and self if the described neuronal and psychodynamic mechanisms would not be activated and recruited.

Why are the neuronal and psychodynamic mechanisms of difference-based coding and self- and object constitution nevertheless recruited and activated? I speculate that this is so because the mechanisms themselves, difference-based coding and self- and object-constitution, are central and existential mechanisms without which the brain's functioning and the own existence are called into question. Let me be more specific. I assumed difference-based coding to be *the* neural code of the brain according to which all information be it of neural, interoceptive, exteroceptive or cognitive origin, is processed. This means that if difference-based coding ceases meaning that no differences are neurally coded and processed anymore, the brain ceases to function–difference-based coding must thus not only be regarded central but also existential to the brain's function meaning that conceptually the latter is defined by the former.

What does this mean for the neuronal context of the brain and our mental context of subjective perception and experience? To maintain difference-based coding is consequently to maintain one's brain's function and its existence (and, psychologically speaking, one's cognitive states, and philosophically speaking, the transformation of one's neuronal states into mental states, i.e. neuronal–mental transformation; see Chapter 5 on cathexis and mental states). While the loss of this battle for the brain's function and existence (and one's cognitive and mental states) results in the subjective perception and experience of a dead or altered brain (and/or body) which is exactly what many psychotic patients, for instance Schreber perceive and experience.

What does this entail for the psychodynamic context? Analogously so in the case of self- and object-constitution. The constitution of self and objects is considered not only central but existential within the psychodynamic context. If there is no constitution of a self, one can no longer subjectively perceive and experience the own self as existent; since the constitution of the self is very much dependent on the constitution of objects, the existence of the self is also at stake when the constitution of objects is threatened. Hence, to preserve the subjective perception and experience of the existence of the own self, the constitution of objects needs to be maintained by all means. To maintain the constitution of objects and subsequently of the subjective existence of the own self is thus not only central but existential, a matter of life and death. If one loses the battle

view taken within here establish neural context, concern the absence of continuous exteroceptive input in dreams while in psychosis there is continuous sensory input thus requiring parallel processing between true rest-extero and rest-rest as rest-extero. Without being able to go into details, I postulate that this difference at the beginning leads to dire consequences at the end, more specifically in the coupling of aCMS and pCMS with the subsequent distortions of the own self. I consequently assume that while the first and second neuropsychodynamic hypotheses in psychosis may somehow apply to dreams too, the third and the fourth one do not apply to dreams. Unlike in psychosis, the dreaming person is therefore not being exposed to the existential dilemma as the mutually exclusive choice between either self- or object constitution. This entails, as I assume that brain–self and brain–object differentiation are maintained and well sustained in dreams while they are threatened in psychosis and it is because of that dreams are not as existential as psychosis.

for the constitution of objects and self, the price one pays for that consists in the (subjective experience and perception) of the death of the own self and its identity. And this is exactly the way many psychotic patients perceive and experience their own self, i.e. as lost and dead.

As demonstrated previously, the brain's difference-based coding is strongly related to the constitution of self and objects and even more specifically it is supposed to enable and predispose to what I called brain–object differentiation (see Part I psychosis and the Appendix here). This entails that the "neuronal struggle" to maintain difference-based coding at any price no matter what neuronal difference must be considered first and foremost an "existential struggle." The brain's "neuronal struggle" (metaphorically speaking) is not purely neuronal but also existential because by maintaining neuronal differences it enables and predisposes the existence of the self in subjective perception and experience. In other words, the "neuronal struggle" for neuronal differences in difference-based coding is transformed into an "existential struggle" as soon as it enables and predisposes the differentiation of the own self from both brain and objects, i.e. brain–self, brain–object, and self–object differentiation.

Loss of the "neuronal struggle" has the dire consequence that one loses the existential struggle and thus the existence of the own self (in subjective perception and experience) and this is exactly what many psychotic patients experience and perceive. Hence, to maintain any kind of neuronal difference (whatever its origin) in the brain's difference-based coding becomes a matter of life and death for the (subjective perception and experience of) existence of the own self. If one loses the "neuronal struggle," one also loses the "existential struggle" for the self and consequently perceives and experiences the own self as mere organic matter or as the philosopher Schopenhauer said as the "gray pulpy mass of our brain." And this is exactly how many psychotic patients as for instance Schreber perceive and experience their own self; this may psychodynamically correspond to the moment where brain–object differentiation and consecutively brain–self and self–object differentiation break down (in subjective perception and experience).

Appendix

What Can We Learn From Depression and Psychosis? A Transdisciplinary and Neuroexistential Account

Background

Freud was an excellent clinician who not only treated his patients but also used their symptoms, including their perceptions, experience, and histories, to learn something about how the healthy mind and its psychic apparatus work and function. He considered his patients' clinical presentations as the entrance door and road map to how the psychic apparatus must function in general, including its basic principle that is at work in both the "normal" and the "pathological" context. The guiding question here was: How must the psychic apparatus be structured and organized and thus function in order to be able to bring forth the kind of symptoms and experiences that can be observed in pathological cases?

What does this imply for our cases of depression and psychosis within the context of neuropsychoanalysis? I postulated in Parts II and III of this book that the brain's neural mechanisms such as rest–stimulus and stimulus–rest interaction, difference-based coding, intero-exteroceptive linkage, affective assignment, and self-related processing may enable and predispose to the constitution of both self and objects as distinct from the brain (i.e. brain–self and brain–object differentiation) (see Figures A1 and A2). However, lacking direct experimental support, this hypothesis stands on rather shaky ground and may therefore be considered by many, including both neuroscientists and (neuro)psychoanalysts, as speculative (see Epilogue). How can I provide further and better evidence for my neuropsychodynamic hypotheses of these neuronal mechanisms that enable and predispose to brain–self and brain–object differentiation? One way is to gather direct empirical support by conducting the respective experimental investigations which, at least at this point, has potential for the future. Another way is to adopt Freud's approach and to assemble some indirect support, as for instance from clinical cases, which is exactly what I intend to do here.

Depression

Let me be more specific. I characterized depression neuropsychodynamically as a disorder of brain–self differentiation (see Chapter 10). If the brain's resting-state activity and subsequent stimulus–rest interaction are indeed enabling and predisposing neural conditions of possible brain–self differentiation, one would expect them to be altered in depression. This is exactly what I described in Chapter 10. Depression can indeed be characterized by an abnormally elevated resting-state activity, which in turn may change and alter subsequent stimulus–rest interaction and ultimately how the depressed patient subjectively perceives and experiences their own self in relation to their brain (and body) and other objects. Thus the case of depression may be considered to provide some indirect support for the neuropsychodynamic hypothesis that the brain's

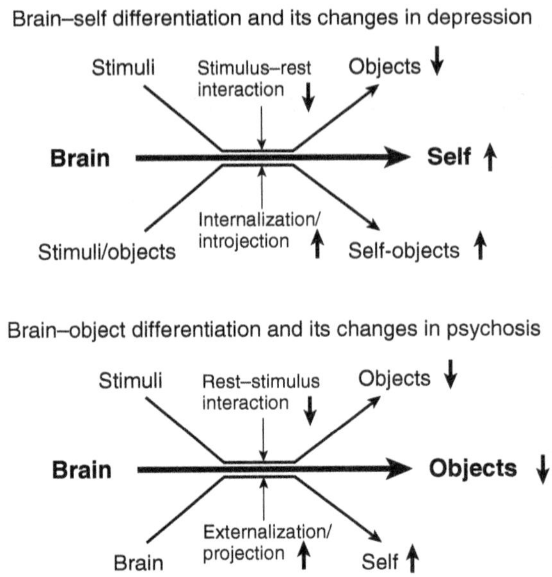

Fig. A1 Brain–self and brain–object differentiation.

resting-state activity and subsequent stimulus–rest interaction may enable and predispose to brain–self differentiation.

However, we need to be more specific. Depression is characterized by symptoms such as extremely negative affects/emotions and bodily symptoms, as well as by an increased self-focus. How is this related to my neuropsychodynamic hypotheses? I described intero-exteroceptive linkage and affective assignment as enabling and predisposing features in the constitution of self-objects, which in turn are central in constituting and stabilizing the self (see Chapter 7). Due to negative affects and bodily symptoms, depressed patients may be characterized by abnormally negative affective assignment and altered intero-exteroceptive linkage in the constitution of their self-objects. I demonstrated that both of these characteristics may be related to the abnormally elevated resting-state activity and thereby to reduced stimulus–rest interaction at the neural level and increased self-focus at the psychological level. The symptomatic abnormalities may thus be closely related to their abnormal self- and self-object constitution, and ultimately to altered

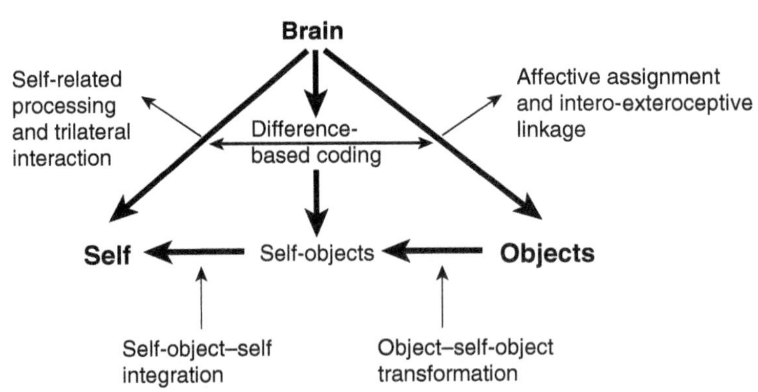

Fig. A2 Balance and mutual dependence between brain–self and brain–object differentiation via difference-based coding in healthy subjects.

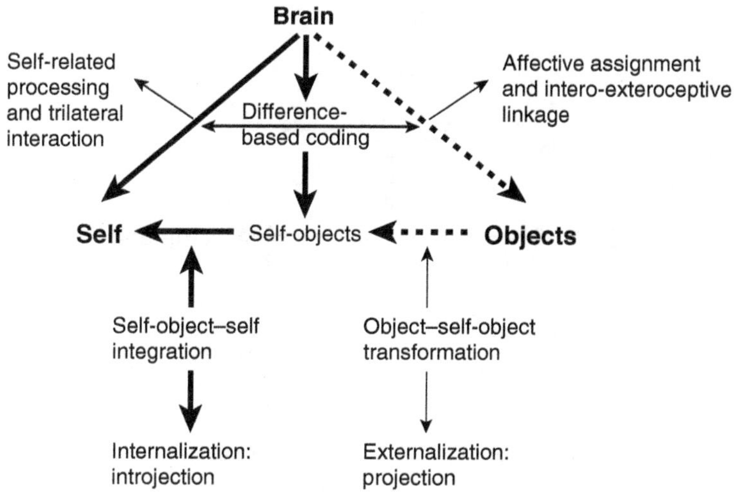

Fig. A3 Shift in balance between brain–self and brain–object differentiation in depression.

brain–self differentiation as the main overarching characteristic of depression. Therefore the case of depression may lend some indirect support to my neuropsychodynamic hypothesis that the constitution of the self and self-objects and their subsequent self–self-object integration may be enabled and predisposed to by intero-exteroceptive linkage and affective assignment (see Figure A3).

Finally, the case of depression also lends some indirect support to another neuropsychodynamic hypothesis of mine, namely that the constitution of self and objects and hence internalization and externalization are mutually dependent on each other. Both internalization (i.e. self) and externalization (i.e. objects) are supposed to balance each other, with changes in one affecting the other (see Chapter 6). Due to its increased self-focus, depression may be a paradigm for studying this hypothesis of the balance between internalization and externalization and their mutual dependence. Depression can be characterized as a case where increased introjection (as a form of internalization) is accompanied by decreased projection (as a form of externalization).

Most importantly, I postulated the balance between internalization and externalization to be mediated neuronally by the balance between rest–stimulus and stimulus–rest interaction and psychologically by self-related processing. If this holds, one would expect imbalance in either of them, which is exactly what seems to be the case in depression as it is supported by empirical data. Taking all of these threads together, the case of depression and its underlying neuronal mechanisms may lend some indirect support to my hypothesis that the brain's resting-state activity and subsequent rest–stimulus and stimulus–rest interaction enable and predispose to the constitution of self-objects (via affective assignment and intero-exteroceptive linkage), the internalization–externalization balance, and thus, more generally, brain–self differentiation.

Psychosis

How about psychosis? I characterized psychosis as a loss of object relations, including lack of (self-stabilizing) self-objects. As such psychosis may lend some indirect support to my hypothesis that brain–object differentiation is enabled and predisposed to by difference-based coding and subsequent rest–stimulus interaction. If so, one would expect abnormal difference-based coding and rest–stimulus interaction in psychosis. This has been demonstrated in psychosis where, on the basis of empirical findings, I assumed the neural coding of false-positive neural differences in

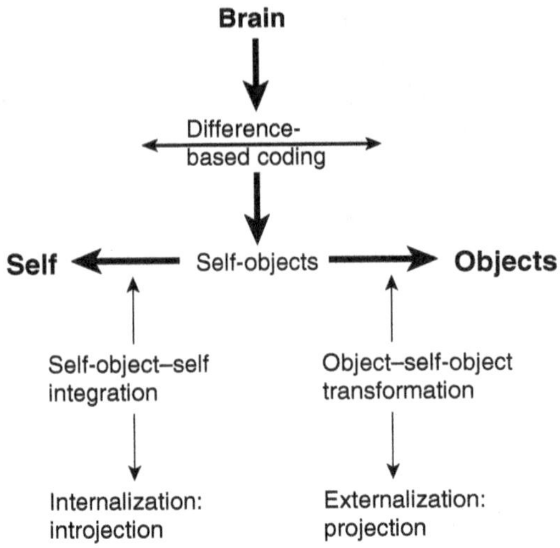

Fig. A4 Dilemma instead of balance between brain–self and brain–object differentiation in psychosis.

sensory cortex, insula and anterior cortical midline structures to enable abnormal object relations. The case of psychosis may thus lend some indirect support to my hypothesis that difference-based coding in general and the neural coding of the "right" neural differences in difference-based coding in particular may enable and predispose to the constitution of objects and thus brain–object differentiation (see Figure A4).

Psychosis may also lend some indirect support to the hypothesis that the neural dynamics of the brain are determined by distributing and balancing energy (i.e. resting-state activity). The withdrawal of energy from objects (i.e. the decathexsis of objects) is accompanied by hypercathexis of the self. This shift in the investment of energy (i.e. cathexis) may correspond neuronally to the increased resting-state activity and hyperconnectivity in anterior cortical midline structures that are observed in psychosis, which seems to correspond psychologically to increased self-related processing. Thus the case of psychosis lends indirect support to the hypothesis that the distribution and balancing of energy (i.e. resting-state activity) may enable and predispose to a certain balance between self and object and thus be crucial for maintaining the balance between brain–self and brain–object differentiation.

Most importantly, the case of psychosis lends support to the mutual dependence between self- and object constitution by confronting the patient with an "existential dilemma" where they encounter an all-or-nothing choice between self and object constitution. Based on empirical findings, I assume that this may be related to altered (i.e. increased) connectivity between anterior and posterior cortical midline structures, which again demonstrates the reciprocal and mutual dependence between self and object constitution. This is because without such reciprocal and mutual dependence, the dilemma that the psychotic patient encounters would not be existential and would thus not concern the existence of their own self when trying to resume (and compensate for lacking) object constitution. Therefore the case of psychosis lends some indirect support to the neuropsychodynamic hypothesis that the balance between anterior and posterior cortical midline structures enables and predisposes to the mutual dependence between self and object constitution.

Finally, the case of psychosis provides some indirect support for the biopsychosocial nature of our brain that appears to be located right at the border between biological, psychological, and social domains. More specifically, I postulated that difference-based coding and thus what I called "constitutive context dependence" (see Chapters 4 and 5) enable and predispose to the apparent biopsychosocial nature of the brain. Changes in difference-based coding and constitutive context dependence should then lead to major changes in the relationship between brain and world, and thus in what neuropsychodynamically is called brain–object differentiation. This is exactly the case in psychosis, where early biopsychosocial trauma in infancy may lead to major changes in later object constitution and hence brain–object differentiation.

Neuroexistential account

Taken together, the neuropsychiatric disorders of depression and psychosis lend some indirect support to my general neuropsychodynamic hypothesis that the brain's resting-state activity and difference-based coding enable and predispose to brain–object and brain–self differentiation. Neuropsychodynamically, depression and psychosis may thus be regarded as instances of border violations. Depression violates the boundary between brain and self, whereas psychosis disrupts the boundary between brain and objects. As such these disorders lend indirect support to my hypothesis that the brain's resting-state activity and difference-based coding may enable and predispose to the constitution of both kinds of boundaries.

More generally, depression and psychosis may be regarded as violations of the boundaries between the brain and the world. And the dramatic nature of their symptoms and our difficulties in understanding and revealing their underlying neuronal mechanisms tells us not only about depression and psychosis, but also about ourselves as humans and about our brains, by touching upon the brain–world boundary. As has already been pointed out, psychosis in particular but also depression show that the brain–world boundary and thus the brain–self and brain–object boundaries are intrinsically biopsychosocial, and that their characterization as either biological or social is doomed to fail. Therefore the brain–world boundary requires a truly biopsychosocial approach that crosses the boundaries between the respective disciplines in order to properly account for it.

Moreover, the alterations in psychotic and depressed patients' possible knowledge of themselves and the world tells us that the brain–world boundary is also epistemically and conceptually relevant, determining what we in principle can and cannot know about ourselves and the world. This makes it clear that the brain–world boundary may also be a target for epistemology in particular and philosophy in general, as was discussed in Part I of this book. For instance, the brain–world boundary as enabled and predisposed to by difference-based coding may prevent us from directly perceiving and experiencing our own brain and its neuronal states as such, which I referred to in Chapter 1 as autoepistemic limitation.

Conceptually, the brain–world border may represent the border between the different concepts of the brain, namely the brain as functioning, the brain as observed, and the brain as experienced. It is here where these three distinct concepts of the brain may have their common and ultimate origin, and it is here where our concepts converge with our epistemic abilities and inabilities, as well as with out empirical observations. Thus the brain–world boundary may be not only a biopsychosocial interface but also a conceptual–epistemic–empirical node point. This makes it clear that the brain–world boundary must also be considered to be truly transdisciplinary, as it cannot be located in one particular discipline, but rather it is situated in between different disciplines such as neuroscience, philosophy, psychiatry, sociology, anthropology, and psychoanalysis, among others.

Finally, the cases of psychosis and depression show that the brain–world boundary is truly existential, as otherwise they would not cause such a threat to the patient's own existence in their subjective experience and perception. It is here at the brain–world boundary where the subjective experience and perception of our own self and its existence is located, which I demonstrated to be enabled and predisposed to by our brain and its specific neural coding (i.e. difference-based coding). This makes our brain, including its difference-based coding and subsequently the brain–world boundary, existentially relevant and thus what I here describe as "neuroexistential." How do we know that the brain–world boundary is not merely neuronally but rather neuroexistentially relevant? This is the lesson that we can learn from the depressed and psychotic patients who pay the price for the neuroexistential nature of the brain–world boundary in their existential sufferings.

Epilogue

The Beauty of Transdisciplinary Failure–A Trialogue

In lieu of a Conclusion, there now follows a dialog between a psychoanalyst (PS), a neuroscientist (NS), and a philosopher (PH) about neuropsychoanalysis.

> Philosophy is not opposed to science, it behaves itself as if it were a science, and to a certain extent it makes use of the same methods; but it parts company with science, in that it clings to the illusion that it can produce a complete and coherent picture of the universe, though in fact that picture must fall to pieces with every new advance in our knowledge. Its methodological error lies in the fact that it over-estimates the epistemological value of our logical operations, and to a certain extent admits the validity of other sources of knowledge, such as intuition.
> (Freud, 1933)

Transcendental and empirical views of the brain

PH: While I am not really entitled and able to judge the second and third empirical parts of the book, I am certainly qualified enough to comment on the theoretical chapters. The transcendental view of the brain still smells like a category error to me.

NS: What do you mean by that?

PH: People and their minds acquire knowledge. This is an epistemic feature. The author's error lies in the fact that he relates such an epistemic feature to the brain and its neuronal states as a mere empirical feature. However, this is to confuse people with brains, and thus to confuse epistemic and empirical categories. Isn't that an error?

PS: But wouldn't the person be unable to acquire knowledge at all without a brain? People are their minds—and minds are brains. Therefore the person is nothing but his brain. What you call different categories thus implode once one considers reality.

PH: You don't know how much it hurts to see an intelligent person like yourself making such a simple mistake. Of course you are right in saying that we could not acquire any knowledge without the brain. How though do you describe your observations? You use concepts, don't you? The observations themselves are not the same as the concepts that you use to describe them. Observations refer to the brain, concepts refer to the person.

NS: Don't they both refer to the brain? Concepts merely describe what we observe in the brain. Isn't that a purely artificial distinction?

PH: No. It becomes relevant in the distinction between the author's transcendental and empirical views of the brain. The transcendental view describes knowledge acquisition, an epistemic feature that can only be associated with people, whereas the empirical view concerns mere observations of the brain's neuronal states.

Since the transcendental view refers to people rather than brains, it is senseless to speak of a transcendental view of the brain. The only transcendental view we can take is of people. The author confuses people and brains when he claims a transcendental view of the brain—a clear instance of a category error!

PS: Objection! You may be right in diagnosing a category error when the empirical view of the brain is associated with a person's knowledge acquisition. But does that hold also for the transcendental view of the brain?

NS: Can anyone finally explain me what is meant by "transcendental view of the brain"?

PS: As I understand it, the transcendental view of the brain concerns those features of the brain that originate from within the brain itself—the "brain's input," as the author says. But there is another input yet—the observer's input, what the observer himself puts into his own observation of the brain. Our observation of the brain in the empirical view is thus a combination of the brain's input and the observer's input. How can we grasp the brain's input itself as distinct from the observer's input? For that, the author argues, we need to take a transcendental view of the brain that will lay bare the brain's input itself.

NS: Nice chimerical fantasies. What on earth is this "brain's input" if not the neuronal states that we observe? What I observe is nothing but the brain itself and its neuronal states. I do not put anything into the observation of the brain by myself. There is no observer's input that is distinct from the brain's input. If, however, there is nothing but the brain's input in our observation of the brain, we do not need any transcendental view of the brain. The empirical view tells us everything there is. What the author calls "transcendental view of the brain" turns out to be a mere fantasy.

PH: Well reasoned, but flawed nevertheless, Mr. Neuroscientist. Kant argued that a transcendental view of the mind allows us to reveal the mind's input to our cognition and knowledge. His concern was a truly epistemic one, that asks for the conditions of our knowledge of the mind and the world. Following Kant, Schopenhauer replaced the mind by the brain when he characterized the brain rather than the mind as the subject of cognition. However, this is to confuse not only empirical and epistemic concerns but also the subject and object of cognition. The brain can only be an object of cognition—it is an object of our knowledge—whereas the brain cannot be a subject of cognition, since only people but not brains can acquire knowledge. Our brain is a mere gray pulpy mass, as Schopenhauer himself called it once, and nothing else. Do you really think that such a gray pulpy mass can acquire knowledge and lead dialogs like this?

PS: You mean then that a transcendental view of the brain is impossible because that would be to presuppose the brain as subject of cognition?

PH: Yes, that is exactly what I mean. The brain cannot be the subject of our knowledge. The brain is only an object of our knowledge, which precludes any transcendental view of it. Kant got it right. Schopenhauer and our poor author got it wrong. They confused the object and the subject of cognition. Our brain's gray pulpy mass is nothing but an object of cognition and can never be its subject.

PS: Impressive reasoning. I concede, though only in this case.

NS: Impressive reasoning. Yes certainly so. But is it really relevant? Who cares whether the brain is the subject or the object of cognition? That is pure logic, play with concepts. Reality is what counts. You cannot infer reality from your concepts. Reality is not logical reality. Reality is what you observe out there in the world, reality is empirical reality. What I observe in empirical reality is the starting point for acquiring knowledge about the world! To start otherwise with mere concepts is to infer empirical reality from the logical world of concepts. That will give you knowledge only about your concepts and their logical rules, whereas it will not provide any knowledge about the world itself. Who cares about your concepts if my observation of the brain tells me everything there is?

Brain, mind, and the psychic apparatus

PS: I have another objection to the category error. The error does not concern the confusion between empirical and epistemic domains, but between neuronal and psychodynamic categories. This is most apparent in the author's concepts of brain–self and brain–object differentiation. He here confuses the neuronal category of the brain with the psychodynamic one of self and objects. How can brain and self/objects differentiate from each other if they belong to different categories? The concepts of brain–self and brain–object differentiation are thus as nonsensical as the transcendental view of the brain.

PH: You learned your introductory philosophical lesson rather well and quickly. Yes, the concepts of brain–self and brain–object differentiation do certainly suffer from a category error. What you describe as confusion between self/object and brain surfaces as confusion between mind and brain in the philosophical context. Analogous to your case, the concept of brain–mind differentiation is a rather hybrid concept: On the one hand, the concept of the brain is purely mechanical and thus scientific, and as such does not carry any meaning in itself. On the other hand, the concept of mind refers to meaning when describing meaningful contents, so is hermeneutical rather than scientific.

NS: Well described. But that does not prevent the brain from constituting the very meaning you exclusively associate with the mind or the self and its objects.

PH: I again feel sensations of mental pain. Are you blind? How can you confuse the meaningful with the meaningless? How can you reduce hermeneutics to science? These are two fundamentally different categories. To reduce hermeneutics to science is to replace the meaningful with the meaningless.

PS: I don't understand. Could you be a little more concrete?

PH: Differentiation presupposes some kind of common origin, as otherwise differentiation would be impossible. By assuming brain–self and brain–object differentiation, our author implicitly presupposes some kind of common origin.

PS: Now, I understand what you mean by mental pain. Freud would have indeed shaken his head. Self and objects do not differentiate from the brain. Instead, self and objects differentiate from what Freud called the "psychic apparatus." Our author got it wrong. He needs to replace his concepts of brain–self and brain–object differentiation with those of "psychic apparatus–self differentiation" and "psychic apparatus–object differentiation."

NS: You are always so quick to sweep over the facts with one glance. How foolish can one be in not seeing that what Freud called the "psychic apparatus" is nothing but the brain and its neural apparatus? The psychic apparatus is the brain. Our author got it right here. Freud's psychoanalysis needs to be traced back to neuroscience.

PH: Strangely, this may be the first time in my life I have ever agreed with a neuroscientist. Our author claims that Freud's psychic apparatus corresponds to what he calls the "brain as functioning" as distinct from the "brain as observed." And, as I understand him, he assumes that brain–self and brain–object differentiation concern the differentiation between the "brain as observed" and self/objects with the "brain as functioning" as their common origin.

PS: My dear philosophical and neuroscientific colleagues, don't you see empirical reality? Brain remains brain, whatever concept (and view) you use to describe it. Since any concept of the brain cannot but refer to the brain, the "brain as functioning" is just a conceptual disguise for the "brain as observed." Neither can account for the meaningful contents associated with psychodynamic concepts such as self and objects and the psychic apparatus. Self and objects do not differentiate from the brain – they differentiate from each other as well as from the psychic apparatus. Whereas the author's concepts of brain–self and brain–object differentiation need to be discarded.

PH: I agree with this. Brain is brain, whether it is described as brain as observed or brain as functioning. Since the brain remains the brain in either concept, it is also impossible to associate the brain with any kind of hermeneutics. The brain remains nothing but a purely physical and material organ, and thus meaningless by itself. The brain is nothing but a gray pulpy mass! Such a gray pulpy mass has nothing to do with either the epistemic or hermeneutical domain.

NS: Surprisingly so, but I agree here, though for different reasons. Brain is brain, no one can dispute that. However, I disagree that mind, as the philosopher seems to call it, or psychic apparatus, as the psychoanalyst prefers to say, are largely different from the brain. Brain is brain, which means that mind or psychic apparatus is nothing but the brain, too. To put it philosophically, the concept of the brain implies all of the features and meanings originally attributed to mind or psychic apparatus, as you call your chimerical entities. The concepts of brain–self and brain–object differentiation should be replaced by those of "self–brain and object–brain reduction." To assume otherwise is as superfluous as adding a fifth wheel to your car. Or doesn't your car already drive well on four wheels?

PS: Now I do not only understand but even feel sensations of pain in the same way as Mr. Philosopher expressed before. How can one sweep away the meaningful contents of objects and the self that were so richly and colorfully described by Freud and replace them with the mere gray pulpy mass of the brain? What a brutal act of scientific arrogance!

NS: You confuse me with reality. I am not brutal—I merely describe reality. Reality itself is brutal!

PS: No, it is you who confuse things. You confuse the hybrid nature of the psychic apparatus with the brain. Freud regarded the concept of the psychic apparatus as truly hybrid by its very nature. On the one hand, the psychic apparatus describes purely mechanical and meaningless processes, while on the other it constitutes meaningful contents as objects and self. To reduce the psychic apparatus to the brain is to disregard the latter, the meaningful contents, in favor of the purely mechanical and meaningless processes. However, that would be like moving a bridge from across the river and placing it to one side of the river. Wouldn't that defeat the purpose of the bridge, if you are no longer able to cross to the other side? Now you know what our author does to the psychic apparatus if he reduces it to the brain.

PH: I disagree, Miss Psychoanalysis. Yes, the concept of the psychic apparatus is hybrid, but it is hybrid not by its very nature but due to your conceptual confusion. Hermeneutics and science are two different categories. They are not like two sides of the same river that can be bridged. They are not like two different centres which can be connected by airplanes. Instead, they are two different universes that cannot be connected. Your hybrid concept of the psychic apparatus is thus nothing but conceptual confusion. The concept of the psychic apparatus is indeed as superfluous as a fifth wheel on a car. Our author is therefore right in replacing the concept of the psychic apparatus with the one of the brain. He even presents an interesting alternative to the mind–brain problem. The current mind–brain discussion focuses on revising the concept of mind to make it amenable with the one of the brain. One may call this the "mind path." By discussing different concepts of the brain, our author tackles the mind–brain problem from the opposite end of the brain, a "brain path" as one may want to call it.

PS: "Brain path" or "mind path"? You don't need any of these concepts. All you need is the psychic apparatus. The psychic apparatus is neither mind nor brain. The psychic apparatus is psychological reality. You cannot start with some concepts like mind or brain and then infer psychological reality from them. You have to start with psychological reality itself. You have to take the "psychic apparatus path." Your "mind path" and "brain path" are utterly flawed conceptual fantasies that you imprint on the psychological reality of the poor psychic apparatus. Nonsense, sheer conceptual nonsense.

Knowledge of the brain

PS: As much as it hurts to admit it, our confused author converges with Freud on another point. Both argue that we may suffer from an epistemic limitation with regard to either the psychic apparatus (Freud) or the brain (the author). We may not be able to completely and directly know the brain as functioning, which therefore may not be identical to the brain as observed. What's up, Mr. Neuroscientist?

NS: I decline to comment. Why comment on something that is nonsense? We observe neuronal processes which make the brain functioning. The brain as observed is the brain as functioning. End of story! To assume otherwise is to assume some non-neuronal properties of the brain that cannot be observed. However, there is nothing more to the brain than what we can observe. The author's doubling of the concept of the brain is consequently utterly flawed and nonsensical.

PS: That's not a very friendly comment. We have to understand, though. Who likes to admit borders to a kingdom, the kingdom of neuroscience, that claims to cover the whole universe and thus to be without borders? Freud already knew. The narcissistic fantasies of grandeur do not like to be spoiled by something as trivial and contingent as borders.

NS: I am almost certain that this is nothing but a projection on your side, Miss Psychoanalysis.

PS: Now he is getting personal.

NS: Who is projecting here upon whom? Aren't you projecting your own fantasies about the psychic apparatus upon me as neuroscientist? You should be thankful that I serve as the projection surface behind which you can hide your own narcissism. Maybe I should demand a psychotherapeutic fee?

PH: Let us be serious again. Yes, the emphasis on epistemic limitation in our possible knowledge of the brain is indeed interesting. That reminds me a little of the claims of philosophers like Colin McGinn, who argue that we suffer from a primary inability to fully know the brain. However, they assume some additional ontological properties as distinct from both mental and physical properties. The exact nature of these properties remains unclear, though, which makes them rather mysterious. Our author seems to be prone to the same problem. How does he characterize the brain as functioning? Either he assumes some additional properties, which makes him rather mysterious, or he refrains from assuming additional properties, in which case his brain as functioning can no longer be distinguished from the brain as observed.

PS: Cold logic, nothing but cold logic and empty concepts, Mr. Prosecutor. Yes or no. All or nothing. That is all you can think. The concept of the psychic apparatus is exactly what you are looking for. It does not assume any additional rather mysterious properties because it refers to nothing but psychological reality. While at the same time it is more than we can observe when describing the psychological structure and organization rather than specific psychological contents. You describe such a concept of the psychic apparatus as hybrid and logically flawed. I describe it as psychological reality and empirical.

NS: Strange, very strange. You accept empirical reality as hybrid in the case of the psychic apparatus, whereas you reject it for the brain. The psychic apparatus can integrate and link mechanical meaningless processes and meaningful contents, but the brain cannot. You accept that psychological reality is hybrid but you reject neuronal reality. Strange, very strange. Our author does exactly the same as you. The only difference is that he extends your beloved psychological reality of the psychological apparatus to the neuronal reality of the brain. How can you reject the latter while at the same time embracing the former?

PS: It's very simple. The psychic apparatus can be characterized by mental states while it does not include any neuronal states. Doesn't that provide enough evidence that there may be some transition between mind and psychic apparatus while the latter is not continuous with the brain? Even worse, you go beyond your own knowledge boundaries. How can you know that the psychic apparatus is nothing but the brain if you cannot have complete and direct knowledge of the psychic apparatus itself? You seem to stubbornly resist the assumption that you may suffer from a limitation in your knowledge of the brain. The blind spot in your knowledge of the brain is in exactly the same place as the psychic apparatus is located.

NS: You and your psychic apparatus. Enshrine it and pray at least five times a day to it. You suffer from a limitation, not me. Your limitation is that you do not see your own illusion. Your concept of a psychic apparatus is nothing but an illusion. What you call a limitation is based on an illusion that there is more to empirical reality than you can observe. The psychic apparatus is nothing but the brain as we can observe, nothing less and nothing more. To deny that is not only to ignore the neuroscientific data but, even worse, to deny the brain and reality altogether.

PH: Naivety, pure naivety, the mentality of a two-year-old. How could I ever have expected a neuroscientist to see the difference between observation and knowledge, between empirical and epistemic domains? Rather than rejecting epistemic limitation, he only illustrates to me how limited we are in our knowledge.

NS: Of course, not being used to reality in his ivory tower, the philosopher cannot do other than simply invent something mysterious. One needs to understand. If he was to admit empirical reality, his ivory tower would be demolished. He must be mysterious in order to survive!

PS: Well analyzed. But let us go back to our author. The limitation he is talking about does not concern some mysterious properties, as our detached philosopher says. Rather it refers to what Freud called the psychological structure and organization. Our author now extends this to the brain. He argues that the brain's neural structure and organization cannot be accessed and known directly as such. He speaks of autoepistemic limitation. Since he assumes that the brain's neural structure and organization constitute mental states, autoepistemic limitation prevents us from accessing and knowing the relationship between neural and mental states. Most interestingly, he associates such "neuronal–mental transformation" with neuronal mechanisms such as a specific neural code.

PH: Is that true? This provides a novel philosophical approach to the perennial mind–brain problem which motivates me to get back into it . . .

NS: Are you sure? Maybe I need to read the book again . . .

Subjectivity and the brain

PS: Another interesting issue is the method suggested here, namely neuropsychodynamic iterativity. The author attempts to out psychodynamic concepts within the context of neuroscientific concepts, and vice versa.

PH: To be honest there is nothing innovative. Worse this method is flawed and naive. The author is apparently not aware of the basic distinction between hermeneutics and science which was long ago established. This is the bedrock foundation for the distinction between humanities and science. How can he mix the two in such a naive and childish way?

PS: Mr. Philosopher, you are always so quick to denounce anything that does not fit your pre-established categories. Everything is nothing but conceptual confusion and category error. Have you ever thought of questioning the categories themselves?

PH: I'm appalled. How can you doubt the bedrock foundation of our disciplinary divisions? Are you self-destructive? Your own discipline, psychoanalysis, would not exist if hermeneutics could not be distinguished from science. You could not even get started with your psychoanalytic confusions if hermeneutics could not be distinguished from science. How can you saw off the very branch that you are sitting on?

PS: This shows me even more clearly that you have no sensitivity to context. The psychological context is different from your conceptual–logical one. Take, for instance, the concept of cathexis, which Freud himself introduced to describe the discharge of psychic energy. This concept is clearly bound to the psychological context, whereas it would make no sense in your conceptual–logical context. To understand the concept of cathexis, you thus need to consider its specific context. You cannot detach concepts from their specific contexts, including the categories associated with the respective context.

NS: Why not simply replace the concept of psychic energy with the concept of neuronal energy? You simply change the context from a psychological one to a neuroscientific one. What Freud described as cathexis is then nothing but the brain's resting state, its intrinsic activity.

PH: My dear Mr. Neuroscientist, I'm amazed how naive you are. You cannot simply translate one concept into another in a one-to-one manner. The philosopher Quine argued that one-to-one translation of one concept into another is doomed to fail because of their contextual differences. And there you go and just claim to translate the psychodynamic concept of cathexis into a neuroscientific concept.

NS: Why not? Where is the problem with that?

PH: The problem is that the psychological context of the psychodynamic concept of cathexis as psychic energy is different from the neuroscientific context of your concept of the brain's resting-state activity.

NS: I don't understand. In both cases the concept of cathexis refers to discharge of energy. Does it matter at all whether the energy refers to psychic energy or neuronal energy?

PS: You cannot imagine how much that hurts, your conceptual naivite disturbs me. Cathexis is psychic energy, not neuronal energy. We should go back to Freud himself and his intentions. Freud aimed to describe subjective processes by using objective terms. Processes are subjective in that they can be accessed only by an individual person from their particular first- (or second-) person perspective. For instance, the concept of the id is an objective third-person-based concept that aims to refer to highly subjective processes from a first-person perspective, such as sexual desire. In short, psychodynamic concepts describe first-person processes in third-person terms.

PH: Yes, now it is clearer to me. That is exactly why they are conceptually confused. You cannot link first- and third-person aspects within one concept. Either you deal with facts that are observable from third-person perspective as in neuroscience, which excludes meaning, context and first-person perspective. This is what science and more specifically neuroscience is all about. Or you deal with concepts rather than facts. However, this brings context, meaning and the first-person perspective into play. This is accounted for by hermeneutics. Do you really want to deny this primary difference when saying that Freud aimed to link first- and third-person perspectives in his concepts?

PS: Be care, Mr. Philosopher. Linkage is not denial. Linking first- and third-person perspectives does not mean that I deny their distinction. You seem to confuse linkage with reduction. To make you aware again, I am the psychoanalyst, not the neuroscientist.

NS: Linkage is not denial? I don't care. Facts are relevant, not concepts. The recent neuroscientific facts tell me that what you call first-person perspective can well be accounted for by specific neuronal states as observed from third-person perspective. Your heralded difference between first- and third-person perspectives is thus relevant only on paper, conceptually as you say, whereas it has no empirical relevance. Why then consider this difference at all?

PS: I again feel a sensation of mental pain. How can someone as intelligent as you completely disregard all of the personal meanings from first-person perspective that Freud tried to capture in his ingenuous descriptions of his patients? You will never get the same information from mere observation from third-person perspective.

PH: Let us be clear. The neuroscientist aims to colonize the first-person perspective with the third-person perspective. How can he be so naive? These are different ways of accessing reality and do therefore require different methods. To link both in the discipline of neuropsychoanalysis is to simple ignore these boundaries and to confuse psychoanalysis as hermeneutics with neuroscience as science.

PS: Bravo! This needed to be stated clearly. Thank you so much.

NS: You must be emotionally unstable. Now you suddenly side with the philosopher whom you were fighting just a couple of minutes ago. Apart from that, I still do not understand the fuss about the first-person perspective. If it can be explained in terms of the third-person perspective, why not embrace the latter? Why leave a broken wheel on the car when you have already replaced it with a new one?

PS: You are at least as blind as I am emotionally unstable. You do not see that the difference between first- and third-person perspectives goes along with a much deeper difference, the one between subjectivity and objectivity. Psychoanalysis targets subjective processes that involve a hermeneutics of mind, whereas neuroscience targets only objective processes and thus a science of the brain. If you reduce the former to the latter, you eradicate any trace of subjectivity. However, that means to completely abandon the basic subjective nature of our human existence.

NS: Now you're becoming really dramatic. I can tell you, though, the real drama is what happens in the brain. What you call subjective is nothing but the work of the brain's objective neuronal processes. To make it even more dramatic, those neuronal processes can well be observed from third-person perspective.

PH: Mr. Certainty, you seem to know everything. How can you know that the brain's objective neuronal processes bring forth subjective processes such as mental states? You must know everything to know that. You must be God.

NS: I think you are confusing yourself in your ivory tower with God. No wonder you don't see reality. The windows and doors of your ivory tower are closed, which prevents you from seeing the empirical reality. That empirical reality is nothing but neuronal reality. What you call mental is neuronal. What you call subjective is objective. To assume that the mental states as subjective must be traced back to something that is subjective by itself is logically flawed. You can access the mental states only in a subjective way from first-person perspective. Yes, you may be right about that. But does this mean that the neuronal states as accessible from third-person perspective must be subjective themselves? I don't need to tell you. That means to confuse access and states or, to put it in terms of your own categories, to confuse epistemic and empirical domains.

PH: Oh, now the neuroscientist is waxing philosophical. Did I ever argue for subjective processes to be described in objective terms? No, psychoanalysis does that. Psychoanalysis is a bad mixture of subjectivity and objectivity. But don't get complacent too soon. You do exactly the same when you claim that the objective neuronal processes cause subjective mental states. Subjectivity is not a matter of mere access, as you try to sideline it. Subjectivity is a basic dimension of human nature.

PS: When do I ever step into an ivory tower? Unlike you, Mr. Philosopher, I step out of it and see psychological reality. This is what psychoanalysis is all about—describing subjectivity in psychological reality. This is why Freud introduced the concept of the psychic apparatus. This is why he abandoned neuroanatomy.

PH: I have to distance myself from such confused attempts at colonization. To associate subjectivity with any kind of empirical reality whatsoever, be it psychological or neuronal, is to lose it. Subjectivity is not empirical, it is epistemic. Therefore subjectivity cannot be linked either with the psychic apparatus or with the brain, as our desperate author suggests. For the same reasons you cannot link the first-person perspective with observations from third-person perspective. This means that all attempts to do so—however sophisticated – are doomed to fail.

NS: Strange. You two seem to be the only ones who still adhere to the rather mysterious claim that subjectivity has nothing to do with the brain. Meanwhile everybody else knows this all too well. Changes in my brain go along with changes in my subjectivity. How can you deny something so obvious? I'm sorry, I was forgetting that you are both sitting in ivory towers called psychoanalysis and philosophy.

Localization and the brain

PS: Another reason why the extension of psychoanalysis into neuropsychoanalysis must fail is the peculiar nature of psychological functions. Although Freud was initially a neuroanatomist, he considered the localization of psychological functions in specific regions of the brain to be impossible.

NS: Why exactly did he abandon neuroanatomy and venture into the rather mysterious territory of psychoanalysis?

PS: He was frustrated with the neuroscience of his time. He aimed for a brain-based scientific psychology in his famous writing from 1895. However, he observed in that writing that the complex psychological functions cannot be localized in specific regions of the brain. This led him to abandon neuroanatomy and to develop psychoanalysis.

NS: Why did he deem any localization of psychological functions in the brain to be impossible?

PS: Because of the complexity of the psychological contents that are subjective, personally relevant, and individual. Something as subjective, personally relevant, and individual as psychological contents cannot be localized in something as objective, personally irrelevant, and non-individual as the neural activity of a specific brain region. Or do you think that your current emotional feelings of contempt for the philosopher can be found in a specific region of your brain?

NS: Wait! That may have been true for the neuroscience in Freud's time. Now we know better. We know, for instance, which regions are active during the feeling of contempt. Even more complex functions like consciousness and the self can be associated with specific regions in the brain.

PS: No, no, it is not only a matter of time. We cannot trace the personal relevance, individuality, and subjectivity of your feeling of contempt to the activity in certain brain regions. You will only see neural activity changes in the brain's regions, but no personal relevance, individuality, and subjectivity. Or are you suffering from hallucinations?

NS: Apparently you are not fully up to date with the latest breakthroughs in neuroscience. The self has been shown convincingly in several brain imaging studies to recruit specific regions in the midline of the brain—the so-called cortical midline structures. And you tell me that we cannot locate the self in the brain? You are wrong, I don't suffer from hallucinations. You, however, must be diagnosed with ignorance.

PS: You are conceptually confused. You talk about the self when I mean the ego. Ego and self are not identical. Freud considered the concept of the self to be only part of the ego.

NS: This is a mere conceptual playground, I know. Our author cites another psychoanalyst, Kohut, who regards the self as central. I hope you enjoy playing with concepts. Do these concepts matter? No! In the future, I am sure, erroneous psychodynamic concepts like self and ego will be replaced by neuronal ones like the cortical midline structures. Do you refer to some mysterious entity when you talk about your digestive problems? Would you? I am almost certain that you don't. Instead, you speak of your stomach, don't you? Why then are you still using mysterious concepts like self and ego instead of cortical midline structures?

PH: I do not need to say that I hate to side with a conceptually confused person such as a psychoanalyst, but here I must agree with her. You are not only talking about different concepts—the concept of the ego and the concept of the self. You also presuppose a rather erroneous fallacy which has recently been termed "mereological fallacy" by Bennett and Hacker. The discipline of mereology describes the relationship between parts and wholes, as for instance between brains and people. The brain is a part of the person as a whole. If you now trace back the features of the person, their self and ego, to the brain and its cortical midline structures, you explain the whole by one of its parts. This, though, is erroneous in very much the same way as when you explain a car in terms of one of its wheels.

PS: Mr. Neuroscientist, isn't this clear? Everybody knows that a car cannot be explained by one of its wheels. Apparently you do not. Otherwise you would not localize psychological functions like ego and self in specific regions of the brain.

NS: How naive do you think we as neuroscientists are?

PS: Very naive, and blinded by neuronal reality.

NS: We do not associate a concept as complex as the ego with the activity of one region. There are multiple connections between different regions. There are so-called re-entrant connections where regions return their information. And there is a global operational workspace in the brain that allows recruitment and operation of different regions at the same time. Localization nowadays does not mean the same as it did in Freud's time. In his time, localization meant one specific region as one part of the brain, whereas nowadays localization refers to complex neural networks encompassing multiple regions, if not the whole of the brain. And you still accuse me of part–whole confusion?

PH: No matter how sophisticated your concept of localization may be, to consider the brain as a whole by itself is to simply ignore the fact that the brain is part of your body and that the body is only part of you as a person. This would amount to the claim that you are nothing but your brain, "I am my brain," which is utterly absurd, Mr. Neuroscientist. You have to accept your limits and boundaries. There is no escape. Best give up and surrender. Unless you claim that you yourself are nothing but your brain. Although I was not aware that I was talking to a brain . . .

Brain and environment

NS: Who tells you that there is any difference between people and brains? Isn't a person what their brain is? Am I not nothing but my brain?

PS: You are so refreshingly naive. Psychoanalysis showed very convincingly that there is more to the person. There is the environment and more specifically attachment to specific people or objects within that environment. Without the environment you are nothing. You need the environment in order to flourish in the same way that you need air for breathing in order to survive. To claim that people are nothing but their brains is to completely disregard the environment. Arrogance, ignorance, and all that because of some gray pulpy mass in your skull.

NS: This is another instance of projection. You are too embarrassed to admit your lack of knowledge about current neuroscience but, being a psychoanalyst, you of course know how to hide it. The best shelter is still behind the back of other people, and that is why you attribute your own ignorance and arrogance to me. To bring it back into your own awareness. You ignore all of the neuroscientific data about neural plasticity, the adaptive changes of the brain's cells and neural activity in response to challenges from the environment. The activity in your cortical midline structures may, for instance, change when you are exposed to plenty of stimuli that are highly personally relevant and meaningful to you.

PS: Well reported. But that still does not capture what Freud meant. He argued that the constitution of the psychological structure and organization is itself dependent on the environment. Only when we interact with the environment are we able to develop the skills and abilities that are necessary to distinguish between self and objects (i.e. self–other differentiation), as they are for instance attributed to the ego. I have no doubt that the brain and its neuronal states can be modulated and influenced by the environment. But the idea that neuronal states themselves are organized and structured by the environment? No, that remains impossible, and it holds only for the psychic apparatus.

NS: Isn't that exactly what the findings concerning the neural plasticity of the brain show when, for example, demonstrating the impact of social interaction or the use of certain skills on our brain's structure and function?

PS: No, they only show that the neuronal states can be modulated by the environmental context. Following our author, one may distinguish between "constitutive context dependence" and "modulatory context dependence." Contrary to the author, though, I assume that constitutive context dependence holds only for the psychic apparatus, but not for the brain.

NS: Holy concepts. Another instance of a conceptual difference that remains empirically irrelevant.

PH: You are wrong again. The concept of constitutive context dependence seems to describe a causal and necessary effect of the environmental context on the brain's neural structure and function, while the concept of modulatory context dependence does not attribute a causal and necessary role to the environmental context in shaping and constituting the brain's neural structure and function. Isn't that empirically relevant?

NS: It is conceptual sophistication, nothing else. Every modulatory impact is causal, otherwise it would not lead to changes in the brain. Your distinction between modulatory and constitutive context dependence is therefore only conceptually relevant, not empirically so. But what can one expect from a philosopher whose discipline moved into an ivory tower 3000 years ago and has stayed there ever since?

PH: Before being finally banished to my ivory tower, I want to step out one last time. Doesn't the author himself claim that he bases his distinction between modulatory and constitutive context dependence on empirical data?

NS: Yes, it would appear so. But all the evidence that he cites could equally well support modulatory context dependence. Isn't that proof enough that there is no difference between modulatory and constitutive context dependence? Our author imposed a conceptual distinction upon the empirical data. Instead of imposing his concepts, he would have done better to have started with the actual data. This example again demonstrates that we need fewer concepts and more facts. And these facts will tell us that the mind is nothing but the brain. The more facts, the more brain, and the less mysterious concepts like mind and psychic apparatus.

PH: I cannot but pull up the drawbridge to my philosophical ivory tower when I hear that.

PS: Completely devoid of any traces of meaning and subjectivity. What an oblivious view of people and the world—and, I'm afraid, of brains.

Neural predisposition and difference-based coding

NS: There is yet another interesting distinction, namely that between neural predisposition and neural correlates.

PH: Yet another one of those neuroscientific artifacts, but this time a conceptual one.

NS: Dear Mr. Philosopher, I would be more than delighted to explain this conceptual difference using your own terminology. Could that help to make you understand? Neural correlates concern those neural conditions that are sufficient to induce a specific psychological content or mental state. However, as you may well know, not every sufficient condition is also necessary. In addition to the neural correlates, the author therefore speaks of neural predispositions that describe the necessary but non-sufficient neural conditions of mental states.

PH: Welcome to my ivory tower. You got the distinction between necessary and sufficient conditions right. You successfully passed the first entrance door.

PS: How long did you spend in the philosopher's ivory tower? Now it's clear to me why you were infected by the philosophical virus and sound as abstract and lofty as the philosopher.

NS: To bring it down to your level, Miss Psychoanalysis. Neural correlates concern those neuronal mechanisms that underlie specific psychological contents as objects or self. Whereas neural predispositions refer to those structures and functions that, like the ego, enable and predispose to the constitution of psychological contents.

PH: Oh my goodness, you have a point. Freud indeed saw the ego as psychological structure and organization that enables and predisposes one to acquire specific psychological contents. He may have spoken thus of psychological predisposition and distinguished it from mere psychological correlates.

NS: You finally see my point?

PS: Don't get too carried away yet. You again disregard the fact that the ego refers to the psychic apparatus rather than the brain. One can speak of psychological predispositions, but neural predispositions? No, that is impossible. One cannot replace the psychic apparatus by the brain.

NS: You accept the predispositions but not the neural predispositions? That tells me that you really are locked into a prison, the prison of the psychic apparatus. Like all inmates of any kind of prison, you are not exposed to reality any more. No wonder that you do not notice that your beloved psychic apparatus has been replaced by the brain in the mean time. Now it is also becoming clear to me why you constantly speak of Freud. You have been imprisoned for quite some time and have never been released. I assume therefore that you must be dangerous.

PH: Happy world. He puts me in an ivory tower and you in a prison. But he himself is speaking out of a cave, the cave of the brain. Caves are dark, with at best only occasional rays of sun glinting through. All he then sees are shadows, and he mistakes them for the brain. Let us be serious again. Mr. Neuroscientist, you are the expert on the brain. Can you tell me in detail what the author's neural predisposition looks like?

NS: What a majestic visitor to my cave. The philosophical emperor has arrived in person. Let me prepare the tea. Neural predispositions refer to the specific way in which the brain codes all incoming information.

PH: Could I have more sugar please? What does this specific method of neural coding look like?

NS: The author shows that the brain's neural activity results from coding differences. What is coded in neural activity is thus the difference between the brain's intrinsic activity and intero- and exteroceptive stimuli from body and environment. Neural activity thus mirrors difference-based coding.

PH: Ha, there it is—the logical flaw. How could I ever have expected a neuroscientist to be logically coherent? How can a neuron code the difference between different stimuli if it does not code the single stimuli by themselves prior to that? How can the difference between different stimuli be calculated except on the basis of the values for each stimulus independent of the others? His suggestion of difference-based coding is circular and thus logically flawed.

NS: I hear the gloating tone in your voice. However, you forget that you are in my cave and not in your ivory tower. You forget the distinction between your own access and the brain itself. Yes, you are right, you can access differences only as differences between two values. Difference-based coding is then indeed circular and logically flawed. These are the logical–conceptual foundations upon which your ivory tower is built. But you forget that my cave is built on a different foundation, one where access and concepts are replaced by the brain itself and its empirical foundation. The brain itself tells us the truth, not you and your specific way of accessing the brain in terms of concepts.

PS: Now I am curious. What does the brain tell us?

PH: The brain does not tell us anything. The person who calls himself Mr. Neuroscientist tells us something about the brain. Or does your brain speak to us?

NS: I'll ignore these comments that are supposed to be amusing. What do the data tell us? The data do indeed show some support for neural activity being coded on the basis of neural differences between neural, interoceptive, and exteroceptive stimuli. Even more interesting, the author shows quite convincingly that the neural differences between these different stimuli predict the occurrence of the respective associated mental states.

PH: Does that amount to an empirical hypothesis about the relationship between difference-based coding and mental states?

NS: That is what the author seems to have in mind. For instance, he postulates that difference-based coding enables and predisposes to what he calls "neuronal–mental transformation."

PS: That has major implications for psychoanalysis. Self and objects can be characterized by mental states. If so, both self and objects should be enabled and predisposed to by the brain's difference-based coding.

NS: Exactly. The author devoted a whole chapter to how self and objects can be constituted on the basis of the brain's difference-based coding.

PH: Miss Psychoanalysis, before you agree enthusiastically, can I make you aware of a flaw inherent in that hypothesis? If self and objects are enabled and predisposed to by difference-based coding, they must be located in the brain. Do you really want that? Didn't you reject that idea earlier?

PS: Of course, that is terrible. No way. I am totally confused. This author seems to make me agree and pull me into something, and then I end up in a territory that I never intended to venture into. He seems to be like an airplane. You booked the flight and chose the destination. The take-off goes well, but then suddenly the airplane changes direction and drops you off in the very country that you did not want to enter. You are scared.

NS: What do you mean? You still have your beloved self and objects. The author only shows that they are enabled and predisposed to by the brain's difference-based coding. What makes it so scary?

PS: Do you know how it feels if you lose the ground under your feet, so that you feel as if you no longer have any control? Maybe not, because you don't seem to have ever left your cave!

NS: The opposite is true. You finally get some ground under your feet—the brain and how it codes all incoming information. Doesn't that stabilize you?

PS: That gray pulpy mass? How can that stabilize me? It's nothing but gray and dead! How can that account for the vividness and color of objects and the self?

PH: Let us all return to our own places where we feel secure and stable. I shall go back to my ivory tower, Mr. Neuroscientist can enjoy his cave, and Miss Psychoanalyst will have to remain locked in prison.

NS: Is that all there is?

PS: It at least shelters us from the colonizing invasions of the dangerous cavemen. But before we close our doors again, we need to decide what to do with the author of this book. Should we recommend his book as showing how disastrous the consequences can be when one does not keep to the boundaries of disciplines?

NS: No, that is too dangerous. Sometimes viruses multiply unexpecte vdly. Neutralization is a better strategy. Let us send him back to the drawing board of neuroscience, the basic neuroscience classes.

PH: Happy disagreement, then. I shall send him back to the drawing board of philosophy, and some introductory courses.

PS: And guess where I shall send him? Our author suffers from the most severe narcissistic fantasies of grandeur. I sense his inner need to extend my beloved Freud to the brain. He urgently needs to be sent to the psychotherapeutic couch . . .

Acknowledgments

I am very grateful to Oliver Lyttelton for encouraging me to include this trialogue as the Epilogue to this book, and to overcome my own doubts. I also thank him for his helpful comments on earlier drafts of the text.

References

Addis, D. R., Wong, A. T. and Schacter, D. L. (2007). Remembering the past and imagining the future: common and distinct neural substrates during event construction and elaboration. *Neuropsychologia*, **45**(**7**), 1363–1377.

Akins, K. (2001). Of sensory systems and the "aboutness" of mental states. In: Bechtel, W., Mandik, P., Mundale, J. and Stuffelebeam, R. S., eds. *Philosophy and the Neurosciences*, pp. 156–178. Oxford: Blackwell Publishers.

Albert, N. B., Robertson, E. M. and Miall, R. C. (2009). The resting human brain and motor learning. *Current Biology*, **19**, 1023–1027.

Alcaro, A. *et al.* (2010). Is subcortical-cortical midline activity in depression mediated by glutamate and GABA? A cross-species translational approach. *Neuroscience and Biobehavioral Reviews*, **34**(**4**), 592–605.

Alcaro, A., Huber, R. and Panksepp, J. (2007). Behavioral functions of the mesolimbic dopaminergic system: an affective neuroethological perspective. *Brain Research Reviews*, **56**(**2**), 283–321.

Aleman, A. *et al.* (2005). The functional neuroanatomy of metrical stress evaluation of perceived and imagined spoken words. *Cerebral Cortex*, **15**(**2**), 221–228.

Allen, P., Aleman, A. and McGuire, P. K. (2007a). Inner speech models of auditory verbal hallucinations: evidence from behavioural and neuroimaging studies. *International Review of Psychiatry*, **19**(**4**), 407–415.

Allen, P. *et al.* (2007b). Neural correlates of the misattribution of speech in psychosis. *British Journal of Psychiatry*, **190**, 162–169.

Allen, P., Laroi, F., McGuire, P. K. and Aleman, A. (2008). The hallucinating brain: a review of structural and functional neuroimaging studies of hallucinations. *Neuroscience and Biobehavioral Reviews*, **32**(**1**), 175–191.

Allison, H. E. (1983). *Kant's Transcendental Idealism*. New Haven, CT: Yale University Press.

Ameriks, K. (2000). *Kant and the Fate of Autonomy: problems in the appropriation of the critical philosophy*. Cambridge: Cambridge University Press.

Arieli, A. *et al.* (1996). Dynamics of ongoing activity: explanation of the large variability in evoked cortical responses. *Science*, **273**, 1868–1871.

Baars, B. J. (2005). Global workspace theory of consciousness: toward a cognitive neuroscience of human experience. *Progress in Brain Research*, **150**, 45–53.

Bacal, H. A. (1987). British object-relations theorists and self psychology: some critical reflections. *International Journal of Psychoanalysis*, **68**(**1**), 81–98.

Barry, R. J. *et al.* (2007). EEG differences between eyes-closed and eyes-open resting conditions. *Clinical Neurophysiology*, **118**, 2765–2773.

Bechara, A. (2004). The role of emotion in decision-making: evidence from neurological patients with orbitofrontal damage. *Brain and Cognition*, **55**(**1**), 30–40.

Beckmann, C. F., DeLuca, M., Devlin, J. T. and Smith, S. M. (2005). Investigations into resting-state connectivity using independent component analysis. *Philosophical Transactions of the Royal Society of London. Series B, Biological Sciences*, **360**, 1001–1013.

Benedetti, G. (1983). *Todeslandschaften der Seele*. Göttingen: Vandenhoeck and Ruprecht.

Benes, F. M. (2007). Searching for unique endophenotypes for schizophrenia and bipolar disorder within neural circuits and their molecular regulatory mechanisms. *Schizophrenia Bulletin*, **33**(**4**), 932–936.

Benes, F. M. *et al.* (2008). Circuitry-based gene expression profiles in GABA cells of the trisynaptic pathway in schizophrenics versus bipolars. *Proceedings of the National Academy of Sciences of the United States of America*, **105**, 20935–20940.

Ben-Simon, E. *et al.* (2008). Never resting brain: simultaneous representation of two alpha related processes in humans. *PLoS One*, **3(12)**, e3984.

Bennett, A. O. M. (2008). Consciousness and hallucinations in schizophrenia: the role of synapse regression. *Australian and New Zealand Journal of Psychiatry*, **42(11)**, 915–931.

Bennett, M. and Hacker, P. M. (2003). *Philosophical Foundations of Neuroscience*. New York: Basic Books.

Bermpohl, F. *et al.* (2009). Attentional modulation of emotional stimulus processing in patients with major depression—alterations in prefrontal cortical regions. *Neuroscience Letters*, **463(2)**, 108–113.

Berretta, S., Gisabella, B. and Benes, F. M. (2009). A rodent model of schizophrenia derived from postmortem studies. *Behavioural Brain Research*, **204(2)**, 363–368.

Berridge, K. C. and Winkielman, P. (2003). What is an unconscious emotion? (The case for unconscious "liking"). *Cognition and Emotion*, **17(2)**, 181–211.

Berze, J (1914). *Die Primare Insuffizienz der Psychischen Aktivitat*. Leipzig: Franz Deuticke.

Bickle, J. and Mandik, B. (2006). *The Philosophy of Neuroscience*. The Stanford Encyclopedia of Philosophy. http://plato.stanford.edu/entries/neuroscience/.

Bin Kimura, B. (1997). Cogito et je. *L'Evolution Psychiatrique*, **62**, 335–348.

Birbaumer, N., Hinterberger, T., Kübler, A. and Neumann, N. (2003). The thought-translation device (TTD): neurobehavioral mechanisms and clinical outcome. *IEEE Transactions on Neural Systems and Rehabilitation Engineering*, **11(2)**, 120–123.

Birnbacher, D. (2005). Schopenhauer und die moderne Neurophilosophie. In: Birnbacher, D. and Koßler, M., eds. *Die Bedeutung Schopenhauers für das moderne Bild des Menschen*, pp. 34–55. Würzburg: Schopenhauer-Jahrbuch.

Blakemore, S. and Frith, C. (2003). Disorders of self-monitoring and symptoms of schizophrenia. In: Kircher, T. and David, A., eds. *The Self in Neuroscience and Psychiatry*, pp. 407–425. Cambridge: Cambridge University Press.

Blakemore, S. J., Goodbody, S. J. and Wolpert, D. M. (1998). Predicting the consequences of our own actions: the role of sensorimotor context estimation. *Journal of Neuroscience*, **18**, 7511–7518.

Blankenburg, W. (1969). Ansatze zu einer Psycholpathologie des "common sense." *Confinia Psychiatrica*, **12**, 144–163.

Blass, R. B. and Carmeli, Z. (2007). Reply to Drs Mancia and Pugh. *International Journal of Psychoanalysis*, **88**, 1068–1070.

Blatt, S. J. (1998). Contributions of psychoanalysis to the understanding and treatment of depression. *Journal of the American Psychoanalytic Association*, **46**, 723–752.

Bleichmar, H. B. (1996). Some subtypes of depression and their implications for psychoanalytic treatment. *International Journal of Psychoanalysis*, **77**, 935.

Bleichmar, H. (2010). Rethinking pathological mourning: multiple types and therapeutic approaches. *Psychoanalytic Quarterly*, **79(1)**, 71–93.

Bleuler, E. (1911). *Dementia Praecox or the Group of Schizophrenias*. New York: International Universities Press.

Bleuler, E. (1916). *Lehrbuch der Psychiatrie*. Berlin: Springer.

Böker, H. and Northoff, G. (2005). Desymbolisierung in der schweren Depression und das Problem der Hemmung: Ein neuropsychoanalytisches Modell der Störung des emotionalen Selbstbezuges Depressiver. *Psyche – Zeitschrift für Psychoanalyse und ihre Anwendungen*, **59**, 964–989.

Böker, H. and Northoff, G. (2010). Die Entkopplung des Selbst in der Depression: Empirische Befunde und neuropsychodynamische Hypothesen. *Psyche – Zeitschrift für Psychoanalyse und ihre Anwendungen*, **64**, 934–976.

Böker, H. *et al.* (2000). [Self concept and object relations of patients with affective disorders —individual centered diagnosis with the repertory-grid technique] [article in German]. *Psychotherapie, Psychosomatik, Medizinische Psychologie*, **50(8)**, 328–334.

Boly, M. *et al.* (2007). Baseline brain activity fluctuations predict somatosensory perception in humans. *Proceedings of the National Academy of Sciences of the United States of America*, **104(29)**, 12187–12192.

Boly, M. *et al.* (2008a). Intrinsic brain activity in altered states of consciousness: how conscious is the default mode of brain function? *Annals of the New York Academy of Sciences*, **1129**, 119–129.

Boly, M. *et al.* (2008b). Consciousness and cerebral baseline activity fluctuations. *Human Brain Mapping*, **29**(**7**), 868–874.

Braun, C. M. J. *et al.* (2003). Brain modules of hallucination: an analysis of multiple patients with brain lesions. *Journal of Psychiatry and Neuroscience*, **28**(**6**), 432–449.

Brockman, R. (2002). Self, object, neurobiology. *Neuropsychoanalysis*, **4**, 87–99.

Brook, A. (1994). *Kant and the Mind*. Cambridge: Cambridge University Press.

Brook, A. (1998). Neuroscience versus Psychology in Freud. *Annals of the New York Academy of Sciences*, **843**(**1**), 66–79.

Brook, A. (2003). *Kant and Freud*. Ottawa: Carleton University.

Broyd, S. J. *et al.* (2009). Default-mode brain dysfunction in mental disorders: a systematic review. *Neuroscience and Biobehavioral Reviews*, **33**(**3**), 279–296.

Buckner, R. L., Andrews-Hanna, J. R. and Schacter, D. L. (2008). The brain's default network: anatomy, function, and relevance to disease. *Annals of the New York Academy of Sciences*, **1124**, 1–38.

Busch, N. A., Dubois, J. and VanRullen, R. (2009). The phase of ongoing EEG oscillations predicts visual perception. *Journal of Neuroscience*, **29**, 7869–7876.

Buzsaki, G. (2006). *Rhythms of the Brain*. New York: Oxford University Press.

Buzsaki, G. and Draguhn, A. (2004). Neuronal oscillations in cortical networks. *Science*, **304**, 1926–1929.

Cain, N. M., Pincus, L. A. and Ansell, E. B. (2008). Narcissism at the crossroads: phenotypic description of pathological narcissism across clinical theory, social/personality psychology, and psychiatric diagnosis. *Clinical Psychology Review*, **28**, 638–656.

Calhoun, V. D., Maciejewski, P. K., Pearlson, G. D. and Kiehl, K. A. (2008). Temporal lobe and "default" hemodynamic brain modes discriminate between psychosis and bipolar disorder. *Human Brain Mapping*, **29**(**11**), 1265–1275.

Camerer, C. F. and Fehr, E. (2006). When does "economic man" dominate social behavior? *Science*, **311**, 47–52.

Canli, T. *et al.* (2004). Brain activation to emotional words in depressed vs healthy subjects. *Neuroreport*, **15**, 2585–2588.

Canli, T. *et al.* (2005). Amygdala reactivity to emotional faces predicts improvement in major depression. *Neuroreport*, **16**(**12**), 1267–1270.

Carhart-Harris, R. L. and Friston, K. J. (2010). The default-mode, ego-functions and free-energy: a neurobiological account of Freudian ideas. *Brain*, **133**, 1265–1283.

Carhart-Harris, R. L., Mayberg, H. S., Malizia, A. L. and Nutt. D. (2008). Mourning and melancholia revisited: correspondences between principles of Freudian metapsychology and empirical findings in neuropsychiatry. *Annals of General Psychiatry*, **24**, 7–9.

Christoff, K. *et al.* (2009). Experience sampling during fMRI reveals default network and executive system contributions to mind wandering. *Proceedings of the National Academy of Sciences of the United States of America*, **106**, 8719–8724.

Churchland, P. (1986). *Neurophilosophy*. Cambridge, MA: MIT Press.

Churchland, P. S. (2002). Self-representation in nervous systems. *Science*, **296**, 308–310.

Clark, A. (2003). *Natural-Born Cyborgs: minds, technologies, and the future of human intelligence*. Oxford: Oxford University Press.

Collins, A. C. (1999). *Possible Experience: understanding Kant's* Critique of Pure Reason. London: University of California Press.

Corlett, P. R., Krystal, J. H., Taylor, J. R. and Fletcher, P. C. (2009). Why do delusions persist? *Frontiers in Human Neuroscience*, **3**, 12.

Craig, A. D. (2002). How do you feel? Interoception: the sense of the physiological condition of the body. *Nature Reviews. Neuroscience*, **3**(**8**), 655–666.

Craig, A. D. (2003). Interoception: the sense of the physiological condition of the body. *Current Opinion in Neurobiology*, **13**, 500–505.

Craig, A. D. (2004). Human feelings: why are some more aware than others? *Trends in Cognitive Sciences*, **8**(**6**), 239–241.

Craig, A. D. (2005). Forebrain emotional asymmetry: a neuroanatomical basis? *Trends in Cognitive Sciences*, **9**(**12**), 566–571.

Craig, A. D. (2009). How do you feel—now? The anterior insula and human awareness. *Nature Reviews. Neuroscience*, **10**(**1**), 59–70.

Cramer, P. (2000). Defense mechanisms in psychology today. Further processes for adaptation. *American Psychologist*, **55**(**6**), 637–646.

Crick, F. (1994). *The Astonishing Hypothesis: the scientific search for the soul*. New York: Simon & Schuster.

Crick, F., Koch, C., Kreiman, G. and Fried, I. (2004). Consciousness and neurosurgery. *Neurosurgery*, **55**(**2**), 273–81; discussion 281–282.

Critchley, H. D. (2005). Neural mechanisms of autonomic, affective, and cognitive integration. *Journal of Comparative Neurology*, **493**(**1**), 154–166.

Critchley, H. D. *et al.* (2004). Neural systems supporting interoceptive awareness. *Nature Neuroscience*, **7**(**2**), 189–195.

Critchley, H. D. *et al.* (2005). Activity in the human brain predicting differential heart rate responses to emotional facial expressions. *NeuroImage*, **24**(**3**), 751–762.

Czisch, M. *et al.* (2004). Functional MRI during sleep: BOLD signal decreases and their electrophysiological correlates. *European Journal of Neuroscience*, **20**(**2**), 566–574.

Dalgleish, T. (2004). The emotional brain. *Nature Reviews. Neuroscience*, **5**(**7**), 583–589.

Damasio, A. (1999). *The Feeling of What Happens: body and emotion in the making of consciousness*. Orlando, FL: Harcourt Brace & Company.

Damasio, A. (2003a). Feelings of emotion and the self. *Annals of the New York Academy of Sciences*, **1001**, 253–261.

Damasio, A. (2003b). Mental self: the person within. *Nature*, **423**, 227.

Damoiseaux, J. S. *et al.* (2006). Consistent resting-state networks across healthy subjects. *Proceedings of the National Academy of Sciences of the United States of America*, **103**, 13848–13853.

Dannlowski, U. *et al.* (2010). Emotion-specific modulation of automatic amygdala responses by 5-HTTLPR genotype. *NeuroImage*, **53**, 893–898.

D'Argembeau, A. *et al.* (2005). Self-referential reflective activity and its relationship with rest: a PET study. *NeuroImage*, **25**(**2**), 616–624.

Davidson, R. J. *et al.* (2003). The neural substrates of affective processing in depressed patients treated with venlafaxine. *American Journal of Psychiatry*, **160**(**1**), 64–75.

deCharms, R. C. and Zador, A. (2000). Neural representation and the cortical code. *Annual Review of Neuroscience*, **23**, 613–647.

de Greck, M. *et al.* (2008). Is our self based on reward? Self-relatedness recruits neural activity in the reward system. *NeuroImage*, **39**(**4**), 2066–2075.

de Greck, M. *et al.* (2010). Decreased neuronal activity in reward circuitry of pathological gamblers during processing of personal relevant stimuli. *Human Brain Mapping*, **31**, 1802–1812.

Demertzi, A. *et al.* (2008). Is there anybody in there? Detecting awareness in disorders of consciousness. *Expert Review of Neurotherapeutics*, **8**(**11**), 1719–1730.

Dennett, D. (1978). *Brainstorms: philosophical essays on mind and psychology*. Cambridge, MA: MIT Press.

Desseilles, M. *et al.* (2009). Abnormal neural filtering of irrelevant visual information in depression. *Journal of Neuroscience*, **29**(**5**), 1395–1403.

Di Cristo, G. (2007). Development of cortical GABAergic circuits and its implications for neurodevelopmental disorders. *Clinical Genetics*, **72**(**1**), 1–8.

Dierks, T. *et al.* (1999). Activation of Heschl's gyrus during auditory hallucinations. *Neuron*, **22**(**3**), 615–621.

Dimaggio, G., Fiore, D., Salvatore, G. and Carcione, A. (2007a). Dialogical relationship patterns in narcissistic personalities: session analysis and treatment implications. *Journal of Constructivist Psychology*, **20**, 1–29.

Dimaggio, G. *et al.* (2007b). Poor metacognition in narcissistic and avoidant personality disorders: four psychotherapy patients analysed using the Metacognition Assessment Scale. *Clinical Psychology and Psychotherapy*, **14**, 386–401.

Dimaggio, G. *et al.* (2008a). Know yourself and you shall know the other . . . to a certain extent: multiple paths of influence of self-reflection on mindreading. *Consciousness and Cognition*, **17**, 778–789.

Dimaggio, G. *et al.* (2008b). States of minds in narcissistic personality disorder: three psychotherapies analyzed using the grid of problematic states. *Psychotherapy Research*, **18**(**4**), 466–480.

Dimaggio, G. *et al.* (2009a). Impaired decentration in personality disorder: a series of single cases analysed with the Metacognition Assessment Scale. *Clinical Psychology and Psychotherapy*, **16**(**5**), 450–462.

Dimaggio, G. *et al.* (2010). A rational model for maximizing the effects of therapeutic relationship regulation in personality disorders with poor metacognition and over-regulation of affects. *Psychology and Psychotherapy*, **83**, 363–384.

Dimaggio, M. D. (2009b). Impaired self-reflection in psychiatric disorders among adults: a proposal for the existence of a network of semi independent functions. *Consciousness and Cognition*, **18**, 653–664.

Donahue, M. J. *et al.* (2009). Cerebral blood flow, blood volume, and oxygen metabolism dynamics in human visual and motor cortex as measured by whole-brain multi-modal magnetic resonance imaging. *Journal of Cerebral Blood Flow and Metabolism*, **29**(**11**), 1856–1866.

Drevets, W. C., Price, J. L. and Furey, M. L. (2008a). Brain structural and functional abnormalities in mood disorders: implications for neurocircuitry models of depression. *Brain Structure and Function*, **213**, 93–118.

Drevets, W. C., Savitz, J. and Trimble M. (2008b). The subgenual anterior cingulate cortex in mood disorders. *CNS Spectrums*, **13**(**8**), 663–681.

Dümpelmann, M. (2002a). Psychosen und affektive Störungen nach Traumatisierung. In: Böker, H. and Hell, D., eds. *Therapie der affektiven Störungen*, pp. 66–90. Stuttgart: Schattauer.

Dümpelmann, M. (2002b). Trauma und Psychose – Trauma oder Psychose? In: Ardjomandi, M. E., Berghaus, A. and Knauss, W., eds. *Jahrbuch der Gruppenanalyse*, pp. 195–218. Göttingen: Vandenhoeck und Ruprecht.

Dümpelmann, M. (2004). Wie lässt sich praktisch anwenden, was psychoanalytische Augen erfassen? In: Machleidt, W., Garlipp, P. and Haltenhof, H., eds. *Schizophrenie – Behandlungspraxis zwischen speziellen Methoden und integrativen Konzepten*, pp. 148–154. Stuttgart: Schattauer.

Dümpelmann, M. (2009). Kontingenzerfahrung und Affektentwicklung. In: Wuensch, S., ed. *Entwicklungspsychologische Ansätze in der Psychotherapie von Psychosen*, pp. 78–96. Hamburg: Dokumentation zum 10jährigen Jubiläum der RPK Hamburg.

Dümpelmann, M. (2010). Zur Bedeutung der Affektentwicklung für die Behandlung von Psychosen. In: Böker, H., ed. *Psychoanalyse im Dialog mi den Nachbarwissenschaften*, pp. 481–501. Giessen: Psychosozial-Verlag.

Edelman, G. M. (2003). Naturalizing consciousness: a theoretical framework. *Proceedings of the National Academy of Sciences of the United States of America*, **100**(**9**), 5520–5524.

Edelman, G. M. and Tononi, G. (2000). *A Universe of Consciousness: how matter becomes imagination.* New York: Basic Books.

Elliott, R. *et al.* (1998). Abnormal neural response to feedback on planning and guessing tasks in patients with unipolar depression. *Psychological Medicine*, **28**, 559–571.

Elliott, R., Rubinsztein, J. S., Sahakian, B. J. and Dolan, R. J. (2002). The neural basis of mood-congruent processing biases in depression. *Archives of General Psychiatry*, **59**(**7**), 597–604.

Ellison-Wright, I. and Bullmore, E. (2009). Meta-analysis of diffusion tensor imaging studies in schizophrenia. *Schizophrenia Research*, **108**, 3–10.

Engel, A. K., Fries, P. and Singer, W. (2001). Dynamic predictions: oscillations and synchrony in top–down processing. *Nature Reviews. Neuroscience*, **2**, 704–716.

Enzi, B. *et al.* (2009). Is our self nothing but reward? Neuronal overlap and distinction between reward and personal relevance and its relation to human personality. *PLoS One*, **4**, e8429.

Ersner-Hershfield, H., Wimmer, G. E. and Knutson, B. (2009). Saving for the future self: neural measures of future self-continuity predict temporal discounting. *Social Cognitive and Affective Neuroscience*, **4**(**1**), 85–92.

Eshel, N. and Roiser, J. P. (2010). Reward and punishment processing in depression. *Biological Psychiatry*, **68**(**2**), 118–124.

Fair, S. B. Gilbert, J. A., Matthys, D. R. and Lindner, C. H. (2000). Development of a phase-displacement equation for panoramic interferometry. *Applied Optics*, **39**, 3289–3294.

Fan, Y. *et al.* (2010). The narcissistic self and its psychological and neural correlates: an exploratory fMRI study. *Psychological Medicine*, **13**, 1–10.

Federn, P. (1952). *Ego Psychology and the Psychoses*. New York: Basic Books.

Fehr, E. and Camerer, C. F. (2007). Social neuroeconomics: the neutral circuitry of social preferences. *Trends in Cognitive Sciences*, **11**, 419–427.

Feinberg, I. and Guazzelli, M. (1999). Schizophrenia – a disorder of the corollary discharge systems that integrate the motor systems of thought with the sensory systems of consciousness. *British Journal of Psychiatry* **174**, 196–204.

Feinberg, T. E. (2001). *Altered Egos: how the brain creates the self.* New York: Oxford University Press.

Feinberg, T. E. (2009). *From Axons to Identity: neurological explorations of the nature of the self.* Norton Series on Interpersonal Neurobiology. New York: W. W. Norton and Company.

Feinberg, T. E. (2010). Neuropathologies of the self: a general theory. *Neuropsychoanalysis*, **12**, 133–158.

Fiorillo, C. D. (2008). Towards a general theory of neural computation based on prediction by single neurons. *PLoS One*, **3**(**10**), e3298.

Fiser, J., Chiu, C. and Weliky, M. (2004). Small modulation of ongoing cortical dynamics by sensory input during natural vision. *Nature*, **431**, 573–578.

Fitzgerald, P. B. *et al.* (2006). An analysis of functional neuroimaging studies of dorsolateral prefrontal cortical activity in depression. *Psychiatry Research*, **148**(**1**), 33–45.

Fitzgerald, P. B. *et al.* (2007). A functional magnetic resonance imaging study of the effects of low frequency right prefrontal transcranial magnetic stimulation in depression. *Journal of Clinical Psychopharmacology*, **27**(**5**), 488–492.

Fletcher, P. C. and Frith, C. D. (2009). Perceiving is believing: a Bayesian approach to explaining the positive symptoms of schizophrenia. *Nature Reviews. Neuroscience*, **10**, 48–58.

Fliessbach, K. *et al.* (2007). Social comparison affects reward-related brain activity in the human ventral striatum. *Science*, **318**, 1305–1308.

Fonagy, P. (2003). Psychoanalysis today. *World Psychiatry*, **2**(**2**), 73–80.

Fonagy, P., Gergely, G. and Target, M. (2007). The parent–infant dyad and the construction of the subjective self. *Journal of Child Psychology and Psychiatry*, **48**, 288–328.

Ford, J. M. *et al.* (2002). Reduced communication between frontal and temporal lobes during talking in schizophrenia. *Biological Psychiatry*, **51**(**6**), 485–492.

Ford, J. M. *et al.* (2007). Synch before you speak: auditory hallucinations in schizophrenia. *American Journal of Psychiatry*, **164**(**3**), 458–466.

Ford, J. M. *et al.* (2008). Out-of-synch and out-of-sorts: dysfunction of motor-sensory communication in schizophrenia. *Biological Psychiatry*, **63**(**8**), 736–743.

Ford, J. M. *et al.* (2009). Tuning in to the voices: a multisite FMRI study of auditory hallucinations. *Schizophrenia Bulletin*, **35**(**1**), 58–66.

Fosshage, J. L. (2009). Some key features in the evolution of self psychology and psychoanalysis. *Annals of the New York Academy of Sciences*, **1159**, 1–18.

Fox, M. D. and Raichle, M. E. (2007). Spontaneous fluctuations in brain activity observed with functional magnetic resonance imaging. *Nature Reviews. Neuroscience*, **8**(**9**), 700–711.

Fox, M. D. *et al.* (2005). The human brain is intrinsically organized into dynamic, anticorrelated functional networks. *Proceedings of the National Academy of Sciences of the United States of America*, **102**, 9673–9678.

Fox, M. D. *et al.* (2006). Coherent spontaneous activity accounts for trial-to-trial variability in human evoked brain responses. *Nature Neuroscience*, **9**, 23–25.

Fransson, P. (2005). Spontaneous low-frequency BOLD signal fluctuations: an fMRI investigation of the resting-state default mode of brain function hypothesis. *Human Brain Mapping*, **26**(**1**), 15–29.

Fransson, P. and Marrelec, G. (2008). The precuneus/posterior cingulate cortex plays a pivotal role in the default mode network: evidence from a partial correlation network analysis. *NeuroImage*, **42**(**3**), 1178–1184.

Freud, S. (1891). *Zur Auffassung der Aphasien*. Leipzig: Franz Deuticke.

Freud, S. (1894). *The Neuro-Psychoses of Defence*. Standard Edition of the Complete Psychological Works of Sigmund Freud. London: Hogarth Press.

Freud, S. (1895). *Project for a Scientific Psychology*. Standard Edition of the Complete Psychological Works of Sigmund Freud. London: Hogarth Press.

Freud, S. (1900). *Die Traumdeutung*. Leipzig: Franz Deuticke.

Freud, S. (1905). *Three Essays on the Theory of Sexuality*, tr. J. Strachey. New York: Basic Books (1962 edition).

Freud, S. (1910). *Leonardo da Vinci and a Memory of his Childhood*. Standard Edition. New York: W. W. Norton.

Freud, S. (1911). *Psychoanalytic Notes on an Autobiographical Account of a Case of Paranoia*. Standard Edition, Volume 12. London: Hogarth Press.

Freud, S. (1914). Zur Einführung des Narzissmus. *Jahrbuch der Psychoanalyse*, **6**, 1–24.

Freud, S. (1915). *The Unconscious*. Standard Edition, Volume 14. London: Hogarth Press.

Freud, S. (1917). *Mourning and Melancholia. On the History of the Psycho-Analytic Movement. Papers on Metapsychology and Other Works*. Standard Edition, Volume 14.

Freud, S. (1921). *Massenpsychologie und Ich-Analyse*. Wien: Internationaler Psychoanalytischer Verlag.

Freud, S. (1923a). *Das Ich und das Es*. Leipzig: Internationaler Psychoanalytischer Verlag.

Freud, S. (1923b). Encyclopedia articles. In: *Psychoanalysis: an empirical science*. London: Vintage.

Freud, S. (1924). *Neurosis and Psychosis*. Standard Edition, Volume 19. London: Hogarth Press.

Freud, S. (1926). *Inhibitions, Symptoms and Anxiety*. Standard Edition, Volume 20. London: Hogarth Press.

Freud, S. (1933). *New Introductory Lectures on Psycho-Analysis*. London: L. Woolf and V. Woolf at the Hogarth Press, and the Institute of Psycho-analysis.

Freud, S. (1940). *An Outline of Psychoanalysis*. Standard Edition, Volume 23. London: Hogarth Press.

Freud, S. (1946). *The Ego and the Mechanisms of Defence*. New York: International Universities Press (originally published 1936).

Fries, P. *et al.* (2001). Modulation of oscillatory neuronal synchronization by selective visual attention. *Science*, **291**, 1560–1563.

Fries, P., Nikolic, D. and Singer, W. (2007). The gamma cycle. *Trends in Neurosciences*, **30**, 309–316.

Friston, K. (2005). Disconnection and cognitive dysmetria in schizophrenia. *American Journal of Psychiatry*, **162**(**3**), 429–432.

Friston, K. and Kiebel, S. (2009). Predictive coding under the free-energy principle. *Philosophical Transactions of the Royal Society of London. Series B, Biological Sciences*, **364**, 1211–1221.

Frith, C. (1999). How hallucinations make themselves heard. *Neuron*, **22**(**3**), 414–415.

Frith, C. D. (1992). *The Cognitive Neuropsychology of Schizophrenia*. Hove, Sussex: Lawrence Erlbaum Associates.

Frith, C. D. and Done, D. J. (1988). Towards a neuropsychology of schizophrenia. *British Journal of Psychiatry*, **153**, 437–443.

Frith, C. D. and Frith, U. (2008). Implicit and explicit processes in social cognition. *Neuron*, **60**(**3**), 503–510.

Frith, C. D., Friston, K. J., Liddle, P. F. and Frackowiak, R. S. (1992). PET imaging and cognition in schizophrenia. *Journal of the Royal Society of Medicine*, **85**(**4**), 222–224.

Fu, C. and McGuire, P. K. (2003). Hearing voices or hearing the self in disguise? Revealing the neural correlates of auditory hallucinations in schizophrenia. In: Kircher, T. and David, A., eds. *The Self in Neuroscience and Psychiatry*, pp. 425–436. Cambridge: Cambridge University Press.

Fu, C. H. *et al.* (2004). Attenuation of the neural response to sad faces in major depression by antidepressant treatment: a prospective, event-related functional magnetic resonance imaging study. *Archives of General Psychiatry*, **61**(**9**), 877–889.

Fu, C. H. *et al.* (2006). An fMRI study of verbal self-monitoring: neural correlates of auditory verbal feedback. *Cerebral Cortex*, **16**(**7**), 969–977.

Gabbard, G. O. (1992). Psychodynamic psychiatry in the "decade of the brain." *American Journal of Psychiatry*, **149**(**8**), 991–998.

Gabbard, G. O. (2005a). *Psychodynamic Psychiatry in Clinical Practice*. Arlington, VA: American Psychiatric Press, Inc.

Gabbard, G. O. (2005b). Mind, brain, and personality disorders. *American Journal of Psychiatry*, **162**(**4**), 648–655.

Gallagher, H. L. and Frith, C. D. (2003). Functional imaging of "theory of mind." *Trends in Cognitive Sciences*, **7**(**2**), 77–83.

Gallagher, S. (2000). Philosophical conceptions of the self: implications for cognitive science. *Trends in Cognitive Sciences*, **4**(**1**), 14–21.

Garfield, S. L. (1995). *Psychotherapy: an eclectic-integrative approach*. 2nd edn. New York: John Wiley & Sons, Inc.

Garrido, M. I. *et al.* (2009). The mismatch negativity: a review of underlying mechanisms. *Clinical Neurophysiology*, **120**(**3**), 453–463.

Gazzaniga, M. S., Ivry, R. B. and Mangun, G. R. (2008). *Cognitive Neuroscience: the biology of the mind*. New York: W. W. Norton & Company, Inc.

Gehrie, M. J. (2009). The evolution of the psychology of the self: toward a mature narcissism. *Annals of the New York Academy of Sciences*, **1159**, 31–50.

Georgopoulos, A.P. *et al.* (1989). Mental rotation of the neuronal population vector. *Science*, **243**, 234–236.

Gilbert, C. *et al.* (2000). Interactions between attention, context and learning in primary visual cortex. *Vision Research*, **40**, 1217–1226.

Gillihan, S. J. and Farah, M. J. (2005). Is self special? A critical review of evidence from experimental psychology and cognitive neuroscience. *Psychological Bulletin*, **131**, 76–97.

Goel, V. and Dolan, R. J. (2003). Reciprocal neural response within lateral and ventral medial prefrontal cortex during hot and cold reasoning. *NeuroImage*, **20**(**4**), 2314–2321.

Golomb, J. D. *et al.* (2009). Enhanced visual motion perception in major depressive disorder. *Journal of Neuroscience*, **29**, 9072–9077.

Gonzalez-Burgos, G. and Lewis, D. A. (2008). GABA neurons and the mechanisms of network oscillations: implications for understanding cortical dysfunction in schizophrenia. *Schizophrenia Bulletin*, **34**(**5**), 944–961.

Grammont, F. and Riehle, A. (2003). Spike synchronization and firing rate in a population of motor cortical neurons in relation to movement direction and reaction time. *Biological Cybernetics*, **88**(**5**), 360–373.

Green, A. (2001). Advice to psychoanalysts: "cognitive psychology is good for you." *Neuropsychoanalysis*, **3**, 16–19.

Greicius, M. (2008). Resting-state functional connectivity in neuropsychiatric disorders. *Current Opinion in Neurology*, **21**(**4**), 424–430.

Greicius, M. D. *et al.* (2004). Default-mode network activity distinguishes Alzheimer's disease from healthy aging: evidence from functional MRI. *Proceedings of the National Academy of Sciences of the United States of America*, **101**, 4637–4642.

Greicius, M. D. *et al.* (2009). Resting-state functional connectivity reflects structural connectivity in the default mode network. *Cerebral Cortex*, **19**(**1**), 72–78.

Grimm, S. *et al.* (2008). Imbalance between left and right dorsolateral prefrontal cortex in major depression is linked to negative emotional judgment: an fMRI study in severe major depressive disorder. *Biological Psychiatry*, **63**(**4**), 369–376.

Grimm, S. *et al.* (2009). Altered negative BOLD responses in the default-mode network during emotion processing in depressed subjects. *Neuropsychopharmacology*, **34**, 932–843.

Grimm, S. *et al.* (2011). Reduced negative BOLD responses in the default-mode network and increased self-focus in depression. *World Journal of Biological Psychiatry*. [Epub ahead of print]

Grunebaum, M. F. *et al.* (2005). Symptom components of standard depression scales and past suicidal behavior. *Journal of Affective Disorders*, **87**(**1**), 73–82.

Gur, R. E. *et al.* (1995). Resting cerebral glucose metabolism in first-episode and previously treated patients with schizophrenia relates to clinical features. *Archives of General Psychiatry*, **52**(**8**), 657–667.

Halari, R. and Kumari, V. (2005). Comparable cortical activation with inferior performance in women during a novel cognitive inhibition task. *Behavioural Brain Research*, **158**(**1**), 167–173.

Hartmann, H. P. (2009). Self and systems explorations in contemporary self psychology. *Annals of the New York Academy of Sciences*, **1159**, 86–105.

Hartwich, P. (2006). Schizophrenie. Zur Defekt- und Konfliktinteraktion. In: Böker, H., ed. *Psychoanalyse und Psychiatrie*, pp. 45–69. Berlin: Springer.

Hasler, G. *et al.* (2009). Reward processing after catecholamine depletion in unmedicated, remitted subjects with major depressive disorder. *Biological Psychiatry*, **66**(**3**), 201–205.

He, B. J. and Raichle, M. E. (2009). The fMRI signal, slow cortical potential and consciousness. *Trends in Cognitive Sciences*, **13**(**7**), 302–309.

He, B. J. *et al.* (2008). Electrophysiological correlates of the brain's intrinsic large-scale functional architecture. *Proceedings of the National Academy of Sciences of the United States of America*, **105**(**41**), 16039–16044.

Heimer, L. (2003). A new anatomical framework for neuropsychiatric disorders and drug abuse. *American Journal of Psychiatry*, **160**, 1726–1739.

Heinzel, A. *et al.* (2006). Self-related processing in the sexual domain: a parametric event-related fMRI study reveals neural activity in ventral cortical midline structures. *Social Neuroscience*, **1**(**1**), 41–51.

Heinzel, A. *et al.* (2009). Segregated neural representation of psychological and somatic-vegetative symptoms in severe major depression. *Neuroscience Letters*, **456**(**2**), 49–53.

Heller, A. S. *et al.* (2009). Reduced capacity to sustain positive emotion in major depression reflects diminished maintenance of fronto-striatal brain activation. *Proceedings of the National Academy of Sciences of the United States of America*, **106**, 22445–22450.

Henningsen, P. (1998). [Self-recognition in the mirror of another? On the significance of cognitive neuroscience for psychoanalysis] [article in German]. *Psychotherapie, Psychosomatik, Medizinische Psychologie*, **48** (**3–4**), 78–87.

Henningsen, P. and Kirmayer, L. J. (2000). Mind beyond the net: implications of cognitive neuroscience for cultural psychiatry. *Transcultural Psychiatry*, **37**(**4**), 467–494.

Herculano-Houzel, S. *et al.* (1999). Precisely synchronized oscillatory firing patterns require electroencephalographic activation. *Journal of Neuroscience*, **19**(**10**), 3992–4010.

Hesselmann, G. *et al.* (2008). Spontaneous local variations in ongoing neural activity bias perceptual decisions. *Proceedings of the National Academy of Sciences of the United States of America*, **105**, 10984–10989.

Hobson, J. A. (2009). REM sleep and dreaming: towards a theory of protoconsciousness. *Nature Reviews. Neuroscience*, **10**(**11**), 803–813.

Hoffer, P. T. (2005). Reflections on cathexis. *The Psychoanalytic Quarterly*, **74**, 1127–1135.

Hoffman, R. E. (2007). A social deafferentiation hypothesis for induction of active schizophrenia. *Schizophrenia Bulletin*, **33**(**5**), 1066–1070.

Hoffman, R. E. (2010). Revisiting Arieti's "listening attitude" and hallucinated voices. *Schizophrenia Bulletin*, **36**(**3**), 440–442.

Hoffman, R. E. *et al.* (2007). Probing the pathophysiology of auditory/verbal hallucinations by combining functional magnetic resonance imaging and transcranial magnetic stimulation. *Cerebral Cortex*, **17**(**11**), 2733–2743.

Hoffman, R. E. *et al.* (2008). Time course of regional brain activation associated with onset of auditory/verbal hallucinations. *British Journal of Psychiatry*, **193**(**5**), 424–425.

Holstege, G., Bandler, R. and Saper, C. B. (1996). The emotional motor system. *Progress in Brain Research*, **107**, 3–6.

Holt, R. R. (1962). A critical examination of Freud's concept of bound vs. free cathexis. *Journal of the American Psychoanalytic Association*, **10**, 475–525.

Hopfinger, J. B., Buonocore, M. H. and Mangun, G. R. (2000). The neural mechanisms of top-down attentional control. *Nature Neuroscience*, **3**(**3**), 284–291.

Hoptman, M. J. *et al.* (2010). Amplitude of low-frequency oscillations in schizophrenia: a resting state fMRI study. *Schizophrenia Research*, **117**(**1**), 13–20.

Hubl, D. *et al.* (2004). Pathways that make voices: white matter changes in auditory hallucinations. *Archives of General Psychiatry*, **61**(**7**), 658–668.

Hunter, M. D. *et al.* (2006). Neural activity in speech-sensitive auditory cortex during silence. *Proceedings of the National Academy of Sciences of the United States of America*, **103**, 189–194.

Husserl, E. (1989). *Ideas Pertaining to a Pure Phenomenology and to a Phenomenological Philosophy*. Dordrecht: Kluwer Academic Publishers.

Ikemoto, S. and Panksepp, J. (1999). The role of nucleus accumbens dopamine in motivated behavior: a unifying interpretation with special reference to reward-seeking. *Brain Research Reviews*, **31**(**1**), 6–41.

Ingram, R. E. (1990). Self-focused attention in clinical disorders: review and a conceptual model. *Psychological Bulletin*, **107**(**2**), 156–176.

Isomura, Y. *et al.* (2006). Integration and segregation of activity in entorhinal-hippocampal subregions by neocortical slow oscillations. *Neuron*, **52**(**5**), 871–882.

Jafri, M. J. *et al.* (2008). A method for functional network connectivity among spatially independent resting-state components in schizophrenia. *NeuroImage*, **39**(**4**), 1666–1681.

James, W. (1884). What is an emotion? *Mind*, **9**, 185–205.

Jaspers, K. (1962a). *Kant*. San Diego, CA: Harcourt Brace & Company.

Jaspers, K. (1962b). *General Psychopathology*, tr. J. Hoenig and M. W. Hamilton. Manchester: Manchester University Press.

Javitt, D. C. (2008). Phenomenology, aetiology and treatment of schizophrenia. *Novartis Foundation Symposium*, **289**, 4–16; discussion 17–22, 87–93.

Javitt, D. C. (2009). When doors of perception close: bottom-up models of disrupted cognition in psychosis. *Annual Review of Clinical Psychology*, **5**, 249–275.

Jeannerod, M. (1999). To act or not to act: perspectives on the representation of actions. *Quarterly Journal of Experimental Psychology: Human Experimental Psychology* **52**(**1**), 1–29.

Kandel, E. R. (1998). A new intellectual framework for psychiatry. *American Journal of Psychiatry*, **155**(**4**), 457–469.

Kandel, E. R., Schwartz, J. H. and Jessell, T. M. (2000). *Principles of Neural Science*. New York: McGraw-Hill Companies.

Kant, G. J. and Meyerhoff, J. L. (1977). Release of endogenous norepinephrine from rat hypothalamus in vitro. *Life Sciences*, **20**(**1**), 149–153.

Kant, I. (1783). *Prolegomena to Any Future Metaphysics*. Indianapolis, IN: Hackett Publishing Company (1972 edition).

Kant, I. (1998). Critique of Pure Reason Translated into English In: Guyer, P. and Wood, A., eds. *Critique of Pure Reason*. Cambridge Edition of the Works of Immanuel Kant. Cambridge: Cambridge University Press.

Kaplan-Solms, K. and Solms, M. (2000). *Clinical Studies in Neuro-Psychoanalysis: introduction to a depth neuropsychology*. London: Karnac Books.

Kaufmann, C. *et al.* (2006). Brain activation and hypothalamic functional connectivity during human non-rapid eye movement sleep: an EEG/fMRI study. *Brain*, **129(3)**, 655–667.

Keedwell, P. A. *et al.* (2005). The neural correlates of anhedonia in major depressive disorder. *Biological Psychiatry*, **58(11)**, 843–853.

Keedwell, P. A. *et al.* (2010). Subgenual cingulate and visual cortex responses to sad faces predict clinical outcome during antidepressant treatment for depression. *Journal of Affective Disorders*, **120(1–3)**, 120–125.

Keenan, J. P. *et al.* (2003). Self-face processing in a callosotomy patient. *European Journal of Neuroscience*, **18(8)**, 2391–2395.

Kelley, W. M. *et al.* (2002). Finding the self? An event-related fMRI study. *Journal of Cognitive Neuroscience*, **14(5)**, 785–794.

Kenet, T. *et al.* (2003). Spontaneously emerging cortical representations of visual attributes. *Nature*, **425**, 954–956.

Kernberg, O. (1966). Structural derivatives of object relationships. *International Journal of Psychoanalysis*, **47(2)**, 236–253.

Kernberg, O. F. (1984). *Severe Personality Disorders*. New Haven, CT: Yale University Press.

Kernberg, O. F. (1998). Pathological narcissism and narcissistic personality disorder: theoretical background and diagnostic classification. In: Ronningstam, E. F., ed. *Disorders of Narcissism: diagnostic, clinical, and empirical implications*, pp. 67–95. Washington, DC: American Psychiatric Press, Inc.

Kihlstroem, J. E. (1987). The cognitive unconscious. *Science*, **237**, 1445–1452.

Kihlstroem, J., Barnhardt, T. M. and Tataryn, D. (1992). The psychological unconscious: found, lost, and regained. *American Psychologist*, **47(6)**, 788–791.

Kim, D. I. *et al.* (2009). Dysregulation of working memory and default-mode networks in schizophrenia using independent component analysis, an fBIRN and MCIC study. *Human Brain Mapping*, **30(11)**, 3795–3811.

Kipper, L. *et al.* (2004). Brazilian patients with panic disorder: the use of defense mechanisms and their association with severity. *Journal of Nervous and Mental Disease*, **192(1)**, 58–64.

Kircher, T. and David, A., eds (2003). *The Self in Neuroscience and Psychiatry*. Cambridge: Cambridge University Press.

Kirshner, L. A. (1991). The concept of the self in psychoanalytic theory and its philosophical foundations. *Journal of the American Psychoanalytic Association*, **39(1)**, 157–182.

Kitcher, P. (1990). *Kant's Transcendental Psychology*. New York: Oxford University Press.

Kitcher, P. (1992). The naturalists return. *Philosophical Review*, **101(1)**, 53–114.

Klein, M. (1935). A contribution to the psychogenesis of manic-depressive states. *International Journal of Psychoanalysis*, **16**, 145–174.

Klein, M. (1955). *On Identification*. New York: Basic Books.

Knutson, B. *et al.* (2008a). Neural antecedents of the endowment effect. *Neuron*, **58(5)**, 814–822.

Knutson, B. *et al.* (2008b). Nucleus accumbens activation mediates the influence of reward cues on financial risk taking. *NeuroReport*, **19(5)**, 509–513.

Koch, C. (2004). *The Quest for Consciousness: a neurobiological approach*. Greenwood Village, CO: Roberts & Company Publishers.

Koch, G. *et al.* (2005). Modulation of excitatory and inhibitory circuits for visual awareness in the human right parietal cortex. *Experimental Brain Research*, **160(4)**, 510–516.

Kohut, H. (1971). *The Analysis of the Self: a systematic approach to the psychoanalytic treatment of narcissistic personality disorders*. New York: International Universities Press.

Kohut, H. (1977). *The Restoration of the Self.* New York: International Universities Press.

Kohut, H. (1984). *How Does Analysis Cure?* Chicago: University of Chicago Press.

Kohut, H. and Wolf, E. S. (1978). The disorders of the self and their treatment. *International Journal of Psychoanalysis*, **59**, 414–425.

Kraepelin, E (1913). *Ein Lehrbuch fur Studierende und Arzte.* Leipzig: Barth.

Krueger, F., Grafman, J. and McCabe, K. (2008). Neural correlates of economic game playing. *Philosophical Transactions of the Royal Society of London. Series B, Biological Sciences*, **363**, 3859–3874.

Kumar, P. *et al.* (2008). Abnormal temporal difference reward-learning signals in major depression. *Brain*, **131**(**8**), 2084–2093.

Kumari, V. *et al.* (2003). Neural abnormalities during cognitive generation of affect in treatment-resistant depression. *Biological Psychiatry*, **54**(**8**), 777–791.

Kumari, V. *et al.* (2008). Uncontrollable voices and their relationship to gating deficits in schizophrenia. *Schizophrenia Research*, **101**(**1–3**), 185–194.

Lambie, J. A., and Marcel, A. J. (2002). Consciousness and the varieties of emotion experience: a theoretical framework. *Psychological Review*, **109**(**2**), 219–259.

Larson-Prior, L. J. *et al.* (2009). Cortical network functional connectivity in the descent to sleep. *Proceedings of the National Academy of Sciences of the United States of America*, **106**(**11**), 4489–4494.

Lawrence, N. S. *et al.* (2004). Subcortical and ventral prefrontal cortical neural responses to facial expressions distinguish patients with bipolar disorder and major depression. *Biological Psychiatry*, **55**(**6**), 578–587.

LeDoux, J. (2002). *Synaptic Self: how our brains become who we are.* New York: Viking.

Legrand, D. (2005). Being a body (book review of Shaun Gallagher: *How the Body Shapes the Mind*). *Trends in Cognitive Sciences*, **9**(**9**), 413–414.

Legrand, D. (2007). Pre-reflective self-as-subject from experiential and empirical perspectives. *Consciousness and Cognition*, **16**, 583–599.

Legrand, D. and Grammont, F. (2005). A matter of facts (commentary on Bickle *Philosophy and Neuroscience*). *Phenomenology and the Cognitive Sciences*, **4**(**4**), 249–257.

Legrand, D. and Ruby, P. (2009). What is self-specific? Theoretical investigation and critical review of neuroimaging results. *Psychological Review*, **116**(**1**), 252–282.

Lemogne, C. *et al.* (2009). In search of the depressive self: extended medial prefrontal network during self-referential processing in major depression. *Social Cognitive and Affective Neuroscience*, **4**(**3**), 305–312.

Lemogne, C. *et al.* (2010). Self-referential processing and the prefrontal cortex over the course of depression: a pilot study. *Journal of Affective Disorders*, **124**(**1–2**), 196–201.

Lennox, B. R. *et al.* (1999). Spatial and temporal mapping of neural activity associated with auditory hallucinations. *Lancet*, **353**, 644.

Levins, R. (2003). Whose scientific method? Scientific methods for a complex world. *New Solutions*, **13**(**3**), 261–274.

Lewis, C. M. *et al.* (2009). Learning sculpts the spontaneous activity of the resting human brain. *Proceedings of the National Academy of Sciences of the United States of America*, **106**(**41**), 17558–17563.

Lewis, D. A. (2009). Neuroplasticity of excitatory and inhibitory cortical circuits in schizophrenia. *Dialogues in Clinical Neuroscience*, **11**(**3**), 269–280.

Lewis, D.A. and Levitt, P. (2002). Schizophrenia as a disorder of neurodevelopment. *Annual Review of Neuroscience*, **25**, 409–432.

Lewis, D. A. and Gonzalez-Burgos, G. (2008). Neuroplasticity of neocortical circuits in schizophrenia. *Neuropsychopharmacology*, **33**(**1**), 141–165.

Lewis, D. A., Hashimoto, T. and Volk, D. W. (2005). Cortical inhibitory neurons and schizophrenia. *Nature Reviews. Neuroscience*, **6**(**4**), 312–324.

Lichtenberg, J. D. and Wolf, E. (1997). General principles of self psychology: a position statement. *Journal of the American Psychoanalytic Association*, **45**(**2**), 531–543.

Liddle, P. F. (1992). Regional brain abnormalities associated with specific syndromes of persistent schizophrenic symptoms. *Clinical Neuropharmacology* **15(Suppl. 1, Part A)**, 401A–402A.

Lin, S. H., Yu, C. Y. and Pai, M. C. (2006). The occipital white matter lesions in Alzheimer's disease patients with visual hallucinations. *Clinical Imaging*, **30(6)**, 388–393.

Liotti, M. *et al.* (2002). Unmasking disease-specific cerebral blood flow abnormalities: mood challenge in patients with remitted unipolar depression. *American Journal of Psychiatry*, **159(11)**, 1830–1840.

Lipska, B. K. and Weinberger, D. R. (2000). To model a psychiatric disorder in animals: schizophrenia as a reality test. *Neuropsychopharmacology*, **23(3)**, 223–239.

Lisman, J. E. *et al.* (2008). Circuit-based framework for understanding neurotransmitter and risk gene interactions in psychosis. *Trends in Neurosciences*, **31(5)**, 234–242.

Llewellyn, S. (2009). In two minds? Is schizophrenia a state "trapped" between waking and dreaming? *Medical Hypotheses*, **73**, 572–579.

Llinás, R. R. (1988). The intrinsic electrophysiological properties of mammalian neurons: insights into central nervous system function. *Science*, **242**, 1654–1664.

Llinás, R. R. (2001) *I of the Vortex: from neurons to self.* Cambridge, MA: MIT Press.

Llinás, R. and Ribary, U. (2001). Consciousness and the brain. The thalamocortical dialogue in health and disease. *Annals of the New York Academy of Sciences*, **929**, 166–175.

Llinás, R. *et al.* (1998). The neuronal basis for consciousness. *Philosophical Transactions of the Royal Society of London. Series B, Biological Sciences*, **353**, 1841–1849.

Lurija, A. R. (1973). *The Working Brain.* New York: Basic Books.

Maandag, N. J. *et al.* (2007). Energetics of neuronal signaling and fMRI activity. *Proceedings of the National Academy of Sciences of the United States of America*, **104**, 20546–20551.

Macaluso, E., Frith, C. D. and Driver, J. (2000). Modulation of human visual cortex by crossmodal spatial attention. *Science*, **289**, 1206–1208.

Macdiarmid, D. (1989). Self-cathexis and other-cathexis. Vicissitudes in the history of an observation. *British Journal of Psychiatry*, **154**, 844–852.

McDowell, J. E. *et al.* (2002). Neural correlates of refixation saccades and antisaccades in normal and schizophrenia subjects. *Biological Psychiatry*, **51(3)**, 216–223.

McGinn, C. (1991). *The Problem of Consciousness: essays towards a resolution.* Oxford: Blackwell Publishers Ltd.

McGinn, C. J. and Kinsella, T. J. (1991). Combined modality therapy. *Current Opinion in Oncology*, **3**, 1049–1054.

McGlashan, T. H. (2009). Psychosis as a disorder of reduced cathectic capacity: Freud's analysis of the Schreber case revisited, *Schizophrenia Bulletin*, **35(3)**, 476–481.

McGuire, P. K. *et al.* (1996a). The neural correlates of inner speech and auditory verbal imagery in schizophrenia: relationship to auditory verbal hallucinations. *British Journal of Psychiatry*, **169(2)**, 148–159.

McGuire, P. K., Silbersweig, D. A. and Frith, C. D. (1996b). Functional neuroanatomy of verbal self-monitoring. *Brain*, **119(3)**, 907–917.

McIntosh, D. (1986). The ego and the self in the thought of Sigmund Freud. *International Journal of Psychoanalysis*, **67(4)**, 429–448.

McIntosh, D. (1993). Cathexes and their objects in the thought of Sigmund Freud. *Journal of the American Psychoanalytic Association*, **41(3)**, 679–709.

McKiernan, K. A., D'Angelo, B. R., Kaufman, J. N. and Binder, J. R. (2006). Interrupting the "stream of consciousness": an fMRI investigation. *NeuroImage*, **29(4)**, 1185–1191.

Maciag, D. *et al.* (2010). Reduced density of calbindin immunoreactive GABAergic neurons in the occipital cortex in major depression: relevance to neuroimaging studies. *Biological Psychiatry*, **67(5)**, 465–470.

Mäki, P. *et al.* (2005). Predictors of schizophrenia—a review. *British Medical Bulletin*, **73–74**, 1–15.

Malancharuvil, J. M. (2004). Projection, introjection, and projective identification: a reformulation. *American Journal of Psychoanalysis*, **64(4)**, 375–382.

Mancia, M. (2004). The dream between neuroscience and psychoanalysis. *Archives Italiennes de Biologie*, **142(4)**, 525–531.

Mancia, M., ed. (2006a). *Psychoanalysis and Neuroscience.* Milan: Springer.

Mancia, M. (2006b). Implicit memory and early unrepressed unconscious: their role in the therapeutic process (how the neurosciences can contribute to psychoanalysis). *International Journal of Psychoanalysis*, **87(Pt 1)**, 83–103.

Mancia, M. (2006c). Introduction: how the neurosciences can contribute to psychoanalysis. In: Mancia, M., ed. *Psychoanalysis and Neuroscience*, pp. 1–15. Milan: Springer.

Mannell, M. V. *et al.* (2010). Resting state and task-induced deactivation: a methodological comparison in patients with schizophrenia and healthy controls. *Human Brain Mapping*, **31(3)**, 424–437.

Marcel, A. J. and Lambie, J. A. (2004). How many selves in emotion experience? Reply to Dalgleish and Power (2004). *Psychological Review*, **111(3)**, 820–826.

Mareschal, D. *et al.* (2007). *Neuroconstructivism. Volume 1. How the brain constructs cognition.* Oxford: Oxford University Press.

Martinez, A., Hillyard, S. A., Dias, E. C. *et al.* (2008). Magnocellular pathway impairment in schizophrenia: evidence from functional magnetic resonance imaging. *Journal of Neuroscience*, **28**, 7492–7500.

Matthias, E. *et al.* (2009). On the relationship between interoceptive awareness and the attentional processing of visual stimuli. *International Journal of Psychophysiology*, **72(2)**, 154–159.

Mayberg, H. (2002). Depression, II: Localization of pathophysiology. *American Journal of Psychiatry*, **159**, 1979.

Mayberg, H. S. (2003a). Positron emission tomography imaging in depression: a neural systems perspective. *Neuroimaging Clinics of North America*, **13(4)**, 805–815.

Mayberg, H. S. (2003b). Modulating dysfunctional limbic-cortical circuits in depression: towards development of brain-based algorithms for diagnosis and optimised treatment. *British Medical Bulletin*, **65**, 193–207.

Mayberg, H. S. (2009). Targeted electrode-based modulation of neural circuits for depression. *Journal of Clinical Investigation*, **119(4)**, 717–725.

Mayberg, H. S. *et al.* (1999). Reciprocal limbic-cortical function and negative mood: converging PET findings in depression and normal sadness. *American Journal of Psychiatry*, **156(5)**, 675–682.

Mechelli, A. *et al.* (2007). Misattribution of speech and impaired connectivity in patients with auditory verbal hallucinations. *Human Brain Mapping*, **28(11)**, 1213–1222.

Meerbote, R. (1989). Kant's functionalism. In: Smith, J. C., ed. *Historical Foundations of Cognitive Science*, pp. 65–78. Dordrecht: Kluwer Academic Publishers.

Meissner, W. W. (2003a). Mind, brain, and self in psychoanalysis: III. Psychoanalytic perspectives after Freud. *Psychoanalysis and Contemporary Thought*, **26**, 345–386.

Meissner, W. W. (2003b). Mind, brain, and self in psychoanalysis: I. Problems and attempted solutions. *Psychoanalysis and Contemporary Thought*, **26**, 279–320.

Meissner, W. W. (2007). Mind, brain, and self in psychoanalysis: therapeutic implications of the mind–body relation. *Psychoanalytic Psychology*, **24**, 333–354.

Mentzos, S. (1991). *Psychodynamische Modelle der Psychiatrie.* Göttingen: Vandenhoeck & Ruprecht.

Mentzos, S. (1992). *Psychose und Konflikt.* Göttingen: Vandenhoeck & Ruprecht.

Merleau-Ponty, M. (1962). *Phenomenology of Perception.* London: Routledge.

Merleau-Ponty, M. (1965). *Phanomenologie der Wahrnehmung.* Berlin: Walter de Gruyter.

Mesulam, M. M. (2000). *Principles of Behavioral and Cognitive Neurology*, 2nd edn. New York: Oxford University Press.

Metzinger, T. (2003). *Being No One.* Cambridge, MA: MIT Press.

Metzinger, T. and Gallese, V. (2003). The emergence of a shared action ontology: building blocks for a theory. *Consciousness and Cognition*, **12(4)**, 549–571.

Milch, W. (2001). *Lehrbuch der Selbstpsychologie.* Stuttgart: Kohlhammer.

Milrod, D. (2002a). The superego. Its formation, structure, and functioning. *The Psychoanalytic Study of the Child*, **57**, 131–148.

Milrod, D. (2002b). The concept of the self and the self representation. *Neuropsychoanalysis*, **4(1)**, 7–23.

Mishara, A. L. (2010). Klaus Conrad (1905–1961): delusional mood, psychosis, and beginning schizophrenia. *Schizophrenia Bulletin*, **36**, 9–13.

Montague, P. R. (2007). Neuroeconomics: a view from neuroscience. *Functional Neurology*, **22**, 219–234.

Montague, P. R. and Berns, G. S. (2002). Neural economics and the biological substrates of valuation. *Neuron*, **36**, 265–284.

Moore, G. E. (1903). *Principia Ethica*. Cambridge: Cambridge University Press.

Moran, J. M. *et al.* (2006). Neuroanatomical evidence for distinct cognitive and affective components of self. *Journal of Cognitive Neuroscience*, **18(9)**, 1586–1594.

Morcom, A. M. and Fletcher, P. C. (2007). Does the brain have a baseline? Why we should be resisting a rest. *NeuroImage*, **37**, 1073–1082.

Morgane, P. J. and Mokler, D. J. (2006). The limbic brain: continuing resolution. *Neuroscience and Biobehavioral Reviews*, **30(2)**, 119–125.

Morgane, P. J., Galler, J. R. and Mokler, D. J. (2005). A review of systems and networks of the limbic forebrain/limbic midbrain. *Progress in Neurobiology*, **75(2)**, 143–160.

Muthukumaraswamy, S. D. *et al.* (2009). Resting GABA concentration predicts peak gamma frequency and fMRI amplitude in response to visual stimulation in humans. *Proceedings of the National Academy of Sciences of the United States of America*, **106(20)**, 8356–8361.

Naatanen, R. *et al.* (2007). The mismatch negativity (MMN) in basic research of central auditory processing: a review. *Clinical Neurophysiology*, **118(12)**, 2544–2590.

Nagel, T. (1998). Conceiving the impossible and the mind-body problem. *Philosophy*, **73**, 337–352.

Nase, G. *et al.* (2003). Features of neuronal synchrony in mouse visual cortex. *Journal of Neurophysiology*, **90(2)**, 1115–1123.

Niedenthal, P. M. (2007). Embodying emotion. *Science*, **316**, 1002–1005.

Nieuwenhuys, R. (1996). The greater limbic system, the emotional motor system and the brain. *Progress in Brain Research*, **107**, 551–580.

Nieuwenhuys, R., Veening, J. G. and van Domburg, P. (1988). Core and paracores; some new chemoarchitectural entities in the mammalian neuraxis. *Acta Morphologica Neerlando-Scandinavica*, **26(2–3)**, 131–163.

Nir, Y. and Tononi, G. (2010). Dreaming and the brain: from phenomenology to neurophysiology. *Trends in Cognitive Sciences*, **14(2)**, 88–100.

Northoff, G. (2001). *Personale Identitaet und operative Eingriffe in das Gehirn*. Paderborn: Schoening.

Northoff, G. (2004a). Why do we need a philosophy of the brain? *Trends in Cognitive Sciences*, **8(11)**, 484–485.

Northoff, G. (2004b). *Philosophy of the Brain: the brain problem*. Philadelphia, PA: John Benjamins Publishing Co.

Northoff, G. (2007). Psychopathology and pathophysiology of the self in depression—neuropsychiatric hypothesis. *Journal of Affective Disorders*, **104(1–3)**, 1–14.

Northoff, G. (2008). Are our emotional feelings relational? A neurophilosophical investigation of the James–Lange theory. *Phenomenology and the Cognitive Sciences*, **7**, 501–527.

Northoff, G. (2010). Humans, brains, and their environment: marriage between neuroscience and anthropology? *Neuron*, **65(6)**, 748–751.

Northoff, G. and Bermpohl, F. (2004). Cortical midline structures and the self. *Trends in Cognitive Sciences*, **8(3)**, 102–107.

Northoff, G. and Boeker, H. (2006). Principles of neuronal integration and defense mechanisms: neuropsychoanalytic hypothesis. *Neuropsychoanalysis*, **8**, 69–84.

Northoff, G. and Musholt, K. (2006). How can Searle avoid property dualism? Epistemic-ontological inference and autoepistemic limitation. *Philosophical Psychology*, **19(5)**, 1–17.

Northoff, G. and Panksepp, J. (2008). The trans-species concept of self and the subcortical-cortical midline system. *Trends in Cognitive Sciences*, **12**(**7**), 259–264.

Northoff, G. *et al.* (2000). Functional dissociation between medial and lateral prefrontal cortical spatiotemporal activation in negative and positive emotions: a combined fMRI/MEG study. *Cerebral Cortex*, **10**(**1**), 93–107.

Northoff, G. *et al.* (2004). Reciprocal modulation and attenuation in the prefrontal cortex: an fMRI study on emotional-cognitive interaction. *Human Brain Mapping*, **21**(**3**), 202–212.

Northoff, G. *et al.* (2006). Self-referential processing in our brain—a meta-analysis of imaging studies on the self. *NeuroImage*, **31**(**1**), 440–457.

Northoff, G., Bermpohl, F., Schoeneich, F. and Boeker, H. (2007a). How can we bridge between psychoanalysis and neuroscience? First-person neuroscience. *Psychotherapy and Psychosomatics*, **76**, 141–153.

Northoff, G. *et al.* (2007b). GABA concentrations in the human anterior cingulate cortex predict negative BOLD responses in fMRI. *Nature Neuroscience*, **10**(**12**), 1515–1517.

Northoff, G. *et al.* (2007c). How does our brain constitute defense mechanisms? First-person neuroscience and psychoanalysis. *Psychotherapy and Psychosomatics*, **76**(**3**), 141–153.

Northoff, G. *et al.* (2008). Differential parametric modulation of self-relatedness and emotions in different brain regions. *Human Brain Mapping*, **30,** 369–382.

Northoff, G. *et al.* (2009). Differential parametric modulation of self-relatedness and emotions in different brain regions. *Human Brain Mapping*, **30**(**2**), 369–382.

Northoff, G., Qin, P. and Nakao, T. (2010). Rest-stimulus interaction in the brain: a review. *Trends in Neurosciences*, **33**, 277–284.

Northoff, G. *et al.* (2011). The 'resting-state hypothesis' of major depressive disorder—a translational subcortical-cortical framework for a system disorder. *Neuroscience and Biobehavioral Reviews*, [E-pub ahead of print].

Nunberg, H. and Federn, E. (1977). *Protokolle der Wiener Psychoanalytischen Vereinigung*. Frankfurt: M. Fischer.

Ornston, D. G. (1985). The invention of "cathexis" and Strachey's strategy. *International Review of Psychoanalysis*, **12**, 391–398.

Pacherie, E., Green, M. and Bayne, T. (2006). Phenomenology and delusions: who put the "alien" in alien control? *Consciousness and Cognition*, **15**, 566–577.

Pally, R. (2006). The predicting brain: psychoanalysis and repeating the past in the present. *International Journal of Psychoanalysis*, **88**(**4**), 861–881.

Panksepp, J. (1998). *Affective Neuroscience: the foundations of human and animal emotions*. New York: Oxford University Press.

Panksepp, J. (2003). At the interface of the affective, behavioral, and cognitive neurosciences: decoding the emotional feelings of the brain. *Brain and Cognition*, **52**(**1**), 4–14.

Panksepp, J. (2005). Affective consciousness: core emotional feelings in animals and humans. *Consciousness and Cognition*, **14**(**1**), 30–80.

Panksepp, J. (2007). The neuroevolutionary and neuroaffective psychobiology of the prosocial brain. In: Dunbar, R. I. M. and Barrett, L., eds. *The Oxford Handbook of Evolutionary Psychology*, pp. 145–162. Oxford: Oxford University Press.

Panksepp, J. and Northoff, G. (2009). The trans-species core SELF: the emergence of active cultural and neuro-ecological agents through self-related processing within subcortical-cortical midline networks. *Consciousness and Cognition*, **18**, 193–215.

Park, I. H. *et al.* (2009). Medial prefrontal default-mode hypoactivity affecting trait physical anhedonia in schizophrenia. *Psychiatry Research*, **171**(**3**), 155–165.

Parnas, J. (2003). Self and schizophrenia: a phenomenological perspective. In: Kircher, T. and David, A., eds.: *The Self in Neuroscience and Psychiatry*, pp. 217–242. Cambridge: Cambridge University Press.

Parnas, J. and Handest, P. (2003). Phenomenology of anomalous self-experience in early schizophrenia. *Comprehensive Psychiatry*, **44**(**2**), 121–134.

Parnas, J. *et al.* (1998). Self-experience in the prodromal phases of schizophrenia. *Neurology, Psychiatry and Brain Research*, **6**, 97–106.

Parnas, J. *et al.* (2001). Visual binding abilities in the initial and advanced stages of schizophrenia. *Acta Psychiatrica Scandinavica*, **103**(**3**), 171–180.

Parnas, J. *et al.* (2003). Anomalies of subjective experience in schizophrenia and psychotic bipolar illness. *Acta Psychiatrica Scandinavica*, **108**(**2**), 126–133.

Paulus, M. P. and Stein, M. B. (2006). An insular view of anxiety. *Biological Psychiatry*, **60**(**4**), 383–387.

Peled, A. (2008). *NeuroAnalysis: the missing link between psychoanalysis and neuroscience*. New York: Routledge.

Peled, N. *et al.* (2008). Increased erythrocyte adhesiveness and aggregation in obstructive sleep apnea syndrome. *Thrombosis Research*, **121**(**5**), 631–636.

Perry, J. C. and Cooper, S. H. (1989). An empirical study of defense mechanisms. I. Clinical interview and life vignette ratings. *Archives of General Psychiatry*, **46**(**5**), 444–452.

Phan, K. L. *et al.* (2002). Functional neuroanatomy of emotion: a meta-analysis of emotion activation studies in PET and fMRI. *NeuroImage*, **16**(**2**), 331–348.

Phan, K. L. *et al.* (2004). Neural correlates of individual ratings of emotional salience: a trial-related fMRI study. *NeuroImage*, **21**(**2**), 768–780.

Phillips, M. L. *et al.* (2003). Neurobiology of emotion perception I: The neural basis of normal emotion perception. *Biological Psychiatry*, **54**(**5**), 504–514.

Pizzagalli, D. A. *et al.* (2009). Reduced caudate and nucleus accumbens response to rewards in unmedicated individuals with major depressive disorder. *American Journal of Psychiatry*, **166**(**6**), 702–710.

Pollatos, O. *et al.* (2007a). Interoceptive awareness mediates the relationship between anxiety and the intensity of unpleasant feelings. *Journal of Anxiety Disorders*, **21**(**7**), 931–943.

Pollatos, O., Schandry, R., Auer, D. P. and Kaufmann, C. (2007b). Brain structures mediating cardiovascular arousal and interoceptive awareness. *Brain Research*, **1141**, 178–187.

Pollatos, O., Herbert, B. M., Matthias, E. and Schandry, R. (2007c). Heart rate response after emotional picture presentation is modulated by interoceptive awareness. *International Journal of Psychophysiology*, **63**(**1**), 117–124.

Pollatos, O., Gramann, K. and Schandry, R. (2007d). Neural systems connecting interoceptive awareness and feelings. *Human Brain Mapping*, **28**(**1**), 9–18.

Pollatos, O. *et al.* (2008). Reduced perception of bodily signals in anorexia nervosa. *Eating Behaviors*, **9**(**4**), 381–388.

Pollatos, O., Traut-Mattausch, E. and Schandry, R. (2009). Differential effects of anxiety and depression on interoceptive accuracy. *Depression and Anxiety*, **26**(**2**), 167–173.

Pomarol-Clotet, E. *et al.* (2008). Failure to deactivate in the prefrontal cortex in schizophrenia: dysfunction of the default mode network? *Psychological Medicine*, **38**, 1185–1193.

Powell, C. T. (1990). *Kant's Theory of Self-Consciousness*. Oxford: Oxford University Press.

Price, J. L. and Drevets, W. C. (2010). Neurocircuitry of mood disorders. *Neuropsychopharmacology*, **35**, 192–216.

Prinz, J. (2000). A neurofunctional theory of visual consciousness. *Consciousness and Cognition*, **9**(**2 Part 1**), 243–259.

Pugh, G. (2006). Cooperation not incorporation. In: Mancia, M., ed. *Psychoanalysis and Neuroscience*, pp. 54–56. Milan: Springer.

Pulver, S. E. (1970). Narcissism. The term and the concept. *Journal of the American Psychoanalytic Association*, **18**(**2**), 319–341.

Pyka, M. *et al.* (2009). Impact of working memory load on FMRI resting state pattern in subsequent resting phases. *PLoS One*, **4**(**9**), e7198.

Qin, P. and Northoff, G. (2011). How is our self related to midline regions and the default-mode network? *NeuroImage*, [E-pub ahead of print].

Qin, P. *et al.* (2008). Mismatch negativity to the patient's own name in chronic disorders of consciousness. *Neuroscience Letters*, **448**(**1**), 24–28.

Qin, P. *et al.* (2010). Anterior cingulate activity and the self in disorders of consciousness. *Human Brain Mapping*, **31**(**12**), 1993–2002.

Qin, P. *et al.* (2011). Dissociation between anterior and posterior cortical regions during self-specificity and familiarity: a combined fMRI–meta-analytic study. *Human Brain Mapping*, in press.

Raichle, M. E. (2009). A brief history of human brain mapping. *Trends in Neurosciences*, **32**(**2**), 118–126.

Raichle, M. E. and Mintun, M. A. (2006). Brain work and brain imaging. *Annual Review of Neuroscience*, **29**, 449–476.

Raichle, M. E. *et al.* (2001). A default mode of brain function. *Proceedings of the National Academy of Sciences of the United States of America*, **98**, 676–682.

Raij, T. T. *et al.* (2009). Reality of auditory verbal hallucinations. *Brain*, **132**, 2994–3001.

Rangel, A., Camerer, C. and Montague, P. R. (2008). A framework for studying the neurobiology of value-based decision making. *Nature Reviews. Neuroscience*, **9**, 545–556.

Raskin, R. and Novacek, J. (1991). Narcissism and the use of fantasy. *Journal of Clinical Psychology*, **47**(**4**), 490–499.

Revonsuo, A. (2005). The self in dreams. In: Feinberg, T. E. and Keenan, J. P., eds. *The Lost Self: pathologies of the brain and identity*, pp. 206–219. New York: Oxford University Press.

Revonsuo, A. (2006). *Inner Presence: consciousness as a biological phenomenon.* Cambridge, MA: MIT Press.

Reznikova, T. N. *et al.* (2004). [Positron emission tomography of the human brain in relation to psychological defense mechanisms in patients with multiple sclerosis] [article in Russian]. *Fiziologiya Cheloveka*, **30**(**4**), 25–31.

Richards, A. D. (1982). The superordinate self in psychoanalytic theory and in the self psychologies. *Journal of the American Psychoanalytic Association*, **30**, 938–957.

Richter, A., Grimm, S. and Northoff, G. (2010). Lorazepam modulates orbitofrontal signal changes during emotional processing in catatonia. *Human Psychopharmacology*, **25**(**1**), 55–62.

Rilling, J. K. *et al.* (2007). A comparison of resting-state brain activity in humans and chimpanzees. *Proceedings of the National Academy of Sciences of the United States of America*, **104**(**43**), 17146–17151.

Rimes, K. A. and Watkins, E. (2005). The effects of self-focused rumination on global negative self-judgements in depression. *Behaviour Research and Therapy*, **43**(**12**), 1673–1681.

Ritter, K. and Lammers, C. H. (2007). [Narcissism—variable of personality and personality disorder] [article in German]. *Psychotherapie, Psychosomatik, Medizinische Psychologie*, **57**(**2**), 53–60.

Rolls, E. T., Tovee, M. J. and Panzeri, S. (1999). The neurophysiology of backward visual masking: information analysis. *Journal of Cognitive Neuroscience*, **11**(**3**), 300–311.

Rose, E. J., Simonotto, E. and Ebmeier, K. P. (2006). Limbic over-activity in depression during preserved performance on the n-back task. *Neuroimage*, **29**(**1**), 203–215.

Rosenbaum, B. and Harder, S. (2007). Psychosis and the dynamics of the psychotherapy process. *International Review of Psychiatry*, **19**(**1**), 13–23.

Rosenbaum, R. S. *et al.* (2005). "Where to?" Remote memory for spatial relations and landmark identity in former taxi drivers with Alzheimer's disease and encephalitis. *Journal of Cognitive Neuroscience*, **17**, 446–462.

Rotarska-Jagiela, A. *et al.* (2010). Resting-state functional network correlates of psychotic symptoms in schizophrenia. *Schizophrenia Research*, **117**(**1**), 21–30.

Sadaghiani, S., Ugurbil, K. and Uludag, K. (2009). Neural activity-induced modulation of BOLD poststimulus undershoot independent of the positive signal. *Magnetic Resonance Imaging*, **27**(**8**), 1030–1038.

Sajonz, B. *et al.* (2010). Delineating self-referential processing from episodic memory retrieval: common and dissociable networks. *NeuroImage*, **50**(**4**), 1606–1617.

Sanacora, G. *et al.* (1999). Reduced cortical gamma-aminobutyric acid levels in depressed patients determined by proton magnetic resonance spectroscopy. *Archives of General Psychiatry*, **56(11)**, 1043–1047.

Sanfey, A. G. *et al.* (2006). Neuroeconomics: cross-currents in research on decision-making. *Trends in Cognitive Sciences*, **10(3)**, 108–116.

Sapir, A. *et al.* (2004). Parietal lobe lesions disrupt saccadic remapping of inhibitory location tagging. *Journal of Cognitive Neuroscience*, **16(4)**, 503–509.

Sass, L. A. (1996). "The catastrophes of heaven": modernism, primitivism, and the madness of Antonin Artaud. *Modernism/Modernity*, **3**, 73–92.

Sass, L. A. (2000). Schizophrenia, self-experience, and the so-called "negative symptoms." In: Zahavi, D, ed.: *Exploring the Self: philosophical and psychopathological perspectives on self-experience*, pp. 149–182. Amsterdam: John Benjamins Publishing.

Sass, L. A. (2003). Self-disturbance in schizophrenia: hyperreflexivity and diminished self-affection. In: Kircher, T. and David, A., eds. *The Self in Neuroscience and Psychiatry*, pp. 242–272. Cambridge: Cambridge University Press.

Sass, L. A. and Parnas, J. (2001). Phenomenology of self-disturbances in schizophrenia: some research findings and directions. *Philosophy, Psychiatry, and Psychology*, **8**, 347–356.

Sass, L. A. and Parnas, J. (2003). Schizophrenia, consciousness, and the self. *Schizophrenia Bulletin*, **29(3)**, 427–444.

Savitz, J. B. and Drevets, W. C. (2009). Imaging phenotypes of major depressive disorder: genetic correlates. *Neuroscience*, **164(1)**, 300–330.

Scarone, S. *et al.* (2008). The dream as a model for psychosis: an experimental approach using bizarreness as a cognitive marker. *Schizophrenia Bulletin*, **34(3)**, 515–522.

Schacter, D. L. and Addis, D. R. (2007a). The cognitive neuroscience of constructive memory: remembering the past and imagining the future. *Philosophical Transactions of the Royal Society of London. Series B, Biological Sciences*, **362**, 773–786.

Schacter, D. L. and Addis, D. R. (2007b). Constructive memory: the ghosts of past and future. *Nature*, **445**, 27.

Schaefer, R. (1972). Internalization: process or fantasy? *Psychoanalytic Study of the Child*, **27**, 411–436.

Scharfetter, C. (1996). The self-experience of schizophrenics. In: *Empirical Studies of the Ego/Self in Schizophrenia, Borderline Disorders and Depression*, 2nd edn, pp. 34–54. Zurich: University of Zurich.

Scharfetter, C (2003). The self-experience of schizophrenics. In: Kircher, T. and David, A., eds.: *The Self in Neuroscience and Psychiatry*, pp. 272–293. Cambridge: Cambridge University Press.

Schlicht, T. (2007). *Erkenntnistheoretischer Dualismus. Das Problem der Erklärungslücke in Geist-Gehirn-Theorien*. Paderborn: Mentis-Verlag.

Schmidt, K. E., Singer, W. and Galuske, R. A. (2004). Processing deficits in primary visual cortex of amblyopic cats. *Journal of Neurophysiology*, **91(4)**, 1661–1671.

Schmidt-Hellerau, C. (1995). [The birth of metapsychology. On the current interpretation of "Entwurf einer Psychologie" (1895)] [article in German]. *Psyche*, **49(12)**, 1156–1195.

Schneider, F. *et al.* (2008). The resting brain and our self: self-relatedness modulates resting state neural activity in cortical midline structures. *Neuroscience*, **157(1)**, 120–131.

Schneider, K. (1959). *Allgemeine Psychopathologie*. Stuttgart: Thieme Verlag.

Schneider, P. (2006). Psychoanalyse und Neurowissenschaft: Inkompatbilitaet! In: Böker, H., ed. *Psychoanalyse und Psychiatrie*, pp. 293–300. Heidelberg: Springer.

Schopenhauer, A. (1966). *The World as Will and Representation. Volume II*. New York: Dover Publications, Inc.

Schuessler, G. (2002). Aktuelle Konzepte des Unbewussten. *Zeitschrift für Psychosomatische Medizin und Psychotherapie*, **48**, 192–214.

Searle, J. (2004). *Mind: a brief introduction*. New York: Oxford University Press.

Sedikides, C. *et al.* (2004). Are normal narcissists psychologically healthy?: Self-esteem matters. *Journal of Personality and Social Psychology*, **87(3)**, 400–416.

Sellars, W. (2007). *Selected Works*. Cambridge, MA: Harvard University Press.

Seth, A. K. *et al.* (2006). Theories and measures of consciousness: an extended framework. *Proceedings of the National Academy of Sciences of the United States of America*, **103(28)**, 10799–10804.

Shapley, R., Hawken, M. and Ringach, M. D. L. (2003). Dynamics of orientation selectivity in the primary visual cortex and the importance of cortical inhibition. *Neuron*, **5(38)**, 689–699.

Shedler, J. and Westen, D. (2004). Refining personality disorder diagnosis: integrating science and practice. *American Journal of Psychiatry*, **161(8)**, 1350–1365.

Sheline, Y. I. *et al.* (2009). The default mode network and self-referential processes in depression. *Proceedings of the National Academy of Sciences of the United States of America*, **106**, 1942–1947.

Shergill, S. S. *et al.* (2000a). Functional anatomy of auditory verbal imagery in schizophrenic patients with auditory hallucinations. *American Journal of Psychiatry* **157(10)**, 1691–1693.

Shergill, S. S. *et al.* (2000b). Mapping auditory hallucinations in schizophrenia using functional magnetic resonance imaging. *Archives of General Psychiatry*, **57(11)**, 1033–1038.

Shergill, S. S. *et al.* (2003). Engagement of brain areas implicated in processing inner speech in people with auditory hallucinations. *British Journal of Psychiatry*, **182**, 525–531.

Shergill, S. S. *et al.* (2004). Temporal course of auditory hallucinations. *British Journal of Psychiatry*, **185**, 516–517.

Shergill, S. S. *et al.* (2007). A diffusion tensor imaging study of fasciculi in schizophrenia. *American Journal of Psychiatry*, **164(3)**, 467–473.

Shin, S. E. *et al.* (2005). Segmented volumes of cerebrum and cerebellum in first episode schizophrenia with auditory hallucinations. *Psychiatry Research*, **138(1)**, 33–42.

Shore, A. (1996). The experience-dependent maturation of a regulatory system in the orbital prefrontal cortex and the origin of developmental psychopathology. *Development and Psychopathology*, **8,** 59–87.

Shore, A. N. (2003). *Affect Regulation and the Repair of the Self.* New York: W.W. Norton.

Shore, S. E., El Kashlan, H. and Lu, J. (2003). Effects of trigeminal ganglion stimulation on unit activity of ventral cochlear nucleus neurons. *Neuroscience*, **119(4)**, 1085–1101.

Shulman, G. L. *et al.* (1997a). Common blood flow changes across visual tasks: I. Increases in subcortical structures and cerebellum but not in nonvisual cortex. *Journal of Cognititve Neuroscience*, **9(5)**, 624–647.

Shulman, G. L. *et al.* (1997b). Searching for activations that generalize over tasks. *Human Brain Mapping*, **5(4)**, 317–322.

Shulman, R. G. *et al.* (2004). Energetic basis of brain activity: implications for neuroimaging. *Trends in Neurosciences*, **27(8)**, 489–495.

Shulman, R. G., Hyder, F. and Rothman, D. L. (2009). Baseline brain energy supports the state of consciousness. *Proceedings of the National Academy of Sciences of the United States of America* **106,** 11096–11101.

Simpson, J. R. Jr *et al.* (2001). Emotion-induced changes in human medial prefrontal cortex: I. During cognitive task performance. *Proceedings of the National Academy of Sciences of the United States of America*, **98(2)**, 683–687.

Smith, A. T., Singh, K. D. and Greenlee, M. W. (2000). Attentional suppression of activity in the human visual cortex. *Neuroreport*, **11(2)**, 271–277.

Solms, M. (1995). Is the brain more real than the mind? *Psychoanalytic Psychotherapy*, **9(2)**, 107–120.

Solms, M. (1996). [What are affects?] [article in German]. *Psyche*, **50(6)**, 485–522.

Solms, M. (1997). What is consciousness? *Journal of the American Psychoanalytic Association*, **45(3)**, 681–703.

Solms, M. (1998). Psychoanalytische Beobachtungen an vier Patienten mit ventromesialen Frontalhirnlasionen. *Psyche*, **52**, 919–962.

Solms, M. (1999). The deep psychological functions of the right cerebral hemisphere. *Bulletin of the British Psychoanalytical Society*, **35(1)**, 9–29.

Solms, M. (2000). Dreaming and REM sleep are controlled by different brain mechanisms. *Behavioral and Brain Sciences*, **23(6)**, 843–850; discussion 904–1121.

Solms, M. (2004). Freud returns. *Scientific American*, **290(5)**, 82–88.

Solms, M. (2006). "Freud" and Bullitt: an unknown manuscript. *Journal of the American Psychoanalytic Association*, **54**(**4**), 1263–1298.

Solms, M. and Saling, M. (1986). On psychoanalysis and neuroscience: Freud's attitude to the localizationist tradition. *International Journal of Psychoanalysis*, **67**(**4**), 397–416.

Solms, M. and Turnbull, O. (2002). *The Brain and the Inner World: an introduction to the neuroscience of subjective experience*. New York: Other Press.

Solms, M. *et al.* (1998). Rotated drawing: the range of performance and anatomical correlates in a series of 16 patients. *Brain and Cognition*, **38**(**3**), 358–368.

Somers, D. C. *et al.* (1999). Functional MRI reveals spatially specific attentional modulation in human primary visual cortex. *Proceedings of the National Academy of Sciences of the United States of America*, **96**(**4**), 1663–1668.

Sommer, I. E. C. *et al.* (2008). Auditory verbal hallucinations predominantly activate the right inferior frontal area. *Brain*, **131**, 3169–3177.

Sommerfield, C. *et al.* (2006). Predictive codes for forthcoming perception in the frontal cortex. *Science*, **314**, 1311–1314.

Stanghellini, G. (2009). Embodiment and schizophrenia. *World Psychiatry*, **8**(**1**), 56–59.

Stanghellini, G. and Ballerini, M. (2007). Criterion B (social dysfunction) in persons with schizophrenia: the puzzle. *Current Opinion in Psychiatry*, **20**(**6**), 582–587.

Stanghellini, G. and Ballerini, M. (2008). Qualitative analysis. Its use in psychopathological research. *Acta Psychiatrica Scandinavica*, **117**(**3**), 161–163.

Steele, J. D. and Lawrie, S. M. (2004). Segregation of cognitive and emotional function in the prefrontal cortex: a stereotactic meta-analysis. *NeuroImage*, **21**(**3**), 868–875.

Stephan, K. E., Friston, K. J. and Frith, C. D. (2009). Dysconnection in schizophrenia: from abnormal synaptic plasticity to failures of self-monitoring. *Schizophrenia Bulletin*, **35**(**3**), 509–527.

Stephane, M., Barton, S. and Boutros, N. N. (2001). Auditory verbal hallucinations and dysfunction of the neural substrates of speech. *Schizophrenia Research*, **50**, 61–78.

Stern, D. (1985). *Die Lebenserfahrung des Sauglings*. Stuttgart: Klett-Cotta.

Stewart, J. L. *et al.* (2010). Attentional bias to negative emotion as a function of approach and withdrawal anger styles: an ERP investigation. *International Journal of Psychophysiology*, **76**(**1**), 9–18.

Stolorow, R. D. (1993). An intersubjective view of the therapeutic process. *Bulletin of the Menninger Clinic*, **57**(**4**), 450–457.

Stolorow, R. D. and Atwood, G. F. (1997). Deconstructing the myth of the neutral analyst: an alternative from intersubjective systems theory. *Psychoanalytic Quarterly*, **66**(**3**), 431–449.

Strachey, J. (1967). An unknown review by Freud. *International Journal of Psychoanalysis*, **48**(**2**), 319–320.

Strawson, G. (1994). *Mental Reality*. Cambridge, MA: MIT Press.

Stuss, D. T., Gallup, G. G. Jr and Alexander, M. P. (2001). The frontal lobes are necessary for "theory of mind." *Brain*, **124**(**2**), 279–286.

Sullivan, H. S. (1962). *Schizophrenia as a Human Process*. New York: W. W. Norton.

Surguladze, S. A. *et al.* (2005). Recognition accuracy and response bias to happy and sad facial expressions in patients with major depression. *Neuropsychology*, **18**, 212–218.

Svare, E. (2006). *Kant and Embodiment*. Oslo: Oslo University Press.

Talvitie, V. and Ihanus, J. (2006). The psychic apparatus, metapsychology, and neuroscience. *Neuropsychoanalysis*, **8**, 85–98.

Tauber, A. I. (2009). Freud's dreams of reason: the Kantian structure of psychoanalysis. *History of the Human Sciences*, **22**(**4**), 1–29.

Taylor, C. B. and Chang, V. Y. (2008). Issues in the dissemination of cognitive-behavior therapy. *Nordic Journal of Psychiatry*, **62**(**Suppl. 47**), 37–44.

Thomaes, S. *et al.* (2009). What makes narcissists bloom? A framework for research on the etiology and development of narcissism. *Development and Psychopathology*, **21**(**4**), 1233–1247.

Thompson, E. (2007). *Mind in Life: biology, phenomenology, and the sciences of mind*. Cambridge, MA: Harvard University Press.

Thompson, J. C. *et al.* (2007). Common and distinct brain activation to viewing dynamic sequences of face and hand movements. *NeuroImage*, **37**(**3**), 966–973.

Tienari, P. (1991). Interaction between genetic vulnerability and family environment. *Acta Psychiatrica Scandinavica*, **84**, 460–465.

Tononi, G. (2008). Consciousness as integrated information: a provisional manifesto. *Biological Bulletin*, **215**(**3**), 216–242.

Tononi, G. and Koch, C. (2008). The neural correlates of consciousness: an update. *Annals of the New York Academy of Sciences*, **1124**, 239–261.

Tregellas, J. R. *et al.* (2009). Increased hippocampal, thalamic, and prefrontal hemodynamic response to an urban noise stimulus in schizophrenia. *American Journal of Psychiatry*, **166**(**3**), 354–360.

Treynor, W., Gonzalez, R. and Nolen-Hoeksema, S. (2003). Rumination reconsidered: a psychometric analysis. *Cognitive Therapy and Research*, **27**, 247–259.

Tsodyks, M. *et al.* (1999). Linking spontaneous activity of single cortical neurons and the underlying functional architecture. *Science*, **286**, 1943–1946.

Turetsky, B. I. *et al.* (2007). Neurophysiological endophenotypes of schizophrenia: the viability of selected candidate measures. *Schizophrenia Bulletin*, **33**(**1**), 69–94.

Turk, D. J. *et al.* (2002). Mike or me? Self-recognition in a split-brain patient. *Nature Neuroscience*, **5**(**9**), 841–842.

Turk, D. J. *et al.* (2003). Out of contact, out of mind: the distributed nature of the self. *Annals of the New York Academy of Sciences*, **1001**, 65–78.

Tye, M. (1995). *Ten Problems of Consciousness*. Cambridge, MA: MIT Press.

Uddin, L. Q., Iacoboni, M., Lange, C. and Keenan, J. P. (2007). The self and social cognition: the role of cortical midline structures and mirror neurons. *Trends in Cognitive Sciences*, **11**(**4**), 153–157.

Uhlhaas, P. J. and Singer, W. (2010). Abnormal neural oscillations and synchrony in schizophrenia. *Nature Reviews. Neuroscience*, **11**(**2**), 100–113.

Uhlhaas, P. J. *et al.* (2009a). The development of neural synchrony reflects late maturation and restructuring of functional networks in humans. *Proceedings of the National Academy of Sciences of the United States of America*, **106**(**24**), 9866–9871.

Uhlhaas, P. J. *et al.* (2009b). Neural synchrony in cortical networks: history, concept and current status. *Frontiers in Integrative Neuroscience*, **3,** 17.

Uhlhaas, P. J. *et al.* (2010). Neural synchrony and the development of cortical networks. *Trends in Cognitive Sciences*, **14**(**2**), 72–80.

Ulman, J. S. (2005). Introduction: Becoming aware of the new unconscious. In: Hassin, G. and Ulman, J. S., eds. *The Unconscious*, pp. 1–12. Cambridge, MA: MIT Press.

Vaillant, G. (1992). *Ego Mechanisms of Defence*. Washington, DC: American Psychiatric Press.

Vaillant, G. E. (1993). Is alcoholism more often the cause or the result of depression? *Harvard Review of Psychiatry*, **1**(**2**), 94–99.

van de Ven, V. G. *et al.* (2005). The spatiotemporal pattern of auditory cortical responses during verbal hallucinations. *NeuroImage* **27**(**3**), 644–655.

van Eijsden, P. *et al.* (2009). Neurophysiology of functional imaging. *NeuroImage*, **45**, 1047–1054.

Varela, F. and Shear, J. (1999). First-person methodologies: what, why, how? *Journal of Consciousness Studies*, **6**(**2–3**), 1–14.

Varela, F. *et al.* (2001). The brainweb: phase synchronization and large-scale integration. *Nature Reviews. Neuroscience*, **2**(**4**), 229–239.

Vincent, J. L. *et al.* (2007). Intrinsic functional architecture in the anaesthetized monkey brain. *Nature*, **447**, 83–86.

Vincent, J. L. *et al.* (2008). Evidence for a frontoparietal control system revealed by intrinsic functional connectivity. *Journal of Neurophysiology*, **100(6)**, 3328–3342.

Vogeley, K. (2003). Schizophrenia as disturbance of the self-construct. In: Kircher, T. and David, A., eds. *The Self in Neuroscience and Psychiatry*, pp. 361–380. Cambridge: Cambridge University Press.

Vogeley, K. and Fink, G. R. (2003). Neural correlates of the first-person perspective. *Trends in Cognitive Sciences*, **7(1)**, 38–42.

Volkan, V. D. (1976). *Primitive Internalized Object Relations: a clinical study of schizophrenic, borderline, and narcissistic patients*. New York: International Universities Press.

Volkan, V. D. (1981). *Linking Object Relations and Linking Phenomena*. New York: International Universities Press.

Wagner, C. and Stucki, J. W. (2002). Construction of an associative memory using unstable periodic orbits of a chaotic attractor. *Journal of Theoretical Biology*, **215(3)**, 375–384.

Wahlberg, K. E. *et al.* (1997). Gene–environment interaction in vulnerability to schizophrenia: findings from the Finnish Adoptive Family Study of Schizophrenia. *American Journal of Psychiatry*, **154**, 355–362.

Walker, M. P. (2009a). The role of sleep in cognition and emotion. *Annals of the New York Academy of Sciences*, **1156**, 168–197.

Walker, M. P. (2009b). The role of slow wave sleep in memory processing. *Journal of Clinical Sleep Medicine*, **5(2 Suppl.)**, S20–26.

Walker, M. P. and van der Helm, E. (2009). Overnight therapy? The role of sleep in emotional brain processing. *Psychological Bulletin*, **135(5)**, 731–748.

Walter, H. (2002). *Neurophilosophy of Free Will*. Cambridge, MA: MIT Press.

Walter, M. *et al.* (2008). Preceding attention and the dorsomedial prefrontal cortex: process specificity versus domain dependence. *Human Brain Mapping*, **30(1)**, 312–326.

Wang, K. *et al.* (2008). Spontaneous activity associated with primary visual cortex: a resting-state FMRI study. *Cerebral Cortex*, **18(3)**, 697–704.

Wehrle, R. *et al.* (2005). Rapid eye movement-related brain activation in human sleep: a functional magnetic resonance imaging study. *Neuroreport*, **16(8)**, 853–857.

Wehrle, R. *et al.* (2007). Functional microstates within human REM sleep: first evidence from fMRI of a thalamocortical network specific for phasic REM periods. *European Journal of Neuroscience*, **25(3)**, 863–871.

Weinryb, R. M. and Rossel, R. J. (1991). Karolinska Psychodynamic Profile. KAPP. *Acta Psychiatrica Scandinavica*, **83(Suppl. 363)**, 1–23.

Weinstein, E. A. and Kahn, R. L. (1959). Symbolic reorganization in brain injuries. In: Arieti, S., ed. *American Handbook of Psychiatry. Volume 1*, pp. 34–54. New York: Basic Books.

Weissman, D. H. *et al.* (2006). The neural bases of momentary lapses in attention. *Nature Neuroscience*, **9(7)**, 971–978.

Westen, D. (1999). The scientific status of unconscious processes: is Freud really dead? *Journal of the American Psychoanalytic Association*, **47(4)**, 1061–1106.

Whitfield-Gabrieli, S. *et al.* (2009). Hyperactivity and hyperconnectivity of the default network in schizophrenia and in first-degree relatives of persons with schizophrenia. *Proceedings of the National Academy of Sciences of the United States of America*, **106(4)**, 1279–1284.

Wible, C. G. *et al.* (2009). fMRI activity correlated with auditory hallucinations during performance of a working memory task: data from the FBIRN consortium study. *Schizophrenia Bulletin*, **35(1)**, 47–57.

Wicker, B., Ruby, P., Royet, J. P. and Fonlupt, P. (2003). A relation between rest and the self in the brain? *Brain Research Reviews*, **43(2)**, 224–230.

Wiebking, C. *et al.* (2010). Abnormal body perception and neural activity in the insula in depression: an fMRI study of the depressed "material me." *World Journal of Biological Psychiatry*, **11(3)**, 538–549.

Wiebking, C., de Greck, M., Tempelmann, C. and Northoff, G. (2011). Are emotions associated with activity during rest or interoception? An exploratory fMRI study in healthy subjects. *Neuroscience Letters*, **491**(**1**), 87–92.

Williamson, P. (2007). Are anticorrelated networks in the brain relevant to schizophrenia? *Schizophrenia Bulletin*, **33**(**4**), 994–1003.

Wink, P. (1991). Self- and object-directedness in adult women. *Journal of Personality*, **59**(**4**), 769–791.

Winnicott, D. W. (1965). *The Maturational Processes and the Facilitating Environment*. Madison, CT: International Universities Press.

Winnicott, D. W. (1971). *Playing and Reality*. New York: Basic Books.

Winnicott, D. W. (1975). *Through Paediatrics to Psychoanalysis*. New York: Basic Books.

Wittgenstein, L. (1994). *On Certainty*. Cambridge: Cambridge University Press (originally published 1969).

Wolf, E. S. (1980) On the developmental line of self-object relations. In: Goldberg, A., ed. *Advances in Self Psychology*, pp. 117–130. New York: International Universities Press.

Wolf, E. S. (1988) *Treating the Self*. New York: Guilford.

Wolf, N. S., Gales, M., Shane, E. and Shane, M. (2000). Mirror neurons, procedural learning, and the positive new experience: a developmental systems self psychology approach. *Journal of the American Academy of Psychoanalysis*, **28**(**3**), 409–430.

Womelsdorf, T. *et al.* (2006). Gamma-band synchronization in visual cortex predicts speed of change detection. *Nature*, **439**, 733–736.

Woo, T. U. *et al.* (2008). N-methyl-D-aspartate receptor and calbindin-containing neurons in the anterior cingulate cortex in schizophrenia and bipolar disorder. *Biological Psychiatry*, **64**(**9**), 803–809.

Woodruff, C. C., Uncapher, M. R. and Rugg, M. D. (2006). Neural correlates of differential retrieval orientation: sustained and item-related components. *Neuropsychologia*, **44**(**14**), 3000–3010.

Yao, Z. *et al.* (2009). Regional homogeneity in depression and its relationship with separate depressive symptom clusters: a resting-state fMRI study. *Journal of Affective Disorders*, **115**(**3**), 430–438.

Yoo, S. S. *et al.* (2003). Neural substrates of tactile imagery: a functional MRI study. *Neuroreport*, **14**(**4**), 581–585.

Yoo, S. S. *et al.* (2007). The human emotional brain without sleep—a prefrontal amygdala disconnect. *Current Biology*, **17**, R877–878.

Yoon, J. H. *et al.* (2010). GABA concentration is reduced in visual cortex in schizophrenia and correlates with orientation-specific surround suppression. *Journal of Neuroscience*, **30**(**10**), 3777–3781.

Young, C. and Brook, A. (1994). Schopenhauer and Freud. *International Journal of Psychoanalysis*, **75**, 101–118.

Zahavi, D. (2005). *Subjectivity and Selfhood: investigating the first-person perspective*. Cambridge, MA: MIT Press.

Zahn R. *et al.* (2008). The neural basis of human social values: evidence from functional MRI. *Cerebral Cortex*, **19**, 276–283.

Index

Printed in the USA/Agawam, MA
January 15, 2019

695462.014